Carbon Capture and Storage

Physical, Chemical, and Biological Methods

SPONSORED BY

Carbon Capture and Storage Task Committee of the Technical Committee on Hazardous, Toxic, and Radioactive Waste Engineering of the Environmental Council of the Environmental and Water Resources Institute of ASCE

EDITED BY
Rao Y. Surampalli
Tian C. Zhang
R. D. Tyagi
Ravi Naidu
B. R. Gurjar
C. S. P. Ojha
Song Yan
Satinder K. Brar
Anushuya Ramakrishnan
C. M. Kao

Published by the American Society of Civil Engineers

Library of Congress Cataloging-in-Publication Data

Carbon capture and storage : physical, chemical, and biological methods / sponsored by Carbon Capture and Storage Task Committee of the Environmental Council, Environmental and Water Resources Institute (EWRI) of the American Society of Civil Engineers ; edited by Rao Y. Surampalli [and 9 others].
pages cm
Includes bibliographical references and index.
ISBN 978-0-7844-1367-8 (pbk.) -- ISBN 978-0-7844-7891-2 (e-book PDF)
1. Carbon sequestration. 2. Sequestration (Chemistry) I. Surampalli, Rao Y., editor. II. Environmental and Water Resources Institute (U.S.). Carbon Capture and Storage Task Committee, sponsoring body.
TP156.S5C37 2015
628.5'32--dc23

2014038868

Published by American Society of Civil Engineers
1801 Alexander Bell Drive
Reston, Virginia, 20191-4382
www.asce.org/bookstore | ascelibrary.org

Errata: Errata, if any, can be found at http://dx.doi.org/10.1061/9780784413678

ISBN 978-0-7844-1367-8 (print)
ISBN 978-0-7844-7891-2 (E-book PDF)
Manufactured in the United States of America.

20 19 18 17 16 15 1 2 3 4 5

Contents

Chapter 5 CO_2 Sequestration and Leakage

Chapter 6 Monitoring, Verification, and Accounting of CO_2 Stored in Deep Geological Formations

Chapter 7 Carbon Reuses for a Sustainable Future

Chapter 8 Carbon Dioxide Capture Technology for the Coal-Powered Electricity Industry

Chapter 9 CO_2 Scrubbing Processes and Applications

Chapter 10 Carbon Sequestration via Mineral Carbonation: Overview and Assessment

Chapter 11 Carbon Burial and Enhanced Soil Carbon Trapping

Chapter 12 Algae-Based Carbon Capture and Sequestrations

Chapter 13 Carbon Immobilization by Enhanced Photosynthesis of Plants

Chapter 14 Enzymatic Sequestration of Carbon Dioxide

Chapter 15 Biochar

Chapter 16 Enhanced Carbon Sequestration in Oceans: Principles, Strategies, Impacts, and Future Perspectives

Preface

Currently, three climate change mitigation strategies are being explored: a) increasing energy efficiency, b) switching to less carbon-intensive sources of energy, and c) carbon capture and sequestration (CCS). As a strong option to achieve the large-scale reductions in CO_2, CCS technology allows the continuous use of fossil fuels and provides time to make the changeover to other energy sources in a systematic way. Therefore, CCS technology is certainly necessary both globally and nationally in order to mitigate climate change.

The ASCE's Technical Committee on Hazardous, Toxic and Radioactive Waste has identified CCS technology as an important area for mitigation of climate change and sustainable development, and thus, made an effort to work with the contributors to put this book together in the context of a) the basic principles of CCS focusing on the physical, chemical and biological methods (see chapters 1–7); and b) applications and research development related to CCS (see chapters 8-17). This structure reflects the historical evolution and current status of CCS technology as well as the major issues/challenges/the path forward for CCS technology.

Many factors decide CCS applicability worldwide, such as technical development, overall potential, flow and shift of the technology to developing countries and their capability to apply the technology, regulatory aspects, environmental concerns, public perception and costs. In this book, the term CCS is defined as any technologies/methods that are to a) capture, transport and store carbon (CO_2), b) monitor, verify and account the status/progress of the CCS technologies employed, and c) advance development/uptake of low-carbon technologies and/or promote beneficial reuse of CO_2. As a reference, the book will provide readers in-depth understanding of and comprehensive information on the principles of CCS technology, different environmental applications, recent advances, critical analysis of new CCS methods and processes, and directions toward future research and development of CCS technology. We hope that this book will be of interest to students, scientists, engineers, government officers, process managers and practicing professionals.

The editors gratefully acknowledge the hard work and patience of all the authors who have contributed to this book. The views or opinions expressed in each chapter of this book are those of the authors and should not be construed as opinions of the organizations they work for. Special thanks go to Ms. Arlys Blakey at the University of Nebraska-Lincoln for her thoughtful comments and invaluable support during the development of this book.

– RYS, TCZ, RDG, RN, BRG, CSPO, SY, SKB, AR, CMK

Contributing Authors

Indrani Bhattacharya, *INRS, Universite du Quebec, Quebec, QC, Canada*

Satinder K. Brar, *INRS, Universite du Quebec, Quebec, QC, Canada*

Munish K. Chandel, *Indian Institute of Technology Roorkee, Roorkee, India*

Stéphane Godbout, *INRS, Universite du Quebec, Quebec, QC, Canada*

W. S. Huang, *National Sun Yat-Sen University, Kaohsiung, Taiwan*

B. R. Gurjar, *Indian Institute of Technology Roorkee, Roorkee, India*

Wenbiao Jin, *Shenzhen Key Laboratory of WRUEPC, Shenzhen, China*

Rojan P. John, *INRS, Universite du Quebec, Quebec, QC, Canada*

C. M. Kao, *National Sun Yat-Sen University, Kaohsiung, Taiwan*

L. Kumar, *INRS, Universite du Quebec, Quebec, QC, Canada*

Archana Kumari, *INRS, Universite du Quebec, Quebec, QC, Canada*

P. N. Mariyamma, *INRS, Universite du Quebec, Quebec, QC, Canada*

T. T. More, *INRS, Universite du Quebec, Quebec, QC, Canada*

Klai Nouha, *INRS, Universite du Quebec, Quebec, QC, Canada*

C. S. P. Ojha, *Indian Institute of Technology Roorkee, Roorkee, India*

Joahnn Palacios, *INRS, Universite du Quebec, Quebec, QC, Canada*

Frédéric Pélletier, *INRS, Universite du Quebec, Quebec, QC, Canada*

Anushuya Ramakrishnan, *University of Nebraska-Lincoln, Lincoln, NE, USA*

Guobin Shan, *University of Nebraska-Lincoln, Lincoln, NE, USA*

Rao Y. Surampalli, *University of Nebraska-Lincoln, Lincoln, NE, USA*

R. D. Tyagi, *INRS, Universite du Quebec, Quebec, QC, Canada*

Mausam Verma, *INRS, Universite du Quebec, Quebec, QC, Canada*

P. P. Walvekar, *Indian Institute of Technology Roorkee, Roorkee, India*

S. S. Yadav, *INRS, Universite du Quebec, Quebec, QC, Canada*

Song Yan, *INRS, Universite du Quebec, Quebec, QC, Canada*

Z. H. Yang, *National Sun Yat-Sen University, Kaohsiung, Taiwan*

Tian C. Zhang, *University of Nebraska-Lincoln, Lincoln, NE, USA*

Xiaolei Zhang, *INRS, Universite du Quebec, Quebec, QC, Canada*

CHAPTER 1

Introduction

Rao Y. Surampalli, B. R. Gurjar, Tian C. Zhang, and C. S. P. Ojha

This book on Carbon Capture and Storage (CCS) mainly includes the Physical, Chemical and Biological Methods. The book starts with a broad overview of CCS in chapter 2 by Gurjar et al. In this chapter, the authors mainly focus on need and importance of CCS so as to control the greenhouse gases (GHGs) emissions and its consequences on climate change. This chapter reveals an overview of CCS, mentioning CCS as a transitional strategy until renewable and nuclear energies can displace fossil fuel energy.

Further, this book reveals its contents sequentially in two parts. The first part deals with the basic principles of CCS, and it is spread over in 5 chapters (chapter 3 to 7). The second part includes applications and research development related to carbon capture and storage and it is covered in 10 chapters (chapters 8 to 17).

Chapter 3 by Verma et al. sheds light on physical/chemical technologies of CCS. This chapter explains various types of existing carbon capture technologies, application schemes, and their possible future improvements and modifications. The present technology utilizes chemical/physical solvents and sorbents, membranes, enzymes, and innovative processes to capture CO_2 at pre-, post-, or oxy-fuel combustion stages. There are numerous other techniques that are under investigation such as physical solvents/sorbents, molecular sieve, activated carbon, membranes, cryogenic fractionation, chemical-looping combustion, and combination processes. In the end, authors insist for the need of research to investigate best strategies for application of suitable CO_2 capture technique at pre-, post-, or oxy-fuel combustion stages.

However, there is considerable upcoming research regarding the several biological methods for efficient sequestration of CO_2. Chapter 4 by Nouha et al. starts with the discussion about the biological processes for carbon capture, and then provide a state-of-the-art review on biological processes and technologies for CCS, including the major biological processes, approaches and alternatives to i) capturing and ii) sequestrating CO_2, iii) advanced biological processes for CCS, an iv) comparison between biotic and abiotic CCS concerning their merits and limitations. Most of the natural methods are slow and need attention on advanced biological techniques for CO_2 reduction. It is emphasized in this chapter that the efficient utilization of biological methods in all over the world can change the fate of our environment to a stable condition.

The next chapter 5 by Mariyamma et al. focuses principally on carbon sequestration and also discuss about the major disposal initiatives of carbon sequestration namely, physical, chemical and biological process. In this chapter, CO_2 sequestration including ocean, geological, and terrestrial sequestration of CO_2 and leakage is discussed. Finally, the authors conclude that relying on a single method for carbon sequestration will prove to be ineffective in the long run to sequester carbon.

In chapter 6, Ramakrishnan et al. overviews monitoring, verification and accounting of CO_2 stored in deep geologic formations. In general, monitoring and verification features are common to onshore or offshore sites. According to Ramakrishnan et al., there is a need of risk management plan which outlines remediation measurements to the monitoring and verification program throughout the project life. This chapter describes various aspects of baseline surveys, chemical tracers and numerous geophysical techniques, direct observations of the reservoir interval. In all, authors suggest that further developments of sea-floor water-gas chemistry and flux monitoring systems be required before fully operational systems will be available for offshore storage areas.

The first part of the book ends with chapter 7 by Verma et al. in which the focus is on current trends of CO_2 utilization and the concept of carbon minimum economy with examples. This chapter presents a detailed description of reuse as fuel (e.g., methanol made from CO_2 and H_2), reuse as raw materials for plastics and low carbon economy. In this chapter, authors also mention that utilisation of CO_2 for the production of synthetic fuels, chemical feedstock, polymers, and polycarbonates are some exemplary steps. However, authors do not forget to mention that risks associated with CCS in deep ocean and geological formations are significant and pose challenge to the implementation of low carbon economy on a global basis.

To start with the second part of the book, Kao et al. provides information about application and research developments of CO_2 capture technologies for the coal-powered electricity industries in chapter 8. Kao et al. looks into the difficulties and challenges regarding implementation of CCS technologies in coal powered electricity industries. In general, choosing the most promising sorbent and the CO_2 capture technology may not be possible due to the fact that multiple parameters would affect the overall process performance and economics. Retrofitting of CCS in coal-based thermal power plants is a key issue. This is due to the fact that the size and space required for

CO_2 capture process facilities are greater than the size and space for conventional air pollution controls.

Although CO_2 separation and capture from point and nonpoint sources is one of the big challenges, CO_2 scrubbing is the most promising technology due to its wild conditions, low costs, easier regeneration and faster loading. Chapter 9 by Jin et al. deals with the process overview to post-combustion CO_2 scrubbing technologies, followed by discussing advantages and disadvantages, scrubber materials, and applications of CO_2 scrubbing processes. According to Jin et al., research on functionalizing solid supports with amine functional groups for CO_2 capture has reached various stages of development; however, sorbents-based systems still have challenges, such as high heat of reaction and long-term stability.

In chapter 10, Verma et al. illustrates overview and assessment of carbon sequestration via mineral carbonation. This chapter includes a detailed process of mineral carbonation and compared with other methods of carbon sequestration. Authors also discuss about the future research directions, considering advantages and disadvantages of this method. Authors conclude the chapter stating that magnesium can be a better choice as a mineral carbonation agent.

Carbon burial is one of the unique techniques being developed over the period of time to neutralize or reduce the deposits of CO_2 released into the atmosphere from the burning of gases, coal, oil, etc. In chapter 11, Bhattacharya et al. discuss in detail about this technique along with enhanced soil carbon trapping. Carbon entrapping in the soil helps in the crop growth and development, and the cycle of carbon returning back to the atmosphere and from the atmosphere to the soil as burial of carbon continues in the similar manner. Finally, authors summarize that choosing the right kind of crop and plant enhances the soil with deposits of carbon, which eventually gets lost over the period of time.

In chapter 12, Zhang et al. explains the algae-based carbon capture and sequestrations. The authors compared the efficiency of algae with other vegetations and state that algae are superior to others in carbon sequestration among all the vegetation, due to their fast growth rate and possibility of using them for producing green energy such as biodiesel, protein, etc. This chapter also deals with the principle and carbon cycle of algae-based carbon dioxide sequestration, influence factors, and applications of algae-based carbon sequestration followed by a brief cost estimation given at last. In the end, authors remind that algae-based CCS is still not a matured technology and calls for much more efforts to achieve high carbon dioxide sequestration efficiency with low cost.

Kumari et al. present enhanced photosynthesis as a carbon immobilization technique in chapter 13. As forest resources can provide long-term national economic benefits, reforestation and preventing deforestation can be better options for carbon immobilization. Authors also focus on genetic engineering which consists of modifying RuBisCO genes in plants as well as increasing the earth's proportion of C4 carbon fixation photosynthesis plants. Also, authors conclude that better understanding of gene expression in chloroplasts and how to manipulate it predictably will also be beneficial.

In chapter 14, Zhang et al. magnify the enzymatic sequestration of carbon dioxide. Enzymatic sequestration of carbon dioxide is a way to sequester carbon dioxide through transforming carbon dioxide into bicarbonate/carbonate ions, which can be collected and converted into secondary chemicals as raw material for the use by industry. In this chapter, a detailed explanation is given about the type of enzymes used and the mechanisms of using enzyme for carbon dioxide sequestration. Authors also discuss the difficulties to scale up the application of enzymatic carbon dioxide sequestration along with the solutions. Finally, the chapter concludes that it is worth to study in this field in order to find a proper method for carbon dioxide sequestration.

In chapter 15, Bhattacharya et al. introduce biochar as one of the most important CCS technologies. Biochar is produced by a process called pyrolysis, which is the direct thermal decomposition of biomass in the absence of oxygen to obtain an array of solid (biochar), liquid (bio-oil), and gas (syngas) products. This chapter reviews topics related to the biochar for carbon sequestration, including certain biochar production methods and its properties, biochar amendment in soil, the effect of biochar on crop productivity and economy, biochar's capacity for mitigating climate change, and biochar as bioenergy lifecycle. Biochar processes take the waste material from food crops, forest debris, and other plant material, and turn it into a stable form that can be buried away permanently as charcoal. Sustainable use of biochar could reduce the global net emissions of CO_2, methane, and nitrous oxide.

It should be accepted that ocean sequestration is a major natural method for carbon dioxide control in the atmosphere. In chapter 16, Mariyamma et al. throw a light on use of ocean iron/urea fertilization application for sequestering carbon. Authors clearly explain that ocean sequestration of carbon dioxide will help to lower the atmospheric carbon dioxide content on a global scale, their rate of increase and in turn will reduce the detrimental effects of climate change and chance of catastrophic events. This chapter ends with the demand for expensive research to develop techniques to monitor the carbon dioxide plumes, their biological and geochemical behavior in terms of long duration and on a large scale.

In chapter 17, authors address the issues related to modeling and uncertainty analysis of CCS technologies and their performance. In general, CO_2 pipe transport could be modeled by using standard hydraulic equation of flow in which CO_2 is mostly assumed to be transported in dense phase. Authors also focus on different multi-dimensional models such as TOUGH2, ECLIPSE, STOMP, NUFT, LLNL to study the CO_2 sequestration in the reservoirs. A hybrid modeling approach can be applied where detailed numerical models are applied as needed and simpler models are applied in other regions. Also, this chapter takes into account important risk associated with the CO_2 sequestration, i.e., possibility of CO_2 leakage from the saline aquifers into the groundwater and to the atmosphere.

In the end, Zhang et al. discuss the major issues, challenges and the path forward for CCS in chapter 18. This chapter covers cost and economics issues, legal and regulatory issues, social acceptability issues, technical issues along with concerned uncertainty and scalability. Authors insist to overcome the technical, regulatory, financial and social barriers. Deployment of large-scale demonstration CCS projects within a few years will be critical to gain the experience necessary to reduce cost, improve efficiency, remove uncertainties, and win public acceptances of CCS. Finally, it is concluded that wide range of research is needed in the future for CCS development.

CHAPTER 2

Carbon Capture and Storage: An Overview

B. R. Gurjar, C. S. P. Ojha, RaoY. Surampalli, Tian C. Zhang, and P. P. Walvekar

2.1 Introduction

With the advent of Industrial revolution around 1750's, human race entered an era of enhanced industrial activity with the introduction of machines in the production cycle. With unprecedented use of machines there rose a sharp demand for energy to sustain this development, which forces human beings to utilize the most viable available source of energy–the fossil fuels.

However, constantly increased exploitation of the carbon-based energy resources in the last century has led to a substantial change in the atmosphere in the form of increased greenhouse gas (GHG) concentrations. According to fourth assessment report of Intergovernmental Panel on Climate Change (IPCC), carbon emissions from fossil fuel combustion, industrial processes and land use change has increased the ambient CO_2 concentrations, resulting in acidification of world oceans, global warming and climate change (Royal Society 2005; IPCC 2007). It is anticipated that, by 2035, the CO_2 level of 450 ppm, the commonly adopted definitions of a dangerous level of climate change, will be reached with a 77–99% chance of exceeding 2 °C warming. This global challenge could be even more severe because the rate of growth in CO_2 emissions between 2000 and 2005 exceeds the worst case scenario (Gough et al. 2010).

The long-term solution of reducing GHG emissions is to uncouple energy use and CO_2 release. To deal with this issue, an energy technology revolution and energy systems transformation are required, involving superior energy efficiency, increased renewable energies and the decarbonisation of fossil fuel based power generation (Oh, 2010; Dangerman and Schellnbuber 2013). However, the crucial questions is whether a swift transition to sustainable energy systems, based on renewable sources (e.g., biomass, hydro, nuclear, solar, wind, geothermal and tidal energy), can be achieved (Oh 2010; Dangerman and Schellnbuber 2013).

However, it is unlikely that in the near future the alternate energy sources and technologies can fully substitute fossil fuels. Fossil fuel usage is expected to continue to dominate global energy supply as the principle indigenous energy resource. Hence, carbon capture and storage (CCS) is being investigated as a mitigation measure for carbon dioxide emissions and climate change. Such a measure is appearing as a transition until renewable and nuclear energies can replace fossil fuel energy (Williams 2006; Surridge and Cloete 2009).

The current technology options available for mitigation of climate change include improved fuel economy, reduced reliance on cars, more efficient buildings, improved power plant efficiency, decarbonisation of electricity and fuels, substitution of natural gas for coal, CCS, nuclear fission, wind electricity, photovoltaic electricity, and biofuels (Pacala and Socolow 2004). CCS is mentioned as a strong option to achieve the large-scale reductions in CO_2 that are required during this century (IPCC 2005). According to a recent analysis, the emissions of CO_2 will be reduced by approximately 350 Mt CO_2/yr by 2030, if CCS is used extensively after 2020 in the US power sector alone (EPRI 2007). CCS allows the continuous use of fossil fuels by reducing CO_2 releases and also provides time to make the changeover to other energy sources in a systematic way. In a recent European Union (EU) survey, a majority of the energy experts believed that CCS is certainly necessary both globally and nationally in order to mitigate climate change (Alphen et al. 2007). However, many factors decide CCS applicability worldwide, such as technical development, overall potential, flow and shift of the technology to developing countries and their capability to apply the technology, regulatory aspects, environmental concerns, public perception and costs (IPCC 2005).

CCS issues have been addressed/reviewed since the early 1990s (e.g., Riemer et al. 1993; USDOE 1999; Herzog 2001; Anderson and Newell 2003; IPCC 2005; IEA 2009; Lackner and Brennan 2009; CCCSRP 2010; Zhang and Surampalli 2013). However, still there is a need to review CCS technologies because new information is now being generated at a faster pace. Particularly, this chapter serves as an overview chapter to introduce CCS technologies, with major issues (e.g., concerns, constrains, and major barriers) and future perspectives being discussed.

2.2 CCS Technologies

CCS is a course of methodologies consisting of the separation of CO_2 from industrial and energy-related sources, compressing this CO_2, transport to a storage location and long-term isolation from the environment (Fernando et al. 2008). Many of these components are already used in other settings and working together to prevent CO_2 from entering the atmosphere (Oh, 2010; Zhang and Surampalli 2013). This section provides a brief overview of the major CCS technologies currently used.

2.2.1 CO_2 Capture

Capture technologies can be categorized based on whether a) carbon capture is from concentrated point sources or from mobile/distributed point- or non-point sources; and b) the technique involves physical/chemical or biological processes (Zhang and Surampalli 2013). Major technologies are briefly described below.

Category a). Mobile/distributed sources like cars, on-board capture at affordable cost would not be feasible, but are still needed. However, industries have used technologies for CO_2 capture from concentrated point sources for very long time, which is mainly to remove or separate out CO_2 from other gases that are produced in the generation process when fossil fuels are burnt (IEA 2009). This can be done in at least three different ways: 'post-combustion', 'pre-combustion' and 'oxy-fuel combustion (see Fig. 2.1).

Post-combustion Capture. This involves CO_2 capture from the exhaust of a combustion process. The methods for separating CO_2 include high pressure membrane filtration, adsorption, desorption processes and cryogenic separation. Among all these methods, the more established method is solvent scrubbing. Currently, in several facilities, amine solvents are used to capture CO_2 significantly (IEA 2009). The absorbed CO_2 is then compressed for transportation and storage.

Pre-combustion Capture. Fuel in any form is first converted to a mixture of hydrogen and carbon dioxide by gasification process and then followed by CO_2 separation to yield a hydrogen fuel gas. The hydrogen produced in this way may be used for electricity production and also in the future to power our cars and heat our homes with near zero emissions. The pre-combustion capture technology elements have already been proven in various industrial processes other than large power plants (IPCC 2005).

Oxy-fuel Combustion Systems. In oxy-fuel combustion, the recycled flue gas enriched with oxygen (separated from air prior to combustion) is used for combusting the fuel so as to produce a more concentrated CO_2 stream for easier purification. This process confirms high efficiency levels and offers key business opportunities. This method has been demonstrated in the steel manufacturing industry at plants up to 250 MW in capacity (IEA 2009).

In general, for power generation projects, most studies estimate CO_2 capture will account for up to 75% of the total cost of CCS, measured in cost per tonne stored. Part of this cost is due to the energy required by the capture process itself. Finally, CO_2 can also be captured in restricted quantities from industrial practices that do not involve fuel combustion, such as natural gas purification (Fernando, et al., 2008).

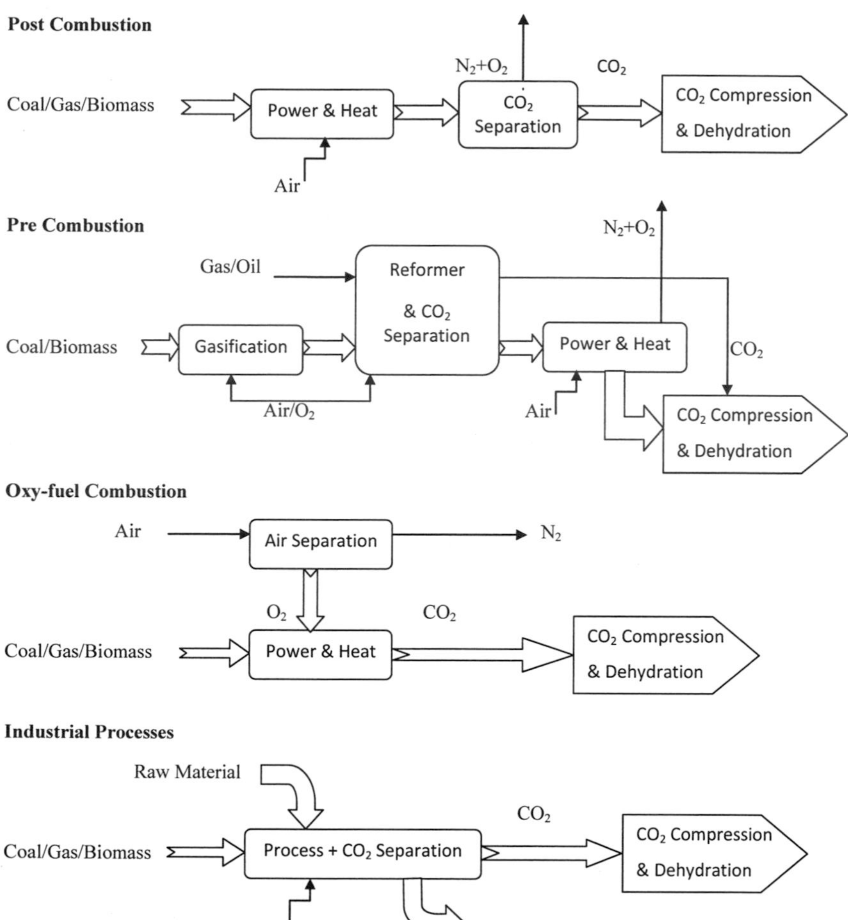

Figure 2.1. Various types of capture processes (adapted from IPCC 2005)

Category b). Other than the three technologies described in Category a), sorption and membranes are the two major physical/chemical technologies for carbon capture. There are many biological technologies that can be used for carbon capture from either point or non-point sources, such as i) trees and organisms; ii) ocean flora; iii) biomass-fueld power plant, biofuels and biochar; and iv) sustainable practices (e.g., soils, grasslands, peat bogs). Biological methods often combine carbon capture and sequestration together, as shown in Table 2.1 (Zhang and Surampulli 2013).

Table 2.1. Alternative biological technologies for carbon capture/sequestration[a]

Methods	Description
1) Trees/organisms	• Capture CO_2 via photosynthesis (e.g., reforestation or avoiding deforestation); cost range 0.03–8\$/t-CO_2; one-time reduction, i.e., once the forest mature, no capture; release CO_2 when decomposed • Develop dedicated biofuel and biosequestration crops (e.g., switchgrass); enhance photosynthetic efficiency by modifying Rubisco genes in plants to increase enzyme activities; choose crops that produce large numbers of phytoliths (microscopic spherical shells of silicon) to store carbon for thousand years.
2) Ocean flora	• Adding key nutrients to a limited area of ocean to culture plankton/algae for capturing CO_2. • Utilize biological/microbial carbon pump (e.g., jelly pump) for CO_2 storage. • Problems/concerns: a) large-scale tests done but with limited success; b) limited by the area of suitable ocean surface; c) may have problems to alter the ocean's chemistry; and d) mechanisms not fully known.
3) Biomass-fueled power plant, bio-oil and biochar	• Growing biomass to capture CO_2 and later captured from the flue gas. Cost range = 41\$/t-CO_2 • By pyrolyzing biomass, about 50% of its carbon becomes charcoal, which can persist in the soil for centuries. Placing biochar in soils also improves water quality, increases soil fertility, raises agricultural productivity and reduce pressure on old growth forests • pyrolysis can be cost-effective for a combination of sequestration and energy production when the cost of a CO_2 ton reaches \$37 (in 2010, it is \$16.82/ton on the European. Climate Exchange).
4) Sustainable practices, e.g., • Soils/grasslands • peat bogs	• Farming practices (e.g., no-till, residue mulching, cover cropping, crop rotation) and conversion to pastureland with good grazing management would enhance carbon sequestration in soil. • Peat bogs inter ~25% of the carbon stored in land plants and soils. However, flooded forests, peat bogs, and biochar amended soils can be CO_2 sources.
5) Enzymatic sequestration	• CO_2 is transformed, via enzymes as catalysts, into different chemicals, such as i) HCO_3^-/CO_3^{2-}, ii) formate, iii) methanol, and iv) methane.

[a]Adapted from Zhang and Surampulli (2013).

2.2.2 CO_2 Transport

After capturing, the CO_2 must be transported for storage at a suitable site by various means such as pipelines, ships, trucks or trains.

Pipeline. Carbon dioxide is already transported for commercial uses by road trucks, pipeline and ships. Hence local and regional infrastructures of pipelines will eventually be developed. The pipeline transportation technologies are little dissimilar from those used extensively for transporting oil and gas all over the world. In some cases it may be possible certainly to re-use existing pipeline networks. Large networks of CO_2 pipelines are already in use and are confirmed to be safe and reliable. The development and management of CO_2 pipeline networks will be a major international business opportunity for professionals in this area (GCCSI report 2009).

Pipeline transportation of CO_2 has some industry experience, primarily in the oil and gas sector. This is the most economical method of high quantity CO_2 transportation over long distances. CO_2 pipelines are in operation and operated safely for over 30 years in USA and Canada through 6200 km of pipeline network. CO_2 pipelines function at much higher pressure than natural gas pipelines and also, CO_2 pipeline technology has comparatively less developed than oil and gas pipelines (IEA 2009).

Land. When pipeline technology is expensive, and smaller quantities are to be transported over short distance, rail and tankers are the best suitable option for CO_2 transportation (IPCC 2005).

Shipping. This option is possible when the distance between emission source and seaport facilities is adequate to load CO_2 for injection in offshore locations. Since several decades, transportation of liquefied natural gas occurs and further research work is in progress in Norway and Japan to adjust this technology to transport CO_2 by ships (GCCSI report 2009).

2.2.3 CO_2 Storage

There are various options available for CO_2 storage such as deep saline reservoirs, depleted or declining gas and oil fields, enhanced oil and gas recovery, enhanced coal bed methane, basalt formations and others (GCCSI report 2009). From the ecological and economic perspectives, storage in geological formations is currently the most attractive option. Some of these methods are described below.

Enhanced Hydrocarbon Recovery. Apart from pure storage, carbon dioxide can also be used for Enhanced Hydrocarbon Recovery. This includes Enhanced Oil Recovery (EOR), Enhanced Gas Recovery (EGR) and Enhanced Coal-bed Methane Recovery (ECBM). Any oil or gas that is recovered through these methods would otherwise not be extracted and therefore has an economic value which would offset some of the costs of CO_2 sequestration.

EOR. In crude oil extraction, numerous different techniques are used to increase the yield. One of these is the injection of CO_2. The injected CO_2 increases the pressure in the reservoir and diffuses into the crude oil, making it more fluid and therefore easier to extract. Therefore, by using CO_2 for EOR the oil yield can be increased, and at the same time carbon dioxide can be permanently transferred into geological formations and be removed from the atmosphere. The latter applies at least to the portion of the CO_2 that is not mixed with the oil. Owing to the economic incentives CO_2 EOR is often regarded as an attractive way to begin using CCS. However, EOR only generates additional profits in those places where it is possible to establish a cost-effective infrastructure (short pipeline distances, etc.). Enhanced oil recovery through carbon dioxide injection is already being used at various places across the world (e.g. the Weyburn oil field in Canada) and can be regarded as an established technology. On the other hand, there has been no practical experience with the analogous process of Enhanced Gas Recovery (EGR), for which to date there has only been work on simulations (Fischedik et al. 2007).

EGR. EGR can be achieved using CO_2 as it is heavier than natural gas. CO_2 is injected into the base of a depleted gas reservoir and will tend to pool there, causing any remaining natural gas to "float" on top of it. This then drives the natural gas towards the production wells. However, since a high percentage of the natural gas contained in many gas fields can be recovered without using enhanced recovery techniques, the potential target for EGR is small.

ECBM. Coal beds (also known as coal seams) can be reservoirs for gases, due to fractures and micro pores in which natural gas, known as coal-bed methane (CBM), can be found adsorbed onto the surface. However, CO_2 has a greater adsorption affinity onto coal than methane. Thus, if CO_2 is pumped into a coal seam towards the end of a coal-bed methane production project, it displaces any remaining methane at the adsorption sites, allowing methane recovery jointly with CO_2 storage. Experiments have been conducted in the San Juan Basin showing that CO_2 injection does appear to have enhanced CBM production. Smaller field trials of ECBM production using CO_2 are under way in Europe, Canada and Japan. However, there are issues with ECBM; for example the low permeability of seams means that a large number of wells may be needed to inject sufficient amounts of CO_2. Moreover, the methane in coal represents only a small proportion of the energy value of the coal, and the remaining coal could not be mined or gasified underground without releasing the CO_2 to the atmosphere. Finally methane is a far more potent greenhouse gas than CO_2. Therefore, steps would have to be taken to ensure no methane leakage to the atmosphere took place (CCSA 2010).

In contrast to geological storage, industrial utilisation (e.g. production of carbonic acid, dry ice, raw materials for polymer chemistry) will only be possible on a small scale. Furthermore, in these cases the CO_2 is not removed for ever from the atmosphere but in fact released again at a later date. A net effect here is only achieved if the CO_2 used replaces technical production and supply of CO_2 (i.e. specially for the industrial purpose) elsewhere.

The storage of CO_2 in geological formations can be accomplished through many processes and technologies already used in the oil and gas industry and in handling liquid wastes. Drilling and injection processes, monitoring methods and computer simulations about the distribution of the CO_2 in the reservoir would, however, have to be adapted to the specific requirements of CO_2 storage. Here there is still a considerable need for research and development. The EU-funded CO_2 sink project at Ketzin near

Berlin is contributing to resolving these questions through its research into the behaviour and controllability of CO_2 in underground reservoirs (Fischedik et al. 2007).

Other Alternatives. The idea of binding CO_2 in the marine environment either directly (storage in the ocean depths) or indirectly (e.g. algae formation) is currently being pursued only sporadically (mainly in Japan) due to public opposition (the question of permanence of storage, insufficient knowledge of the effects on marine ecosystems) and low efficiency. CO_2 can also be fixed through the deliberate cultivation of biomass (e.g. through forest planting), although this stores CO_2 for only a few decades. Additionally, especially in the United States, processes for binding CO_2 to silicates (mineralisation) are being discussed, but the high energy requirements and large amounts of material to be disposed of are discouraging. This means that from today's perspective the geological storage options are clearly the most realistic ones. Owing to the many uncertainties involved, current estimates of storage potential differ enormously. Ultimately, a case-by-case assessment will be required if we are to gain insights into storage capacity. IPCC estimates put global storage capacity at between 1,678 and 11,100 Gt CO_2, with 2,000 Gt CO_2 classed as technically viable (IPCC 2005). By way of comparison, global CO_2 emissions in 2005 amounted to 27.3 Gt CO_2.

2.3 Current Status of CCS Technology

Maturity of CCS System. Only large scale point sources which produce approximately 50% of CO_2, such as power plants, steel mills, cement plants, refineries, and coal-to-liquid plants, are targets for application of CCS techniques (Fernando et al. 2008). The technical maturity of particular CCS system components such as capture, transport or storage varies significantly and the overall CCS system may not be as mature as some of its components. While many of the component technologies of CCS are relatively mature (see Table 2.2), there are no fully integrated, commercial-scale CCS projects in operation till date (McKinsey, 2008).

Phases and Different Kinds of CCS Project. The asset lifecycle model indicates five phases of the CCS project: as identify, evaluate, define, execute and operate. Planned projects are in identification, evaluation and definition stage. Active projects are in executional or operational stage after having been sanctioned. Delayed projects are those that encompass activities postponed and held up. Cancelled projects are those that have ceased activities without fulfilling their purpose and have no intention of resuming. Completed projects are those that have fulfilled their original purpose and have ceased operation. The majority of the completed projects are relatively small in scale (Fig. 2.2). This may be because the economic, technical, regulatory and public acceptance challenges were smaller or fewer at this scale. As a result, these challenges did not present significant barriers to these projects. No integrated projects have been completed at any scale (GCCSI report, 2009).

Table 2.2. Maturity of CCS system components (adapted from IPCC 2005)

Phase	CCS Component	CCS Technology
Research	Ocean Storage, Mineral Carbonation	Direct Injection Natural silicate minerals
Demonstration	Capture Geological Storage Mineral Carbonation	Oxy-fuel combustion Enhanced Coal Bed Methane recovery (ECBM) Waste materials
Economically feasible under specific conditions	Capture Transport Geological Storage	Post-combustion, Pre-combustion Shipping Gas or oil fields, Saline formations
Mature market	Capture Transport Geological Storage	Industrial separation (natural gas processing, ammonia production) Pipeline Enhanced Oil Recovery (EOR)

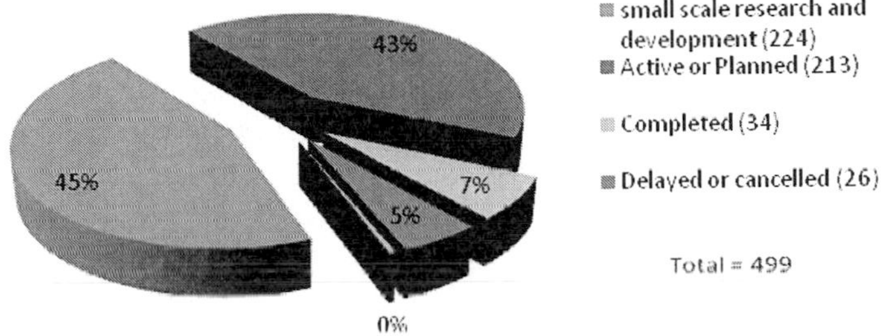

Fig. 2.2. The database of CCS projects (adapted from GCCSI report 2009)

Importance. On 7 October 2008, the European Parliament voted to set an emission limit of 500 g CO_2 per KWh on new plant from 2015, essentially mandating the use of CCS on any new coal-fired power station. In addition the European Parliament also voted to establish a 10 billion dollars fund to support CCS projects (Gough et al., 2010). The IEA has recently weighed up the importance of CCS in achieving required emissions reductions. The resulting roadmap concludes that 100 CCS plants must be operational by 2020, with 38 of these in the power sector and more than 3000 by 2050 (Table 1.3). This needs a huge stepping up from the current condition. An industrial technology is also proposed that captures CO_2 directly from ambient air to target the remaining 50% emissions (Zeman, 2007). CCS deployment will be a function of how policy impacts the power producers' cash flows and in turn how this impacts their least cost compliance strategies (Fernando et al. 2008).

Table 2.3. Global deployment of CCS in 2010–2050 by sector[a]

Year	Number of CCS Projects (Projected)	Power (%)	Industry (%)
2020	100	38	35
2030	850	42	42
2040	2100	47	34
2050	3400	48	32

[a]Adapted from IEA (2009).

Deployment of CCS. CCS has been developed predominantly in Japan, Europe, Australia and North America. However, there is a great potential for the introduction of CCS into China, India and other industrializing countries as the largest growth in CO_2 emissions arises from fast economic growth. In China, major carbon capture opportunities and also CCS enabling technologies exists (Liu and Gallagher 2010). CCS is not currently a priority for the Government of India (GOI) because, as a signatory to the UNFCCC and Kyoto Protocol, there are no existing greenhouse gas emission reduction targets and most commentators do not predict compulsory targets for India in the post 2012 segment (Shackley and Verma 2008; Kapila et al. 2009). CCS could also contribute to energy security and to economic growth, through encouraging technological innovation. Several research and development (R&D) projects on CCS have been initiated in the last few years, and demonstration projects are being implemented all over the world (Fischedik et al. 2007).

CCS can only prudently be applied to large-scale point source emissions. Alongside the power generation as the typical application, this also applies to various industrial applications such as steel industry where carbon-based fuels are used to supply energy or where chemicals like ammonia or fuels are produced. In fact, in industrial applications the conditions may actually be considerably more favourable, because here CO_2 sometimes occurs in higher concentrations than in power generation flue gases. For the many decentralised CO_2 sources outside the power generation sector, CCS is not available for direct application. But indirectly there is potential for CCS to make a contribution here too, through centralised production of low carbon fuels (Fischedik et al., 2007).

Capture technologies are based on those that have been applied in the chemical and refining industries for decades, but the integration of this technology in the particular context of power production still needs to be demonstrated. Transportation of CO_2 over long distances through pipelines has proven successful for more than 30 years in the central US, which has more than 5,000 km of such pipelines for EOR. According to IPCC special report on CCS published in 2005, there have been three commercial projects which concerns CCS. They are offshore Sleipner natural gas processing project in Norway, the Weyburn Enhanced Oil Recovery (EOR) project in Canada and the In Salah natural gas project in Algeria as of mid-2005. 1–2 Mt CO_2 is captured by each project per year. The industry can also build on the knowledge obtained through the geological storage of natural gas, which has been practiced for decades (McKinsey, 2008).

In September 2008, Vattenfall's 30 MW Schwarze Pumpe oxy-fuel pilot capture project in Germany was opened. Several other CCS projects have been announced recently, for example in Germany (RWE's Hürth project), the US (AEP Alstom Mountaineer), Australia (Callide Oxy-fuel) and China (GreenGen). Establishing a first set of such "demonstration" projects is generally considered the next necessary step in CCS development. The purpose of such projects would be to prove that the technology works at scale and in integrated value chains; to get a more accurate picture of the true economics of CCS; to validate storage potential and permanence; to prove transport safety; and to address public awareness and perception issues (McKinsey, 2008).

Numerous other CCS projects (especially demonstration and research projects) are in planning and will play a decisive role for the further development of the technology over the coming 10 to 20 years. They will show whether CCS can fulfil the necessary technical, economic and ecological requirements for its large-scale use and what role CCS can play in national and international energy systems (Fischedik et al., 2007).

2.4 Barriers to CCS

The widespread deployment of CCS projects is not achieved because of major hurdles observed. For scaling up of CCS projects these challenges have to be studied to overcome them. A wide literature is available in which these obstacles are discussed in detail (IPCC 2005; Fernando et al. 2008; McKinsey 2008; GCCSI report 2009; Zhang and Surampalli 2013). Accordingly, some of these barriers are described below.

2.4.1 High Capital Investment

Huge initial investment for CCS projects is a major barrier. Integrated CCS system will have costs attached to the compression, transportation, injection, storage and monitoring of captured CO_2. Initial capital investment is projected to increase approximately 50% for coal power plants with CCS compared with the non-CCS option (McKinsey 2008). The capital cost may be very high for early commercial projects in particular (Fig. 2.3). The subsidy or grants requirements may be as high as $1billion for a 900MW coal power plant (McKinsey 2008). The positive cash flows must be generated by these projects to become commercially viable. But the time horizon required for these projects is longer than the normal because of high capital costs. It is very difficult to guarantee such income streams over long periods as the technology is new with unproven track records (Rai et al. 2010). The firms show reluctance to extend the same performance guarantee for new and unproven technologies in the current construction environment (Fernando et al. 2008). The cost of technology will come down with experience but only incrementally because CCS is not a single technology, rather combination of processes and technological change will occur incrementally with component technologies. Cost reduction opportunities will only arrive through widespread deployment of CCS projects and continuation of R&D to support successive technology improvements (GCCSI report 2009).

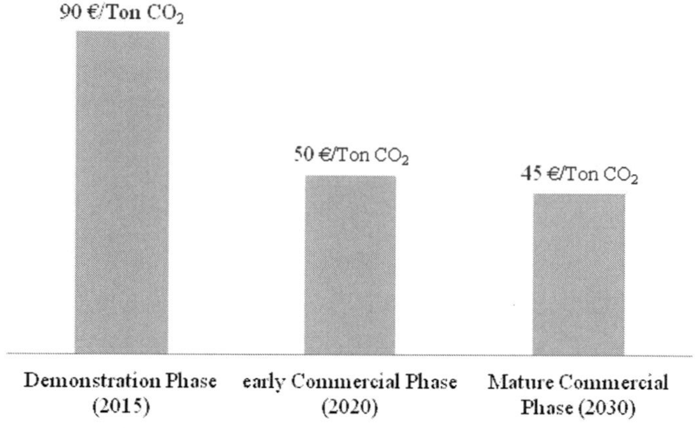

Figure 2.3. Forecast of development of CCS cost (adapted from McKinsey 2008)

The development costs of CCS projects are also high and could be between 10–15% of the total installed capital costs of a project as suggested by industrial experience. This could be a huge amount in hundreds of millions of dollars for CCS projects (GCCSI report 2009). The coal plant construction costs are rising intensively compared to other renewable technologies as escalation in materials cost hit it harder than other technologies. The sources of funding like key funding agencies such as industry groups, national governments and institutions should be identified to support CCS projects. The fundamental role played by governments to reduce project uncertainties and costs will be the key issue for the successful deployment of CCS projects.

2.4.2 Policy Options

The revenue streams from CCS projects depend on regulatory actions to be taken. On the basis of avoided emissions, the cost of CCS ranges from $30–90/tonnes of

CO_2 (Rubin et al. 2007) which turns into 60–80% increase in the cost of electricity (Dalton 2008). The increased cost of electricity has to be paid so that commercial entities are profitable enough to attract continued investment. Policy incentives are given for electricity from renewable energy sources such as mandatory Renewable Portfolio Standards (RPS) as in many states in United States and Feed-In-Tariffs (FIT) in Germany. Such demand-pull scheme does not exist for CCS. The development of CCS systems depends on special government policies applied at broad scale. High risk is associated with undertaking CCS projects without credible schemes in place to ensure cost recovery more broadly (Rai et al. 2010). It is unlikely that commercial developers will invest in CCS projects unless there are supporting governmental policies. The durability of policies and incentives for CCS systems should be over the entire deployment period in order to provide comfort to investors that regulations will not immensely change over time. The selection of CCS as a key component of the compliance strategy depends upon design of climate policy in terms of number of allowances, cap stringency, and the structure of regional electricity markets. The governments must also practise measures that push technologies into marketplace. The crucial factors for the improvement of technologies, reducing costs and mitigating investor risk are performance standards, funding for research and development, and large-scale demonstration projects encompassing the full CCS system (Fernando et al. 2008). It is really critical to provide policy frameworks based on similar incentives as that of other competitive new technologies to develop CCS systems to stay within (GCCSI report 2009).

2.4.3 Uncertainties in Regulations and Technical Performance

There is limited experience with the integrated CCS system that combines power generation with capture, transport and long term storage of CO_2 at scale. The component technologies are at different points of maturity. Although component technologies are not new, the technological and operational experience is almost nil for CCS from power plants. The lack of experience results in higher cost and extreme difficulty in performance predictions (Rai et al. 2010). The uncertainties are also associated with the supporting infrastructure facilities like construction and operation of dedicated CO_2 pipeline system, responsibility of long term storage of CO_2 etc. Mostly, the improvements in CCS technology will be incremental in nature based on first of kind experience of CCS plants and related research activity (Gibbins and Chalmers 2008). Consequently, uncertainties in the technical performance increases affecting long-term viability of investments in the technology.

A strong regulatory regime is required to govern the scaling up of CCS systems. Regulatory and legal framework associated with injection, storage, monitoring and long term liability is needed to ensure that CCS systems are safe and effective enough as a climate change mitigation measure. However regulatory uncertainties allied to scaling up of CCS systems are also high. Liabilities related to leakage and long term storage of CO_2 are most important as enough leakage of the CO_2 would reverse benefits of sequestration and become dangerous to human, water supply and property (Fernando, et al., 2008). A lack of capacity among regulatory agencies is also an important factor which leads to delays in approvals. These regulatory issues are complex because it involves national and international jurisdictions. The existing regulatory systems related to CCS are not yet suited to address some critical issues, such as the need for thorough site characterisation, careful monitoring and long-term stewardship (IRGC, 2008).

The progress in technological and regulatory issues related to CCS systems is mutually dependant on advances in technology and regulation respectively. For the moment, the uncertainties in both, technical and regulatory regimes results in deadlock. Unless there is removal of uncertainty in one area, the other has little chance to move further (Rai et al., 2010).

2.4.4 A Complex Value-Chain

A complex value chain of CCS systems is also a key barrier in scaling up CCS. The component systems of CCS are having completely different risk attitudes, which resemble a major obstacle in scaling up CCS systems. The best example is diversity in risk policies for power generation and geological storage business in the U.S. (Rai et al. 2010). The power generation business is controlled by low risk regulated utilities; on the contrary, geological storage business is controlled by high risk regulated utilities. Accordingly, the informative knowledge about geological storage is not available as it is held by major oil companies based on risk policy. This difference in risk policies in the same value chain leads to investment stalemate as investors face difficulty in managing co-dependent commercial risk. Much experience is not available in complex value chain CCS systems to organise at scale in various contexts but there is enough awareness about solving the complexity of CCS value chain at scale as a most crucial issue.

2.4.5 Public Safety and Support

The support of general public and stakeholders is the prime issue to deploy CCS system at scale. The risk associated and safety measures provided are the key concerns of public in general. This problem concerned to CCS system is relatively isolated in nature and not as acute as of for nuclear power systems (Rai et al. 2010). But according to other study in US the acceptance of CCS system is lower comparative to that of nuclear system (McKinsey 2008). The public concern is mostly about the health and ecosystem risk associated with capture, transport and storage elements, but most importantly with leakage of CO_2 stored. Although demonstration projects show the risk related to CCS systems is low, the public perception plays critical role in deployment of CCS systems at the commercial scale (Fernando et al. 2008). Moreover, several studies show the possible solutions to solve the uncertainties related to public safety and support (Johnsson et al. 2009; Duan 2010; Ashworth et al. 2010).

2.5 Major Issues Related to CCS

The present status of four major issues and related enabling methodologies are discussed in this section.

2.5.1 Costs of Implementation

A massive investment is required to implement CCS technology as a measure of climate change mitigation; running into hundreds of millions of dollars depending on the type of plants. But without CCS, it is just impossible to achieve CO_2 emission targets to be halved by the year 2050 (Oh 2010).

In 2008, McKinsey concluded that the early full commercial scale CCS projects, potentially to be built shortly after 2020, are estimated to cost € 35–50 per tonne CO_2 abated. The initial demonstration projects to be deployed around 2012–15, would typically cost between € 60–90 per tonne CO_2 abated because of their smaller scale, and focus on proving technologies rather than optimal commercial operations. Costs for some projects such as those with large transport distances may even fall outside this range. The later CCS cost after the early commercial stage would depend on several factors including the development of the technology, its economies of scale, the availability of favourable storage sites and the actual roll-out realized. A total CCS cost between € 30–45 per tonne CO_2 abated for new power installations could be reached, assuming a roll-out in Europe of 80–120 projects by 2030. The costs could be lowered roughly by € 5 per tonne CO_2 in case of global roll-out reaching 500–550 projects by 2030 (McKinsey 2008).

CO_2 capture alone will increase the cost of electricity from US $ 43 per MWh to US $ 61–78 per MWh for new power plants and from US $ 17 per MWh to US $ 58–67 per MWh for existing coal plants. Separation and compression typically account for over 75% of the costs of CCS, with the remaining costs attributed to transportation and underground storage (Fig. 2.4). Pipeline transportation costs are highly site-specific as they depend heavily on economy of scale and pipeline length. Costs of underground storage are estimated from US$ 3–10 per tonne CO_2 (Oh 2010).

Individual project costs can vary from the reference case costs, depending on their explicit characteristics such as their location, their scale, and the technologies being experienced. The differences in cost between the three main capture technologies are relatively small today indicating that multiple technologies should be tested at this early stage of development. For a demonstration project, a transportation distance 200 km longer than the reference case would add € 10 per tonne CO_2. As cost of CO_2 capture for new plants is high, new concept of capture ready plants comes into picture. A capture ready plant is a plant which can be retrofitted with CO_2 capture when the necessary regulatory or economic systems are efficient to work (IEAGGP 2007). Retrofitting of existing power plants is likely to be pricier than new installations, and economically viable only for relatively new plants with high efficiencies (McKinsey 2008).

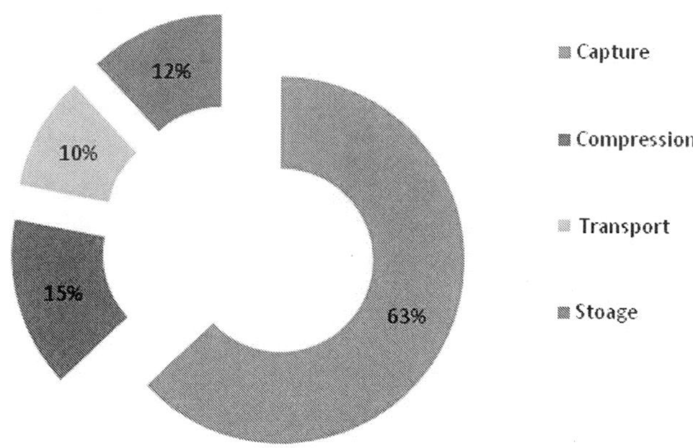

Figure 2.4. Distributions of CCS costs (adapted from Fischedik et al. 2007)

Retrofit requires large additional capital investment which is usually not predictable in the upfront investment decision and may thus make some plants unprofitable before the end of their lifetime. In addition, because of its negative impact on conversion efficiency, it is only suitable for highly efficient plants (Praetorius & Schumacher 2009).

In addition to the high cost of CCS, the energy penalty for capture and compression is also high. The post-combustion, end-of-pipe capture technologies use up to 30% of the total energy produced, thus spectacularly decreasing the overall efficiency of the power plant. Oxy-combustion has a similarly high energy penalty because it requires separation of a pure source of oxygen from air, although eventually, new materials may compensate the penalty by allowing for higher temperature and consequently result in more efficient combustion. Pre-combustion technologies have the potential to lower energy penalties to the range of 10–20%, leading to higher overall efficiency and lower capture costs (Benson 2004).

The reduction of costs and increase in revenue in the short-term is crucial for economic viability and subsequent future deployment of CCS. Vital areas from research perspective include materials and the need to reduce the energy penalty of capture. All stages of the CCS series require cheaper materials, such as new coatings, surface finishes and linings near the point of injection. Costs will also be affected by the global availability of materials. Reduction of the energy penalty of capture is currently hindered by a lack of commercial investment and knowledge gaps (Gough 2010). Significant cost improvements can be expected in CO_2 capture beyond the demonstration phase provided an "industrial scale" roll-out takes place (McKinsey 2008).

Costs of new technologies usually come down as experience is gained by producing and using the product. The share of the market controlled by a new technology plotted against time typically follows an S-curve (Geroski 2000). An observation as 20 percent unit cost reduction for a doubling of cumulative installed capacity is widely used to project future costs of energy technologies (OECD/IEA 2004). But the circumstances for CCS are somewhat different from the usual technology cost curve because it is an integrated process and technological change will occur via incremental improvements to component technologies. Cost reductions of CCS systems should be calculated as the sum of all process cost reductions per level of installed capacity in capture, transport, and storage of CO_2 (Fernando et al. 2008). According to the IEA, cost of capture is estimated to come down 50 percent by 2030 (OECD/IEA 2004), while the IPCC, estimates cost reductions of 20–30% in the next decade (IPCC 2005). For post-combustion capture, research is being conducted to test and develop better solvents that could reduce the energy penalty. Studies suggest that solvents such as chilled ammonia may reduce power diverted for capture to as little as 10%. For pre-combustion capture, researchers are developing membrane technologies for separating the CO_2 from gas, which may have the potential to reduce power requirements by 50% (EPRI 2007). While a significant portion of CCS costs are associated with capture, additional costs will be incurred for transport and storage. Transport costs will largely depend on what type of transportation system is developed. A centralized CO_2 system may develop if the CO_2 is to travel very long distances to localized geologic storage sites. A more decentralized system could also be developed if suitable sequestration sites are located in close proximity to the plant (Fernando et al. 2008). For the reference case of new coal power installations, CCS costs could come down to around € 30–45 per tonne of CO_2 abated in 2030 which is in line with expected carbon prices in that period (McKinsey 2008).

The situation is changing as several governments plan to ramp financial support for CCS demonstration projects. Governments' interest in CCS is generally rooted either in concerns about global warming or in the desire to continue to use coal or unconventional oil reserves even in a carbon constrained world. Concerned governments, notably the US, European Union (EU), Australia, and Canada (Alberta and British Columbia), are gearing up to provide multi-billion dollar support for CCS related R&D projects (Rai 2010).

2.5.2 Health, Safety and Environment Risks

Carbon dioxide is generally regarded as a safe, non-toxic, inert gas and also an indispensable part of the basic biological processes of all living things. Though CO_2 is a physiologically active gas that is essential to both respiration and acid-base balance in all life, the exposure to high concentrations can be harmful and even fatal. Ambient atmospheric concentrations of CO_2 are currently about 370 ppm. Humans can sustain increased concentrations with no physiological effects for exposures up to 1% CO_2 (Benson 2004).

The CCS system achieves a significant reduction of greenhouse gas emissions but has multiple environmental trade-offs. The capture process designed to capture 90% of the CO_2 in the flue gas captures only 75% of the life cycle CO_2 emissions and avoids 70% of total CO_2 emission over the life cycle (Singh et al. 2010). For all capture systems it was found that SO_2, NO_x and PM emissions are expected to be reduced or remain equal per unit of primary energy input compared to power plants without CO_2 capture. For all three capture systems an increase in NO_x emissions per kWh is possible, but is the most likely for the post-combustion capture at coal and gas fired power plants. With pre-combustion CO_2 capture, a reduction in SO_2 emissions per kWh is considered the most probable although an increase is also reported. PM emissions may increase per kWh when applying post- and pre-combustion CO_2 capture. Increase in primary energy input as a result of the energy penalty for CO_2 capture may for some technologies and substances result in a net increase of emissions per kWh output. The emission of ammonia may increase by a factor of up to 45 per unit of primary energy input for post-combustion technologies. Quantitative emission estimates for Volatile Organic Carbons from power plants equipped with CO_2 capture are not available in the scientific literature; although the pertaining literature suggests that VOCs may be emitted during operation of some amine based post-combustion technologies. VOC emission may decrease per primary energy input when implementing oxyfuel combustion and pre-combustion CO_2 capture (Koornneef et al. 2010).

The implementation of CCS reduces the greenhouse gas emissions by 64%, from 459 g CO_2 equiv/kWh to 167 g CO_2 equiv/kWh. This figure is lower than 70% net reduction of CO_2 due to emission of other GHG substances (CH_4, CO, N_2O). With CCS, a major portion of the Global Warming Potential (53%) emanates from the fuel production chain and 28% from the power plant. The transport and storage chain contributes only about 3% to the total GWP impact. However, there is a net increase in all other environmental impact categories, mainly due to the energy penalty, infrastructure development and direct emission from the capture process. CCS causes an increase of 21–167% for other impact categories, with a relatively high increase in all the toxicity potentials. NO_x emission from fuel combustion is largely responsible for increase in most direct impacts other than toxicity potentials and GWP, contributing 69% to direct air pollution. Increased infrastructure requirements contribute most to the increase in human toxicity and terrestrial toxicity. The scenario study of best-case and worst-case CCS shows a decrease of 68–58% in GWP, respectively with significant increases in toxicity impacts (Singh et al. 2010).

Carbon dioxide is regulated by central and State authorities for many different purposes, including working safety and health, ventilation and indoor air quality, confined-space hazard and fire suppression, as a respiratory gas and food additive. Current occupational safety regulations are adequate for protecting workers at CO_2 separation facilities and geologic storage sites (Benson 2004).

Many of the fears and concerns relating to support of international CCS projects are focussed on uncertainty over the environmental performance of the projects over the medium to long term. Environmental reliability can be assured through vital international procedures addressing the selection of storage sites that exhibit excellent trapping mechanisms, assessment and suitable management of the risk of CO_2 leakage, allocation of responsibility for monitoring and reporting, allocation of responsibility for any environmental damage caused (CCSA Position Paper 2009).

Assessment of electricity production with the current technology for post-combustion CO_2 capture and transport indicates that there are considerable adverse environmental interventions of CCS, besides the benefit of reduced global warming potential. The key areas identified to reduce the adverse impacts are technical developments to reduce energy penalty and degradation of toxicity in capture process to reduce the negative impacts (Singh et al. 2010).

The single most important factor for long-term environmental stability is the selection of storage site. CCS experience to date and geology research shows that well-chosen sites would be very unlikely to ever leak CO_2 to the atmosphere or even to the water column. Other commonly quoted fears include unpredictable subsurface CO_2 movement and the possible impact of seismic events. Again, thorough site-selection would minimise unpredictability and effective monitoring of CO_2 plumes would allow for early corrective action in the event of any unexpected CO_2 migration. An internationally accepted Monitoring, Reporting and Verification (MRV) protocol for CCS would provide control for other environmental concerns surrounding the capture, transport and injection aspects of CCS and the possibility of local environmental impacts including specific concentrated CO_2 leaks and general impacts from construction and operation of plant. Existing experience with CO_2 transport and handling, including 30 years of experience of Enhanced Oil Recovery operations, show that these hazards can be avoided through regulation of operating procedures with suitable safety standards (CCSA Position Paper 2009).

The potential public health and environmental risks of CCS are believed to be well understood based on analogous experience from the oil and gas industry, natural gas storage, and the U.S.EPA's Underground Injection Control Program. For CCS, the highest probability risks are associated with leakage from the injection well itself, abandoned wells that provide short-circuits to the surface and inadequate characterization of the storage site–leading to smaller than expected storage capacity or leakage into shallower geologic formations. Potential consequences from failed storage projects include leakage from the storage formation, CO_2 releases back into the atmosphere, groundwater and ecosystem damage. Avoiding these consequences will require careful site selection, environmental monitoring and effective regulatory oversight. Fortunately, for the highest probability risks, that is, damage to an injection well or leakage up an abandoned well, methods are available to avoid and remedy these problems. In fact, many of risks are well understood based on the analogous experience listed above, and over time, practices and regulations have been put in place to ensure that most of these industrial analogues can be carried out safely (Benson 2004).

2.5.3 Legal Issues for Implementing CO_2 Storage

The requirements to build CCS as a climate mitigation measure are more than technological feasibility. The development of incentive and regulatory policies is also required to support business models facilitating extensive implementation. These business models are not yet broadly demonstrated because of having inadequate current policies. As a result, the number of current real projects is small, indicating the public subsidies playing dominant role in pursuing the CCS projects. The most likely projects today are not sufficiently common to support a full scale industry that would store hundreds of millions of tonnes of CO_2 annually (Rai et al. 2010). Accordingly suitable legal framework with effective regulatory oversight is a keystone of effective CCS. Laws must be in place to protect personal property and the environment, and to assign liability for failed storage projects. Regulations must be in place to select and permit storage sites, specify monitoring and verification requirements, and enable constructive engagement with potentially affected citizens and communities (Benson 2004).

The regulatory frameworks shall include items such as the definition or classification of CO_2, access and property rights, intellectual property rights, monitoring and verification requirements, and liability issues (Solomon et al. 2007). The durability of CO_2 storage is one of the key regulatory and performance issues. The concept of "storage effectiveness" has been developed to quantify how much CO_2 must remain underground to avoid compromising the effectiveness of geologic storage. Estimates of the required "storage effectiveness" ranges from about 90% in 100 years to 90% in 10,000 years. The range is explained by differences in assumptions about how much CO_2 is stored, atmospheric stabilization levels, future industrial emissions, economic considerations about the cost of storage, and the effectiveness of the natural carbon cycle as a CO_2 sink. Another approach is that geologic storage will be and should be, for all intents and purposes, permanent. Preference for this approach is determined in part by national attitudes and partly by the belief that geologic structures could provide storage for millions of years. From the perspective of a climate change technology, a storage effectiveness of 90% in 1000 years is acceptable, and in fact, a conservative lower limit to the performance that is needed. Coming to consent on the performance requirements, including the question of durability, for geologic storage is an important issue that must be addressed (Benson 2004).

Although a vigorous regulatory structure is required early on, different systems could be adopted for demonstration projects and commercial scale deployment. Liability is determined with crucial importance, to cover potential leakage both during the active project and in the longer term. Demonstration projects have a central role to play in improving understanding of leakage and hence the extent of long-term liability. An overview of the key regulatory and liability issues associated with CCS can be found in (IRGC 2008).

A reliable effort to mention the major unresolved regulatory issues related to CCS, such as long-term custodian ship of the stored CO_2 is required for rapid implementation of the technology (Oh 2010). There is also no consensus on whether or not adequate regulations are in place for oversight of geologic storage. CCS is sufficiently unique and may be implemented on such a large scale to warrant its own regulatory regime because of the unique physical and geochemical attributes of CO_2 and the long-term storage requirement. A science-based regulatory approach for CCS is required to be soon developed to allow regulatory permitting of upcoming experimental projects and begin to define a set of performance requirements against which projects can be objectively gauged (Benson 2004).

A roadmap study conducted in UK mention that the State has to take ultimate ownership of stored CO_2 but there is a risk that public perception of industry handing over its problems to the UK public sector could suppress CCS. Thus, handover can only take place following adequate prediction and validation of storage performance to ensure that the risk of public liability is extremely low; this could be up to 30 years after storage site closure. Ideally site performance during this interim period would be well-enough understood to be insurable, though lack of insurance will exclude smaller companies from becoming CCS operators (Gough et al. 2010).

With the objective of building a regulatory framework for CCS, the activities should now be undertaken like public engagement and education, development of generalised site selection guidelines, development of GHG accounting protocols for CCS, improvement and standardisation of modelling techniques, development of necessary modifications to existing regulations, negotiation of specialised arrangements for long-term liabilities at a limited, number of early sites and creation of financial incentives to get full-scale demonstration sites up and running (IRGC, 2008). It is good to know that concerted efforts are in progress in the development of national and international level rules and regulations for CCS projects (CSLF 2004).

2.5.4 Public Perception of CCS

As more people are exposed to the concept of CCS, public opinion will be properly shaped, but it is fair to say for now, that the public is generally not aware of the technology and not having any opinion yet (Benson 2004). Awareness of the potential of CCS is the first step towards gaining acceptance for its deployment. If CCS is to be widely accepted, a policy of openness is required. All communication efforts should be based on high quality data. National consultation and regional negotiations are critical to the success of CCS projects since, by its very nature, the technology would require large industrial-sized projects affecting local and regional communities (OECD/IEA 2004).

To facilitate acceptance of CCS by the general public, industry decision-makers, and government policy makers, it will be necessary to develop well-structured education and outreach programmes (Esposito and Locke 2003). The deployment of CCS technologies will require broad understanding and long-term commitment by numerous constituencies including central and local governments, the general public, environmental and non-environmental NGOs, industrial and commercial organisations, academic and scientific institutes, financial institutions, the media, and international organisations (OECD/IEA 2004). Several studies are conducted to know the status of public participation related to CCS technologies. In 2007, Curry et al. report that 5% of respondents in 2006 know about CCS. Miller et al. report that fewer than 18% of Australians had ever heard of CCS in 2007. Consequently, it was also observed that Stakeholders response about CCS in Europe is positive ranging from moderate to strong (Shackley et al. 2007). Several more recent surveys of societal attitudes have focused on public opinion surveys (e.g. Johnsson et al. 2009; Desbarats et al. 2010) and still fewer stakeholder surveys have been conducted. In 2010, Duan report that respondents knew comparatively less about CCS than other renewable energy technologies like solar, wind, nuclear etc. But they also indicated supportive attitude to some extent towards CCS development in China. It is marked that a large share of stakeholders in North America has a clear position on CCS compared to Europe and Japan (Johnsson et al.

2009). Table 1.4 shows the roadmap indicating overview of the various communication activities that have been undertaken.

Table 2.4. Roadmap of CCS communication activities[a]

Month and Year	Communication activity	Country
Oct 02	Citizens Panels	UK (Tyndall)
Apr 03	Survey	Netherland (TUE)
Apr 03	Survey	Japan (RITE)
Jun 03	Survey	US (CMU)
Aug 03	Interviews	Australia (CCSD)
Sep 03	Survey	US (MIT)
Sep 03	Workshop	Europe (CTAP)
Dec 03	Survey	Japan (MIRI)
Mar 04	Focus Groups	Australia (CSIRO)
Apr 04	Survey	Canada (SFU)
May 04	Explorums	Australia (CCSD)
May 04	Workshop	Europe (CTAP)
Aug 04	Survey	UK (Cambridge)
Aug 04	Survey	UK (Tyndall)
Oct 04	Survey	Netherland (CATO)
Dec 04	Survey	Sweden (Chalmers)
Sep 05	Media Track	UK (Tyndall)
Oct 05	Focus Groups	Australia (CLET)
Oct 05	Survey	Spain (CIEMAT)
Jun 06	Survey	Europe (ACCSEPT)
Feb 06	Community Consultation	Australia (CO2CRC)
Mar 06	Survey	Paris (CIRED)
Mar 06	Interviews	Australia (CO2CRC)
May 06	Surveys	Australia (CLET)
Jun 06	Stakeholder Interviews	Australia (CSIRO)
Jul 06	Focus Groups	Australia (CLET)
Jul 06	Survey	Australia (CO2CRC)
Sep 06	Survey	USA (MIT)
Oct 06	Community Consultation	Australia (CO2CRC)
Dec 06	Community Consultation	Australia (Zerogen)
Mar 07	Survey	Netherland (CATO)
Aug 07	Survey	Australia (CSIRO)
Sep 07	Workshop	Canada (C3)
Mar 08	Large Group	Australia (CSIRO)
Apr 08	Interviews	Switzerland (ETH)
Nov 08	Thought Leader Forum	Canada (PEMBINA)

[a]Adapted from Ashworth et al. (2010).

The majority of the demonstration projects are adopting the communication activities related to direct stakeholders like governments, policy makers and NGOs instead of public in general (Ashworth et al. 2010). However, the involvement of public in decision making related to new technologies of CCS has been taken seriously nowadays (Malone et al. 2010). The key stakeholders (policy makers, general public, media and local community) views related to acceptance of specific CCS project and the way in which the project is communicated to public become interesting matter for social sciences (Oltra et al. 2010). Surveys are relatively simple compared to other methodologies because of simplicity to manage and also easy accessibility to a large size of sample and hence the use of surveys as a communication tool is more than 50% as shown in Fig. 2.5. Surveys conducted among more than 90% people having little knowledge of CCS are not at all reliable as a measure of public attitudes and level of acceptance (Malone et al. 2010). The initial dominant attitude towards CCS is motivated by perceived risks, preferences for renewable technologies and the perception that CCS is not viable option for climate change mitigation. Also, high level of trust in safety management and monitoring results in high level of acceptance (Oltra et al. 2010).

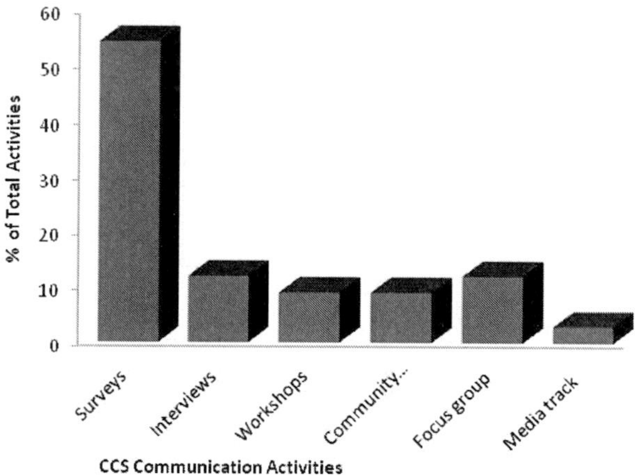

Figure 2.5. Various communication activities conducted from 2002 to 2008 (adapted from Ashworth et al. 2010)

CCS development strategy should be incorporated with an effective education policy to promote stakeholders involvement in CCS public education. CCS demonstration projects may be required to add in public education programs. A special section for public education purposes can be opened by CCS related facilities and some storage sites (Duan 2010). Various communication methodologies can be used to share learning experience and good practice with public by ongoing CCS projects. For the involvement of stakeholders, many already established approaches are available including ongoing focus groups in progress, citizen panels and juries, advisory boards, and specialized techniques, such as Deliberative Polling. Structured interviews and open-ended questions can probe images induced by CCS so as to guide provision of information and inclusion of issues to be addressed (Malone et al. 2010).

Generally, the important forums in the public debate of CCS include retention of CCS as a prominent role in the strategy, raising national awareness of the issues. This debate will be dominated by various issues about economic competitiveness, international trade, policy implements and timing. Non-governmental organizations (NGOs) with an interest in environmental policy will monitor and continue to develop their views on importance of role to be played by CCS in a low-carbon future (Benson 2004). Most importantly, on-the-ground pilot and demonstration projects will draw the interest and concern of the neighbouring communities. The public acceptance of CCS depends on critical outcomes of these debates. The need of CCS and assured safety strategies are to be communicated properly for getting public acceptance and enabling CCS implementation.

2.6 Summary

In this chapter, we introduce the concept of CCS and related technologies. CCS can play a central role in the mitigation of GHG emissions. Currently, CCS technologies are available for large-scale applications, but much more improvements, particularly in CO_2 capture are needed. In addition, there is a gap between what can technically do and what we are doing. High costs, inadequate economic drivers, remaining uncertainties in the regulatory and legal frameworks for CCS deployment, and uncertainties regarding public acceptance are barriers to large-scale applications of CCS technologies in the world (CCCSRP 2010). It is imperative to overcome the technical, regulatory, financial and social barriers.

2.7 References

Alphen, K.V., Ruijven, J.V., Kasa, S., Hekkert, M., and Turkenburg, W. (2008). "The performance of the Norwegian carbon dioxide, capture and storage innovation system." *Energy Policy*, 37, 43–55.

Anderson, S., and Newell, R. (2003). "Prospects for Carbon Capture and Storage Technologies." Discussion Paper published by Resources for the Future,

Washington, DC. Available at <http://rff.org/rff/Documents/RFF-DP-02-68.pdf> (accessed Feb. 2012).

Ashworth, P., Boughen, N., Mayhew, M., and Millar, F. (2010). "From research to action: Now we have to move on CCS communication." *International Journal of Greenhouse Gas Control*, 4, 426–433.

Benson, S.M. (2004). "Carbon dioxide capture and storage in underground geologic formations." In: *Proceeding of the 10–50 solution: Technologies and policies for a low-carbon future*, pp 1–19. The National Commission on Energy Policy, Washington, DC. March 2004.

CCCSRP (California Carbon Capture and Storage Review Panel) (2010). *Findings and Recommendations by the California Carbon Capture and Storage Review Panel.* CCCSRP, Dec. 2010. Available at <www.climatechange.ca.gov/carbon_capture_review_panel/.../2011-01-14...> (accessed Feb. 2012).

CCSA Position Paper (2009). *CCS in an international post-2012 climate change agreement.* Carbon Capture and Storage Association, London. Ref:LW, Sepember 2009. Available at <http://www.ccsassociation.org.uk/news/reports_and_consultations.html> (accessed June 02, 2010).

CCSA (2010). Carbon Capture and Storage Association, London. Available at <http://www.ccsassociation.org.uk/about_ccs/enhanced_hydrocarbon_recovery.html> (accessed June 03, 2010.

Gough, C., Mander, S., and Haszeldine, S., (2010). "A roadmap for carbon capture and storage in the UK." *International Journal of Greenhouse Gas Control*, 4, 1–12.

CSLF, (2004). "Considerations on legal issues for carbon dioxide capture and storage projects." *Report from the Legal, Regulatory and Financial Issues Task Force*, August 2004.

Curry, T.E., Ansolabehere, S., and Herzog, H. (2007). "A Survey of Public Attitudes towards Climate Change and Climate Change Mitigation Technologies in the United States: Analyses of 2006 Results." *Publication No. LFEE 2007-01 WP. Massachusetts Institute of Technology, Laboratory for Energy and the Environment*, Cambridge (http://sequestration.mit.edu/bibliography/).

Dalton, S., (2008). "What are current CO_2 capture and storage technology costs?" *Coal Utilization research council*, Congressional Staff Briefing, 22 May 2008.

Dangerman, A.T.C.J., and Schellnbuber, H.J. (2013). "Energy systems transformation." PANS, Published online Jan. 7, 2013, E549–E558, available at <www.pans.org/cgi/doi/10.1073/pans.1219791110> (accessed March 2013).

Desbarats, J., Upham, P., Riesch, H., Reiner, D., Brunsting, S., Waldhober, Marjolein, D., B., Duetschke, E., Oltra C., Sala, R., and Mclachlan, C. (2010). *Review of the Public Participation Practices for CCS and Non-CCS Projects in Europe.* Institute for European Environmental Policy. January 2010.

Duan, H. (2010). "The public perspective of carbon capture and storage for CO_2 emission reductions in China." *Energy Policy,* 38, 5281–5289.

EPRI (2007). "The power to reduce co_2 emissions: The full portfolio." Discussion Paper, Prepared for EPRI 2007 Summer Seminar, By The EPRI Energy Technology Assessment Centre, August 2007.

Esposito, P.R., and Locke, C.D. (2003). "Education & outreach programs: Important factors in sequestration's future." Paper presented at *the 2nd Annual Conference on Carbon Sequestration*, Alexandria, VA, USA, 5-8 May 2003.

Fernando, H., Venezia, J., Rigdon, C., and Verma, P. (2008). "Capturing king coal - deploying carbon capture and storage systems in the U.S. at scale." *World Resources Institute*, Washington, DC, May 2008.

Fischedick, M., Esken, A., Luhmann, H., Schuwer, D., and Supersberger, N. (2007). "CO_2–capture and geological storage as a climate policy option, technologies, concepts, perspectives." *Wuppertal Institute for Climate, Environment and Energy*, Wuppertal Spezial 35 e, ISBN 978-3-929944-74-7.

GCCSI report (2009). *Strategic Analysis of the Global Status of Carbon Capture and Storage.* Report 5: Synthesis Report, Global CCS Institute, Australia. http://www.ccsassociation.org.uk/news/reports_and_consultations.html accessed July 09, 2010.

Geroski, P.A. (2000). "Models of technology diffusion." *Research Policy*, 29, 603–625.

Gibbins, J., and Chalmers, H. (2008). "Carbon capture and storage." *Energy Policy*, 36, 4317–4322.

Liu, H., and Gallagher, K.S. (2010). "Catalyzing strategic transformation to a low-carbon economy: A CCS roadmap for China." *Energy Policy,* 38, 59–74.

Herzog, H.J. (2001). "What future for carbon capture and sequestration?" *Environ. Sci. Technol.*, 35(7), 148A–153A.

International Energy Agency Greenhouse Gas Programme (IEAGGP) (2007). *CO_2 Capture-ready Plants*. Report number 2007/4, May 2007.

International Energy Agency (IEA) (2009). *Technology Roadmap: Carbon Capture and Storage.* An IEA report, Available at <http://www.iea.org/papers/ 2009/CCS_ Roadmap.pdf> (accessed Feb. 2012).

Intergovernmental Panel on Climate Change (IPCC) (2005). *IPCC Special Report on Carbon Dioxide Capture and Storage.* IPCC Working Group III. Cambridge Univ. Press, Dec., 19, 2005, 431 pp.

Intergovernmental Panel on Climate Change (IPCC) (2007). "Summary for Policymakers." In: *Climate Change 2007: Mitigation. Contribution of Working Group III to the Fourth Assessment Report of the Intergovernmental Panel on Climate Change* [B. Metz, O.R. Davidson, P.R. Bosch, R. Dave, L.A. Meyer (eds)], Cambridge University Press, Cambridge, United Kingdom and New York, NY, USA, May 2007.

IRGC (2008). *Regulation of Carbon Capture and Storage, Policy Brief.* International Risk Governance Council, Geneva, Switzerland.

Johnsson, F., Reinerb, D., Itaokac, K., Herzogd H. (2009). "Stakeholder Attitudes on Carbon Capture and Storage - an international comparison." *Energy Procedia*, 1, 4819–4826.

Kapila, R.V., and Haszeldine, R.S. (2009). "Opportunities in India for carbon capture and storage as a form of climate change mitigation." *Energy Procedia,* 1, 4527–4534.

Koornneef, J., Ramirez, A., Harmelen, T., Horssen, A., Turkenburg, W., and Faaij, A. (2010). "The impact of CO_2 capture in the power and heat sector on the emission of SO_2, NO_x, particulate matter, volatile organic compounds and NH_3 in the European Union." *Atmospheric Environment,* 44, 1369–1385.

Lackner, K.S., and Brennan, S. (2009). "Envisioning carbon capture and storage: expanded possibilities due to air capture, leakage insurance, and C-14 monitoring." *Climatic Change*, 96, 357–378.

Malone, E.L., Dooley, J.J., and Bradbury, J.A. (2010). "Moving from misinformation derived from public attitude surveys on carbon dioxide capture and storage towards realistic stakeholder involvement." *International Journal of Greenhouse Gas Control,* 4, 419–425.

McKinsey (2008). *Carbon Capture & Storage: Assessing the Economics*. McKinsey & Company, September 2008.

Miller, E., Bell, L., and Buys, L. (2007). "Public understanding of carbon sequestration in Australia: socio-demographic predictors of knowledge, engagement and trust." *Australian Journal of Emerging Technologies and Society*, 5(1), 15–33.

OECD/IEA (2004). *Prospects for CO_2 Capture and Storage*. Paris: OECD/IEA.

Oh, T.H. (2010). "Carbon capture and storage potential in coal-fired plant in Malaysia—A review." *Renewable and Sustainable Energy Reviews*, (2010), doi:10.1016/j.rser.2010.06.003

Oltra, C., Sala, R., Sola, R., Masso, M., D., and Rowe, G. (2010). "Lay perceptions of carbon capture and storage technology." *International Journal of Greenhouse Gas Control*, (2010), doi:10.1016/j.ijggc.2010.02.001.

Pacala, S., and Socolow, R. (2004). "Stabilization wedges: Solving the climate problem for next 50 years with current technologies." *Science*, 305, 968–971.

Praetorius, B., and Schumacher, K. (2009). "Greenhouse gas mitigation in a carbon constrained world: The role of carbon capture and storage." *Energy Policy*, 37, 5081–5093.

Rai, V., Victor, D.G., and Thurber, M.C. (2010). "Carbon capture and storage at scale: lessons from the growth of analogous energy technologies." *Energy Policy*, 38, 4089–4098.

Riemer, P., Audus, P., and Smith, A. (1993). *Carbon Dioxide Capture from Power Stations.* Cheltenham, United Kingdom: International Energy Agency Greenhouse Gas R&D Program.

Rubin, E.S., Chen, C., and Rao, A.B. (2007). "Cost and performance of fossil fuel power plants with CO_2 capture and storage." *Energy Policy*, 35, 4444–4454.

Royal Society (2005). *Ocean Acidification due to Increasing Atmospheric Carbon Dioxide*. Royal Society, London.

Shackley, S., Waterman, H., Godfroij, P., Reiner, D., Anderson, J., Draxlbauer, K., and Flach, T. (2007). "Stakeholder perceptions of CO_2 capture and storage in Europe: Results from a survey." *Energy Policy*, 35, 5091–5108.

Shackley, S., and Verma, P. (2008). "Tackling CO_2 reduction in India through use of CO_2 capture and storage (CCS): Prospects and challenges." *Energy Policy*, 36, 3554– 3561.

Singh, B., Stromman, A.,H., and Edgar, H. (2010). "Life cycle assessment of natural gas combined cycle power plant with post-combustion carbon capture, transport and storage." *International Journal of Greenhouse gas Control*, doi: 10.1016/j.ijggc.2010.03.006

Solomon, S., Kristiansen, B., Stangeland, A., Torp, T.A., and Karstad, O. (2007). "A proposal of regulatory framework for carbon dioxide storage in geological formations." In: *International risk governance council workshop*; March 15–16, 2007, Washington, DC.

Surampalli, R., Zhang, T.C., Ojha, C.S.P., Tyagi, R.D., and Kao, C.M. (2013). *Climate Change Modeling, Mitigation and Adaptation,* ASCE, Reston, Virginia, 2013.

Surridge, A.D., and Cloete, M. (2009). "Carbon capture and storage in South Africa." *Energy Procedia*, 1, 2741–2744.

USDOE (U.S. Department of Energy) (1999). *Carbon Sequestration Research and Development.* Washington, DC: Office of Science and Office of Fossil Energy.

Williams, T. (2006). "Carbon capture and storage: Technology, capacity and limitations." Science and Technology Division, the Library of Parliament, Parliament Information and Research Services, March 2006, PRB 05-89E.

Zeman, F. (2007). "Energy and material balance of CO_2 capture from ambient air." *Environment Science and Technology,* 41, 7558–7563.

Zhang, T.C., and Surampalli, R.Y. (2013). "Carbon capture and storage for mitigating climate changes." Chapter 20 in *Climate Change Modeling, Mitigation and Adaptation,* Surampalli, R., Zhang, T.C., Ojha, C.S.P., Gurjar, B.R., Tyagi, R.D., and Kao, C.M. (eds.), ASCE, Reston, Virginia, February 2013.

CHAPTER 3

Carbon Capture and Sequestration: Physical/Chemical Technologies

Mausam Verma, Joahnn Palacios, Frédéric Pélletier, Stéphane Godbout, Satinder K. Brar, R. D. Tyagi, and Rao Y. Surampalli

3.1 Introduction

The gradual upcoming effects of global warming have become increasingly pronounced over the last few decades. This has motivated the political and economic will to minimize the anthropogenic CO_2 emissions by all sectors. Intergovernmental Panel on Climate Change (IPCC) has reported that approximately 75% of the current increase in atmospheric CO_2 is due to fossil fuel uses (IPCC 2001, 2007). Studies conducted over the last several decades suggest that CO_2 has the least potential for global warming among all greenhouse gases (GHGs) but it contributes to about 60% of global warming effects mainly due to its higher proportion in the atmosphere. If all the proven fossil fuel reserves are consumed up, the atmospheric CO_2 levels can increase over 5 folds the pre-industrial era (O'Neill 2002). IPCC (2005) has endorsed that about 60% of the global emission of CO_2 is generated by only 7887 stationary units with $\geq$ 100,000 tonne CO_2/year capacity, including 4942 electric power stations. Mobile carbon sources such as the transportation sector is second and contributes $\geq$ 30% in most of the developed nations and a global average of ~27% to the overall CO_2 emissions (CANSIM 2006; DOT 2013).

The current level of carbon concentrations in the atmosphere is increasing due to accelerated developments in populated countries like India and China. At present, the per capita carbon emission in these developing countries is lower than that of the developed ones. Thus, the future prospects of the global carbon emissions are dreadful once more countries will eventually be developed in few decades from now. As of now, it is gradually being recognized by many researchers that the release of carbon into the atmosphere is unfettered despite several past and ongoing concerted plans and campaigns by international community (e.g., ''United Nations Framework Convention on Climate Change'' initiated efforts) at a global scale (IPCC 2001, 2005, 2007).

Considering the substantial fraction of overall global carbon emissions, it would be pertinent for researchers to start identifying and developing generic as well as specialized technologies to tackle carbon emission from stationary sources, transportation sector, and directly from the atmosphere. In general, CO_2 emission reduction in fossil fuel utilization can be obtained either at pre-combustion, oxy-fuel, or post-combustion stages. However, the most suitable option for CO_2 capture for any emission source will depend upon specific application (Jordal et al. 2004). Figure 3.1 shows different technologies developed for CO_2 separation and capture. At present, almost all of the CO_2 capture technologies are for stationary sources and normally require substantial footprint, which is unsuitable for capturing CO_2 from mobile sources. Solvent- and sorbent-based, and membrane-based CO_2 capture technologies are well documented and attract research and development activities. On the other hand, other technologies such as enzymes, distillation using cryogenics, and hybrid/intermediate technologies are also gaining interest of many researchers (Figure 3.1).

This chapter explains the existing carbon capture technologies, application schemes, and their possible future improvements and modifications.

3.2 Separation with Solvents

In principle, separation with solvents is a two stage process, namely, absorption of CO_2 using absorber solvent followed by desorption using either pressure, temperature, or electric swing or any of the combinations. In general, flue gases from any combustion power plant are at extreme temperatures ($\geq$ few hundreds °C) and need to be cooled down to the optimal absorption temperature level (normally around $\leq$ 40–60°C). In the first stage, flue gas containing CO_2 and inert gases (non-reactive to solvent) enters the absorption solvent chamber, and CO_2 is preferentially separated from the inert gases. The inert gases simply bubbles out from the solvent chamber, and the CO_2 rich solvent is pumped to the desorption chamber (Figure 3.2), where CO_2 is recovered from the solvent using pressure, temperature, or electric swing or any of the combinations. The stripped solvent is then recycled back to the absorber chamber. The energy consumption of the CO_2 capture process is the addition of the energy (pressure, thermal, and/or electric) to pump solvent and flue gases, and to regenerate the solvents. Energy is also required to compress the recovered CO_2 to a specific pressure (10–80 MPa) for storage and transport.

The solvents used for CO_2 absorption can be divided in three categories with respect to the reaction mechanism, e.g., chemical, physical, and intermediate.

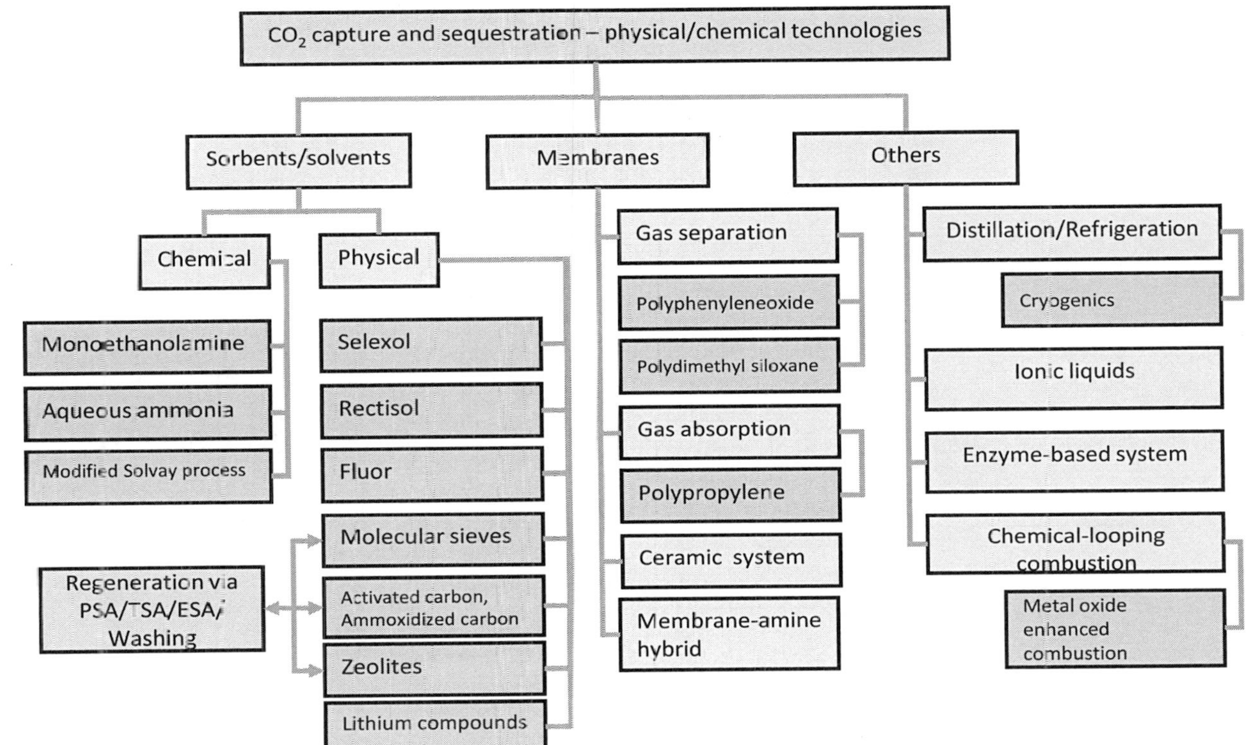

Figure 3.1. Different technologies for CO_2 separation and capture (PSA/TSA/ESA: Pressure swing adsorption/temperature swing adsorption/electric swing adsorption)

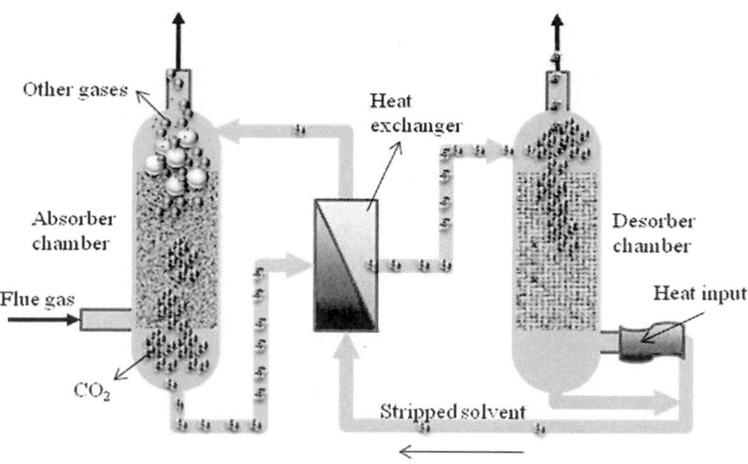

Figure 3.2. CO_2 separation from flue gases by absorption (adapted from CO2CRC 2010)

3.2.1 Chemical Absorption

Chemical adsorption processes for CO_2 have been in use for several decades to sweeten natural gas by chemical industries. Chemical adsorption processes for CO_2 refer to the absorption of CO_2 in a liquid solvent by formation of a chemical bond between CO_2 and the liquid solvent. The general solvent loading profile is a non-linear dependence on partial pressure and is higher at low partial pressures. At the saturation concentrations, the loading of the solvent decreases sharply. For solvent regeneration, heating is necessary and require substantial energy input.

3.2.1.1 Chemical Absorption Based Applications

Amine Absorption. The technology based on amine solution such as monoethanolamine (MEA), is a well-established commercialized technology for more than 60 years used mainly by large-scale chemical industries (Alie 2004; IPCC 2005; Yang et al. 2008). For example, natural gas industry utilizes MEA to selectively absorb CO_2 from natural gas. The MEA absorption processes are widely used for stripping CO_2 from flue gas stream as end-of-pipe application. The CO_2 removal is driven via gas-liquid mass transfer and formation of HCO_3^- ions as per equation (3.1):

$$C_2H_4OHNH_2 + H_2O + CO_2 \Leftrightarrow C_2H_4OHNH_3^+ + HCO_3^- \quad (3.1)$$

The reaction condition is achieved by forced mixing of MEA solution and CO_2 containing flue gases. The CO_2 rich solvent is then regenerated in a separate unit via counter flowing steam at 100–200 °C. The heated mixture of MEA, steam, and CO_2 thus produced is cooled down to around 40–65 °C, resulting into condensation of water vapour and up to about 98% recycle of CO_2 from MEA solution. Thus, highly concentrated CO_2 (≥ 99%) is easily separated from the liquid phase and the MEA, and water mixture is recycled back to the absorption/stripping column (Sander and Mariz 1992; Figueroa et al. 2008). The absorption capacity of MEA solution for CO_2 is very low and requires large volumes, which translates into large equipment size and intensive energy input and lead to unfeasible process economy. In addition to low absorption capacities, degradation of MEA due to light concentrations of SOx and NOx also add to the overall cost. Modifications such as use of mixture of MEA, diethanolamine (DEA) and methyl diethanolamine (MDEA) are reported to enhance heat efficiency and absorption capacities of the absorber (Gray et al. 2005). In the recent past, several attempts have been made to enhance CO_2 absorption capacity and process heat efficiency. It was observed that sterically hindered amines with an amino group attached to a bulky alkyl group, e.g., 2-amino-2-methyl-1-propanol ($NH_2C(CH_3)_2CH_2OH$) allow nitrogen to spontaneously react with CO_2 and improve absorption capacity (Olajire 2010). The overall reaction for this process is according to Equation (3.2):

$$RNH_2 + CO_2 + H_2O \rightarrow RNO_3^+ + HCO_3^- \quad (3.2)$$

Aqueous Ammonia. The aqueous ammonia (ammonia-based wet scrubbing) is similar to amine systems (MEA) in many respects, but it has various advantages over the MEA process. Ammonia and its derivatives can react with CO_2, SO_x, NO_x, HCl and HF, which normally exist in flue gases and may require additional separation steps, otherwise (Olajire 2010). The flue gases other than CO_2 can easily degrade MEA and corrode equipments, which is not true for the aqueous ammonia process. In this process, aqueous ammonia is either atomized and mixed with the flue gas, or simply mixed with the flu gas in a packed bed reactor. The main reaction products of this process are ammonium bicarbonate, ammonium nitrate, and ammonium sulphate as per Equations 3.3–3.5, which have fertilizer application (Xi et al. 1985).

$$2NH_3(l) + CO_2(g) + H_2O(l) \Leftrightarrow NH_4HCO_3(s) \quad (3.3)$$

$$NO_x + SO_x + H_2O \rightarrow HNO_3 + H_2SO_4 \quad (3.4)$$

$$HNO_3 + H_2SO_4 + NH_3 \rightarrow NH_4NO_3 \downarrow + (NH_4)_2SO_4 \quad (3.5)$$

Despite having a number of advantages over amine-based systems, literature on ammonia-based absorption is scarce (Figueroa et al. 2008). The potential for high CO_2

absorption capacity, no degradation during absorption/regeneration, tolerance to oxygen in the flue gas, cost economics, and potential for regeneration using pressure swing are promising features for commercial feasibility. In addition, the thermal energy consumption for the regeneration is expected to be significantly less than the MEA process (Gupta et al. 2003).

Research and development studies have compared the CO_2 removal efficiencies of NH_3 absorbent and MEA absorbent systems. It was concluded that the NH_3 absorbent can reach up to 99% of CO_2 removal efficiency and a CO_2 loading capacity up to 1.20 g CO_2/g NH_3. However, the maximum CO_2 removal efficiency and loading capacity by MEA absorbent were 94% and 0.409 g CO_2/g MEA, respectively, under similar conditions (Bai 1992). Thus, aqueous ammonia based processes can save throughput mass handling cost by up to 3 times in addition to enhanced removal efficiency, which may also reduce the recirculation ratio.

Modified Solvay Process. The Solvay process is also known as dual-alkali approach, where CO_2 and sodium chloride is reacted in the presence of ammonia (primary alkali) as a catalyst under the aqueous environment and produce sodium bicarbonate and ammonium chloride. The commercial process involves saturating brine (aqueous NaCl) with ammonia, followed by mixing with carbon dioxide (in spraying or packed bed reactors). However, the CO_2 removal from sodium bicarbonate is energy intensive; recovery of ammonia consumes lime ($Ca(OH)_2$), a secondary alkali, and requires limestone as source. Use of limestone lead to capture of two moles of CO_2 and liberates one mole of CO_2 for overall reaction steps. The release of additional one mole of CO_2 and energy requirements for regeneration of ammonia from ammonium chloride and release of CO_2 from sodium bicarbonate makes this process inefficient. In order to overcome these challenges, a modified dual-alkali method was developed by replacing ammonia with MEA as primary alkali. Furthermore, MEA can be replaced with methylaminoethanol (MAE) to act as an effective primary alkali by the following overall reaction:

$$CO_2 + NaCl + HOCH_2CH_2(CH_3)NH + H_2O \Leftrightarrow NaHCO_3 \downarrow + HOCH_2CH_2(CH_3)NH{\cdot}HCl \qquad (3.6)$$

This can increase the theoretical CO_2 absorption capacity of the overall process up to 1 mole CO_2/mole of MAE due to an increase in bicarbonate in the products (Xi et al. 1985). However, the researchers did not identify the secondary alkali to regenerate the primary one, i.e., MAE. The regeneration step of the dual-alkali approach can be improved by replacing ($CaCO_3$) limestone with activated carbon (AC) as per the following reaction:

$$NH_4Cl + AC \Leftrightarrow NH_3 + AC{\cdot}HCl \qquad (3.7)$$

Carbonate-Based Systems. Carbonate-based systems for CO_2 stripping from flue gases utilize soluble carbonate to selectively react with CO_2 to form bicarbonate (Equations 3.8 and 3.9).

$$2K^+ + CO_3^{-2} + H_2O + CO_2 \rightarrow 2KHCO_3\downarrow \quad (3.8)$$

$$2KHCO_3 \xrightarrow{\Delta} K_2CO_3\downarrow + CO_2\uparrow + H_2O \quad (3.9)$$

The heat of reaction for soluble carbonate to bicarbonate and for reversion from bicarbonate to carbonate makes this process economically feasible (Rochelle et al. 2006). Researchers at the University of Texas are developing catalyst (piperazine) mediated carbonate-based CO_2 absorption systems to improve absorption efficiency by up to 30% versus a 30% solution of MEA (Figueroa et al. 2008). Studies have indicated that carbonate-based systems can decrease the energy requirements by about 5% and increase the loading capacity up to 10% with respect to MEA. In addition, modifications in reactor design and operational parameters can also provide an additional 5–15% energy savings (Rochelle et al. 2006).

3.2.2 Physical Absorption

Physical solvents selectively absorb CO_2 according to Henry's Law without any chemical interactions. Equation (3.10) is a mathematical expression of Henry's law (at constant temperature), where p is the partial pressure of CO_2 in the gas phase above the physical solvent, c is the concentration of the CO_2 in the physical solvent and k_H is the Henry's law constant which is temperature dependent.

$$p = k_H c \quad (3.10)$$

Physical absorption processes are dependent on the temperature and the partial pressure of CO_2 in the organic solvent (NREL 2006). The higher loading of CO_2 can be achieved by using suitable solvent (with higher CO_2 partial pressures) and lower process temperatures being more favourable for the process economy and efficiency. Physical absorption is weak in comparison to chemical bonding. Therefore, physical solvents provide easier absorption-desorption processes via pressure and temperature swings and are less energy intensive. The major limitation of the physical solvents is the requirements for low temperatures for optimal operation; therefore, the gases should be cooled before absorption (Figueroa et al. 2008). Physical absorption-based applications for carbon capture are briefly introduced below.

Selexol and Rectisol. Selexol and Rectisol are the two most common physical solvent processes. Dimethyl ether of polyethylene glycol is used as the active ingredients of Selexol solvent. Absorption of CO_2 takes place at low temperature (0–5 °C) and desorption of the CO_2-rich Selexol solvent is accomplished either by pressure swing, stripping with air, inert gas, or steam. Sulphur compounds, carbon dioxide, water as well as aromatic compounds can be removed selectively or simultaneously. However, moisture removal is necessary before the Selexol process.

In the case of the Rectisol process, chilled methanol is used as active ingredient. The Rectisol process is mainly used for the efficient treatment of synthesis gas, hydrogen, natural gas, and coal gas. It is normally performed within a temperature range of -1 to -38 °C. The process is not suitable for gas streams containing ethane and heavier components. Selexol and Rectisol are widely used in commercial acid gas removal processes due to several advantages, such as low temperature rise in absorber; moisture removal; low foam; thermal and chemical stability; no degradation problems; low capital cost (use of carbon steel); air stripping (no re-boiler's heat); low operating pressure; and non-aqueous and inert chemical characteristics (less corrosive). Nevertheless, hydrocarbon losses, high pressure, low temperature, and formation of metallic amalgams such as those from mercury are some disadvantages of Selexol and Rectisol based processes and needs consideration before specific application.

Propylene Carbonate (Fluor Process). Propylene carbonate ($C_4H_6O_3$) is commercially used as a polar solvent to strip CO_2 from flue gases. The physical binding of CO_2 with Flour solvent provides energy efficient solvent regeneration. The process is more efficient for CO_2 rich gas streams at high pressure (> 60 psig) (Figueroa et al. 2008; Olajire 2010). In addition, Flour solvent has high affinity and loading capacity for CO_2, no additional water requirement, simple operation, dry gas output and low freezing point. However, high solvent cost, high recirculation requirement, and high affinity for heavy hydrocarbons make this process highly specific for CO_2-rich flue gas streams at high pressure.

Others. Other physical solvents in CO_2 capture applications are methanol, N-methyl-2-pyrrolidone, polyethylene glycol dimethylether, propylene carbonate and sulfolane (Meisen and Shuai 1997). Currently, the processes used for the removal of CO_2 and sulphur compounds from coal syngas are the Shell Sufinol® process and the Amisol® process developed by Lurgi (Gupta et al. 2003).

3.3 Separation with Sorbents

3.3.1 Physical Adsorption

Adsorption of CO_2 on physical sorbents is based on selective intermolecular forces between the molecules of gases and the surfaces of a solid or a liquid sorbent. During the physical adsorption process, the flue gas stream is passed through an adsorbent chamber, where selective separation of CO_2 takes place (Figure 3.3). The degree of selectivity of CO_2 for adsorption is dependent upon the temperature, partial pressure, surface forces, and adsorbent pore sizes. Thus, single or multiple layers of gases can be adsorbed, and the adsorption can be selective to a certain extent. The prevalent methods for regeneration of adsorbent materials are pressure swing, temperature swing, electrical swing, and washing operations, depending on the flue gas composition as well as the adsorption parameters. Pressure swing, temperature swing, and electrical swing operations utilize extreme change in parameters with respect to adsorption and regeneration steps. In the case of pressure swing, the pressure of the adsorption chamber is normally dropped to very low values for regeneration. On the other hand, regeneration via temperature swing requires increase in temperature. Electrical swing is performed by changing the electric current through the adsorbent bed. The washing of the adsorbent bed is also possible using liquid with high affinity for CO_2. However, the adsorption process is not a preferred method for CO_2 capture in large-scale industrial treatment of flue gases because the existing adsorbents are, in general, with low capacity and limited selectivity for CO_2 (Meisen and Shuai 1997).

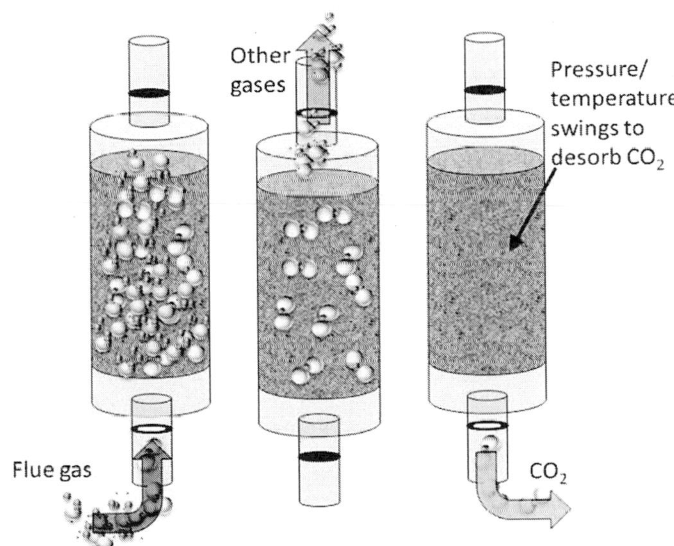

Figure 3.3. CO_2 separation from flue gases by adsorption (adapted from CO2CRC 2010)

3.3.2 Physical Adsorption-Based Applications

Molecular Sieve. Molecular sieves can be naturally occurring materials such as zeolites (aluminosilicate compositions), or non-zeolite, such as aluminophosphates, silico-aluminophosphates, and silica (Scholes et al. 2008), as well as can be tailor-made to desired pore structures to pass gas molecules on selective basis. They differentiate gas molecules based on their kinetic diameter and the relative size of the pore channels.

According to Yang et al. (2008), this technology is believed to be cost-effective and can be adapted to a variety of carbon sequestration schemes. Currently, many researchers have experimented with chemical modifications of the molecular sieve surface to eliminate some limitations of molecular sieves. In particular, adsorbents based on high surface area inorganic supports of basic organic groups, mainly amines, are studied. The CO_2 adsorption capacity was 0.5 mol CO_2/mol surface-bound amine group without the presence of water, and 1.0 mol CO_2/mol surface-bound amine in the presence of water. The mesoporous substrates, such as silica, SBA-1, SBA-15, MCM-41 and MCM-48 are also interesting because of the large enough pores that can be accessed by molecules with amino groups. The porosity and surface functionalized groups have been found to facilitate CO_2 adsorption. Inorganic-organic hybrid adsorbents are also under development, which provide both substantial pore volumes as well as large effective surface area (Chaffee et al. 2007).

Activated Carbon. Activated carbon has extremely porous micro- and meso-structure. The well-developed pore structure and surface chemistry governed by the presence of heteroatoms, such as oxygen, nitrogen, and others of activated carbons are highly suitable for their application in adsorption of numerous compounds. The activated carbon has a particle size between 0.1 and 5 microns (Scholes et al. 2008). The CO_2 adsorption capacity was reported to be 65.7 mg CO_2/g adsorbent for the anthracite activated at 800 °C for 2 h with an effective surface area of 540 m^2/g. On the other hand, the CO_2 adsorption capacity was 40 mg CO_2/g adsorbent from the same anthracite with higher surface area of 1071 m^2/g. Thus, CO_2 adsorption capacity was not only dependent on the surface area but on the surface chemistry as well. Pevida et al. (2008) have indicated that any surface modifications of commercial activated carbons should be carefully selected. For example, the nitrogen functionalities that promote the CO_2 capture characteristics of activated carbon should be implemented without affecting the original structural integrity. In general, the CO_2 capture capacity of activated carbons are lower than zeolites and molecular sieves under low pressure and ambient conditions (Martín et al. 2010), nevertheless, larger CO_2 capture capacities at higher pressures, ease of regeneration, potentially as a low-cost alternative, and tolerance to moisture are some of the strengths of this process.

Zeolite. CO_2 capture can be performed by naturally occurring as well as tailor-made zeolites to meet the specific requirements of a gas separation. There are many reports on separation of CO_2 and H_2 using zeolites of different pore sizes and effective surface areas. In the case of smaller pore size zeolites, relatively smaller gas molecules such as H_2, O_2 and N_2 can pass through, whereas, CO_2 is retained. Alternatively, larger pore sizes with or without surface treatments can selectively retain H_2 and other smaller molecules via Knudsen diffusion, while CO_2 can pass through unhindered (Yan et al. 1997).

Lithium Compounds. Lithium compounds such as lithium zirconate (Li_2ZrO_3), and lithium silicate (Li_4SiO_4) have been reported to be suitable for high temperature CO_2 adsorption (Fauth et al. 2005; Iwan et al. 2009; Nair et al. 2009). The chemical reactions using Li_2ZrO_3, and Li_4SiO_4 to capture CO_2 are as follows:

$$Li_2ZrO_3(s) + CO_2(g) \Leftrightarrow Li_2CO_3(s) + ZrO_2(s) \tag{3.11}$$

$$Li_4SiO_4(s) + CO_2(g) \Leftrightarrow Li_2SiO_3(s) + Li_2CO_3(s) \tag{3.12}$$

The CO_2 capture reactions by Li_2ZrO_3 and Li_4SiO_4 are reversible in the temperature range of 450–590 and around 720 °C, respectively. The adsorption and release of CO_2 can be easily carried out by temperature swing operation. In addition to these compounds, numerous binary and ternary eutectic salt-modified Li_2ZrO_3 were studied for high temperature CO_2 capture applications. The combinations of binary alkali carbonate, binary alkali/alkali earth carbonate, ternary alkali carbonate and ternary alkali carbonate/halide eutectic to Li_2ZrO_3 have been found to increase the CO_2 loading rate and CO_2 capture capacity. The eutectic molten carbonate layer on the outer surface of adsorbent Li_2ZrO_3 particles also facilitates the gaseous CO_2 transfer during the sorption process. High capacity, high absorption rate, wide range of temperature and concentrations of CO_2 and stability are favourable for commercially competitive CO_2 adsorbent. Nevertheless, considering the proven global reserves of lithium, the current and future trends of consumption of lithium compounds in the areas such as electronics, hybrid and electric vehicle production, and many others can jeopardize the economic viability of lithium compounds in the use of CO_2 capture technologies.

3.4 Separation with Membranes

Capture of CO_2 can also be performed via separation with selective membranes, which are specially designed to allow the permeation of desired gas(es) through them. The degree of selectivity of membranes to different gases is intrinsically related to the construction material, whereas the volumetric capacity through the membrane is dependent on the pressure difference across the membrane. Therefore, high-pressure streams are usually preferred for membrane separation (IPCC 2005). Membranes can be used in two different ways: either like a filter (gas separation membranes), or allowing CO_2 to be absorbed into a solvent (gas absorption membrane).

3.4.1 Gas Separation Membranes

Gas separation membranes consist of solid membrane materials and operate on the principle of preferential permeation of desired constituents through the porous structure. The optimal separation efficiency is achieved by selection of membrane material with higher degree of selectivity and higher permeability. A commercial scale membrane separator typically consists of a large number of hollow cylindrical membranes arranged in parallel. The CO_2-rich gas mixture is fed inside the hollow section at elevated pressure. The CO_2 preferentially passes through the membranes and is recovered at reduced pressure on the shell side of the separator as permeate, whereas, the rest of the gas mixture constituents are recovered as retentate (Figure 3.4). Usually, a fraction of retentate is recycled back to the feed stream to improve CO_2 capture efficiency at the expense of operating cost of the process. Herzog and Golomb (2004) discussed several advantages of gas separation using membranes such as high packing density; high flexibility with respect to flow rates and solvent selection; absence of foaming, channelling, entrainment and flooding–common problems associated with packed absorption towers using liquid solvents; transportability; and durability.

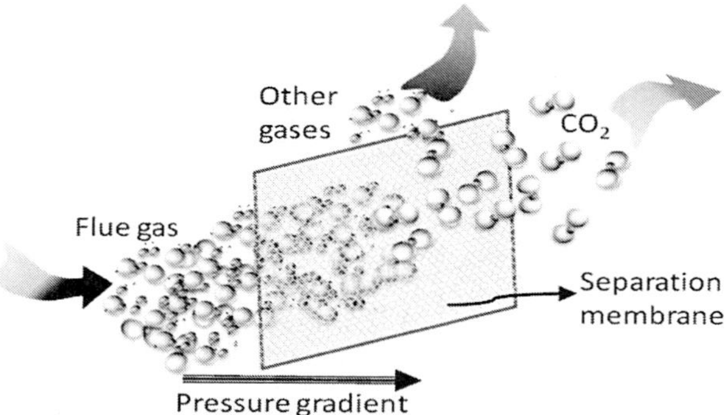

Figure 3.4. Gas separation membrane (adapted from CO2CRC 2010)

Gas Absorption Membrane. Gas absorption membranes consist of highly dense microporous solid structure in contact with an absorbent solvent. The desired gas component to be separated diffuses through the membrane and is then absorbed into and removed by the solvent. The main difference between gas separation membrane and gas absorption membrane is the separation potential. As explained earlier, in a gas separation membrane process, higher permeability usually associates with lower selectivity and vice versa. However, in a gas absorption membrane process, permeability is obtained by the physical presence of the microporous membrane, and selectivity is achieved by the liquid absorbent. This technique favours higher flux of CO_2 as well as more compact equipments than conventional membrane separators (Meisen and Shuai 1997; Ahmad et al. 2010).

Membranes can be employed in pre-combustion; oxyfuel combustion; and post-combustion systems (see below) for the separation of CO_2 from hydrogen, CO, and natural gas; oxygen from nitrogen; and CO_2 from flue gases, respectively. There are many different types of membrane materials (polymeric, metallic, ceramic) that have potential to be used in CO_2 capture systems. Some of the emerging membrane types are discussed in the following.

3.4.2 Membrane Types

Polymeric Membranes. Polymeric membranes are classified as rubbery or glassy based on the glass transition temperature of the polymers used (Plate and Yampol'skii 1994). The rubbery membranes, operating just above the glass transition temperature can easily rearrange and result into low energy adsorption and gas release operations. The gas solubility within the polymer matrix follows Henry's Law and is linearly proportional to the partial pressure (Olajire 2010) as follows:

$$C_D = K_D p \tag{3.13}$$

where, C_D is the gas concentration in the polymer membrane, K_D is the Henry's law constant, and p is the partial pressure of the gas being captured.

On the other hand, glassy membranes operate at temperature below the glass transition and cannot rearrange as effectively as rubbery membranes operating just above the glass transition temperature. This leads to imperfectly packed polymer chains and form excess microscopic voids in the membrane structure, which provide Langmuir adsorption sites. Therefore, glassy membranes have dual-mode sorption capacity as shown by the following equation:

$$C_T = C_D + C_H \tag{3.14}$$

where, C_T is the total adsorbed gas concentration and C_H is the gas concentration due to Langmuir adsorption. Thus, glassy membranes provide higher permeability and lead to lower operating cost. However, permeability is inversely proportional to selectivity and vice versa; therefore, most of the existing polymeric membranes have an optimal trade-off with respect to permeability and selectivity. The performance of the

CO_2-selective polymeric membrane can be achieved either by increasing the solubility of carbon dioxide in the membrane through chemical changes in polymer matrix or, by increasing the diffusion of carbon dioxide by altering the polymer packing within the membrane.

Inorganic Membranes. Inorganic membranes can be classified into porous and non-porous membranes. Porous inorganic membranes consist of a porous thin top layer fixed on a porous metal or ceramic support. The support provides mechanical strength but offers minimum mass-transfer resistance. Some of the commonly used support materials are alumina, carbon, glass, silicon carbide, titania, zeolite, and zirconia membranes. In non-porous membranes, the permeability is mainly due to the atomic interstices, atomic vacancies and dislocations (Bose 2009). The common non-porous inorganic membranes highly selective for hydrogen or oxygen separation consist of a thin layer of metal, such as palladium and its alloys, or solid electrolytes, such as zirconia (Mundschau et al. 2006; Yang et al. 2008).

Evidently, fluxes are higher in the case of porous membranes while higher selectivity can be achieved using non-porous membranes (Mallada and Menéndez 2008). The inorganic membrane systems have ability to operate at high temperatures, which is the most sought-after capability for the separation of carbon dioxide from hydrogen in syngas production (Scholes et al. 2008).

Mixed-Matrix and Hybrid Membranes. Integration of molecular sieves in a polymer membrane (e.g., polymer-zeolite, polyimide-carbon, polyimide-silica, nafion-zirconium oxide, HSSZ-13-polyetherimide, and acrylonitrile butadiene styrene-activated carbon) provides both the permeability of polymers and the selectivity of molecular sieves. However, poor contact at the molecular sieves/polymer interface can hamper the overall performance (Yang et al. 2008). Moore and Koros (2007) demonstrated that the gas sorption in mixed matrix membranes is approximately additive in the absence of other factors such as a contaminant. For example, zeolite 4A can be easily affected by the contaminants from the processing equipments or from the gas mixture itself.

Facilitated Transport Membranes. Facilitated transport membranes (FTM) offer higher selectivity and can handle larger throughputs of gas mixture (Shekhawat et al. 2003). FTMs undergo a reversible complex reaction as well as solution-diffusion mechanism of the polymeric membranes. Higher selectivity in FTM is due to the incorporation of a carrier agent into membrane, which can reversibly react with the desired gas species. The permeating gas dissolves and reacts with the carrier agent inside the membrane to form a complex in the upstream portion of the membrane. The gas species-carrier agent complex diffuses across the membrane and then permeates out of the downstream side of the membrane. The carrier agent is continuously recovered and diffuses back to the feed side (Shekhawat et al. 2003).

Van der Sluijs et al. (1992) concluded that a membrane could be considered as a serious competitor in comparison to other separation techniques for gas separation if it has a selectivity of at least 200 and high permeability. The membrane's best selectivity at that time was 67. At present, the higher selectivity values at the laboratory scale are 160 (Brunetti et al. 2010) and 200 (Duan et al. 2006), showing the potential of the application of this technique.

However, the existing CO_2-selective membranes are not as efficient as some of the others such as H_2-selective membranes (Scholes et al. 2008). Commercially utilized polymeric membranes developed for CO_2 can achieve desired separation at low selectivity. The high CO_2 permeability and selectivity of membranes at low pressure and low temperature make them suitable for post combustion CO_2 capture but less competitive at high pressure and high temperature operations such as integrated gasification combined cycle (IGCC). At present, CO_2-selective membranes for carbon capture from IGCC process are lagging behind other gas separation membranes and require more research efforts.

3.5 Separation with Other Technologies

Cryogenic Distillation. Cryogenic carbon removal methods can capture CO_2 from flue gases in a liquid form by a carefully designed and controlled series of compression, cooling and expansion steps. Except N_2, flue gas components (such as H_2O, NO_x, SO_x and O_2) are removed prior to the cryogenic process. The temperature and pressure are lowered and increased, respectively, in order to liquefy CO_2. In particular, the triple point of CO_2 (216.6 K at 5.11 atm) is attained, where CO_2 condenses, solidifies and remain in gaseous phase simultaneously, while N_2 remains in gaseous form. The main advantage of cryogenic processes is higher recovery of CO_2, provided the CO_2 feed is properly conditioned. The liquefied CO_2 can easily be transported for commercial use or sequestered in the spent oil and gas field reserves. However, the presence of contaminants (SO_X and NO_X) can hinder cryogenic processes (Shackley and Gough 2006) and require the complicated conditioning process of the flue gases. Evidently, the need for pressurization and refrigeration can make cryogenic processes energy intensive and hence expensive depending upon the CO_2 concentration in the flue gases (Meisen and Shuai 1997; Shackley and Gough 2006). Therefore, cryogenic separation is commercially used for the purification of CO_2 from streams with high CO_2 concentrations (typically > 90%) and not normally used for highly diluted streams (Shackley and Gough 2006). A patent filed by Baxter (2009) is pending for a Cryogenic Carbon Capture (CCC) process at Brigham Young University to separate a nearly pure stream of CO_2 from power plant flue gases more cost-effectively and with significantly less energy requirements than current alternatives. The CCC process can be applied in post-combustion systems and is suitable for retrofitting existing power plants with relative ease. Figure 3.5 shows the schematic of the CCC process. In the CCC process, a flue gas stream is dried and cooled, gently compressed and cooled to slightly above the frost point of CO_2. The gas is then adiabatically expanded, which results in further cooling the stream and precipitating solid CO_2. The solid CO_2 is separated from the flue gas by a gas-solid separator, and the pure CO_2 stream is pressurized. The excess cooling of CO_2 and N_2 streams is used to cool incoming flue gas in a heat exchanger. Finally, liquefied CO_2 and a pure gaseous nitrogen stream are obtained separately.

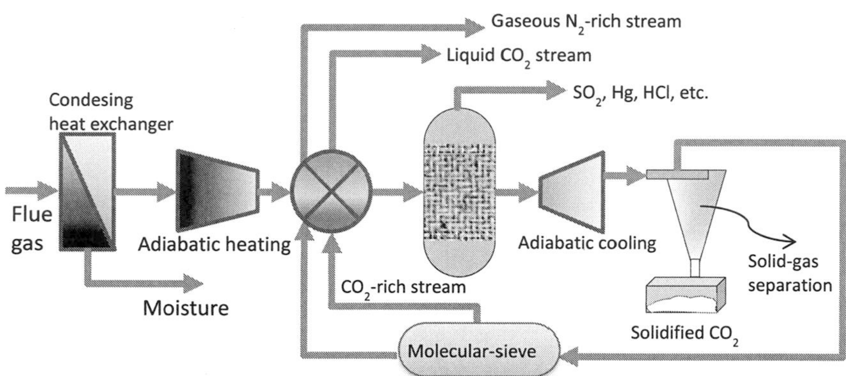

Figure 3.5. Schematic diagram of the cryogenic carbon capture process (adapted from Baxter 2009)

Ionic Liquids. Ionic liquids are organic salts that are liquid near ambient conditions and have been proposed as physical solvents for gas separations to facilitate the sequestration of gases without concurrent loss of the capture agent or solvent into the gas stream. They are stable at temperatures up to several hundred degrees centigrade and can dissolve and retain gaseous CO_2. The temperature stability of ionic liquids is useful for recovering CO_2 from the flue gas without prior cooling. Furthermore, physical binding to CO_2 require ionic liquids minimal heat for regeneration (Figueroa et al. 2008). According to Bates et al. (2002) and Maginn (2005), ionic liquids are regarded as potentially environmentally-benign solvents due to their low volatility.

Enzyme-Based System. Carbonic anhydrase (CA) is currently at the experimental stage to capture CO_2 from flue gases. Enzyme-based systems mimic naturally occurring reactions of CO_2 in living organisms. The enzyme functions as a catalyst to facilitate transformation of CO_2 into carbonic acid in solution (Figueroa et al. 2008; CO_2 Solutions 2010). One CA molecule can catalyze the hydration of 600,000

CO_2 molecules per second in comparison to 0.038 CO_2 molecules per second for an uncatalyzed reaction. Only small-ionized species, principally bicarbonate and carbonate can diffuse across the membrane (Cowan et al. 2003). Immobilized-CA based hollow fibre liquid membrane has demonstrated the potential for up to 90% CO_2 capture followed by regeneration at ambient conditions at the laboratory-scale (Figueroa et al. 2008). The regeneration of CO_2-rich solution is easily possible at ambient temperature with modification to solution's pH in the presence of CA (Salmon et al. 2009). Thus, solution regeneration is less energy intensive. However, surface fouling, loss of enzyme activity, enzyme cost, and scale-up are some of the challenges to this technique.

Chemical-Looping Combustion. Chemical-looping combustion (CLC) involves the use of a metal oxide as an oxygen source for combustion which uptakes oxygen from the air. Thus, the direct contact between fuel and combustion air is prevented. Consequently, the combustion products, e.g., CO_2 and H_2O, are isolated from the rest of the flue gases, e.g., nitrogen and unused oxygen. The net chemical reaction is similar to the normal combustion, and the CO_2 is separated from nitrogen during oxidation of metal in a separate chamber. This is contrary to the known techniques for CO_2 separation from the flue gas, which require large amounts of energy and capital investment (Mattisson and Lyngfelt, 2001). The major developmental challenge of chemical-looping combustion is the metal oxide material that should be able to withstand long-term chemical cycling as well as resistant to physical and chemical degradation from impurities generated from fuel combustion (Thambimuthu et al. 2002). Some of the oxygen carriers are small particles of metal oxides such as Fe_2O_3, NiO, CuO, and Mn_2O_3. Figueroa et al. (2008) reported chemical-looping gasification to gasify coal for syngas (H_2 and CO) production. In this case, a second solid loop is used in a water gas shift reactor, where steam reacts with CO and converts it to H_2 and CO_2. The circulating solid absorbs CO_2 and provide a greater driving force for the water gas shift reaction. The CO_2 can then be released through the calcinations step that produces highly concentrated CO_2 for further compression and sequestration. Nevertheless, both chemical-looping combustion and gasification are under experimental stages.

3.6 Carbon Capture Schemes for Different Sources

Carbon capture schemes for stationary sources such as power generation plants, industrial facilities have been demonstrated via pre-, oxy-fuel, and post-combustion CO_2 capture. There were no reported studies on CO_2 capture from transportation sector (passenger and mass transit systems). However, post-combustion CO_2 capture seems to be a plausible option. Carbon capture from atmosphere involves the separation of CO_2 from exhaust gases emitted by miscellaneous sources; therefore, post-combustion CO_2 capture model can be applicable. The three options for carbon capture schemes are explained in the following sections.

3.6.1 Carbon Capture from Concentrated Sources

Pre-Combustion Capture. Pre-combustion capture involves a solid or liquid fuel such as coal, biomass, or petroleum products reacting with oxygen (gasification) or with air and/or steam, to produce a mixture consisting mainly of carbon monoxide and hydrogen (synthesis gas). The carbon monoxide is reacted with steam in a catalytic reactor, known as shift converter, to provide CO_2 and additional hydrogen. The CO_2 is then separated using a physical adsorption process, resulting in a hydrogen-rich fuel, which can be used in boilers, furnaces, gas turbines, internal combustion engines, and fuel cells. The heat of the exhaust gas from a combustion turbine can be recovered to produce steam for the steam turbine that generates additional power and increases the overall power system efficiency (Figure 3.6). The initial fuel conversion steps in pre-combustion are more elaborate and costly than in post-combustion systems. However, the removal of CO_2 from such gas stream prior to the combustion is often much practical than after combustion due to lower volumetric flow rates, higher pressure, and higher concentrations of CO_2 (typically 15 to 60% by volume on a dry basis) produced by the shift reactor (Marion and Griffin 2001; IPCC 2005). Pre-combustion is an economically feasible process for many cases. The high CO_2 concentration in the high pressure fuel gas, physical solvents such as ethanol or polyethylene glycol, can effectively capture the CO_2 often in combination with sulfur (H_2S) removal. Currently, these state-of-the-art capture technologies are under investigation and will require some more time to be tested at the commercial scale (Figueroa et al. 2008).

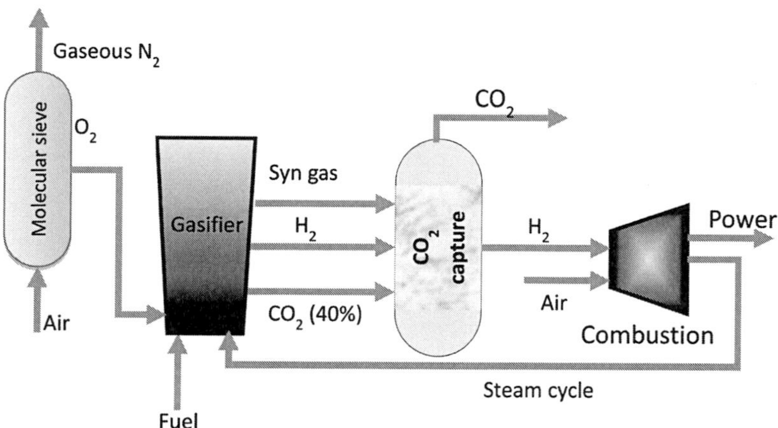

Figure 3.6. Pre-combustion capture system (adapted from Figueroa et al. 2008)

Oxy-Fuel Combustion. In general, fuels are burned in air due to practical reasons; however, the CO_2 concentration in the flue gas is relatively lower due to a higher nitrogen to oxygen ratio in the air. On the other hand, in order to make separations less expensive, higher CO_2 concentrations are necessary. Therefore, oxy-fuel combustion systems use oxygen instead of air for combustion of the primary fuel and produce a flue gas mixture that is mainly water vapour and CO_2. The most common oxy-fuel combustion concept involves a cryogenic air separation unit to supply high purity oxygen to a boiler. The combustion of a fuel in pure oxygen produces very high flame temperatures that exceed the tolerance limits of metals in common boilers. Therefore, a portion of the flue gas containing CO_2 is recycled into the boiler to reduce the combustion temperature (Figure 3.7). The water vapour in the flue gas is removed by cooling and compressing the gas stream, which result in a CO_2 concentration greater than 80% by volume (IPCC 2005; Steeneveldt et al. 2006). The CO_2 rich flue gas must be treated for small amounts of other acid gases (SO_X and NO_X) prior to compression for transport and storage. In most current designs, oxy-fuel combustion requires the oxygen purity $\geq$ 95% (IPCC 2005; Figueroa 2008). As a downside, according to Herzog and Golomb (2004) the air separation unit alone may consume about 15% of a power plant electric output, requiring an increased consumption of fuel for achieving the rated electric output of the plant.

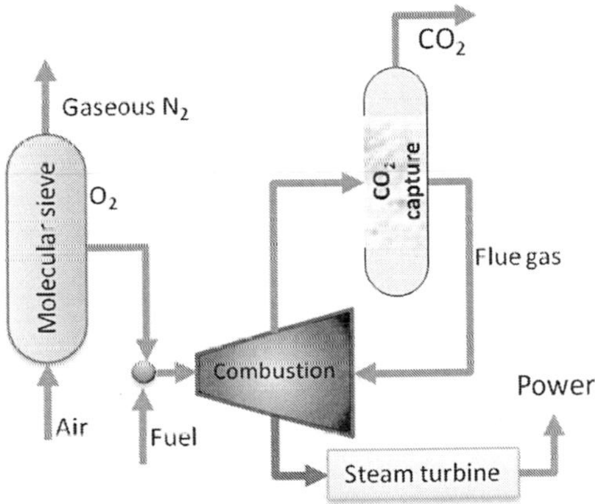

Figure 3.7. Oxy-fuel combustion system (adapted from Figueroa et al. 2008)

The oxy-fuel combustion system is still in the demonstration phase, being the least advanced of all the capture options for power generation. The technology requires further development before the design and construction of a full-scale system (IPCC

2005). The high temperatures encountered during oxy-fuel combustion can be handled either with more advanced materials to enable the direct application of oxy-fuel combustion, or by the use of various diluents to moderate the combustion temperatures (Simmonds et al. 2004). Thus, oxy-fuel combustion faces main challenges such as the high cost of oxygen production, overall design of the boiler and burners, and removal of impurities from CO_2 stream.

Post-Combustion Capture. Post-combustion systems separate CO_2 from the flue gases produced by the combustion of fuels. The lean concentrations of CO_2 in flue gas stream (12–15 v/v% for modern coal fired power plants and 4–8 v/v% for natural gas fired plants) is passed through the equipment which capture and separates most of the CO_2 (Figure 3.8). The CO_2 is then fed to a storage reservoir, and the remaining flue gas is discharged to the atmosphere (IPCC 2005). According to Figueroa et al. (2008) and Brunetti et al. (2010), post-combustion carbon capture has the greatest near-term potential for reducing carbon emissions in terms of industrial sectors (e.g., power, kiln and steel production) because it can be retrofitted to existing units that generate two-thirds of the CO_2 emissions in the power sector. However, the energy requirement and the resulting overall efficiency are not favourable.

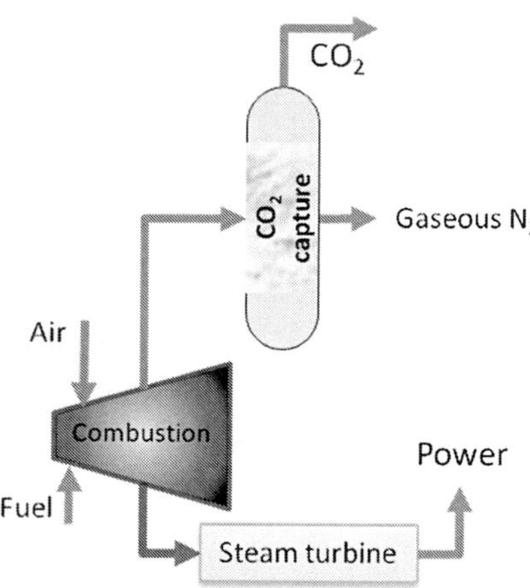

Figure 3.8. Post-combustion capture system (adapted from Figueroa et al. 2008)

Absorption processes based on chemical solvents are currently the preferred option for post-combustion CO_2 capture (IPCC 2005). These systems normally use a liquid solvent to capture the small fraction of CO_2 (typically 3–15% by volume) present in a flue gas stream in which the main constituent is nitrogen (from air). Several potential solvents as described earlier in this chapter are being considered for post-combustion capture, including various types of amines, amino acid salts, ammonia, sodium carbonate solutions and solvent blends, but the most developed post-combustion capture concept is amine separation (Price et al. 2008). However, the presence of oxygen in the flue gas stream can be problematic to the flue gas amine scrubbing, as it can cause degradation of some solvents and corrosion of equipment. At present, the process of scrubbing CO_2 with amines is under the pilot and demonstration scale and can be easily scaled up to actual power plant size (IEA Greenhouse Gas R&D Programme 2007).

Selection of Different Schemes. Several other processes have also been considered to capture CO_2 from power plant and industrial boiler flue gases, e.g. membrane separation, cryogenic fractionation, and adsorption using molecular sieves as described before. Except absorption methods and some membranes, most of these processes are generally less energy efficient and more capital intensive (Gupta et al. 2003; Herzog and Golomb 2004). Research and development on post-combustion capture systems have made it is economically feasible for certain applications. According to IPCC (2005), the main systems of reference for post-combustion capture are the current installed capacity of 2261 GW_e of oil, coal and natural gas power plants. Moreover, the 155 GW_e of supercritical pulverised coal fired plants and the 339 GW_e of natural gas combined cycle plants, both represent the high efficiency power plant technology where post-combustion CO_2 capture can be economically feasible.

In coal-fired power plants, despite CO_2 capture is theoretically less favourable and more energy intensive than from other gas streams, commercial applications exist. However, higher stakes in research and development effort are being undertaken worldwide to develop more efficient and lower cost post-combustion systems. Research and development objectives of most post-combustion research programs are focused on the development of new solvents, membranes, and process integration (such as thermal integration with the power plant) to reduce thermal energy consumption (2 GJ/tonne CO_2) and efficiency loss (Wall 2007).

The different carbon capture/separation technologies can be employed for the carbon capture strategies, namely, post-, pre-, and oxy-fuel combustion captures depending upon the application requirements. Table 3.1 provides a list of the toolbox of both current and emerging technologies for carbon capture for different scenarios.

3.6.2 Carbon Capture from Other Point and Non-point Sources

Besides CO_2 capture from power generation sources, CO_2 could be separated also from the atmosphere as well as stationary sources other than power generation, e.g., integrated steel mills, cement plants, pulp and paper production, processing of heavy oils such as tar sands, and synthetic fuels production. Nevertheless, methods to capture CO_2

at each of these types of facilities depend on their specific production processes, which can be quite complex (Price et al. 2008).

Table 3.1. CO_2 Capture toolbox (adapted from IPCC 2005)

Separation task	**Post-combustion capture**		**Oxy-fuel combustion capture**		**Pre-combustion capture**	
	CO_2/N2		**O_2/N_2**		**CO_2/H_2**	
Separation efficiency	> 80% removal[1]		90–95%[2]		> 80% removal[1]	
Capture Technology	**Current**	**Emerging**	**Current**	**Emerging**	**Current**	**Emerging**
Solvents (Absorption)	**Chemical solvents**	Improved solvents	n. a.	Biomimetic solvents, e.g. hemoglobine-derivatives	**Physical solvents**	Improved chemical solvents
		Novel contacting equipement			**Chemical solvents**	Novel contacting equipment
		Improved design of processes				Improved design of processes
Membranes	Polymeric	Ceramic	Polymeric	Ion transport membranes	Polymeric	Ceramic
		Facilitated transport				Palladium
		Carbon		Facilitated transport		Reactors
		Contactors				Contactors
Solid Sorbents	Zeolites	Carbonates	Zeolites	Adsorbents for O2 /N2 separation	Zeolites	Carbonates Hydrotalcites
				Perovskites		
	Activated carbon	Carbon based sorbents	Activated carbon	Oxygen chemical looping	Activated carbon	Silicates
					Alumina	
Cryogenic	Liquefaction	Hybrid processes	**Distillation**	Improved distillation	Liquefaction	Hybrid processes

Note: Processes shown in bold are commercial processes that are currently preferred in most circumstances. [1] Nominal CO_2 capture levels of 80% is a project specific value and determined by the relationships between the CO_2 partial pressure, capture % and the energy for compression (Wall 2007). In post-combustion capture, a value of 90% is often used as an estimate (Pehnt and Henkel 2009). [2] (Pehnt and Henkel 2009). A range of studies assume a value of 90% (Douglas et al. 2003; Dillon et al. 2004; Sekkappan et al. 2006). However, many studies also expect that an efficiency of nearly 100% could be reached (IPCC 2005; Viebahn et al. 2007).

Carbon capture from mobile sources and atmosphere seems to be a very attractive strategy to mitigate global warming effects as almost 50% CO_2 emission is caused by mobile and unaccounted sources. In fact, many existing techniques for CO_2

capture from industrial and electricity generation plants are suggested, and some are under investigation to capture CO_2 directly from the atmosphere (Stucki and Schuler 1995; Zeman 2007). In addition to existing techniques, wet scrubbing using sodium hydroxide solution is the most common technique for CO_2 capture from ambient air. However, the low CO_2 concentration in ambient air requires recirculation of almost 3,000,000 m^3 of air per ton of CO_2. This is apparent in the lower thermodynamic efficiency of the process with respect to other ''end of pipe'' strategies. Use of alternative energy sources such as wind and solar to power such process can make it economically feasible. On the other hand, direct capture of CO_2 from mobile sources such as automobile, recreational vehicle, mass transit systems is non-existent. This is mainly due to the technical limitation of the present techniques. The size and weight of almost all CO_2 capture systems will not allow them to be integrated with these sources (Hicks et al. 2007). For example, an existing very high capacity adsorbent with 3.1 mmol CO_2/g adsorbent material would require about 854.5 kg of adsorbent material for an average full tank of gasoline (~50 litre). Evidently, existing techniques are not intended for ''end of pipe'' application in mobile sources.

3.7 Conclusions

Current trend of fossil fuel consumption is extremely difficult to curtail, considering the burgeoning world population and emerging economies worldwide. Despite several attempts by international community to check global warming, the problem is getting bigger with time. Unless some reliable and economic alternatives to fossil fuels are established, carbon capture (more precisely, CO_2) from fossil fuel emissions is the most promising approach. There are several carbon capture techniques in place at laboratory as well as commercial-scale power generation plants. These techniques utilize chemical/physical solvents and sorbents, membranes, enzymes, and innovative processes to capture CO_2 at pre-, post-, or oxy-fuel combustion stages. At present, the most commonly used CO_2 capture technique is use of chemical solvents such as MAE, and many proprietary products. However, these processes have higher operating and capital costs, environmental risks, and technical restraints. The CO_2 capture techniques are mainly used by the large scale thermal power generation plants, which add to about 10–20% of the overall fuel consumption. In order to address these problems, several other techniques such as physical solvents/sorbents, molecular sieve, activated carbon, membranes, cryogenic fractionation, chemical-looping combustion, and combination processes have been under investigation. Many of these second generation CO_2 capture techniques are at demonstration and commercial stage experimentation and show good potential. Research is needed to investigate best strategies for application of suitable CO_2 capture technique at pre-, post-, or oxy-fuel combustion stages. However, environmental, geological, and political variables associated with carbon emission sources will play crucial role in determining the best CO_2 capture technique.

3.8 References

Ahmad, A.L., Sunarti, A.R., Lee, K.T., and Fernando, W.J.N. (2010). ''CO_2 removal using membrane gas absorption.'' *Inter. J. Greenhouse Gas Control*, 4, 495–498.

Alie, C.F. (2004). *CO_2 Capture with MEA: Integrating the Absorption Process and Steam Cycle of an Existing Coal-Fired Power Plant.* Thesis presented to the University of Waterloo. Waterloo, Ontario, Canada, 2004.

Bai, H. (1992). *Fundamental study of ammonia-carbon dioxide reactions to form solid particles*. Ph.D. dissertation, University of Cincinnati, USA, 1992.

Bates, E.D., Mayton, R.D., Ntai, I., and Davis, Jr. J.H. (2002). "CO_2 *Capture by a Task-Specific Ionic Liquid.*'' *Journal of the American Chemical Society*, 124, 926–927.

Baxter, L. (2009). ''Cryogenic carbon capture technology." *Carbon Capture Journal*, 10, 18–21.

Bose, A.C. (2009). *Inorganic Membranes for Energy and Fuel Applications*. Springer Science + Business Media. ISBN: 978-0-387-34534-6. U.S. Department of Energy. Pittsbourg, PA, USA.

Brunetti A., Scura, F., Barbieri, G., and Drioli, E. (2010). ''Membrane technologies for CO_2 separation.'' *Journal of Membrane Science*, 359, 115–125.

CANSIM, Statistics Canada (2006). *Computing in the Humanities and Social Sciences (CHASS), Canadian Socio-economic Information Management (CANSIM).* University of Toronto Data Library Service, Toronto, Ontario.

Chaffee, A.L., Knowles, G.P., Liang, Z., Zhang, J., Xiao, P., and Webley P.A. (2007). ''CO_2 capture by adsorption: materials and process development.'' *Int. J. Greenhouse Gas Control*, 1, 11–18.

CO_2 Solutions (2010). <http://www.co2solution.com/en/index.php> (accessed on August, 26 2010).

CO2CRC–The Cooperative Research Centre for Greenhouse Gas Technologies (2010). <http://www.co2crc.com.au/publications/all_factsheets.html> (accessed August 3, 2010).

Cowan, R.M., Ge, J.-J., Qin, Y.-J., McGregor, M.L., and Trachtenberg, M.C. (2003). "CO_2 capture by means of an enzyme-based reactor." *Ann. N.Y. Acad. Sci.* 984, 453–469.

Dillon, D.J., Panesar, R.S., Wall, R.A., Allam, R.J., White, V., Gibbins, J., and Haines, M.R. (2004). "Oxy-combustion process for CO_2 capture from advanced supercritical PF and NGCC power plant." In: 7th International Conference of Greenhouse Gas Control Technologies, Vancouver, Canada, Sept. 5–9, 2004.

DOT (the U.S. Department of Transportation) (2013). Available at <http://climate.dot.gov/ghg-inventories-forcasts/national/us-inventory-structure.html> (accessed Jan. 2013).

Douglas, P.L., Singh, D., Croiset, E., and Douglas M.A. (2003). ''Techno-economic study of CO_2 capture from an existing coal-fired power plant: MEA scrubbing vs. O_2/CO_2 recycle combustion.'' *Energy Conversion & Management*, 44, 3073–3091.

Duan, S., Kouketsu, T., Kazama, S., and Yamada, K. (2006). ''Development of PAMAM dendrimer composite membranes for CO_2 separation.'' *Journal of Membrane Science*, 283(1–2), 2–6.

Fauth D.J., Frommell E.A., Hoffman J.S., Reasbeck R.P., Pennline H.W., (2005). ''Eutectic salt promoted lithium zirconate: novel high temperature sorbent for CO_2 capture.'' *Fuel Process Technol*, 86, 1503–1521.

Figueroa, J. D., Fout, T., Plasynski, S., McIlvried, H., and Srivastava, R. D. (2008). "Advances in CO2 capture technology–The U.S. Department of Energy's Carbon Sequestration Program." *International Journal of Greenhouse Gas Control*, 2(1), 9–20.

Gray, M. L., Soong, Y., Champagne, K. J., Pennline, H., Baltrus, J. P., Stevens Jr, R. W., Khatri, R., Chuang, S. S. C., and Filburn, T. (2005). "Improved immobilized carbon dioxide capture sorbents." *Fuel Processing Technology*, 86(14–15), 1449–1455.

Gupta, M., Coyle, I., and Thambimuthu, K. (2003). ''CO_2 Capture Technologies and Opportuniries in Canada.'' 1st Canadian CC&S Technology Roadmap Workshop, 18–19 September 2003, Calgary, Alberta, Canada.

Herzog, H., and Golomb, D. (2004). ''Carbon Capture and Storage from Fossil Fuel Use.'' Contribution to Encyclopedia of Energy. Available on line: http://sequestration.mit.edu/pdf/enclyclopedia_of_energy_article.pdf

Hicks, J.C., Drese, J.H., Fauth, D.J., Gray, M.L., Qi, G., and Jones, C.W. (2008). ''Designing Adsorbents for CO_2 Capture from Flue Gas-Hyperbranched Aminosilicas Capable of Capturing CO_2 Reversibly.'' *Journal of American Chemical Society*, 130, 2902–2903.

IEA Greenhouse Gas R&D Programme. (2007). *Capturing CO_2*. ISBN : 9781-898373-414.

IPCC (2007). Contribution of Working Group III to the Fourth Assessment Report of the Intergovernmental Panel on Climate Change. B. Metz, O.R. Davidson, P.R. Bosch, R. Dave, and L.A. Meyer (eds). Cambridge University Press, Cambridge, United Kingdom and New York, NY, USA.

IPCC (2005). IEA special report on carbon dioxide capture and storage. Accessed from: www.ipcc.ch

IPCC (2001) Climate change 2001: impacts, adaptation and vulnerability. Contribution of Working Group II to the Third Assessment Report of the Intergovernmental Panel on Climate Change. Cambridge: Cambridge University Press.

Iwan, A., Stephenson, H., Ketchie, W.C., and Lapkin, A.A. (2009). ''High temperature sequestration of CO_2 using lithium zirconates.'' *Chemical Engineering Journal*, 146, 249–258.

Jordal, K., Anheden, M.,Yan, J., and Strömberg, L., (2004). ''Oxyfuel combustion for coal-fired power generation with CO_2 capture-opportunities and challenges.'' In: Proceedings of the 7th International Conferenceon Greenhouse Gas Control Technologies, International Energy Agency Greenhouse Gas Programme.

Maginn, E.J. 2005. Design and Evaluation of Ionic Liquids as Novel CO_2 Absorbents. Quarterly Technical Report. University of Notre Dame. Available on line: http://www.osti.gov

Mallada, R., and Menéndez, M. (2008). Inorganic Membranes. Synthesis, characterization and applications. Membrane Science and Technology Series, 13. ISBN: 978-0-444-53070-7. Elsevier. Oxford, UK.

Marion, J., and Griffin, T. (2001). Controlling power plant CO_2 emissions: A long range view. First National Conference on Carbon Sequestration. May 14–17, 2001.

Martín, C.F., Plaza, M.G., Pis, J.J., Rubiera, F., Pevida, C., and Centeno, T.A. (2010). "On the limits of CO2 capture capacity of carbons." *Separation and Purification Technology*, 74, 225.

Mattisson, T., and Lyngfelt, A. (2001). Applications of chemical-looping combustion with capture of CO_2. Second Nordic Minisymposium on Carbon Dioxide Capture and Storage, Göteborg, October 26, 2001.

Meisen, S., and Shuai, X. (1997). ''Research and Development Issues in CO_2 Capture.'' *Energy Convers. Mgrnt*., 38, 37–42.

Moore, T.T., and Koros, W.J. (2007). ''Gas sorption in polymers, molecular sieves, and mixed matrix membranes.'' *Journal of applied polymer science*, 104, 4053–4059.

Mundschau, M.V., Xie, X., Evenson IV, C.R., and Sammells, A.F. (2006). ''Dense inorganic membranes for production of hydrogen from methane and coal with carbon dioxide sequestration.'' *Catalysis Today*, 118, 12–23.

Nair, B.N., Burwood, R.P., Goh, V.J., Nakagawa, K., and Yamaguchi, T. (2009). ''Lithium based ceramic materials and membranes for high temperature CO_2 separation.'' *Progress in Materials Science*, 54, 511–541.

NREL – National Renewable Energy Laboratory. (2006). Equipment Design and Cost Estimation for Small Modular Biomass Systems, Synthesis Gas Cleanup, and Oxygen Separation Equipment. Task 2.3: Sulfur Primer. San Francisco, California, USA.

O'Neill, B.C. (2002). ''Oppenheimer M. Climate change: dangerous climate impacts and the kyoto protocol.'' *Science*, 296 (5575), 1971–1972.

Olajire, A.A. (2010). "CO_2 capture and separation technologies for end-of-pipe applications – A review." *Energy*, 35(6), 2610–2628.

Pehnt, M., and Henkel, J. (2009). ''Life cycle assessment of carbon dioxide capture and storage from lignite power plants.'' *International Journal of greenhouse gas control*, 3, 49–66.

Pevida, C., Plaza, M. G., Arias, B., Fermoso, J., Rubiera, F., and Pis, J. J. (2008). "Surface modification of activated carbons for CO2 capture." *Applied Surface Science*, 254(22), 7165–7172.

Plate, N., and Yampol'skii, Y.P. (1994). Relationship between structure and transport properties for high free volume polymeric materials. In: Polymeric gas separation membranes. Baton Rouge: CRC Press, p. 115–208.

Price, J., McElligot, S., Price, I., and Smith, B. (2008). ''Carbon Capture and Storage. Meeting the challenge of climate change.'' Bluewave Resources. LLC of McLean, Virginia, USA.

Rochelle, G., Chen, E., Dugas, R., Oyenakan, B., and Seibert, F., (2006). ''Solvent and process enhancements for CO2 absorption/ stripping.'' In: 2005 Annual Conference on Capture and Sequestration, Alexandria, VA, May 8–11, 2006.

Salmon, S., Saunders, P., and Borchert, M. (2009). ''Enzyme technology for carbon dioxide separation from mixed gases.'' IOP Conf. Series: *Earth and Environmental Science* 6 (2009) 172018.

Sander, M.T., and Mariz, C.L. (1992). ''The Fluor Daniel® Econamine™ FG process: Past experience and present day focus.'' *Energy Conversion Management*, 33(5–8), 341–348.

Scholes, C.A., Kentish S.E., and Stevens, G.W. (2008). ''Carbon Dioxide Separation through Polymeric Membrane Systems for Flue Gas Applications.'' *Recent Patents on Chemical Engineering*, 1, 52–66.

Sekkappan, G., Melling, P.J., Anheden, M., Lindgren, G., Kluger, F., Sanchez-Molinero, I., Maggauer, C., and Doukelis, A. (2006). ''Oxyfuel Technology for CO_2 Capture from Advanced Supercritical Pulverized Fuel Power Plants.'' In: 8th International Conference on Greenhouse Gas Control Technologies, Trondheim, Norway, June 19–22, 2006.

Shackley, S., and Gough, C. (2006). ''Carbon capture and its storage.'' Ashgate Publishing Limited. Hampshire, England.

Shekhawat, D., Luebke, D.R., and Pennline, H.W. (2003). A Review of Carbon Dioxide Selective Membranes. A Topical Report. National Energy Technology Laboratory, United States Department of Energy, DOE/NETL-2003/1200.

Simmonds, M., Miracca, I., and Gerdes, K. (2004). ''Oxyfuel technologies for CO_2 capture: a techno-economic overview. In: 7th international conference on greenhouse gas control technologies.'' Vancouver, Canada; September 2004.

Steeneveldt, R., Berger, B., and Torp, T.A. (2006). ''CO_2 Capture and Storage: Closing the Knowing–Doing Gap.'' *Chemical Engineering Research and Design*, 84(A9), 739–763.

Stucki, S., Schuler, A., and Constantinescu, M. (1995). ''Coupled CO_2 recovery from the atmosphere and water electrolysis: feasibility of a new process for hydrogen storage.'' *Int. J. Hydrogen Energy*, 20, 653–663.

Thambimuthu, K., Davison, J., and Gupta, M. (2002). In: IPCC workshop on carbon dioxide capture and storage. Regina, Canada, 18–21 November 2002.

Van der sluijs, J.P., Hendriks, C.A., and Blok, K. (1992). ''Feasibility of polymer membranes for carbon dioxide recovery from flue gases.'' *Energy Convers. Mgmt.*, 33(5–8), 429–436.

Viebahn, P., Nitsch, J., Fischedick, M., Esken, A., Schüwer, D., Supersberger, N., Zuberbühler, U., and Edenhofer, O. (2007). ''Comparison of carbon capture and storage with renewable energy technologies regarding structural, economic, and ecological aspects in Germany.'' *International Journal of Greenhouse Gas Control*, 1, 121–131.

Wall, T.F. (2007). "Combustion processes for carbon capture." *Proceedings of the Combustion Institute*, 31(1), 31–47.

Xi, Z, Shi, X, Liu, M, Cao, Y, Wu, X, and Ru, G. (1985). ''Agrochemical properties of ammonium bicarbonate.'' *Turang Xuebao*, 22(3), 223–232.

Yan, Y., Davis, M.E., and Gavalas, G.R., (1997). ''Use of diffusion barriers in the preparation of supported zeolite ZSM-5 membranes.'' *Journal of Membrane Science*, 126, 53–65.

Yang, H., Xu, Z., Fan, M., Gupta, R., Slimane, R. B., Bland, A. E., and Wright, I. (2008). "Progress in carbon dioxide separation and capture: A review." *Journal of Environmental Sciences*, 20(1), 14–27.

Zeman, F. (2007). "Energy and Material Balance of CO_2 Capture from Ambient Air." *Environmental Science and Technol*ogy, 41(21), 7558–7563.

CHAPTER 4

Carbon Capture and Sequestration: Biological Technologies

Klai Nouha, Rojan P. John, Song Yan, R. D. Tyagi, Rao Y. Surampalli and Tian C. Zhang

4.1 Introduction

Strategies to reduce emissions of carbon dioxide (CO_2) from fossil fuels, and hence mitigate climate change, include energy savings, development of renewable biofuels, and carbon capture and sequestration (CCS). For CCS, several scenarios are being considered (Zhang and Surampalli 2013). One approach is capture of point-source CO_2 from power plants or other industrial sources and subsequent injection of the concentrated CO_2 underground or into the ocean (Benson et al. 2008). An alternative to this point-source CCS method is expansion of biological carbon sequestration of atmospheric CO_2 by measures such as reforestation, changes in land use practices, increased carbon allocation to underground biomass, production of biochar, and enhanced biomineralization (Janssan et al. 2010).

Biological CCS is one of the natural and cost effective technologies for CCS. It includes i) the photosynthetic systems of microorganisms or higher plants (e.g., algae, microbes), ii) sustainable practices (e.g., soil conservation, development of grasslands) and iii) use of biomass/residues (e.g., biomass-fuelled power plant, production of biofuels, and biochar). Currently, considerable studies have been conducted on biological CCS. Many reviews have been published about biotic carbon sequestration, wetlands, soil carbon sequestration (Bruce et al. 1999; Lehmann 2007; Lal 2008; Trumper et al. 2009). However, research and development of CCS technologies has been generating new information on advanced biological processes, genetic and/or protein engineering of microorganisms, plants and biomass to improve/optimize biotic CCS, which warrants rigorous review for wide-range dissimilation. This chapter provides a state-of-the-art review on biological processes and technologies for CCS, including the major biological processes, approaches and alternatives to i) capturing and ii) sequestrating CO_2, iii) advanced biological processes for CCS, an iv) comparison between biotic and abiotic CCS concerning their merits and limitations. Specific attention is paid to the principles and related mechanisms for biological CCS at the ecosystems, organism, and molecular levels.

4.2 Biological Processes for Carbon Capture

The biological processes for CCS normally affect the cycle of carbon in the planet. The carbon cycle is the biogeochemical cycle by which carbon is exchanged among the biosphere, pedosphere (the soil-containing earth surface), geosphere, hydrosphere, and atmosphere of the Earth. Carbon moves from: a) atmosphere (as CO_2) to plants (via photosynthesis) or the oceans and/or other water bodies (via absorption processes), b) plants (or other animals) to animals via food chains, c) died plants/animals to the ground (e.g., fossil fuels formed in millions and millions of years), and d) living things (via respiration) and/or fossil fuels (upon being burned) to the atmosphere (Zhang and Surampalli 2013). When humans burn fossil fuels for energy, most of the carbon quickly enters the atmosphere as CO_2. Each year, 5.5 billion tons of carbon is released by burning fossil fuels; 3.3 billion tons enter the atmosphere and most of the rest is absorbed by the oceans. CO_2 and other greenhouse gases (GHGs) trap heat in the atmosphere to keep the Earth warm and liveable for living beings. However, there is about 30% more CO_2 in the air today (due to human fossil burning activities) than there was about 150 years ago, causing our planet to become warmer (Johnson 2010).

The Carbon Sequestration Regional Partnerships (Litynski et al. 2008) have estimated that 1120 to 3400 billion tonnes of CO_2 can be sequestered in the formations identified so far. CO_2 sequestration in geologic formations shows great promise because of the large number of potential geologic sinks. Also, with higher petroleum prices, there is increased interest in using CO_2 flooding as a means to enhance oil recovery (EOR). With higher gas prices, there will be growing interest in using CO_2 for enhanced coal bed methane production (ECBM). However, none of these activities will be possible unless CO_2 is first captured.

There are many carbon capture technologies (Zhang and Surampalli 2013). These technologies can be categorized as a) physical/chemical and biological technologies and b) technologies for carbon capture from concentrated point sources and mobile/distributed point- or non-point sources. In general, on-site capture is the most viable approach for large sources and initially offers the most cost-effective avenue to sequestration. For mobile and/or distributed sources like cars, on-board capture at affordable cost would not be feasible, but are still needed. It should be noted that none of the currently available CO_2 capture processes are economically feasible on a national implementation scale to capture CO_2 for sequestration, because they consume large amounts of parasitic power and significantly increase the cost of electricity. Thus, improved CO_2 capture technologies are vital if the promise of geologic sequestration, EOR, and ECBM is to be realized.

This section discusses two major biological processes for carbon capture: i) the use of biomass/residuals for biomass energy generation with CO_2 capture, and ii) the photosynthetic systems of microorganisms. The "carbon-negative" energy system concept is also introduced for better understanding of the framework of biotic CCS.

4.2.1 Carbon Capture from Biomass

Major Routes to Biomass Energy Products with CO_2 Capture. Figure 4.1 illustrates the major routes to biomass energy products with CCS, including a) biological processing (e.g., fermentation) with capture of CO_2 by-products to produce liquid fuels and chemical products, and b) biomass combustion to i) produce electricity with CCS either by oxyfuel or pre-combustion CO_2 capture (PCC) routes and ii) biomass gasification with shift and CO_2 separation to produce hydrogen. These basic routes can be combined or integrated, for example, by gasification with CCS of residual biomass from biological processes, or by syngas conversion to liquid fuels with CCS, or by burning hydrogen-rich syngas to produce electricity with CCS.

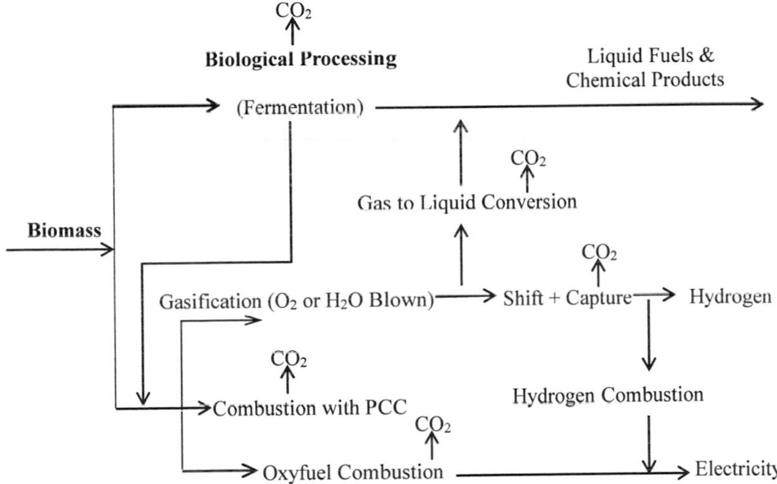

Figure 4.1 Routes to biomass energy products with CO_2 capture (adapted from Rhodes et al. 2005)

Biomass has important similarities with fossil fuels (particularly coal), including conversion technologies and the range of energy products that can be generated, dispatchable, base-load electricity as well as liquid and gaseous fuels. As a result, all three technological routes used for CCS in coal-fired power plants (i.e., post-, pre- and oxyfuel-combustion CO_2 capture) could be applied to biomass energy systems. Biological processes, such as bio-ethanol fermentation, provide additional CCS opportunities. PPC or oxyfuel-combustion CO_2 capture could be integrated with modern biomass boiler technologies or retrofitted to existing plants, though the small-scale and low efficiency of existing biomass boilers would make this relatively inefficient. Alternatively, coal-fired power plants could be retrofitted to co-fire biomass and incorporate CCS such that biomass carbon captured would more than offset incomplete capture of coal carbon (Robinson et al. 2003). With sufficiently stringent emissions controls, such a plant could be retrofitted to burn only biomass. The feasibility of this depends on i) emission controls inducing both a low purchase price for unmodified coal-fired power plants, and ii) financial dominance of large negative emissions over potentially high fuel costs, as well as iii) local access to very large biomass resources. Modern biomass gasification technologies could incorporate pre-combustion sequestration. Syngas dilution with atmospheric nitrogen largely eliminates the benefits of pre-combustion sequestration in air-blown gasification systems. However, indirectly heated, steam-blown systems or oxygen blown systems could effectively leverage pre-combustion sequestration. Oxygen-blown biomass gasification has been demonstrated and offers higher energy efficiencies and carbon capture, though somewhat less operating experience and economic data is available for these systems (Babu 2004).

Biological processes provide additional opportunities for biomass-CCS. CO_2 is a by-product of fermentation in bio-ethanol production, implying that CO_2 available for capture scales with ethanol production and that fuel carbon capture rates scale with conversion efficiency. The retrofit potential of this strategy implies nearly 9 MtC/yr is available at a very low capture cost given global bio-ethanol production of approximately 40 Mm^3 in 2003 (Berg et al. 2004). Bio-ethanol production generally also includes combustion or gasification and combustion of waste biomass, providing further carbon capture opportunities, with additional cost (Mollersten et al. 2003). The ability to generate emissions offsets extends the scope of carbon mitigation with biomass and may provide cost-effective mitigation alternatives across the economy, fundamentally changing the economics of biomass-based mitigation. Negative emissions from biomass-CCS do not, however, offer strict capon mitigation costs since its costs must scale with the biomass supply curve, which may become steep if large-scale bioenergy crops compete for limited land resources. Environmental impacts may further constrain biomass mitigation potential (Kheshgi et al. 2000). However, the extraordinary heterogeneity of emissions sources provides many niches, and integrating CCS will extend the opportunities for biomass-based mitigation.

Concept of "Carbon-Negative" Energy System Technology. Figure 4.2 illustrates the photosynthetic biomass production on Earth (center) that can take CO_2 from the atmosphere. Biomass pyrolysis (upper left) could produce biochar and biofuels (such as H_2), which could be optionally utilized to make NH_4HCO_3 char and/or urea–char fertilizers. Use of biochar fertilizers could store carbon into soil and subsoil earth layers, reduce fertilizer (such as NO_3^-) runoff, and improve soil fertility for more photosynthetic biomass production to provide more win-win benefits, including more forest, more fabric and wooden products, more food and feedstocks. It is also possible to create biochar/energy reserves (bottom right) as "global carbon thermostat" to control global warming. This char-producing biomass-pyrolysis approach essentially employs the existing natural green-plant photosynthesis on the planet as the first step to capture CO_2 from the atmosphere; then, the use of a pyrolysis process converts biomass materials primarily into biofuel and char, a stable form of solid carbon material that is resistant to microbial degradation. The net result is the removal of CO_2 from the atmosphere since the total process capturing CO_2 from the atmosphere and placing it into soils and/or subsoil earth layers as a stable carbon (biochar) while producing a significant amount of biofuel energy through biomass pyrolysis.

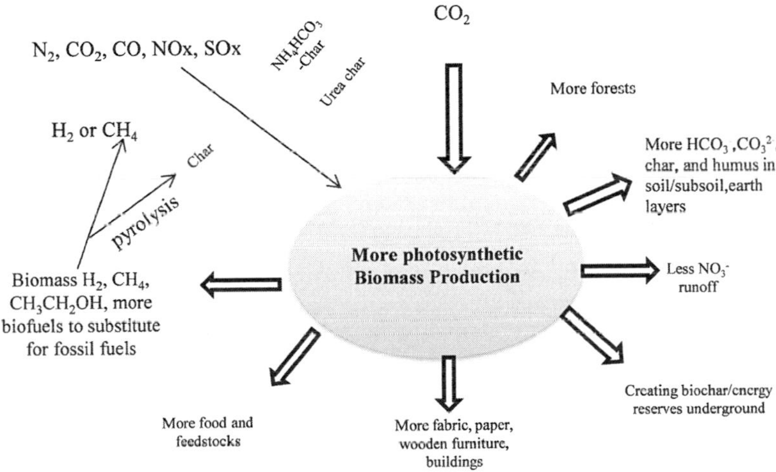

Figure 4.2. The potential benefits of the carbon-negative energy systems technology concept with biomass pyrolysis and ammonia carbonation for CCS (adapted from Lee et al. 2010)

Many studies reported smokeless biomass pyrolysis for biochar and biofuel production for global CCS at scales of gigaton carbon (GtC) (Lee et al. 2010). Each year, land-based green plants capture about 440 GtC from the atmosphere into biomass. That is, about one-seventh of all the CO_2 in the atmosphere (820 GtC) is fixed by photosynthesis (gross primary production) every year. However, biomass is not a stable form of carbon material with nearly all returning to the atmosphere as CO_2 in a relatively short time. Because of respiration and biomass decomposition, there is nearly equal amount of CO_2 (about 120 GtC/yr) released from the terrestrial biomass system back into the atmosphere each year (Sauerbeck et al. 2001). As a result, using biomass for carbon sequestration is limited. Any technology that could significantly prolong the lifetime of biomass materials would be helpful to global CCS. The approach of biomass pyrolysis provides such a possible capability to convert the otherwise unstable biomass into biofuel and, more importantly, biochar, which is suitable for use as a soil amendment and serves as a semi-permanent carbon sequestration agent in soils/subsoil earth layers for hundreds and perhaps thousands of years.

Low-temperature biomass pyrolysis is a process in which biomass such as forest waste woods and/or crop residues (e.g., corn stover) are heated to about 400 °C in the absence of oxygen and, as a result, the biomass is converted to biofuel (bio-oils and syngas) and charcoal (char), a stable form of solid black carbon (C) material. Although its detailed thermo-chemical reactions are quite complex, the biomass-pyrolysis process can be described by the following general equation:

$$\text{Biomass (e.g., lignin cellulose)} \rightarrow \text{char} + H_2O + \text{bio-oils} + \text{syngas} \qquad (4.1)$$

The chemical composition and the yield of biochar depend on the feedstock properties and pyrolysis conditions, including temperature, heating rate, pressure, moisture, and vapor-phase residence time (Antal et al. 2003). With certain refinery processes, the organic volatiles (bio-oils) and syngas (CO, CO_2, and H_2, etc.) from biomass pyrolysis could be used as biofuels for clean energy production. Typically, about 15–50% of the biomass C (carbon) can be converted into biochar, while the remaining C going to the biofuel fraction. Depending on the biomass materials, the low temperature pyrolysis process can be slightly exothermic so that once the process is started it could sustain itself with its own heat. That is, the exothermic heat evolution from biomass pyrolysis can elevate the temperature of the incoming (dry) biomass feedstock sufficiently to initiate the carbonation reactions (Antal et al. 2003). Consequently, once initiated, it is possible to convert large amounts of biomass into biochar and biofuel with minimal exogenous energy cost.

According to a recent study (Das et al. 2010) using a pilot-scale pyrolysis unit at Eprida, pyrolysis of 100 kg biomass (southern yellow pine pellets) at 482 °C can typically produce 26.3 kg of biochar. Based on the energy value calculation, the 26.3 kg biochar contains 528 million joules (MJ) (28%) of the 100 kg biomass energy (1859 MJ). The remaining biomass energy (1331 MJ) exists as pyrolysis vapors (crude biofuel) and heat in the gas phase from the pyrolyzer. The addition of steam and the use of a steam-reforming process (the water gas shift reaction) can convert the pyrolysis vapors and water into syngas (126.6 kg), giving an average composition of 47.6% H_2 (6.66 kg), 18.3% CO_2 (55.9 kg), 2.7% CH_4 (3.00 kg), 13.7% CO (26.6 kg), and 17.7% N_2 (34.4 kg). The significant amount of nitrogen in the syngas was due to the N_2 used to purge the lines for the sensor equipment and for pressuring the biomass feed and char discharge systems. The higher heating value (HHV) of this syngas is calculated to be 1403 MJ and the total energy cost for the pyrolysis and steam-reforming process is 787 MJ, assuming no heat recovery. The net biofuel (syngas) energy production is 616 MJ (per 100 kg biomass), which represents about 33% of the biomass energy (1859 MJ) while the biochar product contains 28% of the biomass energy (1859 MJ). The total combined process energy-conversion efficiency for the production of biochar and biofuel is 61% in this case. With better process designs such as use of heat recovery techniques, the energy conversion efficiency may be further improved. Therefore, this example demonstrates that it is possible to produce both biochar and biofuel from biomass with reasonable energy efficiency.

Therefore, this is a "carbon negative" energy production approach. Currently, the United States can annually harvest over 1.3 gigaton (Gt) of dry biomass, of which about 1.0 Gt is generated from the croplands and over 0.3 Gt is from a fraction of forestlands that are accessible by roads (Perlack et al. 2005). If this amount of biomass (1.3 Gt/yr) is processed through controlled low temperature pyrolysis and assuming 50% conversion of biomass C to stable biochar C and 33% of the biomass energy to biofuels (syngas and bio-oils), it could produce biochar (0.325 GtC/yr) and biofuels (with heating value equivalent to that of 1300 million barrels of crude oil) to help control global warming and achieve energy independence from fossil fuel. In 2008, the US brought in 845 million barrels of crude oil, according to the Energy Information Administration, an arm of the US Department of Energy. The heating value (equivalent to that of 1300 million barrels of crude oil) of the syngas/bio-oils from biomass pyrolysis in this scenario exceeds that of the USA-imported crude oils (845 million barrels). Even if a half of the syngas/bio-oils may be consumed to cover any other energy costs in handling the biomass (such as biomass collection, drying and transport), the remaining syngas/bio-oils (equivalent to that of 650 million barrels of crude oil) is still quite significant as the net biofuel output. Therefore, if a cost-effective biofuel-refinery technology can be developed to convert the syngas/bio-oils from biomass pyrolysis into liquid transportation fuels, use of this approach could significantly help reduce the imports of foreign oil. In the immediate future before such a biofuel refinery technology is available, the syngas/bio-oils from biomass pyrolysis could be used for its heating energy by combustion to replace fossil fuels including coal, natural gas and heating oils.

In addition, application of the 0.325 GtC/yr of biochar products as soil amendment and carbon sequestration agent in soil could stimulate the agriculture economy and achieve major carbon sequestration for the US to control global warming as well.

4.2.2 Carbon Capture by Photosynthetic Microorganisms

Photosynthetic autotrophic organisms and plants utilise this process of carbon fixation as their food source by converting the CO_2 into organic carbon (Stephan et al. 2001). Photosynthesis is the process by which green plants use energy from light in the photo-synthetically active radiation (PRA) range (PAR wavelength = 400–700 nm) to form glucose in their chlorophyll-containing tissues (Stephan et al. 2001). There are two stages in photosynthesis, namely the light dependant stage and the light independent stage. The formation of glucose and O_2 from CO_2 and water in the presence of light is shown in Eq. 4.2 (Petela et al. 2008):

$$6H_2O + 6CO_2 \rightarrow C_6H_{12}O_6 + 6O_2 \tag{4.2}$$

The produced glucose is then converted into starch and cellulose, storing the carbon in cells of the plant, thus mitigating the inorganic carbon by creating organic carbon (Stephan et al. 2001).

4.2.2.1 Microalgae

Photosynthesis is the original process that created the fixed carbon present in today's fossil fuels, and microalgae are the origin of these fuels. Microalgae are among the fastest growing photo-synthetic organisms, using CO_2 as their main building blocks (Kurano et al. 1996). The biomass volumes can be doubled in less than 24 h for most species of microalgae. For a flow rate of 0.3 l/min of air with a 4% CO_2 concentration, a carbon fixation rate of 14.6 g C/m^2-day at a growth rate of 30.2 g can be achieved (Watanabe et al. 1996). This makes microalgae very well suited to carbon mitigation, as their high growth rates can keep up with the continuous flow of CO_2 from the power plant. One of the main challenges with microalgae culturing is the capital cost. The required photobioreactors (PBR) are expensive, and therefore, a government grant scheme may be necessary to encourage utilisation of the technology by power plants. Raceway pond production is less expensive to build and operate; thus, it may be a more viable solution to the economics of the process. In many places (e.g., Ireland), poor sunlight may be a problem especially in the winter months. Artificial lighting may then have to be utilised to ensure survival of the microalgal culture but with the consequence of increasing production costs.

Although culturing microalgae at an industrial scale can be expensive, it has huge potential in producing fuel either from direct combustion, thermo chemical or biochemical processes. These include gasification, pyrolysis, liquefaction, and anaerobic digestion (as shown in Fig. 4.1). Co-firing is an attractive option for converting microalgae into biofuels such as biodiesel. Biodiesel is cleaner than petroleum diesel and is virtually free of sulphur, which eliminates the production of sulphur oxides (Stephan et al. 2001). Microalgal photosynthesis can also result in the precipitation of calcium carbonate, a potentially long-term sink of carbon (Aresta el al. 2005). Microalgal culturing also yields high value commercial products. Sale of these high value products can offset the capital and the operation costs of the process (Oleizola et al. 2004). A cost and energy balance shows that energy production from marine biomass is an attainable target with the currently available technologies. However, in general, the obtained biofuel is too expensive when compared to fossil fuel prices. With the introduction of carbon taxes and the ever increasing price of oil, the cost energy balance will become economically favourable while reducing carbon emissions (Aresta et al. 2005).

4.2.2.2 Macroalgae

Like microalgae, macroalgae are autotrophic aquatic plants using inorganic CO_2 as their food source. Microalgae have received greater attention than macroalgae in the past for CO_2 fixation due to their facile adaptability to grow in ponds or bioreactors and the extended knowledge and research of the many strains used for fish feeding (Aresta et al. 2005). Due to technological and financial hindrances, commercial production of macroalgae is unviable; thus, traditionally it has been harvested from natural basins. Production of algae requires balancing the sea water pH with CO_2 and adding essential nutrients, primarily nitrogen and phosphorus as in microalgal cultivation.

Currently, large scale cultivation of macroalgae is confined mainly to Asia, where commercial CO_2 is used to increase the biomass yield. However, in recent times it is being considered elsewhere as its capacity as a valuable resource becomes more apparent across the world. Macroalgae is very well suited for carbon mitigation due to its high growth rates, satisfactorily utilising CO_2 from the flue gases of the power plant, and, like microalgae, have many value-added by-products. Through gasification, energy yields of up to 11.000 MJ/t dry algae have been achieved compared with 9500 MJ/t for microalgae. The *Gracilaria cornea* (Rhodophyta) strain is produced on a large scale for animal feed using commercial CO_2. Using flue gases as a CO_2 source would greatly reduce the cost of production of macroalgae while increasing the biomass yield (Israel et al. 2005). The use of flue gases containing 12 to 15% CO_2 has been found to maintain the desired pH (Israel et al. 2005). Yields of macroalgal species, such as *Porphyra yezoensis, Gracilaria sp., G. chilensis*, and *Hizikia fusiforme*, were increased 2–3 times when grown at the enhanced levels of CO_2 compared with atmospheric CO_2 cultivation (Wu et al. 2008). A study carried out by Israel et al. (2005) shows that the biomass yield of *Gracilaria cornea* using flue gases was comparable with the achieved yields using commercial CO_2. In addition, macroalgae has many useful applications in the alginate industry, horticulture, cosmetics, biomedicine and their nutritional value in sea vegetables as well as agricultural fertilisers. It may also be co-fired in the power plant to reduce the need for fossil fuels (Ross et al. 2008).

4.2.2.3 Cyanobacteria

Bacteria are fast growing unicellular organisms. Cyanobacteria (also known as blue-green algae) are photoautotrophic bacteria utilising CO_2 as their food source and, therefore, are functional in carbon mitigation. They grow in a temperature range of 50 to 75°C and require light for good growth with hydrogen as a by-product (Eq. 4.3) (Akkerman et al. 2002):

$$CO_2 + H_2O + \text{light energy} \rightarrow C_n(H_2O)_n + O_2 \tag{4.3}$$

This process may be exploited by utilising selected strains of cyanobacteria for direct biological carbon mitigation applications.

The biofixation of CO_2 by cyanobacteria in photobioreactors is considered a sustainable strategy, as CO_2 can be incorporated into the molecular structure of bacterial cells in the form of proteins, carbohydrates and lipids. One strain found to be particularly adept at carbon mitigation is *Synechococcus* and has achieved a CO_2 uptake rate of 0.025 g/l-h or 0.6 g/l-day at a cell mass concentration of 0.286 g/l. If scaling up were plausible, this would equate to a bioreactor of size 4000 m^3 with an average fixation rate of 1 t CO_2/h from emission sources although there could be challenges to overcome. Using *Chlamydomonas reinhardtii*, collection rates of 2 ml/h of hydrogen and 12 ml/h of oxygen were obtained (Stewart and Hessami 2005).

Through photosynthesis and calcification, cyanobacteria have the potential to capture CO_2 from flue gas and store it as precipitated $CaCO_3$. Calcium is abundant in many terrestrial, marine and lacustrine ecosystems. By using halophilic cyanobacteria, seawater or brines (e.g., agricultural drainage water) or saline water produced from petroleum production or geological CO_2 injections can serve as potential calcium sources for the calcification process. Calcification can further be boosted by supplying calcium from gypsum (Mazone et al. 2002) or silicate minerals, possibly in connection with biologically accelerated weathering. However, successful implementation of calcifying cyanobacteria for point-source CCS is met with significant challenges that need to be addressed.

It has been accepted that alkalinization in exopolysaccharide (EPS) or proteinaceous surface layers (S-layer) depends on HCO_3^- import. Therefore, the question arises as to whether calcification in cyanobacteria will occur also under high CO_2 conditions (e.g., when fed CO_2 from a flue gas stream). At high CO_2 levels, the carbon concentrating mechanism (CCM) is not needed and cells will preferentially take up CO_2 rather than HCO_3^-. The conversion of CO_2 during transport to the cytosol produces H^+ (i.e., via $CO_2 + H_2O \rightarrow H^+ + HCO_3^-$) that needs to be neutralized, possibly via export to the medium (Price et al. 2002). This counter balances the subsequent and opposite alkalinization reactions in the carboxysome. Also, rapid infusion of gaseous CO_2 into a cyanobacterial pond will probably lower the ambient pH, impeding alkalinization at the extracellular surface. Cyanobacteria still calcify under elevated CO_2 levels but photosynthesis seems to exert little or no influence on the process (Obst et al. 2009). Furthermore, $CaCO_3$ precipitates were found to be more peripherally located on the extracellular surface and have a different morphology in cells predominantly taking up CO_2 instead of HCO_3^- (Yates and Robbins 1998). It remains to be clarified as for whether reactions such as photosynthesis electron transport (PET) and Ca^{2+} efflux suffice to generate extracellular alkaline microenvironments, to which extent carbonic anhydrase (CA) activities in the EPS are involved or if $CaCO_3$ precipitation during rapid CO_2 uptake becomes a passive process, relying mainly of Ca^{2+} binding and nucleation at the EPS or S-layer. It is important to unravel the mechanisms of calcification and how they are regulated in cyanobacteria growing under flue gas conditions, and in the presence of pulverized gypsum or calcium silicate minerals. Mutant cells grew at high CO_2 levels, but growth was not observed under CO_2-limiting conditions. Another option might be to have the flue gas pass through a CA system so as to convert incoming CO_2 to HCO_3^- before reaching the calcifying cyanobacteria. CA could either be overproduced and secreted as extracellular enzymes directly into the solution by cyanobacteria or other bacteria, or immobilized on solid supports.

Another issue relates to scale. A 500 MW coal-fired power plant emits between 3 and 4 Mt of CO_2 per year (Herzog et al. 2004). To be industrially relevant, ponds (or photobioreactors) with calcifying cyanobacteria have to produce amounts of $CaCO_3$ large enough to make an impact. Only a few attempts have been made at evaluating the rate of calcification in cyanobacteria. Lee et al. (2006) evaluated Bahamain whitings events in the Great Bahama Bank with an average of 70 km^2 and microcosm experiments with the marine *Synechococcus* 8806 (*S.* 8806). They estimated that calcification by *S.* 8806 could account for approximately 2.5 Mt $CaCO_3$ per year. This translates to sequestration of over half of the CO_2 produced from a 500 MW power plant (Lee et al 2006). Robust cyanobacterial strains or consortia need to be designed that exhibit maximized photosynthetic CO_2 uptake and that can fully utilize the plentiful calcium available in silicate minerals or gypsum. Calcification can be enhanced by increasing the number of carboxylate amino acids in the EPS that can be used as nucleation sites, and by increasing CA activities in the EPS. It is also crucial to develop strains that have highly efficient light utilization and photoprotection properties. Cyanobacteria, in general, have low light requirements, but when grown in ponds, cells below the surface will be light-limited while those at the top might experience excessive light intensities. Furthermore, the information gained from studying calcification in cyanobacteria can be used for biomimetic approaches where artificial systems based on CA, EPS, or S-layers are designed for CO_2 capture and biomineralization. Crucial to these efforts is optimizing the long-term stability of the resulting carbonates (Addadi et al. 2003).

In short, employment of cyanobacteria for point-source CCS of flue gas via calcification offers promising strategies for reducing anthropogenic CO_2 emissions. However, much research is urgently needed to further our understanding of the biochemical and physical processes in cyanobacteria that promote calcification and that will allow us to select or design strains with optimized properties for specific applications and conditions using genetic engineering or directed evolution.

4.3 Biological Processes for CO_2 Sequestration

Biotic sequestration is based on managed intervention of higher plants and microorganisms in removing CO_2 from the atmosphere. It differs from management options which reduce or offset emission. Increasing use efficiency of resources (e.g. water, energy) is another option for managing the carbon pool.

4.3.1 Ocean Sequestration

There are several biological processes leading to carbon sequestration in the ocean through photosynthesis. Phytoplankton photosynthesis is one such mechanism (Rivkin and Legendre 2001), which fixes approximately 45 Pg C/yr (Falkowski et al. 2000). Some of the particulate organic material formed by phytoplankton is deposited at the ocean floor and is thus sequestered (Raven and Falkowski 1999). Availability of Fe is one of the limiting factors on phytoplankton growth in oceanic ecosystems. Thus, several studies have assessed the importance of Fe fertilization on biotic CO_2 sequestration in the ocean. Researchers have targeted high-nutrient/low chlorophyll (HNLC) ocean regions, specifically the eastern Equatorial Pacific, the northeastern Subarctic Pacific, and the Southern Ocean. Four major open-ocean experiments have been conducted to test the "iron hypothesis," two in the Equatorial Pacific (Ironexi in 1993 and Ironexii in 1995) and two in the Southern Ocean (Soiree in 1999 and Eisenex in 2000). These experiments, funded through basic science programs (not sequestration programs), show conclusively that phytoplankton biomass can be dramatically increased by the addition of iron. However, although a necessary condition, it is not sufficient to claim iron fertilization will be effective as a CO_2 sequestration option. The proponents of iron fertilization claim very cost effective mitigation, on the order of \$1–10/t C (Herzog et al. 2004), but critical scientific questions remain unanswered. Although iron increases uptake of CO_2 from the atmosphere to the surface ocean, CO_2 needs to be exported to the deep ocean to be effective for sequestration. No experiments have yet attempted to measure export efficiency, which is an extremely difficult value to measure (some people claim that it cannot be measured experimentally). In addition, there are concerns about the effect on ecosystems, such as inducing anoxia (oxygen depletion) and changing the composition of phytoplankton communities.

Many studies reported that the addition of nmol amounts of dissolved iron resulted in the nearly complete utilization of excess NO_3, whereas in the controls without added Fe, only 25% of the available NO_3 was used. They also observed that the amounts of chlorophyll in the phytoplankton increased in proportion to the Fe added. They conclude that Fe deficiency is limiting phytoplankton growth in these major nutrient-rich waters (Falkowski 1997; Boyd et al. 2004). Similar to deep injection, ocean fertilization may also change the ecology of the ocean (Chisholm et al. 2001). However, with the current state of knowledge, the topic of ocean fertilization remains a debatable issue (Johnson et al. 2002).

4.3.2 Soil Sequestration

Anthropogenic perturbations exacerbate the emission of CO_2 from soil caused by decomposition of soil organic matter (SOM) or soil respiration (Schlesinger 2000). The emissions are accentuated by agricultural activities, such as i) tropical deforestation and biomass burning, ii) plowing (Reicosky 2002), iii) drainage of wetlands and low-input farming, and iv) shifting cultivation (Tiessen et al. 2001). In addition to its impact on decomposition of SOM (Trumbore et al. 1996), macroclimate has a large impact on the fraction of the soil organic carbon (SOC) pool (Franzluebbers et al. 2001). Conversion of natural to agricultural ecosystems increases maximum soil temperature and decreases soil moisture storage in the root zone, especially in drained agricultural soils (Lal 1996). Thus, land use history has a strong impact on the SOC pool (Pulleman et al. 2000).

Biomass burning is an important management tool, especially in agricultural ecosystems of the tropics. The process emits numerous gases immediately but also leaves charcoal as a residual material. Charcoal, produced by incomplete combustion, is a passive component, and may constitute up to 35% of the total SOC pool in fire-prone ecosystems (Skjemstad et al. 2002). As the SOC pool declines due to cultivation and soil degradation, the more resistant charcoal fraction increases as a portion of the total C pool (Skjemstad et al. 2001).

Similar to deforestation and biomass burning, cultivation of soil by plowing and other tillage methods also enhances mineralization of SOC and releases CO_2 into the atmosphere. Tillage increases SOC mineralization by bringing crop residue closer to microbes where soil moisture conditions favor mineralization (Gregorich et al. 2001), physically disrupts aggregates and exposes hitherto encapsulated C to decomposition. Both activities decrease soil moisture, increase maximum soil temperature and the exacerbate rate of SOC mineralization. Thus, a better understanding of tillage effects on SOC dynamics is crucial to developing and identifying sustainable systems of soil management for C sequestration. There is a strong interaction between tillage and drainage. Both activities decrease soil moisture, increase maximum soil temperature and the exacerbate rate of SOC mineralization. Conversion from plow till to no-till with residue mulch is a viable option for SOC sequestration.

Nutrient mining, as is the case with low input and subsistence farming practices, is another cause of depletion of SOC pool (Smaling 1993). Negative elemental balance, a widespread problem in sub-Saharan Africa, is caused by not replacing the essential plant nutrients harvested in crop and livestock products by the addition of fertilizer and/or manure. Excessive grazing has the same effect as mining of soil fertility by inappropriate cropping. Uncultivated fallowing, plowing for weed control but not growing a crop so that soil moisture in the profile can be recharged for cropping in the next season, is another practice that exacerbates SOC depletion. In the west central Great Plains of the U.S., this system requires a 14-month fallow period between the harvest and continuous cropping in some instances. Fallowing during summer keeps the soil moist and enhances the mineralization rate. Therefore, elimination of summer fallowing is an important strategy of SOC sequestration (Rasmussen et al. 1998). The objective is to maintain a dense vegetal cover on the soil surface so that biomass C can be added and/or returned to the soil. Consequently, the SOC pool can be maintained or increased in most semi-arid soils if they are cropped every year. Crop residues are returned to the soil, and erosion is minimized.

Generally, the term ''soil C sequestration'' implies removal of atmospheric CO_2 by plants and storage of fixed C as soil organic matter. The strategy is to increase SOC density in the soil, improve depth distribution of SOC and stabilize SOC by encapsulating it within stable micro-aggregates so that C is protected from microbial processes or as recalcitrant C with a long turnover time. In this context, managing agroecosystems is an important strategy for SOC/terrestrial sequestration. Agriculture is defined as an anthropogenic manipulation of C through uptake, fixation, emission and transfer of C among different pools. Thus, land use change, along with adoption of recommended management practices (RMPs), can be an important instrument of SOC sequestration (Post and Kwon 2000). Whereas land misuse and soil mismanagement have caused depletion of SOC with an attendant emission of CO_2 and GHGs into the atmosphere, there is a strong case that enhancing SOC pool could substantially offset fossil fuel emissions (Kauppi et al. 2001).

The SOC sink capacity depends on the antecedent level of SOM, climate, profile characteristics and management. The sink capacity of SOM for atmospheric CO_2 can be greatly enhanced when degraded soils and ecosystems are restored, marginal agricultural soils are converted to a restorative land use or replanted to perennial vegetation, and

RMPs are adopted on agricultural soils. Although generic RMPs are similar (e.g., mulch farming, reduced tillage, integrated nutrient management (INM), integrated pest management (IPM), precision farming), site-specific adaptation is extremely important. With adaptation of RMPs, SOC can accumulate in soils because tillage-induced soil disturbances are eliminated, erosion losses are minimized, and large quantities of root and above-ground biomass are returned to the soil. These practices conserve soil water, improve soil quality and enhance the SOC pool. Incorporation of SOC into the sub-soil can increase SOC's mean residence time (MRT). Converting agricultural land to a more natural or restorative land use essentially reverses some of the effects responsible for SOC losses that occurred upon conversion of natural to managed ecosystems. Applying ecological concepts to the management of natural resources (e.g., nutrient cycling, energy budget, soil engineering by macro-invertebrates and enhanced soil biodiversity) may be an important factor to improving soil quality and SOC sequestration (Lavelle 2000).

Biodiversity is also important to soil C dynamics. It is defined as ''the variability among living organisms" from all sources, including terrestrial, marine ecosystems and other aquatic ecosystems and ecological complexes of which they are part; this includes diversity within species, between species and for ecosystems. It is possible to distinguish between genetic diversity, organism species diversity, ecological diversity and functional diversity'' (UNCBD 1992). A healthy soil is teeming with life, and comprises highly diverse soil biota. The latter comprises representatives of all groups of micro-organisms and fungi, green algae and cyanobacteria, and of all but a few exclusively marine phyla of animals (Lee 1991). With reference to SOC pool and its dynamics, important members of soil biota include earthworms, termites, ants, some insect larvae and few others of the large soil animals that comprise ''bioturbation'' (Lavelle 1997). Activity of these animals have a strong influence on soil physical and biological qualities, especially with regards to soil structure, porosity, aeration, water infiltration, drainage, nutrient/elemental cycling and organic matter pool and fluxes. Soil biodiversity has a favourable impact on soil structure. Activity of soil biota produces organic polymers, which form and stabilize aggregates. Fungal hyphae and polysaccharides of microbial origin play an important role in soil aggregation.

4.3.2.1 Conservation Tillage

Conventional tillage and erosion deplete SOC pools in agricultural soils. Thus, soils can store C upon conversion from plow till to no till or conservation tillage by reducing soil disturbance, decreasing the fallow period and incorporation of cover crops in the rotation cycle. Eliminating summer fallowing in arid and semi-arid regions and adopting no till with residue mulching improves soil structure, lowers bulk density and increases infiltration capacity (Shaver et al. 2002). However, the benefits of no till on SOC sequestration may be soil/site specific, and the improvement in SOC may be inconsistent in fine textured and poorly drained soils (Wander et al. 1998). Similar to the merits of conservation tillage reported in North America, Brazil and Argentina (Sa et al. 2001), several studies have reported the high potential of SOC sequestration in European soils. Smith et al. (1998) estimated that adoption of conservation tillage has the potential to sequester about 23 Tg C/yr in the European Union or about 43 Tg C/yr in the wider Europe including the former Soviet Union. In addition to enhancing SOC pool, up to 3.2 Tg C/yr may also be saved in agricultural fossil fuel emissions. Smith et al. (1998) concluded that 100% conversion to no till agriculture could mitigate all fossil fuel C emission from agriculture in Europe.

4.3.2.2 Cover Crops

The benefits of adopting conservation tillage for SOC sequestration are further enhanced by growing cover crops in the rotation cycle. Growing leguminous cover crops enhances biodiversity, the quality of residue input and SOC pool (Fullen and Auerswald 1998; Singh et al. 1998). It is well established that ecosystems with high biodiversity absorb and sequester more C than those with low or reduced biodiversity. Legume-based cropping systems reduce C and N losses from soil. Franzluebbers et al. (2001) observed that, in Georgia, the USA, improved forage management can enhance the SOC pool. However, the use of cover crops as a short-term green manure may not necessarily enhance the SOC pool. Sainju et al. (2002) also reported that practicing no till with hairy vetch can improve SOC in Georgia, the USA. The beneficial effect of growing cover crops on enhancing SOC pool has been reported from Hungary by Berzseny and Gyrffy (1997), U.K. by Fullen and Auerswald (1998) and Europe by Smith et al. (1998).

4.3.2.3 Nutrient Management

Judicious nutrient management is crucial to SOC sequestration. In general, the use of organic manures and compost enhances the SOC pool more than application of the same amount of nutrients as inorganic fertilizers (Gregorich et al. 2001). The fertilizer effects on SOC pool are related to the amount of biomass C produced/returned to the soil and its humification. Adequate supply of N and other essential nutrients in soil can enhance biomass production under the elevated CO_2 concentration (Van Kessel et al. 2000). Long-term manure applications increase the SOC pool and may improve aggregation (Gilley and Risse 2000); the effects may persist for a century or longer. The potential of conservation tillage to sequester SOC is greatly enhanced whereby soils are amended with organic manures (Hao et al. 2002). Smith and Powlson (2000) reported that 820 million metric tons (MMTs) of manure are produced each year in Europe, and only 54% is applied to arable land and the remainder to non-arable agricultural land. They observed that applying manure to cropland can enhance its SOC pool more than it does on pasture land. Smith and Powlson estimated that if all manure were incorporated into arable land in the European Union, there would be a net sequestration of 6.8 Tg C/yr, which is equivalent to 0.8% of the 1990 CO_2-C emissions for the region.

Beneficial impacts of applying manure to the U.S. cropland were reported by Lal et al. (1998).

4.3.2.4 Forest Soils

Converting degraded soils under agriculture and other land uses into forests and perennial land use can enhance the SOC pool. The magnitude and rate of SOC sequestration with afforestation depends on climate, soil type, species and nutrient management (Lal 2001). Despite its significance, a few studies have assessed the C sink capacity of forest soils (Kimble et al. 2002). In East central Minnesota, an experiment by Johnston et al. (1996) showed an average SOC sequestration rate of 0.8–1.0 Mg/ha-yr. Afforestation, however, may not always enhance the SOC pool. In New Zealand, Groenendijk et al. (2002) reported that afforestation of pastures with radiata pine (*Pinus radiata*) decreased the SOC concentration by 15% to a depth of 12–18 cm. These researchers concluded that afforestation of hill country pasture soils resulted in net mineralization of the SOC pool. In the Cerrado region of Central Brazil, Neufeldt et al. (2002) also observed that reforestation of pasture with pine led to a clear reduction of SOC compared to pasture and eucalyptus plantation. In such cases, agroforestry may be another option of conserving soil and improving the SOC pool. In Europe, Nabuurs et al. (1997) reported that total SOC pool of soils supporting European forests is 12.0 Pg, but did not provide an estimate of the rate of SOC sequestration in forest soils. Afforestation of marginal agricultural soils or degraded soils has a large potential of SOC sequestration. Bouma et al. (1998) observed that in Europe a major change in land use may occur because of technological, socio-economic and political development. For example, adoption of RMPs or technical advances in modern agriculture may produce the same yield on 30–50% of the current agricultural land. That being the case, there is a potential for converting spare agricultural land to forestry. With conversion to a permanent land cover, there is a large potential of SOC sequestration through agricultural intensification.

4.3.3 Microbial Processes for Carbon Sequestration

The microbial contribution to soil C storage is directly related to microbial community's dynamics and the balance between formation and degradation of microbial by-products. Soil microbes indirectly influence the carbon cycle by improving soil aggregation, which physically protects SOM. Consequently, the microbial contribution to C sequestration is governed by the interactions between the amount of microbial biomass, microbial community structure, microbial by-products, as well as soil properties (e.g., texture, clay mineralogy, pore-size distribution) and aggregate dynamics. The capacity of a soil to protect microbial biomass and microbially-derived organic matter (MOM) is directly and/or indirectly (i.e., through physical protection by aggregates) related to the reactive properties of clays. Increasing the potential for agricultural soils to sequester C requires a thorough understanding of the underlying processes and mechanisms controlling soil C levels, for which a great deal of knowledge already exists.

Microbial processes affecting C sequestration in agroecosystems have been extensively reviewed by many researchers. Previous reviews have examined the relationships among microbial communities, SOM decomposition (Scow 1997), and management controls on soil C (Paustian et al. 1997). Organic C taken up by the microbial biomass is partitioned between microbial cell biomass production, metabolite excretion, and respiration (Figure 4.3). The degree to which MOM accumulates in soil depends on a balance between production and decomposition of microbial products, that is: (1) the microbial growth efficiency (MGE), the efficiency with which substrates are incorporated into bacterial and fungal biomass and by-products; (2) the degree of protection of microbial biomass in the soil structure; and (3) the rate at which bacterial and fungal by-products are decomposed by other microorganisms. The proportion of substrate C retained as biomass versus respired as CO_2 depends on MGE and the degree of protection of microbial biomass; the lower the MGE or the less protected the biomass, the more MOM-C is lost as CO_2.

Six et al. (2006) focus specifically on how soil bacteria and fungi may differentially influence the formation and stabilization of different SOM components in agricultural soils via differences in metabolism, the recalcitrance of microbial products, and interactions with soil physical properties (i.e., texture, mineralogy, and structure). However, the stabilization of MOM in the soil is also related to the efficiency with which microorganisms utilize substrate carbon and the chemical nature of the by-products they produce. Crop rotations, reduced or no-tillage practices, organic farming, and cover crops increase total microbial biomass and shift the community structure toward a more fungal-dominated community, thereby enhancing the accumulation of MOM. Thus, Agricultural systems that favour increased levels of microbial biomass include those associated with increased carbon inputs (Schnurer et al. 1985), reduced tillage (Beare et al. 1992; Frey et al. 1999), retention of crop residues rather than removal by burning (Gupta et al. 1994), and integrated farming systems that combine reduced tillage with increased carbon inputs through organic amendments (Hassink et al. 1991).

4.3.3.1 Crop Rotations

Crop rotation affects microbial biomass, activity, and the fungal-to-bacterial biomass ratio. Soils under crop rotation show increased soil enzyme activity when compared with soils under continuous monocropping (Acosta-Martinez et al. 2003). Microbial biomass levels are typically higher when legume cover crops are included in the rotation than when fields are left fallow between cash crops (Lupwayi et al. 1999).

Likewise, microbial biomass and fungal-to-bacterial biomass ratios were higher in a non-tillage (NT) wheat-barley compared with wheat-fallow rotation (Bell et al. 2003). Viable propagules of mycorrhizal fungi (Thompson 1987) and glomalin concentrations (Wright and Anderson 2000) have also been observed to decline in rotations that include a fallow period. However, varying crop rotations that include only mycorrhizal crops have minimal to no effect on mycorrhizal colonization (An et al. 1993). Incorporation of forage species into crop rotations increases hot water-extractable carbohydrates (Haynes and Francis 1993), likely due to higher root inputs and the stimulation of microbial activity that follows (Haynes and Francis 1993).

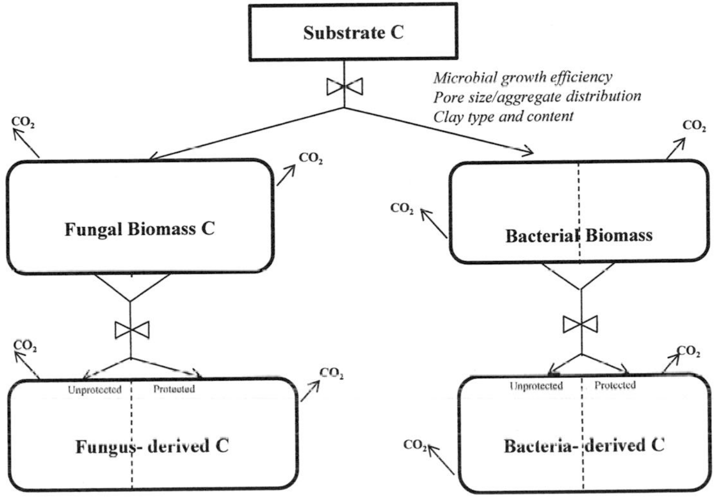

Figure 4.3 Conceptual diagram of the microbial contribution to C sequestration in Agro-ecosystems (adapted from Six et al. 2006).

The total acid hydrolyzable carbohydrate content, on the other hand, is variably affected by crop rotation and forage species (Baldock et al. 1987). Additionally, the proportion of SOM present as acid hydrolyzable carbohydrates was observed to be constant across differing rotational systems (Baldock et al. 1987). In contrast, hot water-extractable carbohydrate content was not a constant proportion of total SOM under different crop rotations (Angers et al. 1993). This indicates that the hot water extractable carbohydrates are more influenced by crop rotations than are acid hydrolyzable carbohydrates. Since hot water extractable carbohydrates are considered more microbially derived than acid hydrolyzable carbohydrates, the microbially-derived carbohydrates seem to be more affected by crop rotations than plant-derived carbohydrates. In contrast, Guggenberger et al. (1995) found that the change in the carbohydrate content in a native savannah converted to pasture was dominated by a change in carbohydrates associated with the plant-derived SOM in the sand fraction. The same was observed when soils under barley versus alfalfa were compared (Angers and Mehuys 1990).

To summarize, crop rotation as a management practice may increase soil C sequestration in comparison with continuous crop management or rotations that include fallow periods. More intensive cropping rotations increase not only soil C input but also microbial activity and biomass, which alter the microbial community composition by increasing the fungi levels in the soil.

4.3.3.2 Tillage

Minimum tillage (MT) and NT systems often exhibit increased C storage compared with conventional tillage (CT) (Six et al. 2002; West and Post 2002); however, this difference disappears with the use of fallow rotations (Peterson et al. 1998), demonstrating the importance of using multiple management practices to enhance soil C storage. The greatest differences between NT and CT soils occur in the top 5 cm, with NT soils having greater fungal-to-bacterial biomass ratios (Beare et al. 1992; Frey et al. 1999), enzyme activity (Acosta-Martinez et al. 2003), macroaggregation (Six et al. 2000), total C and N contents (Feng et al. 2003), and concentrations of bacteria-and-fungus-derived amino sugars (Simpson et al. 2004). Differences in tillage intensity also impact microbial community composition. Beare (1997) examined the fungal-to-bacterial biomass ratios at six long-term tillage comparison experiments located along two climatic gradients; they observed that fungal biomass and fungal-to-bacterial biomass ratios increased in response to reduced tillage at all sites. This suggests that a shift toward fungal dominated microbial communities under NT will be most important for residue decomposition and nutrient cycling processes near the soil surface. However, another study found no difference in the fungal-to-bacterial ratio in surface NT and CT soils and a 10% lower fungal-to-bacterial ratio in NT soil at 6 to 12 cm (Feng et al. 2003). Less intensively managed agroecosystems (e.g., NT) bear the closest resemblance to natural ecosystems, which are fungal-dominated (Yeates et al. 1997; Bailey et al. 2002). These fungal-dominated agroecosystems require fewer inputs to sustain organic matter decomposition and nutrient cycling, that is, these systems show greater self-regulation (Bardgett and McAlister 1999).

However, Allison et al. (2005) concluded that an improved metabolic efficiency due to the increased relative abundance of arbuscular mycorrhizal fungi (AMF), and saprophytic fungi does not promote the stabilization of C on cessation of tillage-based agriculture. Three main factors have been identified as potential controls on bacterial and fungal biomass in NT and CT soils: degree of disturbance, soil moisture content, and residue placement. However, the relative importance of these factors on soil fungal-to-bacterial biomass ratios in NT soil is unknown. Reduced disturbance in NT systems may favor fungal growth and activity due to enhanced establishment and maintenance of extensive hyphal networks (Wardle 1995). Thus, NT systems can accumulate fungal pathogens as well as mycorrhizal fungi (Miller and Lodge 1997). Soil moisture may differentially influence bacteria and fungi either by directly affecting survival and growth or indirectly by shifting substrate availability and microbivore populations. Frey et al. (1999) observed that fungal biomass and fungal-to-bacterial biomass ratios were positively related to soil moisture in both NT and CT soil. Fungal biomass and relative fungal abundance were not significantly different in NT compared with CT when the data were analyzed by analysis of covariance with soil moisture as the covariate, suggesting that observed tillage treatment effects on the microbial community are related to differences in soil moisture (Frey et al. 1999).

Residue placement (surface residues in NT vs. incorporated residues in CT) has also been shown to alter bacterial and fungal populations (Beare et al. 1992). It was hypothesized that the presence of surface residues favors fungal growth because fungi, unlike bacteria, can bridge the soil-residue interface and simultaneously utilize the spatially separated C and N resources by translocating soil inorganic N into the C-rich surface residues (Beare et al. 1992). Frey et al. (2000) demonstrated that fungi do have the potential to translocate significant quantities of soil inorganic N into decomposing surface residues in NT systems, and that this N flow increases fungal proliferation in the surface residues themselves. In addition, reciprocal translocation of C from surface residues to mineral soil via fungal hyphae occurs (Frey et al. 2003). The fungal-to-bacterial ratio can also be greater due to the greater amount and better quality of SOM in the surface layer of NT systems; the latter being induced by the surface residue layer (Paustian et al. 2000) and the increased root growth in the surface layers of NT systems (Qin et al. 2005).

An increased resource availability in soil surface layers has been suggested as the main factor for greater proportions of fungi in the microbial community of the surface layer compared with deeper soil horizons (Fierer et al. 2003). Differential effects of NT and CT on MOM have been observed. Hot water-extractable carbohydrates, acid hydrolysable carbohydrates, and the relative enrichment of SOM in carbohydrates increased under reduced tillage compared with moldboard plowing (Angers et al. 1993). Arshad et al. (1990) reported increased SOM quality under NT compared with CT; SOM in NT soil contained more acid hydrolyzable carbohydrates, amino acids, and amino sugars, and was more aliphatic and less aromatic. The ratio of galactose 1 mannose to arabinose 1 xylose in the whole soil's acid hydrolyzable carbohydrate pool increased under NT compared with CT, indicating a higher microbially-derived carbohydrate C pool under NT. Similar enrichment of microbially-derived carbohydrates under NT versus CT has been reported by Ball et al. (1996) and Beare et al. (1997). Beare et al. (1997) suggested that the greater proportion of microbial-derived compared with plant-derived carbohydrates under NT compared with CT is due to the relatively higher fungal biomass under NT (Frey et al. 1999). This is supported by the observation that the ratio of glucosamine to muramic acid is higher under NT compared with CT, and this is due to a higher enrichment of glucosamine (Simpson et al. 2004). In addition, Wright and Anderson (2000) found a significantly higher glomalin concentration under NT than CT management. Hu et al. (1995) found that inhibition of fungi by the fungicide Captan (N-Trichloromethylthio- 4cyclohexene-1,2-dicarbonximide) only reduced concentrations of soil C and acid hydrolysable carbohydrates in NT soils, not CT soils. The above-mentioned studies clearly indicate that NT induces a higher fungal (saprophytic and mycorrhizal) biomass, which leads to a quantitative and qualitative improvement of SOM. However, the increase in fungal biomass under NT not only leads to an increase in MOM, but may also affect the accumulation of plant-derived C [i.e., particulate organic matter (POM)]. Six et al. (1999) suggested that increased macroaggregate turnover in tilled soil is an important mechanism causing a loss of POM and MOM. Fungi are expected to retard macroaggregate turnover due to their positive influence on aggregate stabilization. However, the link between fungal abundance, macroaggre-gate turnover, and MOM plus POM has not been directly investigated.

4.3.3.3 Organic Farming and Cover Crops

Compared with conventional practices, organic farming practices have been shown to promote higher microbial biomass and to alter microbial community composition (Petersen et al. 1997). In an incubation experiment where conventional and organic soils were amended with organic matter and exposed to similar incubation conditions, no differences in microbial biomass C or substrate-induced respiration were observed (Gunapala et al. 1998). However, differences were observed by the end of the incubation experiment in potentially mineralizable N (higher in organic soils), bacterivorous nematodes (higher in conventional soils), and fungivorous nematodes (higher in organic soils). Fliessbach et al. (2000) found that a biodynamic (organically managed) system had higher microbial biomass than the conventionally managed soils, and suggested that organic soils provide greater protection of microbial biomass. Bossio et al. (1998) found that conventionally managed, organic, and low input management systems all had significantly different microbial communities, and that organic soils had higher fungal-to-bacterial biomass ratios than the conventionally managed soils. Organic farming practices frequently employ cover crops, which can change the soil microbial community and have a variable effect on MOM. Different overall microbial community composition was observed in continuous maize vs. a maize (*Crotalaria grahamiana*)-fallow rotation, with and without P addition (Bunemann et al. 2004). Higher levels of fatty acid biomarkers for fungi and Gram-negative bacteria as well as higher overall microbial biomass were present in the maize *C. grahamiana* fallow rotation. Soil from the maize *C. grahamiana*–fallow rotation also exhibited faster decomposition of added C substrates. A winter cover crop mix of oats and vetch (*Avena sativa* L. and *Vicia sativa* L.) increased microbial biomass C, respiration and N mineralization and changed the microbial community composition compared with a winter fallow treatment (Schutter and Dick 2002).

Roberson et al. (1991) found that a permanent grass cover crop significantly increased the acid hydrolysable heavy fraction carbohydrate content. An investigation of the effect of fertilizer and cover crop N supply on carbohydrates found that the heavy fraction carbohydrates were more microbially derived than the light fraction (LF) carbohydrates (Roberson et al. 1995). A positive effect of vetch (*Vicia dasycarpa* L.) and 168 kg N/ha fertilizer addition on the heavy fraction carbohydrate content was observed. In contrast, the 280 kg N/ha fertilizer rate and the oat (*Avena sativa* L.) cover crop reduced the heavy fraction carbohydrate content compared with a control. A variable effect of cover crops on carbohydrate content was also observed by Kuo et al. (1997). They concluded that the cover crop effect on carbohydrate content was related to the C inputs from the cover crops. However, Roberson et al. (1995) suggested that the N supply more than C supply controls the effect of cover crops on soil carbohydrate content. More research is needed to investigate relationships between the use of organic farming practices (including cover crops) and soil microbial communities. Organic management appears to increase fungal biomass, which would favor increased soil C sequestration (Figure 4.3, Step I). Further studies are also required to attain a greater understanding of the effects of cover crops and nitrogen additions on microbially-derived soil carbohydrates.

4.4 Advanced Biological Processes for CCS

Capture, compression and transportation of CO_2 requires much energy and would increase the fuel needs of a coal-fired plant with CCS. Without further investment and a stable CO_2 price, CCS will not be commercially viable in the energy market at present. Thus, the major challenge in the implementation of CCS is to address the high cost of CCS, particularly for dilute and hot streams such as those from power plants. The progress made in recent research has been mainly aimed at reducing the cost and increasing the efficiency and selectivity of separation and capture of CO_2. Therefore, many advanced processes, particularly biological techniques, have been developed for CCS.

4.4.1 Carbonic Anhydrase Enzymes for Bio-CCS

Biologically-based carbon capture systems are the potential avenue for improvement in CO_2 capture technology. These systems are based upon naturally occurring reactions of CO_2 in living organisms. One of these possibilities is the use of enzymes. An enzyme-based system, which achieves CO_2 capture and release by mimicking the mechanism of the mammalian respiratory system, is under development by Carbozyme. The process, utilizing CA in a hollow fiber contained liquid membrane, has demonstrated at the laboratory scale the potential for 90% CO_2 capture followed by regeneration at ambient conditions (Rishiram et al. 2009). This is a significant technical improvement over the monoethanolamine (MEA) temperature swing absorption process. The CA process has been shown to have a very low heat of absorption that reduces the energy penalty typically associated with absorption processes.

The rate of CO_2 dissolution in water is limited by the rate of aqueous CO_2 hydration, and the CO_2-carrying capacity is limited by buffering capacity. Adding the enzyme CA to the solution speeds up the rate of carbonic acid formation; CA has the ability to catalyze the hydration of 600,000 molecules of carbon dioxide per molecule of CA per second compared to a theoretical maximum rate of 1,400,000 (Trachtenberg et al. 1999). This fast turnover rate minimizes the amount of enzyme required. Coupled with a low make-up rate, due to a potential CA life of 6 months based on laboratory testing, this biomimetic membrane approach has the potential for a step change improvement in performance and cost for large scale CO_2 capture in the power sector. Although the reported laboratory and economic results may be optimistic, the ''carbozyme biomimetic process can afford a 17-fold increase in membrane area or a 17 times lower permeance value and still be competitive in cost with MEA technology'' (Yang and Ciferno 2006). The idea behind this process is to use immobilized enzyme at the gas/liquid interface to increase the mass transfer and separation of CO_2 from flue gas.

Along the same lines, it was recently demonstrated that the most promising biological carbon dioxide sequestration technologies is the enzyme catalyzed carbon dioxide sequestration into bicarbonates, which was endeavored in a study with a purified *C. freundii* SW3 b-carbonic anhydrase (CA) (Rishiram et al. 2009). An extensive screening process for biological sequestration using CA has been defined. Six bacteria with high CA activity were screened out of 102 colonies based on plate assay and the presence of CA in these bacteria was further emphasized by activity staining and Western blot. The identity of selected bacteria was confirmed by 16S rDNA analysis. CA was purified to homogeneity from *C. freundii* SW3 by subsequent gel filtration and ion exchange chromatography which resulted in a 24 kDa polypeptide and this is in accordance with the Western blot results. The effect of concentration of carbon dioxide on carbonic anhydrase is well known as many organisms have been reported to grow well in carbon dioxide rich conditions, and CA is known to be required for organisms existing in these conditions (Kusian et al. 2002). However, there have been contradictory reports of CA being inhibited at higher concentrations of carbon dioxide and is induced at lower concentrations (Bahn et al. 2005). In the study of Rishiram et al. (2009), the enzyme activity was gradually enhanced at concentrations above atmospheric levels, which was maintained as a control, until 5.0% CO_2. However, at high CO_2 concentration of 7.5%, a decrease in activity was observed. At 10% CO_2 concentration, the activity of CA almost halved the highest activity achieved and this may be due to the feedback inhibition by bicarbonate ions or shall be attributed to the decrease in pH with increasing CO_2 concentration, as CO_2 is an acidic gas. This suggests that CA enzyme has the potential to sequester CO_2 even at high CO_2, yet, a feedback inhibition by bicarbonate has to be prevented by precipitating to calcium carbonate. The pH of the medium also has to be maintained to avoid losing CA activity. The study also attenuates a well acclaimed fact that CA has a significant role in regulating the CO_2 concentration in the cell and subsequent metabolism of Dissolved Inorganic Carbon (DIC). The wide distribution and multiple occurrence of the CAs in bacteria also emphasize this key role (Spalding 2008).The effect of host on metal ions, cations and anions, which influence activity of the enzyme in sequestration studies, suggests that mercury and HCO_3^- ion almost completely inhibit the enzyme whereas sulfate ion and zinc enhances carbonic anhydrase activity. Calcium carbonate deposition was observed in calcium chloride solution saturated with carbon dioxide catalyzed by purified enzyme and whereas a sharp decrease in calcium carbonate formation has been noted in purified enzyme samples inhibited by EDTA and acetazolamide.

In conclusion, many studies provide a comprehensive screening methodology for CA in bacteria. It reported the isolation, purification and characterization of carbonic anhydrase enzyme in bacteria. The presence of carbonic anhydrase in diverse heterotrophic bacteria offers much promise for the progress in studies on carbon sequestration as carbonic anhydrase is envisaged to have wide industrial application. The purified enzyme shall be used in an immobilized enzyme reactor to sequester carbon dioxide in the form of mineral carbonates. The sequestration of carbon dioxide by the enzyme proves that the immobilized over-expressed CA reactor may have large implications in the efforts to biologically sequester carbon dioxide into mineral carbonates. In this case, many approaches have been reported using protein engineering.

4.4.1.1 Engineered *Escherichia coli* with Periplasmic Carbonic Anhydrase

Carbonic anhydrase is an enzyme that reversibly catalyzes the hydration of carbon dioxide (CO_2). It has been suggested recently that this remarkably fast enzyme can be used for sequestration of CO_2, a major GHG, making it a promising alternative for chemical CO_2 mitigation. To promote the economical use of enzymes, Buyang et al. (2013) engineered the carbonic anhydrase from *Neisseria gonorrhoeae* (ngCA) in the periplasm of *Escherichia coli*, thereby creating a bacterial whole-cell catalyst. We then investigated the application of this system to CO_2 sequestration by mineral carbonation, a process with the potential to store large quantities of CO_2. Because the cell has a membrane-enclosed structure, and the membrane functions as a selective barrier for the passage of various substances, including ions such as HCO_3^- and CO_3^{2-}, CO_2 hydration activity may not directly correlate with the ability of the whole cell to precipitate $CaCO_3$.

To verify the ability of the constructed periplasmic whole cell system to efficiently sequestrate CO_2 in carbonate mineral, the conversion of CO_2 to $CaCO_3$ was examined. $CaCO_3$ is formed by the reaction between Ca^{2+} and CO_3^{2-}, the latter being formed from HCO_3^-. Because the pK_a for HCO_3^- dissociation is quite high (10.3) (Lower et al. 1999), alkaline buffer with a pH of 11 was used for the conversion reaction. Although CA accelerates the rate of the CO_2 hydration reaction, it does not affect the equilibrium between the different species of carbonate (Uchikaw et al. 2012). In a closed system, the final amount of precipitated $CaCO_3$ does not depend on whether the reaction is catalyzed by CA or not (Favre et al. 2009). Therefore, we focused on the ability of the whole-cell biocatalyst to improve the precipitation (mineralization) rate rather than the measurement of the quantity of the resulting precipitate. ngCA was highly expressed in the periplasm of *E. coli* in a soluble form, and the recombinant bacterial cell displayed the distinct ability to hydrate CO_2 compared with its cytoplasmic ngCA counterpart and the previously reported whole cell CA systems. The expression of ngCA in the periplasm of *E. coli* greatly accelerated the rate of calcium carbonate ($CaCO_3$) formation and exerted a striking impact on the maximal amount of $CaCO_3$ produced under conditions of relatively low pH (8.5).

It was also shown that the thermal stability of the periplasmic enzyme was significantly improved. The cells harbouring periplasmic ngCA exhibited excellent stability compared with purified ngCA. During the 5-h incubation period at 40°C, periplasmic ngCA retained all activity, while free ngCA retained less than 70% of its initial activity. When incubated for 5 h at 50°C, the residual activity (60%) of the periplasmic ngCA was much greater than that (25%) of the free enzyme. These results clearly show that the activity of periplasmic ngCA is protected by some cellular mechanism(s) in the course of high-temperature incubation. Immobilization of ngCA in the periplasm, a hypothetic reason for the resistance to periplasmic release by osmotic shock, may be the responsible factor contributing to the enhanced thermal stability. These results demonstrate that the engineered bacterial cell with periplasmic ngCA can successfully serve as an efficient biocatalyst for CO_2 sequestration. The mechanisms may also include protection by heat shock proteins or an increase in ion permeability caused by the elevated temperature (Guyot et al. 2010). In this study, they expect that finding or engineering a thermostable CA would synergistically improve the thermal stability of the periplasmic whole-cell system for practical application to post combustion capture of industrial CO_2.

In conclusion, for the feasibility of practical application of the periplasmic whole-cell catalyst system, there are two pioneering studies on construction of microbial cells with recombinant CA for CO_2 mitigation and/or mineralization (Fan et al. 2011,

Barbero et al. 2013). In both, whole cells were constructed using a surface display system, which resulted in quite low CA activities. Accordingly, the periplasmic system engineered in the work of Jo et al. (2013) is currently the most efficient whole-cell CA catalyst, exhibiting 2 to 3 orders of magnitude higher activity (1.77 U/ml.OD_{600}) than the other reported systems (6.09 10^{-2} and 5.38 10^{-3} U/ml.OD_{600}, respectively).

4.4.1.2 Low-Cost Biocatalyst for Acceleration of Energy Efficient of CCS

Chemical absorption with regenerable alkaline aqueous solvents is considered the nearest term option for post combustion CCS (Rochelle et al. 2009). In this process, CO_2 is removed from the flue gas stream in the absorber column and then desorbed in a heated stripper column to give relatively pure CO_2 for compression and storage. The challenge facing these capture processes is in energy loss in desorption. Solvents such as MEA tend to bind CO_2 tightly such that the parasitic energy loss in desorbing the CO_2 would almost double the cost of electricity (Ciferno et al. 2009). CA has been shown to facilitate the use of aqueous solvents with a far lower heat of desorption (e.g. hindered and tertiary amines), thus enabling a far lower energy penalty on CCS (Blais et al. 2003). Absorption of CO_2 tends to be slower in these solvents and thus requires the use of an accelerant such as CA. The poor stability and activity of naturally-derived CA under the harsh conditions of these processes (i.e., temperatures from 50 to over 125°C, high concentrations of organic amine, trace contaminants such as heavy metals, and sulfur and nitrogen oxides) have limited their use. Therefore, to overcoming these limitations, the approach of Savile and Lalonde (2011) have included sourcing CAs from thermophilic organisms, using protein engineering techniques to create thermo-tolerant enzymes (Newman et al. 2010), immobilizing the enzyme (for both stabilization and restriction to the cooler process zones) or process modifications such as cooling of the flue gas.

Small molecule analogs of CA have been reported with potentially higher stability than proteins, but their rates of acceleration tend to be orders of magnitude less than the enzymes (Looney et al. 1993). CO_2 Solution (Quebec, Canada) has a patent portfolio on the use of CAs for CO_2 capture from combustions sources. Besides covering carbon capture from fossil fuel combustion (Fradette et al. 2009), they describe the use of CA to accelerate capture in solvents such as MEA, methyldiethanolamine (MDEA), and piperazine (Fradette et al. 2010). The conditions described are at near ambient temperatures and dilute aqueous solvents, presumably due to the sensitivity of the human enzyme. Generally, these CA accelerated capture processes employ immobilized CA in contact with CO_2. CO_2 Solution describes the use of CA in various bioreactor formats including a packed column triphasic bubble reactor (Blais et al. 2003), a tower reactor and a spray absorber (Fradette 2004).

Given the high temperature for regeneration in a solvent-based capture system, thermophilic organisms represent a source of stable CAs. Three CAs from thermophilic organisms have been the primary focus of most biochemical studies, that is, an α-class CA from *Methanosarcina thermophile* (CAM) (Ferry et al. 2010), a β-class CA from *Methanobacterium thermoautotrophicum* (CAB) (Rowlett et al. 2010) and a γ-class CA from *Pyrococcus horikoshii* (Jeyakanthan et al. 2008). CAM shows optimal activity at 55 °C, with k_{cat} values approaching 105/s. CAM is a relatively stable CA, and it shows 50% residual activity after 15 min at 70 °C and is inactivated at 75 °C. By comparison, human CA II shows optimal activity at 37 °C and is inactivated above 50 °C (Alber et al. 1996). However, it is an extremely fast enzyme with a k_{cat} of 106/s (Smith et al. 1999).

Several heat-stable CAs for extraction of CO_2 from CO_2 containing media have been reported (Borchet et al. 2010). An α-class CA from *Bacillus clausii* showed higher thermostability than CAM with 17% residual activity (via the Wilbur-Andersen assay) after 15 min at 80 °C in 1 M sodium bicarbonate with a pH of 8.05. At 0.6 g/L enzyme loading, this CA could extract > 99% of CO_2 from a 15% CO_2 gas stream versus only 33% removal without CA. The CA from the thermophilic organism *Caminibacter mediatlanticus* DSM 16658 had a melting temperature of 109 °C at pH 9 and retained 40% of its original activity after 15 min in 1 M sodium bicarbonate at 100 °C, and also increased the amount of CO_2 extracted with 1 M sodium bicarbonate solutions at pH 9 (Borchert et al. 2011).

While thermophiles represent a good source of thermostable CAs, naturally occurring CAs typically do not show suitable activity or stability in the presence of high concentrations of alkaline capture solvents at high temperatures. However, there are no reports in the literature of CA activity or stability in amine-based captures solvent such as MDEA or 2-amino -2-methyl -1- propanol AMP at process-relevant concentrations (e.g. 50%, v/v) and temperatures (> 45 °C). Savile and Lalonde (2011) generated a soluble CA biocatalyst capable of catalysing CO_2 hydration in capture solvents at absorber temperatures (45–60 °C) and, in an ideal case, surviving the high desorber temperatures (> 100 °C). Codexis is directed evolution technology platform applying to increase the activity and stability of CAs by screening a wide range of genetically-diverse biocatalysts using high throughput (HTP) screens that closely mimic solvent-based Post Combustion Carbon Capture (PCCC) process conditions in MDEA and other amine solvents. Initial evaluation of 50 wild-type CAs identified a CA from a mesophilic organism with higher activity and stability in MDEA than human CA-II (Carbon anhydrase II = short name of enzyme). It was inactivated at temperatures above 40 °C in MDEA, but after four rounds of directed evolution, the temperature of half-inactivation was increased by > 40–82 °C in 5 M MDEA with a pH of 11.8 (Savile and Lalond 2011).

Savile and colleagues used immobilization of CA to both stabilize the enzyme and to limit exposure to denaturing conditions in CCS processes. Immobilization has been reported on a number of solid supports including polyamide (Berzil et al. 2005), *n*-

vinyl formamide (Drevon et al. 2003), chitosan (Sharma et al. 2011), and alkyl sepharose (Azari and Nemat-Gorgani 1999). In the case of Savile et al. (2011), the CA is immobilized on the hollow fiber wall to maximize contact with CO_2 at the gas-liquid interface and facilitate CO_2 uptake into the liquid membrane by rapid conversion to bicarbonate, and can enhance the rate of both absorption and desorption. This technology is applicable for moderate temperature (10–85 °C) gas flows at low to high pressure with CO_2 concentrations from < 1% to > 20% (Tranchtenberg et al. 2008). Continuously, CA is used in the biomineralization process. This biological process has been used as a model for CO_2 fixation in the form of carbonate salts of divalent metals. Lee has recently reviewed the use of CA for CCS and biomineralization (Lee et al. 2010). Very recently, a strategy introducing the use of CA to accelerate in 'bioweathering' of silicates was reported in which CA is used to accelerate to hydration of CO_2 and subsequent precipitation of Mg (II) released from weathering of magnesium silicate rock (Swanson et al. 2010).

4.4.2 Microbial Enhanced CCS

The study of Indrew et al. (2010) suggested the potential of microorganisms for enhancing CCS via mineral-trapping (where dissolved CO_2 is precipitated in carbonate minerals) and solubility trapping (as dissolved carbonate species in solution). The bacterial hydrolysis of urea (ureolysis) was investigated in microcosms including synthetic brine (SB) mimicking a prospective deep subsurface CCS site with variable headspace pressures [$p(CO_2)$] of $^{13}C\text{-}CO_2$. Dissolved Ca^{2+} in the SB was completely precipitated as calcite during microbially-induced hydrolysis of 5–20 g/l urea. The incorporation of carbonate ions from $^{13}C\text{-}CO_2$ ($^{13}C\text{-}CO_3^{2-}$) into calcite increased with increasing $p(^{13}CO_2)$ and increasing urea concentrations, from 8.3% of total carbon in $CaCO_3$ at 1 g/l to 31% at 5 g/l, and 37% at 20 g/l. This demonstrated that ureolysis was effective at precipitating the initial gas [CO_2 (g)] originating from the headspace over the brine. While $CaCO_3$ is precipitated, the moles of calcite precipitated were equal to or less than the moles of urea derived carbonate ions. Thus, no net precipitation of CO_2 (g) in $CaCO_3$ occurred during urea hydrolysis-induced mineral-trapping. Additionally, the consumption of Ca^{2+}, and the likelihood that this will not be replenished via equilibration because of the concurrent rise in pH, may reduce the magnitude of natural carbonation in the aquifer.

However, the precipitation of $CaCO_3$ by ureolysis in the brine provides a number of distinct engineering advantages. First, ureolytic organisms appear to be ubiquitous in surface and subsurface soils (Fujita et al. 2008). *S. pasteurii* has been shown to be ureolytically active at pressures and temperatures relevant to geologic carbon sequestration scenarios (P > 89 bar, T > 32 °C) (Mitchel et al. 2009). Thus, engineering solutions could likely rely on the stimulation of native organisms to induce $CaCO_3$ precipitation. Second, wastewater provides a potential supply of waste urea. If wastewater urea sources can be utilized in this process, it may provide the simultaneous and advantageous degradation of urea waste, and thus a reduction in labile earth surface carbon, and the mineral- and solubility-trapping of injected CO_2. The annual volume of wastewater was 1347 km^3/yr in Europe, North America and Asia in 1995 (WWAP 2003). Wastewater commonly includes urea concentrations of 20 g/l (333 mM) (Rittstieg et al. 2001), which is more than what is required to maintain calcite super saturation in groundwater with chemistries like those of the Powder River Basin, and to precipitate out all available dissolved Ca even in concentrated oilfield brines with Ca concentrations of 125 mM or more (Figure 4.4). Third, the precipitation of $CaCO_3$ in the subsurface provides a means to decrease formation porosity and reduce the potential of CO_2 leakage.

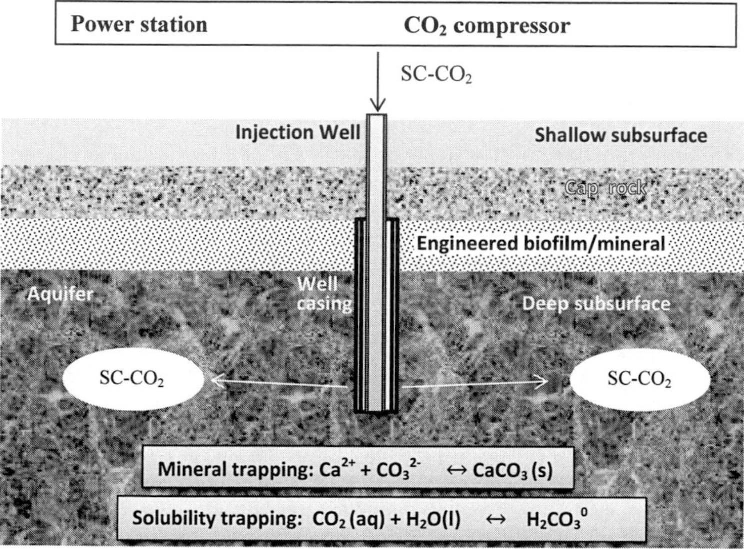

Figure 4.4 Schematic of microbially-enhanced CCS (adapted from Andrew et al. 2009)

For example, subsurface ureolysis-induced carbonate precipitation has been investigated for permeability reduction for enhanced oil recovery and radionuclide contaminant sequestration (Ferris et al. 1988). However, the pH increase induced by bacterial ureolysis generated a net flux of CO_2 (g) into the brine. This reduced the head space concentration of CO_2 by up to 32 mM per 100 mM urea hydrolyzed because the capacity of the brine for carbonate ions was increased, thus enhancing the solubility-trapping capacity of the brine. Together with the previously demonstrated permeability reduction of rock cores at high pressure by microbial biofilms and resilience of biofilms to supercritical CO_2, this suggests that engineered biomineralizing biofilms may enhance CCS via solubility-trapping, mineral formation, and CO_2(g) leakage reduction.

The main potential limitation of microbially-enhanced CCS is the ability of microorganisms to withstand high pressure and supercritical carbon dioxide (SC-CO_2). Planktonic cells (free floating) show limited resistance to SC-CO_2. Twenty-two tested vegetative species of microorganisms reported in the literature were completely deactivated at some combination of pressure and temperature in the presence of SC-CO_2 (Zhang et al. 2006). However, biofilms, which are microorganism assemblages attached to a surface, appear to exhibit a higher resistance to SC-CO_2 (Mitchell et al. 2009). For example, a 19-min exposure to 35 °C, 136 atm SC-CO_2 resulted in a 3 $\log_{10}$ viability reduction of planktonic *Bacillus mojavensis* cells, but only a 1 $\log_{10}$ reduction in viable cell numbers from biofilm cultures. It is hypothesized that the small reduction in the viability of biofilm microorganisms reflects the protective role of the biofilm extracellular polymeric substances (Mitchell et al. 2009). The resilience of microorganisms in biofilm states, to high pressure gaseous and SC-CO_2 suggests microbially-enhanced mineral trapping and solubility-trapping of CO_2 during CCS may be effectively used.

In conclusion, a microbial ureolysis-based approach appears to offer potential for enhancing CCS by (i) increasing the flux of gas into the brine, and the capacity of brine for carbonate ions, and (ii) the formation of carbonate minerals, potentially reducing formation porosity and the potential of CO_2 leakage to the surface. The apparent resilience of biofilm-organisms to SC-CO_2 suggests these and other micobially-mediated processes may offer the ability to enhance the capacity and rates of CO_2 trapping.

4.5 Biotic versus Abiotic CCS

Biotic CCS, based on removal of atmospheric CO_2 through photosynthesis, is a natural process. The magnitude of CO_2 removal through photosynthesis, in woody plants of managed and natural ecosystems, is likely to increase in the future due to the CO_2 fertilization effect. The process can be managed through input of essential nutrients (e.g. N, P, K, Ca, Mg, S, Zn, Cu, Mo) and management of water. There are numerous ancillary benefits of terrestrial/biotic C sequestration as reported by Lal et al. (2008), including: i) improved quality of soil and water resources; ii) decreased nutrient losses from ecosystems; iii) reduced soil erosion; iv) better wildlife habitat; v) increased water conservation; vi) restored degraded soils; and vii) increased use efficiency of input. Soil C sequestration, both as SOC and soil inorganic carbon (SIC), is also a natural process essential for recycling of elements and water. Similar to the terrestrial pool, increase in the SOC pool also has numerous ancillary benefits affecting local, regional and global processes. Principal benefits of SOC sequestration to soil quality are: i) improvement in soil structure; ii) reduction in soil erosion; iii) decrease in non-point source pollution; iv) increase in plant-available water reserves; v) increase in storage of plant nutrients; vi) denaturing of pollutants; vii) increase in soil quality; viii) increase in agronomic productivity of food security; ix) moderation of climate; and x) increase in aesthetic and economic value of the soil. Therefore, the process of biotic C sequestration strengthens and enhances ecosystem services while enhancing agronomic production. The process is cost-effective, and RMPs for adoption on agricultural and forest soils/ecosystem are available for most ecoregions of the world (IPCC 1999). However, the total sink capacity for biotic C sequestration especially that in terrestrial ecosystems is low at 50–100 Pg C over a period of 25–50 year (Lal 2004). Further, C sequestered in soil and biota can be re-emitted with change in soil management (e.g. ploughing) and land use (e.g. deforestation).

In contrast to biotic sequestration, abiotic sequestration is an engineering process. The technology for deep injection in the ocean, geological strata, coal mines and oil wells, etc. are being developed and may be routinely available by 2025 and beyond. At present, these techniques are expensive, and injected CO_2 is prone to leakage. In addition to the high cost, the issues of measurement and monitoring, adverse ecological impacts and regulatory measures need to be developed and implemented. However, the sink capacity of abiotic techniques is extremely large at thousands of Pg C, and often estimated to exceed the fossil C reserves.

Therefore, biotic and abiotic systems are complimentary to one another. Depending upon ecosystem characteristics, there may be site-specific ecological niches for biotic or abiotic CCS options. Biotic CCS options are immediately available. Use of such options buys us time, while C-neutral energy production technologies of alternatives to fossil fuels and techniques of abiotic sequestration take effect.

4.6 Summary

CCS implies transfer of CO_2 into other long-lived global pools including oceanic, pedologic, biotic and geological strata to reduce the net rate of increase in atmospheric CO_2. Engineering techniques of injecting CO_2 into the deep ocean, geological strata, old coal mines and oil wells, and saline aquifers along with mineral carbonation of CO_2 constitute abiotic techniques. While with a large potential to store thousands of Pg C, these abiotic techniques are expensive with leakage risks, but may be available for routine use by 2025 and beyond. In comparison, biotic techniques are natural and cost-effective processes, have numerous ancillary benefits, are immediately applicable but have finite sink capacity. Biological CCS is the major environmentally viable method as compared with other sequestration methods. Therefore, considerable research has been conducted to develop biological technologies and processes that increase the efficiency of capture systems while reducing overall cost. Since plants and soils naturally absorb CO_2, preventing outright deforestation and managing forests and agricultural lands as carbon sinks can remove significant amounts of GHGs. Algal biomass can sequestrate carbon and be part of soil sink for long duration. Higher plants including trees can fix CO_2 into their biomass and prevent it from easily releasing to the environment till the death and degradation. Many of the microbes and some of the higher plants can fix CO_2 in non-photosynthetic pathway. Therefore, biotic communities, water and soil in all ecosystems take part in CCS and other parts of the carbon cycle.

The use of biological systems for the mineralization in the form of crustacean or mollusc shell can fix and store carbon for long time. Carbon in the soil in the form of organic carbon is more than the living vegetation in the terrestrial ecosystem. Prevention of degradation of SOC is a good method for carbon storage and can be achieved by afforestation and reforestration of perennial plants having rich lignin biosynthesis. Most of the natural methods are slow and need attention to utilize these systems artificially such as use of carbonic anhydrase.

Biological CCS using microalgae has many advantages over conventional carbon sequestration methods as the CO_2 is being utilised to produce high value biomass which has a number of applications in energy production, biochemical generation as well as food and fees applications. This is a more desirable method for CCS compared with abiotic CCS technologies which often associate with a high cost for separating CO_2 from flue gases without any economic gains from the process.

Microalgae has high photo synthetic efficiencies, utilising solar energy to convert CO_2 to organic carbon and locking it into their cells. Depending on the application of the produced biomass this carbon may become permanently sequestered. However, even if the biomass were to be co-fired in a fossil fuelled power plant, the CO_2 emissions per unit energy would be significantly reduced as the CO_2 released during algae co-firing is recycled and is reused by the algae. Essentially, the net CO_2 emission for burning produced algae is zero; thus, the net plant emissions remain the same while the energy output is significantly increased. In conclusion, there are several advantages both environmentally and economically for the utilisation of microalgae in biotic CCS from point sources. Most importantly, the reduced use of petroleum-derived fuels by the advanced use of biofuels can balance the release and fixation of carbon to the environment. Many countries already utilized biotic CCS with biofuel production.

The applications such as ocean fertilization have unpredictable effect as there is lack of proper or limited experimentations. Utilization of all these biological methods efficiently in all over the world can change the fate of our environment to a stable condition.

4.7 Acknowledgements

Sincere thanks are due to the Natural Sciences and Engineering Research Council of Canada (Grant A 4984, Canada Research Chair) for their financial support. The views and opinions expressed in this chapter are those of the authors.

4.8 Abbreviations

AMF	Arbuscular mycorrhizal fungi
CCS	Carbon capture and sequestration
CCM	Carbon concentrating mechanism
CA	Carbonic anhydrase
CA	Carbonic anhydrase
CT	Conventional tillage
DIC	Dissolved inorganic carbon
ECBM	Enhanced coal bed methane production
EPS	Exopolysaccharide
GHGs	Greenhouse gases
GT	Gigatonne (109 tonnes)
HTP	High throughput
HHV	Higher heating value
HNLC	High-nutrient/low chlorophyll
IEA	International energy agency
INM	Integrated nutrient management
IPM	Integrated pest management
IPCC	Intergovernmental Planet on Climate Change
LF	Light fraction
MRT	Mean residence time
MGE	Microbial growth efficiency
MOM	Microbially-derived organic matter
MMTs	Million metric tons
MT	Minimum tillage
mm	Millimeter
MMT	Million metric tonnes (10^9 Kilograms)
MEA	Monoethanolamine
NT	Non-tillage
POM	Particulate organic matter
Pg	Petagram (10^{15} grams)
PBR	Photobioreactors
PRA	Photo-synthetically active radiation
PCCC	Post combustion carbon capture

PPC	Pre-combustion CO_2 capture
RMPs	Recommended management practices
SOC	Soil organic carbon
SOM	Soil organic matter
SC-CO_2.	Supercritical carbon dioxide
Tg	Teragram (10^{12} grams)
yr	Year

4.9 References

Acosta-Martinez, V., Zobeck, T.M., Gill, T.E., and Kennedy, A.C. (2003). "Enzyme activities and microbial community structure in semiarid agricultural soils." *Biol. Fertil. Soils,* 38, 216–227.

Addadi, L., Raz, S., and Weiner, S. (2003). "Taking advantage of disorder: Amorphous calcium carbonate and its roles in biomineralization." *Advanced Materials,* 15, 959–970.

Akkerman, I., Janssen, M., Rocha, J., and Wijffels, R.H. (2002). "Photobiological hydrogen production: Photochemical efficiency and bioreactor design." *International Journal of Hydrogen Energy,* 27, 1195–208.

Alabi, A.O., Tampier, M., and Bibeau, E. (2009). *Microalgae technologies & processes for biofuels-bioenergy production in British Columbia: current technology, suitability & barriers to implementation: final report.* Publisher: British Columbia Innovation Council, Vancouver, Canada.

Alber, B.E., and Ferry, J.G. (1996). "Characterization of heterologously produced carbonic anhydrase from *Methanosarcina thermophile.*" *J. Bacteriol.,* 178, 3270–3274.

Allison, V.J., Miller, R.M., Jastrow, J.D., Matamala, R., and Zak, R.D. (2005). "Changes in soil microbial community structure in a tall grass prairie chronosequence." *Soil Sci. Soc. Am. J.,* 69, 1412–1421.

An, Z.Q., Hendrix, J.M., Hershman, D.M., Ferris, R.S., and Henson, G.T. (1993). "The influence of crop rotation and soil fumigation on a mycorrhizal fungal community associated with soybean." *Mycorrhyza.*, 3, 171–182.

Andrew, C., Knud DidEriksen, Lee, H., Spangler, Alfred, B., Cun Ningh, A.M, and Robinger, L. (2010). "Microbially Enhanced Carbon Capture and Storage by Mineral-Trapping and Solubility-Trapping Environ." *Sci. Technol.*, 44, 5270–5276.

Angers, D.A., and Mehuys, G.R. (1990). "Barley and alfalfa cropping effects on carbohydrate contents of a clay soil and its size fractions." *Soil Biol. Biochem.*, 22, 285–288.

Angers, D.A., Bissonnette, N., Legere, L., and Samson, D. (1993). "Microbial and biochemical changes induced by rotation and tillage in a soil under barley production." *Can. J. Soil Sci.,* 73, 39–50.

Antal, M.J., and Gronli, M. (2003). "The art, science, and technology of charcoal production." *Ind. Eng. Chem. Res.,* 42(8), 1619–1640.

Aresta, M., Dibenedetto, A., and Barberio, G. (2005). "Utilization of macro-algae for enhanced CO_2 fixation and biofuels production: Development of a computing software for an LCA study." *Fuel Processing Technology*, 86, 1679–93.

Arshad, M.A., M. Schnitzer, D.A. Angers, and J.A. Ripmeester. (1990). "Effects of till versus no till on the quality of soil organic matter." *Soil Biol. Biochem*., 22, 595–599.

Azari, F., and Nemat-Gorgani, M. (1999). "Reversible denaturation of carbonic anhydrase provides a method for its adsorptive immobilization." *Biotechnol Bioeng.,* 62, 193–199.

Babu, S.P. (2004). "Biomass gasification for hydrogen production – process description and research needs. Des Plaines, IL: Gas Technology Institute, International Energy Agency, Hydrogen Program.

Ball, B.C., Cheshire, M.V., Robertson, E.A.G., Hunter, E.A. (1996). "Carbohydrate composition in relation to structural stability, compatibility and plasticity of two soils in a long-term experiment." *Soil Tillage Res.,* 39, 143–160.

Bahn, Y.S., Cox, G.M., Perfect, J.R., and Heitman, J. (2005). "Carbonic anhydrase and CO_2 sensing during Cryptococcus neoformans growth, differentiation, and virulence." *Curr. Biol.,* 15, 2013–2020.

Bailey, V.L., Smith, J.L., and Bolton, H., Jr. (2002). "Fungal-to-bacterial ratios in soils investigated for enhanced C sequestration." *Soil Biol. Biochem*., 34, 997–1007.

Baldock, J.A., Kay, B.D., and Schnitzer, M. (1987). "Influence of cropping treatments on the monosaccharide content of the hydrolysates of a soil and its aggregate fractions." *Can. J. Soil Sci.,* 67, 489–499.

Barbero, R., Carnelli, L., Simon, A., Kao, A., Monforte, A.D., Ricco, M., Bianchi, D., and Belcher, A. (2013). "Engineered yeast for enhanced CO_2 mineralization." *Energy & Environmental Science*, 6, 660–674. DOI: 10.1039/C2EE24060B.

Bardgett, R.D., and McAlister, E. (1999). "The measurement of soil fungal:bacterial biomass ratios as an indicator of ecosystem self-regulation in temperate meadow grasslands." *Biol. Fertil. Soils,* 29, 282–290.

Beare, M.H., Parmelee, R.W., Hendrix, P.F.,Cheng, W., Coleman, D.C., and Grossley Jr, D.A. (1992). "Microbial and faunal interactions and effects on litter nitrogen and decomposition in agroecosystems." *Ecol. Monogr*., 62, 569–591.

Beare, M.H. (1997). "Fungal and bacterial pathways of organic matter decomposition and nitrogen mineralization in arable soils." 37–70. In L. Brussaard and R. Ferrera-Cerrato (Eds). *Soil Ecology in Sustainable Agricultural Systems*. Lewis Publishers, Boca Raton, Florida, USA.

Bell, J.M., Smith, J.L., Bailey,V.L., and Bolton, H. (2003). "Priming effect and C storage in semi-arid no-till spring crop rotations." *Biol. Fertil. Soils*, 37, 237–244.

Belzil, A., and Parent, C. (2005). "Qualification methods of chemical immobilizations of an enzyme on solid support." *Methodes de Qualification des Immobilisations Chimiques D'une Enzyme sur un Support Solide*, 83, 70–77.

Benson S.M. and Orr, F.M., Jr. (2008). "Carbon dioxide capture and storage." *MRS Bulletin*, 33, 303–305.

Berg, C. (2004). "World fuel ethanol analysis and outlook"; available from: http://www.distill.com/World-Fuel-Ethanol-A&O-2004.html.

Berzseny, Z., and Gyrffy, B. (1997). "Effect of crop rotation and fertilization on maize and wheat yield stability in long-term experiments." *Agrok. Ma. S. Talajtan.*, 46, 377–398.

Blais, R., and Rogers, P. (2003). "Process and apparatus for the treatment of carbon dioxide with carbonic anhydrase." US Patent 6524843.

Bohlin, F., Vinterbäck, J., Wisniewski, J., and Wisniewski, J. (1998). "Solid biofuels for carbon dioxide mitigation." *Biomass Bioenerg.*, 15, 277–81.

Borchert, M., and Saunders, P. (2010). "Heat-stable carbonic anhydrases and their use." US Patent WO/2010/151787.

Bossio, D.A., Scow, K.M., Gunapala, N., and Graham. K.J. (1998). "Determinants of soil microbial communities: Effects of agricultural management, season, and soil type on phospholipid fatty acid profiles." *Microb. Ecol.*, 36, 1–12.

Boyd, P.W., Law, C.S., Wong, C.S., Nojiri, Y., Tsuda, A., Levasseur, M., Takeda, S., Rivkin, R., Harrison, P.J., Strzepek, R., Gower, J., McKay, R.M., Abraham, E., Arychuk, M., Barwell, C.J., Crawford, W., Crawford, D., Hale, M., Harada, K., Johnson, K., Kiyosawa, H., Kudo, I., Marchetti, A., Miller, W., Needoba, J., Nishioka, J., Ogawa, H., Page, J., Robert, M., Saito, H., Sastri, A., Sherry, N., Soutar, T., Sutherland, N., Taira, Y., Whitney, F., Wong, S.K.E., and Yoshimura, T. (2004). "The decline and fate of an iron-induced subarctic phytoplankton bloom", *Nature*, 428, 549–553.

Bruce James, P., Michele, Frome., Eric, Haites., Henry, Janzen., Rattan, Lal., and Keith, Paustian., 1999 "Carbon sequestration in soils." *Journal of Soil and Water Conservation*, 54(1), 382–389.

Bunemann, E.K., Bossio, D.A., Smithson, P.C., Frossard, E., and Oberson, E. (2004). "Microbial community composition and substrate use in a highly weathered soil as affected by crop rotation and P fertilization." *Soil Biol. Biochem.*, 36, 889–901.

Chisholm, S. W., Falkowski, P. G., and Cullen, J. J. (2001). "Discrediting ocean fertilization." *Science*, 294, 309–310.

Christian, Azar., Kristian, Lindgren., Eric, Larson., and Kenneth, Mollersten. (2006). "Carbon capture and storage from fossil fuels and biomass – costs and potential role in stabilizing the atmosphere." *Climatic Change*, 74, 47–79.

Ciferno, J., Fout, T., Jones, A., and Murphy, J.T. (2009). "Capturing carbon from existing coal-fired plants." *Chem Eng Prog.*, 105, 33–41.

Das, K.C., Singh, K., Adolphson, R., Hawkins, B., Oglesby, R., Lakly, D., and Day, D. (2010). "Steam pyrolysis and catalytic steam reforming of biomass for hydrogen and biochar production." *Appl. Eng. Agric.,* 26(1), 137–146.

Drevon, G.F., Urbanke, C., and Russell, A.J. (2003). "Enzyme-containing Michaeladduct- based coatings." *Biomacromolecules,* 4, 675–682.

Falkowski, P. R. J. Scholes, E. Boyle, J. Canadell, D. Canfield, J. Elser, N. Gruber, K. Hibbard, P. Högberg, S. Linder, F. T. Mackenzie, B. Moore III, T. Pedersen, Y. Rosenthal, S. Seitzinger, V. Smetacek, W. Steffen (2000). "The global carbon cycle: a test of our knowledge of earth as a system." *Science, New Series,* 290(5490), 291–296.

Fan, L.H., Liu, N., Yu, M.R., Yang, S.T., and Chen, H.L. (2011). "Cell surface display of carbonic anhydrase on Escherichia coli using ice nucleation protein for CO_2 sequestration." *Biotechnol. Bioeng.,* 108, 2853–2864.

Farrelly, J., Colm, D., Colette, and C., Kevin, P. (2013). "Carbon sequestration and the role of biological carbon mitigation: A review." *Renewable and Sustainable Energy Reviews*, 21,712–727.

Favre, N., Christ, L., and Pierre, A.C. (2009). "Biocatalytic capture of CO_2 with carbonic anhydrase and its transformation to solid carbonate." *J. Mol. Catal. B Enzym.*, 60, 163–170.

Feng, Y., A.C. Motta, D.W. Reeves, C.H. Burmester, E. van Santen, and J.A. Osborne. (2003). "Soil microbial communities under conventional-till and no-till continuous cotton systems." *Soil Biol. Biochem.*, 35, 1693–1703.

Ferris FG, Fyfe WS, and Beveridge TJ. (1988) "Metallic ion binding by Bacillus subtilis: Implications for the fossilization of microorganisms." *Geology,* 16, 149-152.

Fierer, N., A.S. Allen, J.P. Schimel, and P.A. Holden. (2003). "Controls on microbial CO_2 production: A comparison of surface and subsurface soil horizons." *Global Change Biol.,* 9, 322–1332.

Fliessbach A., Paul Mäder and Urs Niggli (2000). "Mineralization and microbial assimilation of ^{14}C-labeled straw in soils of organic and conventional agricultural systems." *Soil Biol. Biochem.,* 32, 1131–1139.

Fradette, S. (2004). "Process and apparatus using a spray absorber bioreactor for the biocatalytic treatment of gases." US Patent WO04056455.

Fradette, S., and Ruel, J. (2009). "Process and a plant for recycling carbon dioxide emissions from power plants." US Patent 7596952. Patent claims the use of carbonic anhydrases for CO_2 capture from fossil fuel derived power generation.

Fradette, S., and Ceperkovic, O. (2010). "CO_2 absorption solution." US Patent US07699910.

Franzluebbers, A.J., Haney, R.L., Honeycutt, C.W., Arshad, M.A., Schomberg, H.H., and Hons, F.M. (2001). "Climatic influences on active fractions of soil organic matter." *Soil Biology and Biochemistry,* 33, 1103–1111.

Frey, S.D., E.T. Elliott, and K. Paustian. (1999). "Bacterial and fungal abundance and biomass in conventional and no-tillage agroecosystems along two climatic gradients." *Soil Biol. Biochem.,* 31, 573–585.

Frey, S.D., J. Six, and E.T. Elliott. (2003). "Reciprocal transfer of carbon and nitrogen by decomposer fungi at the soil-litter interface." *Soil Biol. Biochem.,* 35, 1001–1004.

Frey, S.D., Elliott, E.T., Paustian, K., and Peterson, G.A. (2000). "Fungal translocation as a mechanism for soil nitrogen inputs to surface residue decomposition in no-tillage agroecosystem." *Soil Biol. Biochem.,* 32, 689–698.

Fujita, Y., Taylor, J. L., Gresham, T. L., Delwiche, M. E., Colwell, F. S., McLing, T., Petzke, L., and Smith, R. W. (2008). "Stimulation of microbial urea hydrolysis in groundwater to enhance calcite precipitation." *Environ. Sci. Technol.,* 42, 3025–3032.

Fullen, M.A., and Auerswald, K. (1998). "Effect of grass ley set-aside on runoff, erosion and organic matter levels in sandy soil in east Shropshire, UK." *Soil and Tillage Research,* 46, 41–49.

Gilley, J.E., and Risse, L.M. (2000). "Runoff and soil loss as affected by the application of manure." *Transactions of the ASAE,* 43, 1583–1588.

Gregorich, E.G., Drury, C.F., and Baldock, J.A. (2001). "Changes in soil carbon under long-term maize in monoculture and legume-based rotation." *Canadian Journal of Soil Science,* 81, 21–31.

Guggenberger, G., Zech, W., and Thomas, R.J. (1995). "Lignin and carbohydrate alteration in particle-size separates of an oxisol under tropical pastures following native savanna." *Soil Biol. Biochem.,* 27, 1629–1638.

Gunapala, N., Venette, R.C., Ferris, H., and Scow. K.M. (1998). "Effects of soil management history on the rate of organic matter decomposition." *Soil Biol. Biochem.,* 30, 1917–1927.

Gupta, V.V.S.R., Roper, M.M., Kirkegaard, J.A., and Angus, JF. (1994). "Changes in microbial biomass and organic matter levels during the first year of modified tillage and stubble management practices on a red Earth." *Aust. J. Soil Res.,* 32, 1339–1354.

Guyot, S., Pottier, L., Ferret, E., Gal, L., and Gervais, P. (2010). "Physiological responses of Escherichia coli exposed to different heat-stress kinetics." *Arch. Microbiol.,* 192, 651–661.

Hao, Y., Lal, R., Owens, L.B., Izaurralde, R.C., Post, M., and Hothem, D. (2002). "Effect of cropland management and slope position on soil organic carbon pools in the North Appalachian Experimental Watersheds." *Soil and Till. Res.,* 68, 133– 142.

Hassink, J., G. Lebbink, and J.A. VanVeen. (1991). "Microbial biomass and activity of a reclaimed-polder soil under a conventional and a reduced-input farming system." *Soil Biol. Biochem.*, 23, 507–514.

Haynes, R.J., and Francis, G.S. (1993). "Changes in microbial biomass C, soil carbohydrate composition and aggregate stability induced by growth of selected crop and forage species under field conditions." *J. Soil Sci.* 44, 665–667.

Hu, S., D.C. Coleman, M.H. Beare, and P.F. Hendrix. (1995). "Soil carbohydrates in aggrading and degrading agroecosystems: Influences of fungi and aggregates." *Agric. Ecosyst. Environ.*, 54, 77–88.

IPCC, (1999) *Land use, land use change and forestry. Intergovernment panel on climate change.* Cambridge, UK: Cambridge University Press.

Israel, A., Gavrieli, J., Glazer, A., and Friedlander, M. (2005). "Utilization of flue gas from a power plant for tank cultivation of thered seaweed Gracilaria cornea." *Aquaculture*, 249, 311–316.

Jansson, C., Wullschleger, S.D., Udaya, C.K., and Tuskan, G.A. (2010). "Phytosequestration: carbon biosequestration by plants and the prospects of genetic engineering." *Bioscience*, 60, 685–696.

Jeyakanthan, J., Rangarajan, S., Mridula, P., Kanaujia, S.P., Shiro, Y., Kuramitsu, S.S.Y., and Sekar, K. (2008). "Observation of a calcium-binding site in the [gamma]-class carbonic anhydrase from Pyrococcus horikoshii." *Acta Crystallogr. D Biol. Crystallogr.*, D64, 1012–1019.

Jo, B.H., Kang, D.G., Kim, I.G., Seo, J.H., and Cha, H.J. (2013). "Engineered *Escherichia coli* with periplasmic carbonic anhydrase as a biocatalyst for CO_2 sequestration." *Applied and Environmental Microbiology,* 79, 6697–6705.

Johnston, M.H., Homann, P.S., Engstrom, J.K., and Grigal, D.F. (1996). "Changes in ecosystem carbon storage over 40 years on an old field/forest landscape in east – central Minnesota." *Forest Ecology and Management,* 83, 17–26.

Johnson, K. S., Karl, D. M., Chisholm, S.W., Falowski, P. G. and Cullen, J. J. (2002). "Is ocean fertilization credible and creditable?" *Science,* 296, 467–468.

Johnson, R. (2010). "The carbon cycle." Available at <http://www.windows2 universe.org> (accessed Feb. 2012).

Kadam, K.L. (2001). *Microalgae production from power plant flue gas: environmental implications on a life cycle basis.* Tech. Rep. NREL/TP-510-29417, National Renewable Energy Laboratory, U.S.

Kauppi, P., Sedjo, R., Apps, M., Cerri, C., Fujimoro, T., Janzen, H., Krankina, O., Makundi, W., Marland, G., Masera, O., Nabuurs, G.-J., Razali, W., and Ravindranath, N.H. (2001). "Technological and economic potential of options to enhance, maintain and manage biological carbon reservoirs and geo-engineering." In: Metz, B., Davidson, O., Swart, R., Pan, J. (Eds.), *Climate Change 2001- Mitigation.* Cambridge Univ. Press, UK, pp. 301–343.

Kheshgi, H.S., Prince, R.C., and Marland, G. (2000). "The potential of biomass fuels in the context of global climate change: focuson transportation fuels." *Annual Review of Energy and the Environment*, 25(1), 199–244.

Kimble, J.M., Heath, L.S., Birdsey, R.A., and Lal, R. (2002). *The Potential of U.S. Forest Soils to Sequester Carbon and Mitigate the Greenhouse Effect.* Edited by Kimble, J.M., Heath, L.S., Birdsey, R.A., and Lal, R., CRC Press, Boca Raton, FL. 429 pp. Print ISBN: 978-1-56670-583-7.

Kuo, S., U.M. Sainju, and E.J. Jellum. (1997). "Winter cover crop effects on soil organic carbon and carbohydrate in soil." *Soil Sci. Soc. Am. J.,* 61, 145–152.

Kurano, N., Ikemoto, H., Miyashita, H., Hasegawa, T., Hata, H., Miyachi, S. (1996). "Fixation and utilization of carbon dioxide by microalgal photosynthesis." *Energy Conversion and Management*, 36, 689–92.

Kusian, B., Sultemeyer, D., and Bowien, B. (2002). "Carbonic anhydrase is essential for growth of *Ralstonia eutropha* at ambient CO_2 concentrations." *J. Bacteriol.*, 184(18), 5018–5026.

Lal, R. (1996). "Deforestation and land use effects on soil degradation and rehabilitation in western Nigeria. II: soil chemical properties." *Land Degradation and Development*, 7, 87–98.

Lal, R., Kimble, J.M., Follett, R.F., and Cole, C.V. (1998). *The Potential of U.S. Cropland to Sequester Carbon and Mitigate the Greenhouse Effect.* Ann Arbor Sci. Publ., Chelsea, MI. 128 pp.

Lal, R. (2001). "The potential of soil carbon sequestration in forest ecosystem to mitigate the greenhouse effect." In: Lal, R. (ed.), *Soil Carbon Sequestration and the Greenhouse Effect.* Soil Science Society of America Special Publication, vol. 57.

Lal, R. (2004) "Soil carbon sequestration impacts on global climate change and food security." *Science,* 304, 1623–1627.

Lal, R. (2008). "Carbon sequestration." *Philosophical Transactions of Royal Society B*, 363, 815–830.

Lavelle, P. (1997). "Faunal activities and soil processes: adaptive strategies that determine ecosystem function." *Advances in Ecology Research,* 27, 93–132.

Lavelle, P. (2000). "Ecological challenges for soil science." *Soil Science*, 165, 73–86.

Lee, B.D., Apel, W.A., Walton, and M.R. (2006). *Whitings as a Potential Mechanism for Controlling Atmospheric Carbon Dioxide Concentrations – Final Project Report*, Idaho National Laboratory.

Lee, K.E. (1991). "The diversity of soil organisms." In: Hawksworth, D.L. (ed.), *The Biodiversity of Microorganisms and Invertebrates: Its Role in Sustainable Agriculture*. CAB International, Wallingford, UK, pp. 73–87.

Lee, S.W., Park, S.B., Jeong, S.K., Lim, K.S., Lee, S.H., and Trachtenberg, M.C. (2010). "On carbon dioxide storage based on biomineralization strategies." *Micron,* 41, 273–282.

Lehmann, J. (2007). "A handful of carbon." *Nature*, 447, 143–144.

Litynski, J.T., Plasynski, S., McIlvried, H.G., Mahoney, C., and Srivastava, R.D. (2008). The United States Department of Energy's Regional Carbon Sequestration Partnerships Program: validation phase. *Environ. Int.,* 34(1), 127–38.

Looney, A., Han, R., McNeill, K., and Parkin, G. (1993). "Tris(pyrazoly1)-hydroboratozinc hydroxide complexes as functional models for carbonic anhydrase: on the nature of the bicarbonate intermediate." *J Am Chem Soc.* 115, 4690–4697.

Lower, S.K. (1999). *Chem1 virtual textbook,* p. 1–26. Simon Fraser University, Burnaby, British Columbia, Canada

Lupwayi, N.Z., W.A. Rice, and G.W. Clayton. (1999). "Soil microbial biomass and carbon dioxide flux under wheat as influenced by tillage and crop rotation." *Can. J. Soil Sci.,* 79, 273–280.

Madison, W.I., Groenendijk, F.M., Condron, L.M., and Rijkse, W.C. (2002). "Effects of afforestation on organic carbon, nitrogen and sulfur concentrations in New Zealand hill country soils." *Geoderma.,* 108, 91–100.

Mata, T.M., Martins A.A., and Caetano, N.S. (2010). "Microalgae for biodiesel production and other applications: a review." *Renewable & Sustainable Energy Reviews*, 14, 217–32.

Mazzone, E.J., Guentzel, J.L., and Oliazola, M. (2002). "Carbon removal through algal mediated precipitation of calcium carbonate." *Abstracts of Papers of the American Chemical Society*, 223, U537–U538.

Mitchell, A. C., Phillips, A., Hiebert, R., Gerlach, R.; Spangler, L., and Cunningham, A. B. (2000). "Biofilm enhanced subsurface sequestration of supercritical CO_2." *Int. J. Greenhouse Gas Control,* 3, 90–99.

Molina Grima, E., Belarbi, E.H., Acien Fernandez, F.G., Robles Medina, A., and Chisti, Y. (2003). "Recovery of microalgal biomass and metabolites: process options and economics." *Biotechnology Advances,* 20, 491–515.

Mollersten, K., Yan, J., and Moreira, J.R. (2003). "Potential market niches for biomass energy with CO_2 capture and storage–opportunities for energy supply with negative CO_2 emissions." *Biomass and Bioenergy,* 25(3), 273–85.

Nabuurs, G.J., Pa¨ivinen, R., Sikkema, R., and Mohren, G.M.J. (1997). "The role of European forests in the global carbon cycle – a review." *Biomass Bioenergy,* 13, 345–358.

Neufeldt, H., Resck, D.V.S., and Ayarza, M.A. (2002). "Texture and land use effects on soil organic matter in Cerrado Oxisols, central Brazil." *Geoderma,* 107, 151–164.

Newman L.M., Clark L., Ching C., and Zimmerman S. (2010). Carbonic anhydrase polypeptides and uses thereof. US Patent WO10081007; 2010.

Obst, M., Dynes, J.J., Lawrence, J.R., Swerhone, G.D.W., Benzerara, K., Karunakaran, C., Kaznatcheev, K., Tyliszczak, T., and Hitchcock, A.P. (2009). "Precipitation of amorphous $CaCO_3$ (aragonite-like) by cyanobacteria: a STXM study of the influence of EPS on the nucleation process." *Geochimica et Cosmochimica Acta,* 73, 4180–4198.

Olaizola, M., T., Bridges, S. Flores, L. Griswold, J. Morency, and T. Nakamura (2004). "Microalgal removal of CO_2 from flue gases: CO_2 capture from a coal combustor." The third annual conference on Carbon Capture and Sequestration. Alexandria, VA.

Packer, M. (2009). "Algal capture of carbon dioxide; biomass generation as a tool for greenhouse gas mitigation with reference to New Zealand energy strategy and policy." *Energy Policy,* 37(9), 3428–3437.

Paustian, K., H.P. Collins, and E.A. Paul. (1997). "Management controls on soil carbon." p. 15–49. In E.A. Paul et al. (Eds.) *Soil Organic Matter in Temperate Agroecosystems*. CRC Press, Boca Raton, FL.

Perlack, R.D., Wright, A., Turhollow, A., Graham, R., Stokes, B., and Erbach, D.S. (2005) *Biomass as feedstock for a bioenergy and Bioproducts Industry: the technical feasibility of a billion-ton annual supply*. Report ORNL/TM-2005/66 prepared by Oak Ridge National Laboratory for the US Department of energy.

Petela, R. (2008). "An approach to the energy analysis of photosynthesis." *Solar Energy,* 82, 311–28.

Peterson, G.A., Halvorson, A.D., Havlin, J.l., Jones, O.R., Lyon, D.J., and Tanaka, D.L. (1998). "Reduced tillage and increasing cropping intensity in the Great Plains conserves soil C." *Soil Tillage Res.,* 47, 207–218.

Post, W.M., and Kwon, K.C. (2000). "Soil carbon sequestration and land use change: processes and potential." *Global Change Biology,* 6, 317–327.

Price, G.D., Maeda ,S., Omata, T., and Badger, M.R. (2002). "Modes of active inorganic carbon uptake in the cyanobacterium, Synechococcus sp PCC7942." *Functional Plant Biology*, 29, 131–149.

Pulleman, M.M., Bouma, J., van Essen, E.A., and Meijles, E.W. (2000). "Soil organic matter content as a function of different land use history." *Soil Science Society of America Journal,* 64, 689–693.

Qin, R., P. Stamp, and W. Richner. (2005). "Impact of tillage and banded starter fertilizer on maize root growth in the top 25 centimeters of the soil." *Agron. J.,* 97, 674–683.

Rajagopal, D., and Zilberman, D. (2007). Review of environmental, economic and policy aspects of biofuels (Policy Research Working Paper Series 4341). Washington, DC: The World Bank. The World Bank Development Research Group Sustainable Rural and Urban Development Team, September.

Rasmussen, P.E., Goulding, K.W.T., Brown, J.R., Grace, P.R., Janzen, H.H., and Korschens, M. (1998). "Long-term agroecosystems experiments: Assessing agricultural sustainability and global change." *Science*, 282, 893–896.

Raven, J.A., and Falkowski, P.G. (1999). "Oceanic sinks for atmospheric CO_2." *Plant Cell Environ.,* 22, 741–755.

Reicosky, D.C. (2002). "Long-term effect of moldboard plowing on tillage-induced CO_2 loss." In: Kimble, J.M., Lal, R., Follett, R.F. (eds.), *Agricultural Practices and*

Policies for Carbon Sequestration in Soil. CRC/Lewis, Boca Raton, FL, pp. 87–97.

Rhodes, S., David, W., and Keith, B. (2005). "Engineering economic analysis of biomass IGCC with carbon capture and storage." *Biomass and Bioenergy,* 29, 440–450.

Ramanan, R., Kannan, K., Sivanesan, S.D., Mudliar, S., Kaur, S., Tripathi, A.K., and Chakrabarti, T. (2009). "Bio-sequestration of carbon dioxide using carbonic anhydrase enzyme purified from Citrobacter freundii." *World J. Microbiol Biotechnol.,* 25, 981–987.

Rittstieg, K., Robra, K.H., and Somitsch, W. (2001). "Aerobic treatment of a concentrated urea wastewater with simultaneous stripping of ammonia." *Appl. Microbiol. Biotechnol.,* 56(5), 820–825.

Rivkin, R.B., and Legendre, L. (2001) "Biogenic carbon cycling in the upper ocean: Effects of microbial respiration." *Science,* 291, 2398–2400.

Roberson, E.B., Sarig, S., and Firestone, M.K. (1991). "Cover crop management of polysaccharide-mediated aggregation in an orchard soil." *Soil Sci. Soc. Am. J.,* 55, 734–739.

Roberson, E.B., Sarig, S., Shennan, C., and Firestone, M.K. (1995). "Nutritional management of microbial polysaccharide production and aggregation in an agricultural soil." *Soil Sci. Soc. Am. J.,* 59, 1587–1594.

Robinson, A.L., Rhodes, J.S., and Keith, D.W. (2003), "Assessment of potential carbon dioxide reductions due to biomass-coal cofiring in the United States." *Environmental Science and Technology,* 37(22), 5081–5089.

Rochelle, G.T. (2009). "Amine scrubbing for CO_2 capture." *Science,* 325, 1652–1654.

Rowlett, R.S. (2010). "Structure and catalytic mechanism of the bcarbonic anhydrases." *Biochim. Biophys. Acta (BBA): Proteins Proteomics,* 1804, 362–373.

Sa, J.C.D., Cerri, C.C., Dick, W.A., Lal, R., Venske, S.P., Piccolo, M.C., and Feigl, B.E. (2001). "Organic matter dynamics and carbon sequestration rates for a tillage chronosequence in a Brazilian Oxisol." *Soil Science Society of America Journal,* 65, 1486–1499.

Sainju, U.M., Singh, B.P., and Yaffa, S. (2002). "Soil organic matter and tomato yield following tillage, cover cropping and nitrogen fertilization." *Agriculture Journal,* 94, 594–602.

Sauerbeck, R.D. (2001). "CO_2 emissions and C sequestration by agricultureperspectives and limitations." *Nutr. Cycling Agroecosyst.*, 60(1–3), 253–266.

Savile, C., and James, J. Lalonde (2011). "Biotechnology for the acceleration of carbon dioxide capture and sequestration." *Current Opinion in Biotechnology*, 22, 818–823.

Savile, C., Nguyen, L., Alvizo, O., Balatskaya, S., Choi, G., Benoit, M., Fusman, I., Geilhufe, J., Ghose, S., and Gitin, S. (2011). "Low cost biocatalyst for the acceleration for energy efficient CO_2 capture." ARPA-E Energy Innovation Summit. Washington, DC: ARPA.

Schlesinger, W.H. (2000). "Carbon sequestration in soil: some cautions amidst optimism." *Agricultural Ecosystems and Environment,* 82, 121–127.

Scow, K.M. (1997). "Soil microbial communities and carbon flow in agroecosystems." p. 367–413. In L.E. Jackson (ed.) *Ecology in agriculture.* Academic Press, San Diego, CA.

Sharma, A., Bhattacharya, A., and Shrivastava, A. (2011). "Biomimetic CO_2 sequestration using purified carbonic anhydrase from indigenous bacterial strains immobilized on biopolymeric materials." *Enzyme Microb Technol.,* 48, 416–426.

Shaver, T.M., Peterson, G.A., Ahuja, L.R., Westfall, D.G., Sherrod, L.A., and Dunn, G. (2002). "Surface soil physical properties after twelve years of dryland no till management." *Soil Science Society of America Journal,* 66, 1296–1303.

Schnurer, J., Clarholm, M., and Rosswall, T. (1985). "Microbial biomass and activity in an agricultural soil with different organic matter contents." *Soil Biol. Biochem.,* 17, 611–618.

Simpson, R.T., Frey, S.D., Six, J., and Thiet, R.K. (2004). "Preferential accumulation of microbial carbon in aggregate structures of notillage soils." *Soil Sci. Soc. Am. J.,* 68, 1249–1255.

Singh, B.R., Borresen, T., Uhlen, G., and Ekeberg, E. (1998). "Long-term effects of crop rotation, cultivation practices and fertilizers on carbon sequestration in soils in Norway." In: Lal, R., Kimble, J.M., Follett, R.F., and Stewart, B.A. (eds.), *Management of Carbon Sequestration in Soil.* CRC Press, Boca Raton, pp. 195–208.

Singh, J., and Gu, S. (2010). "Commercialization potential of microalgae for biofuels production." *Renewable and Sustainable Energy Reviews,* 14, 2596–2610.

Six, J., Elliott, E.T., and Paustian, K. (1999). "Aggregate and soil organic matter dynamics under conventional and no-tillage systems." *Soil Sci. Soc. Am. J.,* 63, 1350–1358.

Six, J., Paustian, K., Elliott, E.T., and Combrink, C. (2000). "Soil structure and organic matter: I. Distribution of aggregate-size classes and aggregate-associated carbon." *Soil Sci. Soc. Am. J.,* 64, 681–689.

Six, J., Feller, C., Denef, K., Ogle, K., Sa, M.J.C., and Albrecht, A. (2002). "Soil organic matter, biota and aggregation in temperate and tropical soils-effects of no-tillage." *Agron. Agric. Environ.,* 22, 755–775.

Six, J., Frey, S.D., Thiet, R.K., and Batten, K.M. (2006). "Bacterial and fungal contributions to carbon sequestration in agroecosystems." *Soil Sci. Soc. Am. J.,* 70, 555–569.

Skjemstad, J.O., Dalal, R.C., Janik, L.J., and McGowan, J.A. (2001). "Changes in chemical nature of soil organic carbon in Vertisols under wheat in southeastern Queensland." *Australian Journal of Soil Research,* 39, 343–359.

Skjemstad, J.O., Reicosky, D.C., Wilts, A.R., and McGowan, J.A. (2002). "Charcoal carbon in U.S. agricultural soils." *Soil Science Society of America Journal,* 66, 1255–1949.

Smaling, E.M.A. (1993). "Soil nutrient depletion in sub-Saharan Africa." In: Van Reuler, H., and Prins, W.H., (Eds.), *the Role of Plant Nutrients for Sustainable Food Crop Production in Sub-Saharan Africa*. Vereniging Van Kunstmest-Producenten, Leidschendam, the Netherlands, pp. 53–67.

Smith, K.S., and Ferry, J.G. (1999). "A plant-type (b-class) carbonic anhydrase in the thermophilic methanoarchaeon *Methanobacterium thermoautotrophicum*." *J. Bacteriol.,* 181, 6247–6253.

Smith, P., Powlson, D.S., Glendining, M.J., and Smith, J.U. (1998). "Preliminary estimates of the potential for carbon mitigation in European soils through no-till farming." *Global Change Biology*, 4, 679–685.

Smith, P., Powlson, D.S., Smith, J.U., Falloon, P., and Coleman, K. (2000). "Meeting the U.K.'s climate change commitments: options for carbon mitigation on agricultural land." *Soil Use and Management,* 16, 1–11.

Spalding, M.H. (2008). "Microalgal carbon-dioxide-concentrating mechanisms: Chlamydomonas inorganic carbon transporters." *J. Exp. Bot.,* 59(7), 1463–1473.

Stephan, D.J., Shockey, R.E., Moe, T.A., and Dorn, R. (2001). *Carbon dioxide sequestering using microalgal systems.* Pittsburgh: U.S. Department of Energy.

Stewart, C., and Hessami, M.A. (2005). "A study of methods of carbon dioxide capture and sequestration–the sustainability of a photosynthetic bioreactor approach." *Energy Conversion and Management,* 46, 403–20.

Swanson, E., Patel, T., Banta, S., Brady, P., Davis, K., and Park, A.HA. (2010). "Chemical and biological catalytic enhancement of weathering of silicate minerals." NETL/DOE 2010 CO_2 Capture Technology Meeting. Pittsburgh, PA: NETL.

Thompson, J.P. (1987). "Decline of vesicular-arbuscular mycorrhizae in long fallow disorder of field crops and its expression in phosphorous deficiency of sunflower." *Aust. J. Agric. Res.,* 38, 847–867.

Tiessen, H., Sampaio, E.V.S.B., and Salcedo, I.H. (2001). "Organic matter turnover and management in low input agriculture of NE Brazil." *Nutrient Cycling in Agroecosystems,* 61, 99–103.

Trachtenberg, M.C., Tu, C.K., Landers, R.A., Wilson, R.C., McGregor, M.L., Laipis, P.J., Paterson, M., Silverman, D.N., Thomas, D., Smith, R.L., and Rudolph, F.B. (1999). "Carbon dioxide transport by proteic and facilitated transport membranes." *Life Support Biosph. Sci.,* 6, 293–302.

Trumbore, S.E., Chadwick, O.A., and Amundson, R. (1996). "Rapid exchange of soil C and atmospheric CO_2 driven by temperature change." *Science,* 272, 393–396.

Trumper, K., Bertzky, M., Dickson, B., van der Heijden, G., Jenkins, M., and Manning, P. (2009). *The Natural Fix? The role of ecosystems in climate mitigation.* A UNEP rapid response assessment. United Nations Environment Programme, UNEP-WCMC, Cambridge, UK. ISBN: 978-82-7701-057-1.

UNCBD, (1992). *United Nations Convention on Biological Diversity.* UNCBD, Bonn, Germany.

Van Kessel, C., Horwath, W.R., Hartwig, U., Harris, D., and Luscher, A. (2000). "Net soil carbon input under ambient and elevated CO_2 concentrations: isotopic evidence after 4 years." *Global Change Biology*, 6, 435–444.

Wander, M.M., Bidar, M.G., and Aref, S. (1998). "Tillage impacts on depth distribution of total and particulate organic matter in three Illinois soils." *Soil Science Society of America Journal*, 62, 1704–1711.

Wardle, D.A. (1995). "Impacts of disturbance on detritus food webs in agro-ecosystems of contrasting tillage and weed management practices." *Adv. Ecol. Res.*, 26, 107–185.

Watanabe, Y., and Hall, D.O. (1996). "Photosynthetic CO_2 conversion technologies using a photobioreactor incorporating microalgae–energy and material balances." *Energy Conversion and Management*, 37, 1321–1326.

West, T.O., and Post, W.M. (2002). "Soil organic carbon sequestration rates by tillage and crop rotation: A global data analysis." *Soil Sci. Soc. Am. J.*, 66, 1930–1946.

World Water Assessment Programme (WWAP) (2003). *Water for People: Water or Life–The UN World Water Development Report (The United Nations World Water Development Report)*; Berghahn Books: Oxford, NY.

Wright, S.F., and Anderson, R.L. (2000). "Aggregate stability and glomalin in alternative crop rotations for the central Great Plains." *Biol. Fertil. Soils,* 31, 249–253.

Wu, H.Y., Zou, D., and Gao, K.S. (2008). "Impacts of increased atmospheric CO_2 concentration on photosynthesis and growth of micro-and macro-algae." *Science in China Series C: LifeScience,* 51, 1144–50.

Yang, W.C., and Ciferno, J. (2006). *Assessment of Carbozyme Enzyme-Based Membrane Technology for CO_2 Capture from Flue Gas.* DOE/NETL 401/072606.

Yates, K.K., and Robbins, L.L. (1998). "Production of carbonate sediments by a unicellular green alga." *American Mineralogist,* 83, 1503–1509.

Yeates, G.W., Bardgett, R.D., Cook, R., Hobbs, P.J., Bowling, P.J., and Potter, J.F. (1997). "Faunal and microbial diversity in three Welsh grassland soils under conventional and organic management regimes." *J. Appl. Ecol.,* 34, 453–471.

Zhang, J., Quay, P.D., and Wilbur, D.O. (1995). "Carbon isotope fractionation during gas-water exchange and dissolution of CO_2." *Geochim. Cosmochim. Acta,* 59(1), 107–114.

Zhang, T.C., and Surampalli, R.Y. (2013). "Carbon capture and storage for mitigating climate changes." Chapter 20 in *Climate Change Modeling, Mitigation and Adaptation,* Surampalli, R., Zhang, T.C., Ojha, C.S.P., Gurjar, B.R., Tyagi, R.D., and Kao, C.M. (eds.), ASCE, Reston, Virginia, February 2013.

CHAPTER 5

CO_2 Sequestration and Leakage

P. N. Mariyamma, Song Yan, R. D. Tyagi, Rao Y. Surampalli and Tian C. Zhang

5.1 Introduction

Carbon dioxide, a major greenhouse gas (GHG), which is important to earth's energy balance and climate, has been a major cause of increase in global temperature. The total estimated GHG emissions due to carbon dioxide account to 64%, making it the target for mitigating GHGs (Bachu and Adams 2003). The removal of carbon dioxide is of paramount importance and measures have to be taken to control this in order to ensure our commitment for the preservance of the environment.

One of the most crucial ingredients in mitigating the rising concentrations of anthropogenic GHG emissions is the successful and efficient sequestration of CO_2 in various ecosystems (West and Marland 2002). Carbon sequestration is a technique to mitigate the accumulation of GHGs in the atmosphere. It refers to the process of removing the atmospheric CO_2 and storing it in long lived pools or sinks so that it is not emitted back into the environment. Carbon sequestration is a practical and immediate process, but the sequestration methods vary in their benefits, costs and effectiveness. A variety of schemes have been proposed for carbon sequestration; these schemes can be grouped into three modes as follows (Figure 5.1):

1) physical processes: these include : i) biomass-related physical sequestration methods (e.g., bio-energy with carbon capture and storage, biochar burial, biomass burial), ii) ocean storage and iii) subterranean injection;
2) chemical processes: these include: i) mineral carbonation, ii) ocean-related chemical sequestrations methods (e.g., ocean basalt storage, ocean acid neutralization and ocean hydrogen chloride removal), iii) industrial use (e.g., making cement by absorbing CO_2 from ambient air during hardening), and iv) chemical scrubbers. Chemical sequestration takes it basis from the reactive nature of carbon dioxide and is based on sorption or chemical reaction; and

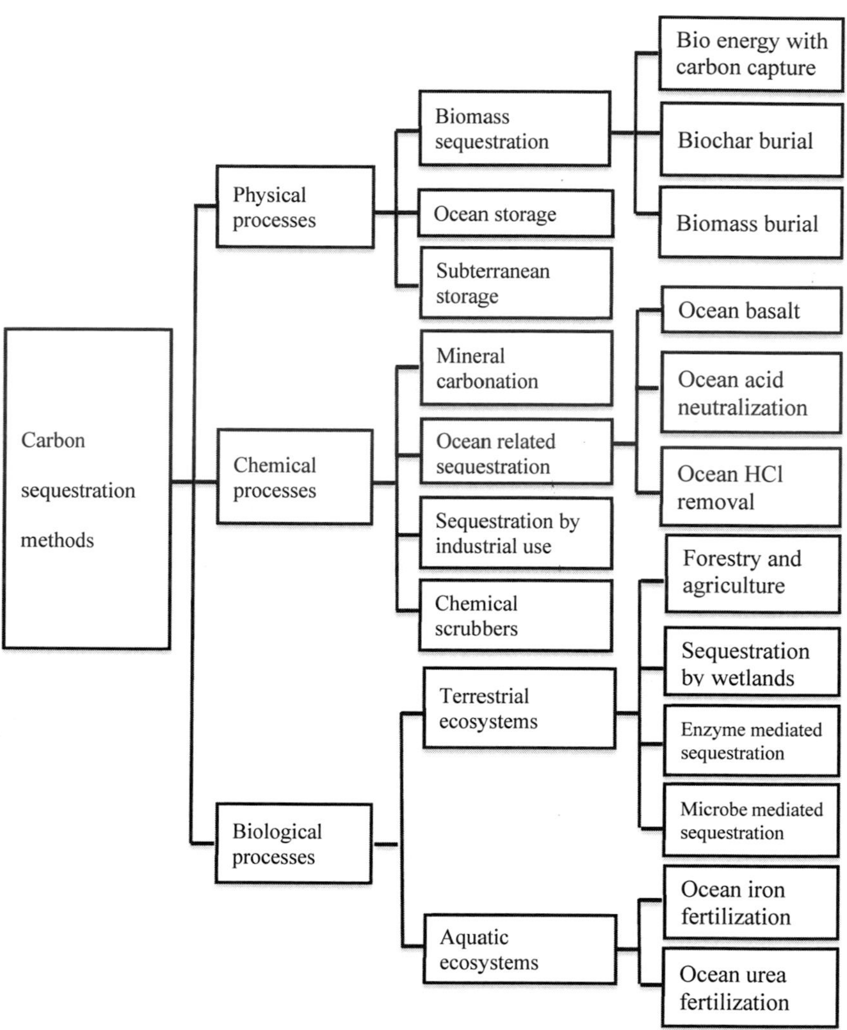

Figure 5.1. The major carbon sequestration methods

3) Biological processes: Biological sequestration (biosequestration) refers to sequestering carbon in biomass, either dead or alive, in terrestrial and/or aquatic ecosystems. Biological processes include forestry and agriculture management practices, reducing emission, enhancing carbon dioxide removal, peat production and sequestration using enzymes. Biosequestration can also be achieved by the use of specific microorganisms in different ecosystems and also by enzyme mediated ways to convert carbon dioxide into bicarbonates. In different ecosystems, biological sequestration can be achieved by ocean iron fertilization, by soil conservation and best management practice in agriculture and forestry, and by soil carbon enhancement.

One of the possible ways to mitigate the GHG effect is to make changes to the global carbon cycle. The primary components of global carbon cycle are ocean, atmosphere, plants and soil and they actively exchange carbon (Prentice et al. 2001). Table 5.1 shows the capacity of different CO_2 sinks. There is a tremendous amount of CO_2 storage capacity in oceans, geological and terrestrial sittings. In this chapter, CO_2 sequestration including ocean, geological, and terrestrial sequestration of CO_2 as well as associated leakage, monitoring and risk management are discussed.

Table 5.1. Capacity of different CO_2 sinks (Hoffman 2009; Zhang and Surampalli 2013)

Sink	Amount (Gt CO_2)
Atmosphere	578 as of 1700 and /766 as of 1999
Ocean ecosystems	
• water column	• 38,000–40,000
• Marine sediments & sedimentary rocks	• 60,000,000–100,000,000
Terrestrial ecosystems	
• Plants	• 540–610
• Soil organic matter	• 1,500–1,600
• Fossil fuel deposits	• 4,000 (not a sink, for comparison purpose only)
Geological storage	
• Saline formations	• > 1,000
• Oil/gas recovery	• 675–900
• Unminable coalbeds	• 3–200

5.2 Ocean Carbon Sequestration (OCS)

Among the various sequestration methods available, oceans have been found to be the most technologically feasible, immediately available and low cost technique for carbon sequestration (see Table 5.1). About 80 percent of carbon dioxide released into the atmosphere could be sequestered in ocean though it may take years to equilibrate with carbonate sediments. In this section, we will discuss a) the principles, b) the most dominant methods (i.e., i) direct injection, ii) carbonate mineral dissolution, and iii) ocean nourishment), and c) impact of OCS.

5.2.1 Principles of OCS

Potential and Capacity. Oceans, which occupy 70 percent of earth's surface, with an average depth of 3800 m, offer the most powerful long term buffer, against the increase in temperature and atmospheric carbon dioxide. The vastness of the ocean provides no practical physical limit to the amount of anthropogenic carbon dioxide that can be stored in the ocean. Comparing the difference between carbon dioxide partial pressure of oceans and the atmosphere, it has been understood that the large amount of carbon dioxide influxes and out fluxes are no longer the same as in preindustrial times. The underlying influx-out flux balance had been skewed by anthropogenic pertubrances to favor influx, where the net ocean uptake is 2 GtC per year, which accounts to 30 percent of total anthropogenic emissions (Herzog et al. 2001). On average, the ocean absorbs 2% more carbon than they emit each year, forming an important sink in the overall carbon cycle.

Principle of Ocean Carbon Sequestration. In ocean dissolved inorganic carbon can be present in any of four forms: dissolved carbon dioxide (CO_2), carbonic acid (H_2CO_3), bicarbonate ions (HCO_3^-) and carbonate ions (CO_3^{2-}). Addition of CO_2 to seawater leads to an increase in dissolved CO_2 (Equation 5.1). Dissolved carbon dioxide reacts with seawater to form carbonic acid (Equation 5.2). Carbonic acid being unstable in seawater rapidly dissociates to form bicarbonate ions (Equation 5.3), which further dissociate to form carbonate ions (Equation 5.4) (IPCC 2005):

$$CO_{2\,(atmos)} \rightarrow CO_{2\,(aq)} \tag{5.1}$$
$$CO_2 + H_2O \rightarrow H_2CO_3 \tag{5.2}$$
$$H_2CO_3 \rightarrow H^+ + HCO_3^- \tag{5.3}$$
$$HCO_3^- \rightarrow H^+ + CO_3^{2-} \tag{5.4}$$

Once in the ocean, CO_2 is transported and/or transformed in two major mechanisms (Zhang and Surampalli 2013):

a) Physical pump. Cold water holds more CO_2 than warm water. Because cold water is denser than warm water, this cold, CO_2-rich water is pumped down by vertical mixing to lower depths. Total dissolved inorganic carbon (DIC) is the sum of carbon contained in H_2CO_3, HCO_3^-, and CO_3^{2-}, but majorly in the form of HCO_3^-. The net results of adding CO_2 to sea water is the generation of H^+ (i.e., lowering pH) and decreases the concentration of CO_3^{2-}.

b) Biological carbon pump (BCP) forcing CO_2 going through the food chain. This is a process whereby CO_2 in the upper ocean is fixed by primary producers and transported to the deep ocean as sinking biogenic particles or as dissolved organic matter. The fate of most of this exported material is remineralization to CO_2, which accumulates in deep waters until it is eventually ventilated again at the sea surface. However, a proportion of the fixed carbon is not mineralized but is instead stored for millennia as recalcitrant dissolved organic carbon (DOC).

The consequence of pathways a) and b) are that ocean surface waters are super-saturated with respect to $CaCO_3$, allowing the growth of corals and other organisms that produce shells or skeletons of carbonate minerals. In contrast, the deepest ocean waters have lower pH and lower CO_3^{2-} concentrations, and are thus under saturated with respect to $CaCO_3$ (Zhang and Surampalli 2013).

Depending on the density of the carbon dioxide in relation to the surrounding water, injected carbon dioxide can either move upward or downward. Drag forces aid in transferring the momentum from carbon dioxide droplets to surrounding water, creating motion in the direction of droplet motion. Eventually carbon dioxide dissolves, making surrounding water denser and then sinks. As the CO_2-enriched water moves, it gets mixed with less CO_2 enriched surrounding water, creating additional dilution and diminishing the density contrast between the CO_2-enriched water and the surrounding water. Carbon dioxide transported by ocean currents undergo mixing and dilution with other water masses along surfaces of constant density, whereas in a stratified fluid, buoyancy forces inhibit vertical mixing (Alendal and Drange 2001).

Although the ocean's biomass represents almost 0.05 percent of the whole, it transforms annually, almost 50 GtC of inorganic carbon to organic. This process is often referred to as biological (carbon) pump, i.e., the aforementioned pathway b). The net effect of pathway b) is that a large amount of carbon is suspended in the water column as dissolved organic carbon (DOC). For example, green, photosynthesizing plankton converts as much as 60 gigatons of carbon per year into organic carbon roughly the same amount fixed by land plants and almost 10 times the amount emitted by human activity. Even though most of DOC is only stored for a short period of time, marine organisms are capable to convert immense amounts of bioavailable organic carbon into difficult-to-digest forms known as *refractory* DOC; this organisms driven conversion has been named the "jelly pump" (Hoffman 2009) and the microbial carbon pump (MCP) (Jiao et al. 2010). Once transformed into "inedible" forms, these DOCs may settle in under saturated regions of the deep oceans and remain out of circulation for thousands of years, effectively sequestering the carbon by removing it from the ocean food chain (Hoffman 2010). There is a tremendous amount of CO_2 storage capacity in marine sediments and sedimentary rocks. It is the natural biological pump that drives the carbon towards the bottom of the ocean. Carbon sequestration methods deal with modifying or accelerating this natural biological pump by adopting various strategies.

5.2.2 *Most Dominant Methods for OCS*

Direct Injection of CO_2. Direct injection of carbon dioxide aims at the artificial acceleration of the natural process of carbon dioxide absorption which occurs through biological pump there by reducing the atmospheric carbon dioxide concentrations. Direct ocean disposal of carbon dioxide refers to the injection of solid, liquid or gaseous carbon dioxide into the mid and deep ocean waters. It was proposed for the first time by Marchetti (1977), for capturing carbon dioxide from combustion of fossil fuels and injecting to the deep ocean.

Properties of CO_2 and Conditions Required for Injecting CO_2. Carbon dioxide will take several forms depending on the characteristics of the injection site and the methodology used for injection. The behavior of carbon dioxide injected into the ocean depends on the physical properties of carbon dioxide and method of release (Song et al. 2005). Carbon dioxide can be injected either as gas, liquid, solid or solid hydrate. Irrespective of the form in which carbon dioxide is injected, it gets dissolved in the sea water with time. The dissolution rate of carbon dioxide is highly variable and depends on certain factors such as the form of carbon dioxide, depth and temperature of disposal and local water velocities.

Carbon dioxide can be potentially released as a gas at a depth shallower than 500 m. In this case, carbon dioxide gas bubbles, being less dense than the surrounding water, will rise to surface at a radial speed of 0.1 cm/hr (Teng et al. 1996). CO_2 can exist as a liquid in ocean at depths roughly deeper than 500 m. At a depth shallower than roughly 2500 m, CO_2 is less dense than sea water, and hence, liquid CO_2 released into the seawater shallower than 2500 m would tend to rise towards the surface. Brewer (2004) observed that a 0.9 cm diameter CO_2 droplet would rise about 400 m in an hour before dissolving completely, and 90% of its mass would be lost in the first 200 m.

Solid CO_2 surfaces being denser than sea water will sink while simultaneously dissolve in the sea water at a speed of about 0.2 cm/hr (Aya et al. 1999). Proportionately, small quantities of solid CO_2 would dissolve completely before reaching the sea floor whereas large masses could potentially reach the sea floor before complete dissolution. CO_2 hydrate refers to a form of CO_2 in which water molecules surrounds each molecule of CO_2. It is normally formed in ocean waters below about 400 m depth. A fully formed crystalline CO_2 hydrate is denser than sea water and dissolves at a speed similar to that of solid CO_2 (about 0.2 cm/hr) (Teng et al. 1999; Rehder et al. 2004). In water colder than 9 °C and at greater depths, a carbon dioxide hydrate film will be formed on the droplet wall, which makes the droplet's radius diminish at a speed of 0.5 cm/hr. Fully formed crystalline CO_2 hydrate being denser than sea water will sink. Liquid carbon dioxide being negatively buoyant at depths greater than 2600 m forms a hydrate skin on water droplet due to ambient temperature and pressure, possess the potential to remove carbon dioxide from the atmospheric reservoir (Haugane and Drange 1992). Pure CO_2 hydrate is a hard crystalline solid and will not flow through a pipe; however a paste-like composite of hydrate and sea water may be extruded, and this will have a dissolution rate intermediate between those of CO_2 droplets and a pure CO_2 (Tsouris et al. 2004). Although formation of solid carbon dioxide hydrate is a dynamic process, the nature of hydrate nucleation in these systems are imperfectly understood (Sloan 1998).

CO_2 diffusers can produce droplets that will dissolve within 100 m of the depth of release. Alternatively CO_2 diffusers can be engineered with nozzles that can produce mm scale droplets, that would produce carbon dioxide plumes that would rise less than 100 m. Hence droplets could be produced that either dissolve completely in the sea water or sink to the sea floor.

There are several techniques for CO_2 Injection. A pure stream of carbon dioxide that has been captured and compressed can be either directly injected to the ocean or deposited on the sea floor. It can also be loaded on ships or transported to fixed platforms and dispersed from a towed pipe to the ocean forming a carbon dioxide lake on the sea floor (Nakashiki 1997). There are several techniques available for the implementation of direct injection of CO_2, such as i) medium-depth (1000–2000 m) sequestration, ii) high-depth (> 3000 m) sequestration, iii) sequestration on the bottom of the ocean, and iv) sequestration at the undersea earth's layer.

The process of switching industrial carbon dioxide emissions directly to the oceanic column below 800 m was studied by Ametistova et al. (2002). The most preferred injection capture would be the dissolution of liquid carbon dioxide using fixed pipeline at depths between 1000–1500 m. Liquid carbon dioxide will be diffused as droplets at this depth. The advantages of this method are that carbon dioxide will be transferred close to the carbonate dissolution boundary at very slow release rates and with reduced environmental impacts.

Carbon dioxide can be sequestered more effectively and for a longer period of time if carbon dioxide is stored in liquid form on the sea floor or in hydrate form below 3000 m depths (Shindo et al. 1995). CO_2 hydrate could be designed to produce a hydrate pile or pool on the sea floor. CO_2 released onto the sea floor deeper than 3 km is denser than the surrounding sea water and is expected to fill topographic depressions, accumulating as a lake of CO_2, over which a thin hydrate layer would form. This hydrate layer would retard dissolution, but it would not insulate the lake from the overlying water, and thus, it would dissolve into the overlying water (Haugan and Alendal 2005). However, the hydrate layer would be continuously renewed through the formation of new crystals (Mori 1998). Laboratory experiments (Aya et al. 1995) and small deep ocean experiments (Brewer et al. 1995) show that deep-sea storage of CO_2 would lead to CO_2 hydrate formation and subsequent dissolution. The time taken for complete dissolution of carbon dioxide in a carbon dioxide lake with an initial depth of 50 m varies from 30 to 400 years, depending on the local sea and the sea floor environment. The dissolution time also depends on the mechanism of carbon dioxide dissolution, properties of carbon dioxide in solution, turbulence characteristics and dynamics of the ocean bottom layers and the depth and complexity of the ocean lake.

Carbonate Mineral Dissolution. One of the methods to accelerate carbonate neutralization is by the natural dissolution of carbonate mineral in sea floor sediments and lands. Sea water acidity caused by the carbon dioxide addition for over thousands of years can be neutralized by this way, which in turn allows oceans to sequester more carbon dioxide from the atmosphere without significant change in the ocean pH and carbonate ion concentration (Archer et al. 1998). Weathering or reaction involving solid carbonate dissolution, in the presence of water and carbon dioxide to form bicarbonate, has the capacity to absorb significant fraction of anthropogenic carbon dioxide input (Sheps et al. 2009). Studies were performed on artificially enhancing mineral carbonate dissolution for carbon dioxide sequestration at an accelerated pace. The largest participants in natural chemical weathering are metal carbonates and their complexes such as calcium carbonate, magnesium carbonate and calcium magnesium carbonate that are especially contained in calcite, limestone and dolomite. It had been observed that enhanced mineral weathering reactions occur in environments with elevated carbon dioxide like decomposing organic rich soils and in the deep ocean.

The partial pressure of carbon dioxide from waste gas streams is several orders higher in magnitude than in the atmosphere. Therefore, it is highly advantageous to place water and mineral carbonate in direct contact with waste gas streams. This would allow the formation of aqueous carbon dioxide and carbonic acid in higher concentrations than in the atmosphere, which in turn elevates the acid concentration and bicarbonate formation, accounting that half to be derived from waste gas carbon dioxide. Rau and Calderia (1999) proposed the reaction of carbon dioxide separated from flue gases with crushed limestone and sea water. The carbonic acid formed as a result of this reaction would accelerate the dissolution of carbonate containing minerals such as calcite, dolomite, limestone and aragonite. The solution containing dissolved inorganic carbon and calcium ions can be released back to the ocean where it would be diluted with additional sea water. Carbonate weathering reactions proceed as follows:

$$CO_{2\,(gas)} \rightarrow CO_{2\,(aq)} \quad (5.5)$$

$$CO_{2\,(aq)} + H_2O \rightarrow H_2CO_{3\,(aq)} \quad (5.6)$$

$$H_2CO_{3\,(aq)} + CaCO_{3\,(solid)} \rightarrow Ca^{2+}_{(aq)} + 2HCO_3^-{}_{(aq)} \quad (5.7)$$

$$\text{Net reaction: } CO_{2\,(gas)} + CaCO_{3\,(solid)} + H_2O \rightarrow Ca^{2+}_{(aq)} + 2HCO_3^-{}_{(aq)} \quad (5.8)$$

Carbonate minerals have been considered as the primary source of alkalinity for the neutralization of carbon dioxide acidity. Kheshgi (1995) suggested promoting the reaction of calcining limestone to form readily soluble CaO since ocean surface waters being over saturated with carbonate minerals. Carbonate neutralization approaches involves the reactions in which limestone reacts with carbon dioxide and water forming calcium and bicarbonate ions in the solution. According to the speciation of dissolved inorganic carbon in sea water, for each mole of calcium carbonate dissolved, there would be 0.8 moles of additional carbon dioxide sequestered in equilibrium with fixed carbon dioxide partial pressure. Adding alkalinity would increase ocean carbon storage in short term and in long term time periods.

Sequestration by Ocean Nourishment. Another method of sequestering carbon in oceans is by ocean nourishment. Ocean fertilization is considered as a potential strategy to mitigate the GHG emissions. Ocean nourishment refers to the introduction of nutrients (e.g., Fe or urea) to the upper ocean so as to increase the marine food chain and to sequester carbon dioxide from the atmosphere. It belongs to the group of geoengineering techniques, which intentionally alters the environment on a planetary scale and thereby mitigates the global warming. Ocean nourishment offers the prospect of reducing the concentration of atmospheric GHGs and of increasing the primary production in ocean. Primary production refers to the process of producing organic compounds from atmospheric or aquatic carbon dioxide through the process of photosynthesis. Majority of the primary production is carried out by microscopic organisms called phytoplanktons and algae. Horiuchi et al. (1995) reported that inorganic carbon concentration in the deep ocean is not in equilibrium with the atmospheric carbon dioxide partial pressure. The surface ocean is considered to be deficient in nutrients since they are consumed rapidly by phytoplanktons. Fertilizing with nutrients promote propagation of phytoplankton and assimilate organic carbon. Consequently this leads to a decrease in the partial pressure of carbon dioxide on the ocean surface, resulting in an increase in drawing carbon dioxide from the atmosphere. The corresponding freshly-added carbon dioxide is fixed by plants like phytoplanktons near the ocean surface through the process of photosynthesis. Once essential nutrients such as N, P and Fe are used up, algal bloom die; they sink, and thereby sequester carbon. Dead phytoplanktons and marine organisms act as carbon dioxide vessels, during the natural biological pump as they sink towards the bottom of the ocean (Fertiq 2004). Microorganisms that feed on this particulate organic matter produces carbon dioxide of which some portion dissolves in ocean and rest ends up as detritus (recalcitrant DOC) that would remain out of circulation for thousands of years.

Ocean Iron Fertilization. Ocean iron fertilization refers to the addition of iron artificially to the water to promote the phytoplankton growth in ocean, which in turn will help in enhancing oceanic carbon dioxide uptake and reducing carbon dioxide in the atmosphere (Denman 2008, Buesseler et al. 2008). Iron is one of the major limiting factors for primary production in oceans. In order to understand the idea of carbon sequestration by ocean, it is good to have knowledge about the marine carbon cycle.

Carbon dioxide is taken by the ocean from the atmosphere and it rapidly dissolves in water forming aqueous carbon dioxide, bicarbonate and carbonate which are referred as dissolved inorganic carbon. In the presence of sunlight and other nutrients like N, P and Fe, phytoplankton will absorb carbon dioxide from seawater to carry out photosynthesis and builds up organic molecules. A portion of organic molecules and dissolved organic carbon will sink from the surface to deep ocean layers by process of mixing and advection (Raven and Falkowski 1991).

Ocean iron fertilization has been suggested as a potential tool to reduce atmospheric carbon dioxide concentrations, since the increase in carbon export due to iron addition corresponds to equivalent storage of carbon dioxide on time scales ranging from few months to thousands of years. The method of applying fertilizers or activated sludge with nutrients will increase the growth of phytoplankton in the ocean and helps in the assimilation of inorganic carbon to produce HCO_3^- and CO_3^{2-}. Martin (1990) published the Iron Hypothesis, suggesting that Fe could be the limiting factor for photosynthesis in ocean, where the concentration of macronutrients is high and chlorophyll is low. The availability of light, nutrients and trace elements are the factors that influence carbon cycling and growth of phytoplankton in ocean. In oceanic region where there is deficiency of Fe, all the macronutrients cannot be used for photosynthesis and hence fertilizing the surface ocean in these regions increases the amount of carbon dioxide used by phytoplanktons for photosynthesis and will increase the primary production and carbon sequestration in the deep ocean. All the carbon taken up by the phytoplankton is not sequestered in the ocean; some portion return to the atmosphere within short time scales. Being the lowest in hierarchy of food chain, phytoplanktons are grazed upon by zooplanktons which in turn are taken by fish and other higher animals. A fraction of carbon goes back to ocean as dissolved inorganic or organic carbon due to the physiological activities of higher animals. Moreover, bacteria also remineralize much of the organic carbon into inorganic carbonate and bicarbonate ions (Denman 2008). It was observed that 50% of exported organic carbon get remineralized during 100 m of sinking, another 2–25% reaches to depths of 1000–1500 m and 1–15% of carbon sinks below 500 m (Powell 2008).

Bakker et al. (2005) explored the changes in the biological carbon uptake and surface water fugacity of carbon dioxide in the iron fertility experiment in Southern Ocean. They studied the effect of carbon dioxide air-sea transfer on dissolved inorganic carbon for patch iron fertilization. Patch fertilization refers to fertilizing an ocean area, measuring a few hundred kilometers for a period of one month to several years. They observed that algal carbon uptake reduced surface carbon dioxide from 4th day onwards. Surface water carbon dioxide decreased at the rate of 3–8 micro atoms per day, thus making iron enriched water a potential sink for atmospheric carbon dioxide. The surface water carbon dioxide and dissolved inorganic carbon decreased at the rate of 32–38 micro atoms after thirteen days. The studies revealed that iron addition made ocean water sinks for atmospheric carbon dioxide, and replenishment of carbon dioxide by air sea exchange was less in comparison to algal carbon uptake. IPCC has predicted that accumulated carbon dioxide emission until the year 2100 would be in the range of 770–2540 Gt. The potential of large scale iron fertilization could be 26–70 GtC for a period of one month. Large scale iron fertilization could share to mitigate carbon dioxide concentration whereas potential for patch fertilization would have only a relative small impact. In situ iron fertility experiments have demonstrated that it will promote the development of algal bloom, buildup of biomass and uptake of inorganic carbon (Boyd et al. 2000; Watson et al. 2000). Four Lagrangian in-situ iron fertilization experiments conducted in Southern Ocean proved that iron addition promoted uptake of inorganic carbon, algal bloom and build-up of biomass (Coale et al. 2004).

Ocean Urea Fertilization. Ocean urea fertilization refers to the process of fertilizing the ocean with urea, the nitrogen rich substance, so as to boost the growth of carbon dioxide absorbing phytoplankton, as a means to combat climate change. It had been proved that efficiency of urea fertilization is dependent on the efficiency of carbon burial and species composition of stimulated bloom (Gilbert et al. 2008). The production of higher phytoplankton biomass can be stimulated by nitrogen fertilization. The desired amount of nutrients required to offset the rising concentration of carbon dioxide in the atmosphere is based on the redfield ratio (PNC ratio: Phosphorous: Nitrogen: Carbon ratio) of the phytoplankton in the ocean. Typical chemical composition of an algal cell is 106 C: 16 N: 1P: 0.0001 Fe. Hence for each unit of Fe added 1,000,000 units of carbon biomass. For each unit of nitrogen that is added to a nitrogen limited region, seven units of carbon biomass can be produced. It was observed that urea uptake was positively correlated with the proportion of phytoplankton composed of cyanobacteria in water. Studies were conducted by Lucas et al. (2007) on rates of phytoplankton production in iron fertilized region in oceans and also the sensitivity of cell size to iron availability through the size fractionated measurements of nitrate, ammonium and urea uptake. In contrast, nitrate uptake is positively correlated with diatom biomass and negatively correlated with urea uptake (Heil et al. 2007; Gilbert et al. 2008). The cellular enzyme urease hydrolyzes urea to ammonium and the enzyme activity is positively correlated with temperature. Diatoms do not excrete inorganic nitrogen, and hence the catabolic end products of urea cycle are returned to the anabolic pathways that produce glutamine and glutamate. A firm response of heterotrophic bacteria to phytoplankton bloom, in terms of biomass production and respiration can be induced by natural iron fertilization (Obernosterar et al. 2008). Urea enrichment leads to enhanced production of cyanobacteria, and picoeukaryotes rather than diatoms (Berg et al. 2001).

5.2.3 Impact of OCS

Changes in Ocean Chemistry due to CO_2 Injection. Anthropogenic carbon dioxide uptake by oceans will lead to severe chemical changes in the ocean surface water environment. The anthropogenic perturbation is greatest in the ocean upper layers where the biological activity is high. Direct injection of liquid carbon dioxide will yield complex dispersion halos due to oceanic density and temperature gradients. Depending on the location, pressure and temperature at the time of injection, carbon dioxide may either get dissolved directly in the sea water or puddle into carbon dioxide puddles or sometimes carbon dioxide hydrate will be formed (Herzog et al. 1997). The most important factor for carbon sequestration is the depth (as described above) that has to be sufficient to keep carbon from surface ocean, and it depends on various factors like ocean current, temperature, weather, patch dissolution and grazing activity (de Baar et al. 2005).

There exists a slow exchange between carbon dioxide and the deep ocean. The surface ocean water with a pH of 8.1 is in equilibrium with atmospheric pCO_2, which is 360 μatm. At a typical seawater pH of 8.1 and salinity of 35, the dominant dissolved inorganic carbon species is HCO_3^- with only 1% in the form of dissolved CO_2. It is the relative proportions of the dissolved inorganic carbon species that control the pH of seawater on short-to-medium timescales. It had been predicted that, with the consumption of fossil fuels at the present rate, by the year 2030, pH of ocean surface water will decrease to 7.8 with the atmospheric carbon dioxide doubling to 700 μatm. However, injection of carbon dioxide of 1300 GtC would decrease the pH by 0.3 units, with a corresponding decrease of pH by 0.5 units in the deep ocean. Ocean general circulation models have been used to predict the changes in ocean chemistry as a result of the dispersion of injected carbon dioxide. It had been predicted that injection of 0.37 Gt CO_2/yr for 100 years would produce a pH change of 0.3 units over a volume of sea water equivalent to 0.01 percent (Wickett et al. 2003).

It had been observed that 80% of carbon dioxide could be sequestered permanently within a residence time of 1000 years; up to 150–300 GtC can be absorbed with a pH change of 0.2–0.4. The outgoing of remaining 20% of injected carbon would occur within a time period of 300–1000 years (Cole et al. 1993). The lateral transport of re-mineralized carbon dioxide originating from shallow layers of ocean represent a powerful route for carbon sequestration in deep sea which needs to be investigated (Hoppemma 2004).

The most obvious consequence of injection of carbon dioxide is that carbon dioxide alters the food web by changing the partitioning of energy between metabolic processes (Angel 1997). The magnitude of ocean sequestration and its impact on environment depends on the duration of exposure, the organism's compensatory mechanism, energy requirement and mode of life (Adams et al. 1997). The low pH is harmful to zooplankton, bacteria, bottom dwelling plants and animals due to limited mobility. The potential effects of liquid carbon dioxide injection on deep sea foraminiferal assemblage on California margin was studied by Ricketts et al. (2009); the results suggested that foraminiferal diversity decreased due to carbon dioxide emplacement. It was also observed that release of liquid carbon dioxide cause an increase in the dissolution of calcareous taxa in sediments directly below the carbon dioxide pool. It was also found to cause significant mortality in benthic foraminifera, since increased CO_2 concentrations cause metabolic changes such as intracellular acidosis and respiration stress.

The effect of OCS to the physiological changes in the marine organisms and to the ecosystems should be taken into consideration while studying the impacts of OCS. The adverse effect on the diverse fauna that resides in the deep ocean and in sediments is also one of the most alarming consequences of ocean carbon dioxide sequestration, which can lead to changes in ecosystem composition and functioning. The dissolution of carbon dioxide can lead to dissolution of calcium carbonate present in the sediments or in the shells of the microorganisms. Changes in the productivity pattern of algal/heterotrophic bacterial species, biological calcification or decalcification and metabolic impacts on zooplankton species are the observed consequences of lowered sea water pH. Changes in the pH of the marine environment will affect the carbonate system, nitrification, speciation and uptake of nutrients (Huesemann et al. 2002). Although it had been stated that CO_2 injection has ecosystem consequences, no controlled ecosystem experiments have been performed in the deep ocean, and hence, no environmental thresholds have been identified.

Impact of Ocean Fertilization Experiments. Ocean fertilization has drawbacks in the sense that it can affect ocean ecosystems in the long run and change the plankton structure. The potential for ocean nourishment experiments for mitigation of GHGs is controversial since the magnitude and direction of carbon stored remains uncertain, and the verification of carbon stored is impossible (Gnanadesikan et al. 2003). The most crucial limitation is the production of methane gas triggered by the sinking of organic matter as a result of large scale iron fertilization. Moreover, it can induce the production of GHGs such as nitrous oxide, methane, dimethylsulphide, alkyl nitrates and halocarbons (Jin and Gruber 2003; Turner et al. 2004). Production of other GHGs and their outgassing could partially offset the carbon drawdown from the atmosphere into the ocean. Changes in the marine ecology and biogeochemical changes are induced by large-scale iron fertilization experiments (Chisholm et al. 2001). Experimental studies have revealed that significant microbial community structural modifications occur in response to 7% increase in carbon dioxide concentration (Sugimori et al. 2002). It also alters the partitioning of energy between metabolic processes. Ocean fertilization can also induce ocean acidification and alter the physical properties of ocean.

Ocean iron fertilization has some negative effects such as the development of toxic algal blooms, unforeseeable changes in food web and ecosystems, anoxia due to remineralization sinking organic matter, increased production of nitrous oxide which can lead to the death of marine life (Denman 2008). More research has to be undertaken before using iron fertilization on a large scale due to substantial uncertainties and effectiveness associated to it (Buesseler et al. 2008). Limitations of ocean sequestration include distraction of energy usage in an efficient way, alternate energy generation from renewable sources and tampering the ecological processes. Urea fertilization is likely to cause eutrophication impacts which include development of hypoxic or anoxic zones and alteration of species leading to harmful algal blooms (Anderson 2004). Sinking of algae to the deep ocean cause hypoxia upon their decomposition and hence is responsible for fish kills. Nitrogen loading can bring forth a shift in the marine community of coral reef directed towards algal overgrowth of corals and ecosystem disruption (McCook et al. 2001). In addition, what is the contribution due to the inedible forms of DOCs or carbonate compounds that are promoted by nutrient addition via biological pumps and the related mechanisms are not fully understood yet, rending more studies about the real contribution of these mechanisms to CO_2 storage.

5.3 Geological Carbon Sequestration (GCS)

Earth's subsurface stores carbon in coals, gas organic rich shales and carbonate rocks. Rocks and fossil fuel deposits such as coal measures, oil reservoirs and oil shales act as a greenhouse sinks which can hold a substantial amount of carbon dioxide so that less carbon is released to the atmosphere. GCS refers to the injection of carbon dioxide captured from industrial sources into porous rocks in earth's crust so as to isolate the CO_2 from the atmosphere. The large underground accumulations of CO_2 provide us with the huge potential of GCS in underground in right geological conditions for millions of years.

Although GCS is not a new technique, considerable research and development of GCS has been performed since the late 1990s. After more than one decade research, a significant amount of information has been accumulated (e.g., see the special issue published in *Environmental Science and Technology* on Jan. 2013). This section serves as a brief overview of GCS, including a) principles and mechanisms of GCS, b) basic requirement for GCS, c) GCS in various geological formations, d) monitoring and verification techniques for GCS, and e) cost associated with GCS.

5.3.1 Principles and Mechanisms of GCS

Geochemical Reactions at GCS sites. Important geochemical reactions under GCS conditions include: a) supercritical CO_2 ($scCO_2$) dissolution into brine, b) acidified brine induced reactions (i.e., dissolution of pre-existing formation rocks and caprock), c) wet $scCO_2$ driven interactions (e.g., precipitation of secondary mineral phases and surface reactions that affect wettability of rocks), d) $scCO_2$-wellbore interactions, and e) partition of organic contaminants in GCS sites (Burant et al. 2013; Jun et al. 2013). All these reactions are induced by the injection of $scCO_2$ because the pH at GCS sites is a function of the distance from and time after the injection. For example, at pH of 3–6 near the injection well, carbonates have a much higher dissolution rates than other minerals, which stimulates the release of several cations (e.g., Ca^{2+}, Mg^{2+}, Fe^{2+}) and increases porosity and permeability of the sites.

These reactions affect fate/transport and behavior of sc- or aqueous-CO_2 and the associated environmental risks of GCS sites (see below and Jun et al. 2013). These reactions are affected by many factors, such as temperature, pressure, the salinity of formation fluids, microbial communities, transport processes of injected $scCO_2$ and impurity gases. Usually, reaction rates increase with temperature and pressure, both of which are a function of injection depth (i.e., 33 °C/km and 99 atm/km). Microbial communities at GCS sites can mediate surface reactions, provide exopolysaccharides (EPS), and affect redox reactions. For example, Fe(III) reducing microorganisms may promote the increase in pH of the brine, leading to the secondary mineral precipitation (amorphous silica, clay minerals, halite, and carbonates), and causing a decrease in permeability at pore throats of the sites.

Trapping Mechanisms. There are mainly three mechanisms by which carbon dioxide can be sequestered in geological formations, namely physical trapping (e.g., due to the capillary effect), chemical (mineral) trapping, and hydrodynamic (solubility) trapping (IPCC 2005; Jun et al. 2013). Physical trapping refers to immobilizing carbon dioxide in gaseous or supercritical phase in geological formations, which can be either static trapping in structural traps or residual gas trapping in porous structures. Chemical trapping in formation fluids by dissolution or by ionic trapping, in which the injected carbon dioxide gets dissolved and react with minerals in formations or get adsorbed to mineral surface (adsorption trapping). In hydrodynamic trapping, upward migration of carbon dioxide at low velocities will lead to trapping in intermediate layers.

It is important to understand the phase behavior of carbon dioxide at specific temperature and pressure for the maximum utilization of available storage space within the reservoir for the injection of carbon dioxide. Buoyancy?? trapping is a mechanism by which carbon dioxide can be injected near the base of the reservoir or in the down deep part of the dipping storage reservoir. In residual gas trapping carbon dioxide while migrating through the reservoirs get entrapped as small bubbles of gas either due to buoyancy drive or injection pressure. In solubility trapping, when carbon dioxide gets dissolved in water becomes heavier than native brine in aquifers and thus in turn allows dissolved carbon dioxide to sink by advection. CO_2 can also be locked up in a solid mineral or dissolved bicarbonate mineral phase preventing its migration to the surface. Abandoned mines offer potential for carbon dioxide storage with added benefit of adsorption of carbon dioxide into coal remaining in mined out area (Piessens and Dusar 2004).

5.3.2 Basic Requirement for GCS

Site Selection and Characteristics for Geological Repositories. The most important step to ensure the efficient GCS for a long period of time is the characterization of potential storage sites. Once the site is selected for GCS, it is mandatory to check for the feasibility of GCS operation based on geological models. Particular attention should be given for location features, faults, thickness, quality, distribution of cap and reservoir rocks. Behavior of carbon dioxide plume, designing, positioning and management of injection bore holes have to be studied using model and plume simulations. Risk assessment associated with features, events and processes that might occur within or external to the storage complex need to be evaluated. Measurement and monitoring of the function of the system along with verification should be carried out in line with established baseline conditions prior to carbon dioxide injection. Any variations from expected site performance pose a risk such as leakage or induce damage or disturbances to other resources. Geological storage sites in general should have adequate capacity and injectibility, sealing cap rock and geological environments to contain the integrity of the storage site. Basin characteristics like basin suitability, basin resources, industry maturity and infrastructure should be considered while assessing criteria for basin suitability (Bachu 2003).

Properties of CO_2 Required for GCS. Carbon dioxide has to be compressed to a dense fluid state so as to store in geological formations. Depending on geothermal gradient, carbon density will increase with depth. Carbon dioxide has the density to remain in subsurface through physic-chemical immobilization reactions. In geological formations open fractures, pore spaces and cavities which are filled with fluid are displaced with carbon dioxide by injection of carbon dioxide either by reacting with fluids or with mineral grains. Hydrocarbons, gases and fluids containing carbon dioxide can remain trapped in oil gas fields for millions of years (Bradshaw et al. 2005). The rate of fluid flow in an injection well depends on many factors such as fluid phases present in formations. A single miscible fluid of natural gas and carbon dioxide are formed if carbon dioxide is injected into a gas reservoir where it forms a supercritical dense liquid phase in deep saline formations.

5.3.3 Carbon Sequestration in Various Geological Formations

Carbon dioxide is stored in geological formations by physical or geochemical trapping reactions depending on formation types and fluid properties. Geological repositories such as aquifers, petroleum fields and carbon deposits can be considered as the first long range opportunity for massive carbon sequestration. In hydrocarbon reservoirs with little of water encroachment, pore volume previously occupied by oil or natural gas will be occupied by injected carbon dioxide. However in open hydrocarbon reservoirs, pore space available for carbon dioxide storage will be less due to capacity reduction caused by capillarity and local effects (Stevens et al. 2001). Carbon plunge and coal bed methane formations give the potential for carbon sequestration. High depth water formations and saline water formations are possible storage mechanisms for carbon dioxide storage. Carbonate rocks such as limestones and chalks also represent a sink of carbon on earth's surface. Geological structural and stratigraphic traps have demonstrated their ability to seal and store hydrocarbon liquids; they also pose the ability to store carbon dioxide and other non-hydrocarbon gases. GCS in different geo-formations are briefly summarized below.

Carbon Storage in Oil and Gas Reservoirs. Carbon dioxide can be stored in depleted oil and gas reservoirs since they do not have leakage paths. Moreover, the geological structure and physical properties are understood due to the oil and gas excavation industries. Abandoned oil and gas fields have more advantages for geological storage due to various reasons. The oil and gas that have already been trapped proves the integrity and safety along with the extensively studied and characterized geological and structural properties. Movement, displacement behavior and trapping of hydrocarbon can be predicted using already developed computer models. The global estimate of oil reservoir storage capacity varies from 120–400 Gt carbon dioxide, whereas it to accounts to 800 Gt carbon dioxide for gas reservoirs (Freund 2003).

Carbon Sequestration with Enhanced Oil Recovery. Carbon dioxide injection for enhanced oil recovery in production wells is the second largest enhanced oil recovery technique after steam flooding (IEA 2005). The mass objective of enhanced oil recovery is the production of maximum oil with minimum carbon dioxide. Carbon dioxide is an ideal gas for accessing oil that cannot be produced under natural pressure or pumping. Carbon dioxide is an excellent solvent for hydrocarbons in dense and liquid phase. Various technical and economic variables such as oil density, viscosity, minimum miscibility pressure, microscopic sweep effects and formation of vertical and lateral heterogenecities should be taken into consideration before the selection of enhanced oil recovery technologies. The global capacity for geological storage of carbon dioxide by carbon dioxide accounts to 61-123 Gt carbon dioxide (Nguyen 2003, IPCC 2005).

Carbon dioxide is injected into the reservoir where it is miscible or nearly miscible pressure with respect to oil, which in turn makes the oil to swell and become less viscous. Although carbon dioxide enhanced oil recovery is effective for lighter oils when combined with thermal techniques it can improve the production from heavy oils. Enhanced oil recovery mainly consists of three steps. The primary production step is the one in which carbon dioxide is injected into reservoirs. In the secondary recovery phase, water is injected to push out oil out of production wells after primary production. In the tertiary recovery phase recoverable oil after the water flood phase is targeted by carbon dioxide. The water alternating gas technology makes use of recovering maximum amount of oil by guiding injected carbon dioxide through parts of the field where recoverable oil remains. The final step of water alternating gas method is the injection of water so as to flush out recoverable oil and carbon dioxide. However, carbon dioxide enhanced oil recovery is limited to oil fields deeper than 600 m, where 20–30% of the original oil is recovered and where primary and secondary production methods have been applied (Goldberg and Slagle 2009; Matter et al. 2009). An incremental oil recovery of 7-23% of the original oil in place can be obtained by using carbon dioxide in enhanced oil recovery (Moritis 2002).

Depending on characteristics of hydrocarbon and reservoir performance, enhanced oil recovery can increase total recovery of an average field as much as fifty percent. Recovery of oil ranges from 25-100% depending on geology of the oil field and oil type. The incremental recovery of oil will be high if the hydrocarbon is light. It has been suggested that carbon dioxide enhanced oil recovery can increase long term conventional oil supply substantially (Mathiassen 2003). To estimate the potential benefits of carbon dioxide enhanced oil recovery projects, detailed field-by-field assessments are needed. Carbon dioxide storage in miscible enhanced oil recovery ranges from 2.4-3 tonnes CO_2 per tonne of oil produced. Project cost for enhanced oil recovery by carbon dioxide depends on various factors such as size of the field, spacing pattern, location and existing facilities. Although carbon dioxide enhanced oil recovery is a mature technology, the main techno-economic challenges include improving sweep efficiency in the case of formation heterogenicities, handling off shore environment, retrofitting surface facilities to handle corrosive fluids, and developing infrastructure to minimize the cost of carbon dioxide delivered for various projects (Gozalpour et al. 2005).

GCS with Enhanced Gas Recovery. Depleted fields can be repressurised by injecting carbon dioxide to increase gas recovery and reduce drain down related subsidence. Carbon dioxide being denser than natural gas flows downward, leading to gravity stabilized displacement. Carbon dioxide being less mobile than methane allows a stable displacement and being more soluble than methane delays break through. The characteristics that make depleted oil and gas fields suitable for GCS are their readily availability and extensive geologic and hydraulic assessment from oil and gas operations. The existing infrastructure like wells, surface facilities and pipelines make it practically feasible for adoption of this technology. The carbon thus sequestered can be stored for extended periods of time in these reservoirs due to the presence of sealing mechanisms. Carbon dioxide can be stored in a dense phase in natural geological formations at the depth of > 600 m. At depths below 800–1000 m, carbon dioxide exists either in liquid or supercritical state due to ambient temperature and pressure, which occupies smaller volume than gaseous state and hence efficient utilization of underground storage space.

The worldwide storage capacity of depleted oil and gas fields range between 675–1200 Gt carbon dioxide. Initial screening of depleted gas fields suggests that a worldwide storage potential of 800 Gt in carbon dioxide at a cost of US dollars of 120 per tonne of carbon dioxide (Stevens et al. 2000. Approximately 0.035–0.05 tonne of methane can be recovered for each tonne of carbon dioxide injected. Carbon dioxide enhanced gas recovery has not yet become a demonstrated technology and hence needs efforts before it becomes established. The only carbon dioxide enhanced gas recovery project that has been undertaken is K12B injection offshore project in Netherlands (Dreux 2006).

GCS with Enhanced Coal Methane Recovery. The technique of enhanced coal methane recovery utilizes the recovery of methane released by mining of gassy coals. The mechanical shock caused by the drilling allows the increased permeability of the coal fraction and also desorption of weakly bound methane, the latter is used for power generation. In unmineable coals water is injected to wells drilled into coal seams at high pressure so as to fracture and mechanically shock coal to release methane. In coal methane recovery, injection of nitrogen gas is practiced to displace any additional methane present in micropores and fractures. In carbon dioxide enhanced coal bed methane recovery nitrogen is replaced with carbon dioxide due to the reason that chemical bonding of carbon dioxide with coal will stimulate methane gas production along with inhibition of displaced methane to production well due to reduced permeability and increased plasticity (Korre et al. 2009; Freidmann et al. 2009). Coal can absorb 2 moles of carbon dioxide for every mole of methane that is initially contained.

Coal beds contain methane absorbed in its pores. Injection of carbon dioxide to unmineable coals, i.e., the coal seams that cannot be commercially exploited due to higher depth or too thin nature, enhance production of coal methane recovery and provide alternative storage mechanism for carbon dioxide. The development of technology for coal bed methane recovery depends on depth, coal rank, permeability and configuration of geological layers. Although permeability of formations varies from few millidarcies to thousands of millidarcies, lighter fractions require hydraulic fractioning for commercial production of methane. The geological factor that has to be considered for meeting the coal bed reservoirs for enhanced coal bed methane recovery are depth of coal seam, pressure and temperature parameters, composition, ash content, local hydrology, dewatering ability, lateral continuity, thickness of coal seam and minimum folding (Shi and Durucan 2005). The main technology gaps that have been addressed by the ongoing research and development projects include interaction between carbon dioxide and coal, chemical interaction of carbon dioxide with in-situ water, impact of heterogenicities, cap rock integrity, monitoring technologies and field wide cross well and well bore monitoring technologies. According to IEA-GHG (2006), the most important criteria that has to be taken into consideration while selecting area for carbon dioxide enhanced coal bed methane recovery are adequate permeability, coal geometry, simple structure, homogeneous and confined coal seam, depth (down to 1500 m), suitable gas saturation conditions and ability to dewater the formations.

Carbon Storage in Sedimentary Rocks. Rocks can be considered as a potential option for GCS, since the open pores within the rocks can be filled with gas, water or oil. Among the different types of rocks, available sedimentary rocks can be considered as the most promising hosts for storing carbon dioxide. Most of the world's oil and gas fields and underground water supplies are hosted by them. Sedimentary rocks are formed from the sediments by erosion from rocks, biogeochemical deposition or combination of both processes, and these sediments get accumulated and buried over time. Permeability of rocks refers to the interconnectivity between pores which permits the fluids (gas, water, oil) to migrate through the rocks. Permeability is an important characteristic of the reservoir, which is very crucial for the successful storage of carbon dioxide. Carbon dioxide can also be injected into igneous and metamorphic rocks such as basalts, serpentines and ophiolites, which contain minerals reactive to carbon dioxide.

GCS in Deep Saline Aquifers. Deep sedimentary rocks are rocks that are saturated with formation water or brines containing high concentration of dissolved salts. Aquifers are layers of sedimentary rocks that can be either open or confined. An aquitard refers to a layer of shale rocks from which no water can be produced but has enough porosity so that it allows water to flow on a geological time scale. Water in deep sedimentary basins is confined by overlaying or overlying aquitards with a high content of dissolved solids making it unsuitable for human consumption. These confined aquifers with favorable alternative applications have been proposed for carbon dioxide storage. Injected carbon dioxide can be trapped in saline aquifers during severed phases, either as a plume at the top of the aquifer in stratigraphic and structural traps in its free phase or as bubbles trapped in pore space, dissolved in aquifer water, or as precipitated carbonate mineral from reaction between carbon dioxide and aquifer water and rocks. It had been studied that 29% of injected carbon dioxide with a density being lower than brine dissolves in brine, the remainder floats on the top of brine and accumulates below the cap rock. Carbon dioxide can be stored over a period of thousands of years (Bachu 2000). Modeling studies reveals that carbon dioxide would flow and spread below aquifer cap rock, which may extend to 10−100 s of square kilometers depending on aquifer porosity, permeability of cap rock and volume of carbon dioxide injected (Saripalli and McGrail 2002). The main technology challenges associated with this cap rock integrity and up scaling of seal characteristics while injecting large volumes of carbon dioxide are developing accurate simulation models, geochemical and geomechanical modeling of reactive transport of carbon dioxide. Although there is no adequate literature to support the geological storage capacity of deep saline formation, the IPCC (2007) assessed the capacity to be at least 1000 Gt carbon dioxide.

Other GCS Options. There exist some other GCS options which are under preliminary stage of research, such as mineral carbonation, acid gas injections and storage in salt caverns. The principle behind mineral carbonation is the reaction between grounds Mg/Ca silicate so as to form solid carbonates. Periodites and serpentines are the most preferred rocks for mineral carbonation because of their worldwide occurrence and content of calcium and magnesium. It had been estimated that 1.6-3.7 tonnes of Silica need to be mined for each tonne of carbon dioxide sequestered by this approach. Acid gas injections provide a commercial GCS analogue. Acid gas, which is a mixture of hydrogen sulphide and carbon dioxide along with minute amounts of hydrocarbons from petroleum production and processing facilities, has been injected into geological formations. Although the main purpose is to dispose hydrogen sulphide, simultaneously significant quantities of carbon dioxide are also sequestered. Basalts also possess some potential for mineral trapping of injected carbon dioxide in which carbon dioxide may react with silicates in the basalt to form carbonate minerals (McGrail et al. 2003). Salt cavern storage is a mature technology for underground gas storage; however they loss volume due to salt creep in course of time. Though salt caverns have to purge of stored contents when decommissioned they can be used as a temporary buffer store. One of the main limitations of the salt caverns, despite their high injectivity, is the low capacity and shallow depth. Storage of carbon dioxide in salt cavern is advantageous in that it include high capacity per volume, efficiency and injection flow rate.

5.3.4 Monitoring and Verification Techniques for GCS

Monitoring is important for addressing various issues related to carbon sequestration. It is important to measure injection well conditions, injection rates, well head and formation pressures as well as to verify the quantity of injected carbon dioxide stored by various mechanisms. It is also highly important for optimizing the efficiency of storage projects, utilization of storage volume, detecting any leakage or any seepage associated with mitigation actions. Appropriate monitoring techniques demonstrate that carbon dioxide remains contained in the intended storage formations. It is also essential for detecting microseismicity associated with storage projects, measuring surface fluxes of carbon dioxide and designing and monitoring remediation activities (Benson et al. 2004). Baseline parameters for storage site can be established by monitoring to ensure that carbon dioxide induced changes are recognized with calibrating and confirming performance assessment in detail (Wilson and Monea 2004).

5.3.5 Cost Estimates for GCS

The major elements for costs of geological storage are drilling wells, infrastructure and project management. The cost for geological storage is site specific and leads to high degree of variability. Cost mainly depends on the type of storage option, location, depth and characteristics of storage reservoir formation. Onshore storage cost depends on location and other geographic factors. The other factors that add to carbon dioxide storage are in field pipeline establishment costs, facilities for handling produced oil and gas, remediation costs for abandoned wells, manpower, maintenance and field costs, costs for licensing, geological, geophysical and engineering feasibility study for site selection, reservoir characterization and evaluation before reservoir storage. Characterization cost varies according to the site, depending on the prevailing data available, geological complexity of storage formations and risks of leakage.

5.4 Terrestrial Carbon Sequestration (TCS)

IPCC predicts that the terrestrial carbon sink will continue to sequester up to 5–10 Gt C per year by the end of the twenty-first century (Houghton et al. 2001). Usually, TCS uses photosynthesis–part of the natural carbon cycle–to create organic matter that is stored in vegetation and soils, which differs from CO_2 mitigation technologies that focus on capturing and permanently storing human-generated emissions. There are a variety of options for TCS, e.g., restoring mined lands, afforestation, reforestation, rangeland improvement, improved tillage practices, and wetlands restoration (NETL 2010). This section describes TCS by terrestrial ecosystems (e.g., agricultural soils, wetlands, forests) and microbes/enzymes in these ecosystems.

5.4.1 Carbon Sequestration by Agricultural Management Practices

Agriculture occupies almost 35% of global land area (Betts et al. 2007). Agricultural soils are being advocated as a possible sink to sequester atmospheric carbon dioxide in terrestrial biosphere to partially offset fossil fuel emissions. Soil carbon sequestration refers to the process of transferring atmospheric carbon dioxide into soil carbon pool, either by humification of photosynthetic biomass or formation of secondary carbonates, where it is held in a relatively permanent form. Carbon dioxide emissions from agriculture results from factors that affect changes in soil carbon reserve like oxidizing soil organic carbon due to soil disturbances, carbon dioxide used by the use of fossil fuels for the production of fertilizer and pesticides, and machinery used for cultivating land. The breaking up of agricultural lands in most regions leads to the depletion of carbon stocks in soil.

Of the several approaches existing for increasing the carbon sequestration in soils, most feasible can be by either increasing the carbon input or decreasing the decay. Appropriate and improved management practices to improve the organic matter content of top soil and to reduce the decomposition rates can turn soil into potential carbon sinks. Adoption of good management practices such as reduction in left over land fallows, use of direct drilling, incorporating legumes and grasses in crop rotations, use of high intensity short term and rotation grazing, planting of windbreaks and conversion of marginal farmlands to perennial grasses helps in increasing the long term carbon restoration capacity of the soil (Forge 2001). Introduction of cover crops in arable lands as a means of increasing plant inputs helps in increasing the overall carbon content and also increasing fertility of soil. Cropping intensity with rotation of winter crops can add to the amount of carbon biomass returned to soil, when compared to monoculture and hence can increase soil organic carbon sequestration (Franzluebber et al. 1994).

TCS in agricultural soils can be attained by changing the tillage practices (Kern and Johnson 1993; Reeves 1997). Tillage breaks up the soil aggregates and exposes organomineral complexes to decomposition. No tillage will reduce the release of soil carbon, by exposing young and labile organic matter to microbe decomposition and reducing turnover of soil aggregates (Kern and Johnson 1993; Paustian et al. 2000; Freibauer et al. 2004). Conversion of croplands into grasslands is another most effective option for carbon mitigation which also implies putting surplus arable land into long term alternative land use suitable for climate change abatement (Smith et al. 2001; Vleeshouwers and Verhagen 2002). Carbon sequestration in soils can be accomplished by changes in agricultural practices like effective use of pesticides, fertilizers, farm machinery and conservation tillage, which may lead to an increase in soil organic carbon, yield and organic matter addition to the soil (West and Marland 2002; NETL 2010). The reclamation of degraded and poorly managed lands for carbon sequestration with use of fossil fuels, biosolids, and organic waste from sewage treatment facilitates to improve soil quality (Palumbo et al. 2004). However, it also increases the soil microbial activity and contributes towards GHG production.

5.4.2 Carbon Sequestration by Wetlands

Wetlands can be considered as GHG sinks as carbon dioxide removed from the atmosphere is stored in soil carbon pool. Wetlands occupy 5% of the earth's surface area and are characterized by waterlogged or standing water conditions, during one part of the year. The unstable water level makes them dynamic ecosystems with high productivity. The carbon pool of wetlands accounts to 770 GtC, which overweighs the total carbon pool of farms, temperate and rain forests. The presence of elevated water table, higher productivity and lower decomposition rates are characteristics of these soils, which enable them to store significant amounts of carbon. The slow diffusion of oxygen and low temperature in these soils will offer a reduced environment which facilitates long term storage of carbon dioxide. The carbon sequestration potential of different ecosystems is presented in Table 5.2. The functioning of wetlands either as a GHG source or sink is dependent on the difference between the greenhouse equivalents of carbon dioxide taken up and methane released (Rudd et al. 1993). However, methane emissions from coastal and estuarine wetlands, pocasins and playas have been found to be lower.

A variety of management practices will help in improving the ability of wetlands to sequester carbon, such as allowing the growth of natural vegetation, controlling fires and deep burns, controlling drainage, land and water management practices that lead to dewatering of wetlands and oxidation. The carbon sequestration potential of lakes and swamps and restoration of wetlands was studied by Duan et al. (2008). It was observed that mangroves possessed the highest carbon sequestration rate than coastal salt marshes, freshwater and peatlands. Coastal marshes and mangrove ecosystem have greater potential to sequester carbon at higher rates, due to the organic sediments they accumulate continuously over a long period of time. Hence restoration and protection of these ecosystems should be given due importance.

Table 5.2. Carbon sequestration potential of different ecosystems

Ecosystem	Location	C-sequestration potential	Reference
Permafrost peatlands	Canada	5.5 Tg C/ha/yr	Tarnocai et al. 2005
Permafrost peatlands	N. America	6.6 Pg C/ha/yr	Tarnocai et al. 2005
Permafrost and non permafrost peatlands	Global	55 Pg C/ha/yr	Maltby and Immirzi 1993
Freshwater mineral soil	Canada	4.6 Pg C/ha/yr	Tarnocai et al. 1998
Freshwater mineral soil	Global	39 Pg C/ha/yr	Bridgham et al. 2006
Tidal marsh	Canada	0.09 Pg C/ha/yr	Bridgham et al. 2006
Tidal Marsh	N. America	4.8 Pg C/ha/yr	Bridgham et al. 2006
Tidal marsh	Global	4.6 Pg C/ha/yr	Chmura 2003
Freshwater marsh	China	811.23 TgC/a	Zhao et al. 2002
Peatlands	China	1345.85 Tg C/a	Ma et al. 1996
Forest wetlands	China	1137.79 Tg C/a	Tan and Zhang 1997
Lakes in Eastern China	China	1056.49 Tg C/a	Duan et al. 2008
Temperate forests	Global	12000 g C/m^2	Oelbermann et al. 2004
Tropical forests	Global	8300 g C/m^2	Oelbermann et al. 2004
Boreal forests	Global	15000 g C/m^2	Oelbermann et al. 2004
Tundra	Global	12750 g C/m^2	Oelbermann et al. 2004
Wetlands	Global	72000 g C/m^2	Oelbermann et al. 2004
Deserts	Global	10500 g C/m^2	Oelbermann et al. 2004

5.4.3 Carbon Sequestration by Forests

Forests possess great potential for carbon sequestration since their biomass accumulates carbon over decades. Forests occupy almost 30% of land area of earth's surface and stores almost 120 GtC. Forests account for 90% of the annual exchange of carbon between the atmosphere and the land (IPCC 2000). Forests add to the reduction of carbon dioxide from the atmosphere as long as there is net productivity. The four components of carbon storage in forest ecosystem are plants, trees growing on forest floor, detritus and leaf litter on forest floor and forest soils. Trees absorb carbon dioxide, which they utilize for photosynthesis in presence of light, the major part of which goes for the formation of cellulose. Carbon accumulated in the leaves may return to the atmosphere, after a period of time when they fall and decompose, whereas carbon is stored in the woods for a longer period of time. The period, for which carbon remains locked in the woods depend on the tree species, growing conditions and forest management practices. Once the life cycle is over, biomass becomes a component of food chain and eventually enters soil as soil organic carbon. Incineration of biomass returns a portion of carbon dioxide back to the atmosphere and it enters the carbon cycle. One of the possible methods to mitigate the accumulating carbon dioxide in the atmosphere is the collection and storage of carbon in growing trees, reforestation and afforestration or by recycling carbon through biomass fuels.

Carbon sink can be maintained or increased by reforestation, afforestration and more environment friendly logging operations, since carbon is locked up in the forest during the growth process. Rehabilitation of degraded or logged over forests not only increase carbon sequestration but serve as potential source of timber. Land use changes, afforestration and forestry activities are widely recognized as strategies to mitigate GHG emissions (Moulton and Richards 1990; Graham et al. 1992). A carbon sequestration strategy was proposed by Zeng (2008) in which dead or live trees are harvested, either by collection or selective cutting and buried in trenches or stowed away in above ground shelters. The sufficiently thick layer of soil provides a largely anaerobic condition which prevents the decomposition of buried wood. Since a large flux of carbon dioxide is constantly being assimilated by photosynthesis into world's forests, cutting of its return pathway to the atmosphere form an effective carbon sequestration strategy. The agroforestry system has carbon storage capability in trees and soil and is an environment friendly strategy to offset immediate effect to GHG emission associated with deforestation and shifting cultivation (Dixon 1995; Nair and Nair 2002).

Biochar is created by the pyrolysis of biomass, and is under investigation as a method of carbon sequestration. It is a novel approach to establish a significant, long-term, sink for atmospheric carbon dioxide in terrestrial ecosystems, i.e. the application of biochar. Biofuel production using modern biomass can produce a biochar by-product through pyrolysis which results in 30.6 kg C sequestration for each GJ of energy produced (Lehmann et al. 2006).

5.4.4 Sequestration by Microorganisms and Enzymes

Active sequestration of carbon dioxide can be accomplished by microorganisms in subterranean formations, by introducing exogenous methanogenic microorganism which is capable of methanogenesis. Subterranean formations comprises of mixed consortium of microorganisms which are capable of methanogenesis, which can convert sedimentary organic matter directly to methane or over long geological time periods depending upon the prevailing environmental conditions. Methanotropic archae in partnership with sulfate reducing organisms are capable of converting methane produced during methanogenesis into carbon dioxide and water. Microorganisms that convert carbon dioxide to acetate also exist in subterranean formations. The methanogenic microorganism which is capable of hydrogen oxidizing and carbon dioxide reducing, will aid in the metabolic conversion of carbon dioxide to methane when introduced into the subterranean formation and methane gas obtained can be recovered and used as fuel (Converse et al. 2003).

Enzyme catalyzed carbon dioxide sequestration into bicarbonates have been studied as one of the most promising biological carbon dioxide sequestration technologies by Ramanan et al. (2009). An extensive screening for biological sequestration by carbonic anhydrase purified from *Citrobacter freundii* was explored. Carbonic anhydrase, a zinc mettaloenzyme, reported to be present in microorganisms, plants and animals has the ability to catalyze the conversion of carbon dioxide to bicarbonates and protons. Bacteria with high carbonic anhydrase activity were isolated from environmental samples such as high strength waste water from polluting industries, sewage, municipal solid wastes and soils. Addition of purified enzyme from the *Citrobacter freundii* to the carbon dioxide saturated calcium chloride reaction mixture resulted in the increased deposition of carbonate and bicarbonate salts. Biological carbon sequestration will have larger impacts if, carbon dioxide can be sequestered as mineral carbonates by using the enzyme in immobilized enzyme reactors. Moreover, the presence of carbonic anhydrase in heterotrophic bacteria provides us a platform to do more investigations on carbon sequestration using carbonic anhydrase enzyme.

A company named Carbon dioxide solution in Québec City, Canada has genetically engineered *an Escherichia coli bacterium* that is capable of producing carbonic anhydrase enzyme which converts carbon dioxide to bicarbonate. The scientists envision the utilization of this enzyme for the core of a bioreactor technology that could be scaled up to capture carbon-dioxide emissions from power plants that run on fossil fuel. A proposal for using the enzyme carbonic anhydrase that works as a catalyst which accelerates carbon dioxide hydration for subsequent fixation into stable mineral carbonates have been proposed by Bond et al. (2001). One of the main concerns regarding the feasibility of using the enzymes in this system is the capability of the enzyme to function in the presence of various chemical species that are present in industrial flue gases. However, research work has demonstrated excellent enzyme activity in presence of SO_2 and NO_2 that are expected to be present in the industrial effluents.

5.4.5 Limitations and Future Prospects

Carbon sequestration is one of the most promising methods to mitigate the rising concentrations of carbon dioxide in the atmosphere. The crucial roles played by microbes and plants in terrestrial ecosystem to sequester carbon and in the global cycling of carbon in the environment make the process of soil carbon sequestration unique and efficient. The major drawback for the practical application of carbon sequestration using agricultural soils is that a single strategy for land use or management may not be suitable for sequestering carbon in all regions. Changes in soil organic carbon can be predicted in a better fashion only with the better understanding of the physical and chemical processes involved in the soil. Long term agricultural trails and knowing the history of the farming system before the conversion to perennial vegetation is crucial for future research. Emphasis should be laid on the increasing the carbon sequestration potential of the above ground and underground systems of the terrestrial ecosystems. Focus should be laid on increasing the long lived soil carbon pool and value added organic products. Reforestation of agricultural lands depends on socio economic factors and land use balance to meet farmers' production demands. Controlling the extraction of peat sources and protecting the wetland functions will help to protect the natural carbon sources in the ecosystem.

5.5 Leakage, MVA and LCRM

CCS is a potential strategy for near term reduction in atmosphere gas emissions. Although CCS appears to be solid from technical perspective, uncertainties remain on scientific and institutional aspects. One of the main concerns of CCS is the underground migration and the possible leakage and escape to surface (IPCC 2005). One of the primary concerns for storage integrity is the leakage of carbon dioxide and brine along faults at GCS sites. Surface release would undermine efforts to minimize atmospheric carbon dioxide concentrations, and the worst case will pose ecological and human health risks. The high pressure that prevails in these storage reservoirs following injection creates an environment for leakage. After injection the leakage slowly subsides or diminishes, since carbon dioxide being increasingly immobilized by residual gas trapping, dissolution and mineral trapping. In the subsurface, the damage caused to natural resources due to carbon dioxide or indirectly by fluids and substances that are immobilized or displaced is of great concern. It also poses risk to the ecosystem and human beings. Natural terrestrial sweeps have a profound negative influence on the growth of plants and microbial responses due to high carbon dioxide exposure. Despite the fact that reservoirs may be well configured to store carbon dioxide, there is chance of leakage in storage sites due to the carbon dioxide being buoyant in geological settings. In this section, we will focus on monitoring, verification and accounting (MVA) of CO_2 storage in GCS reservoirs and associated life cycle risk management (LCRM).

5.5.1 Leakage in Different GCS Reservoirs and Related Processes

One of the hazard common to all injection operations is the widespread pressure perturbation that arises in the formation due to the injection process. The additional mass forced into the injection formation is mainly accommodated by increased fluid pressure, displacing brine at open formation boundaries and uplift to land surface. Open fractures through cap rock could serve to dispose the buoyant phase over large surface areas with non-potable aquifers bounded above by impermeable aquitards, thus promoting dissolution and mineralization. The processes and the pathways that account to the release of carbon dioxide from geological storage sites are complex, including pore systems, openings in cap rock and anthropogenic pathways (Teng and Tondeur 2007). It is inevitable that highly pressurized carbon dioxide will leak to some extent due to the permeable nature of the porous rock which leads to the uncertainties in storability of reservoirs. The scaling induced due to the reaction between carbon dioxide and formation water and rock surface or other formations or damage factors also cause a decrease in the injectivity over time.

Lewicki et al. (2005) developed a strategy to measure the carbon dioxide fluxes or concentration in near surface environment with the help of algorithm, which enhance temporally and spatially correlated leakage signal, while suppressing random background noise. The over ground hydrostatic pressure required for carbon dioxide injection may also open previously closed fracture in reservoirs thereby allowing fluids to drive into faults (Klusman 2003). Saripalli and McGrail (2002) have studied the potential for surface leakage as buoyant carbon dioxide bubble grows in a hydrocarbon reservoir. Hence modeling tools are necessary to predict the leakage rates and pattern in the injection system and potentially leakage wells. Nordbotten et al. (2005) suggested a semianalytic solution for carbon dioxide leakage, in the case of injection of carbon dioxide in supercritical phase in brine saturated deep aquifer, through abandoned well. This approach gives information on carbon dioxide injection plume, leakage rate and plume extent. Free phase carbon dioxide being lighter than formation water due to its increase in buoyancy has more potential for upward leakage. Recent global analyses indicate that the maximum acceptable leakage rates in the range of 0.01%-1.0% (Hepple and Benson 2003; Pacala 2003). Many regulatory agencies and researchers had been studying the issue of potential leakage of hazardous waste through wells into shallow ground water aquifers for the last two decades.

Well bores represent the potential risk of leakage although little is known about the current distribution of potentially leaky wells. Injected gases that are not trapped are in a mobile phase and hence carbon dioxide might escape under typical conditions in wells that do not penetrate cap rock. The wells that are not properly cemented act as a high permeability conduit, through which carbon dioxide can escape (Ide et al. 2006). Well borage can also occur due to the presence of improperly plugged and abandoned wells and loss of integrity due to exposure to high concentration of carbon dioxide environments (Pawar et al. 2009). Carbon dioxide can even leak from properly plugged wells when carbon dioxide dissolution occurs in brines leading to the formation of carbonic acid. Damen et al. (2003) studied the health and environmental issues related to the carbon dioxide injection. A mathematical model for probabilistic risk assessment for carbon sequestration in geological formations was developed after analyzing the risks and uncertainties of unmineable coal beds. The main issues for commercial scale of carbon sequestration in geological reservoirs are uncertainties of geology, risk to environment and inevitably immense financial burden (Xie and Economides 2009).

Formation damage caused by reservoir compaction, precipitation of minerals, oil emulsification and bacterial growth can reduce the permeability and injectivity. Hence, two third of the injected carbon dioxide returns to the surface along with the oil and gas production. One of the potential and most serious environment consequences due to carbon dioxide leakage is the contamination of ground wells. Carbon dioxide being highly effective solvent under supercritical conditions is capable of extracting contaminants from geological materials such as aromatic hydrocarbons. Hence the mobilization of toxic compounds could compromise water quality in nearby aquifers (Stevens et al. 20001. Structural geometry of geological reservoirs has an important role in influencing the direction of migration. Injection of carbon dioxide into aquifers creates an increase in pore pressure which creates subsequent stress in permeability and porosity variation. It can lead to carbon dioxide leakage through fractioned rocks and may give risk to seismicity (Rutqvist and Tsang 2002). Heterogenicity of saline formations controls the mobility ratio and displacement efficiency. The low displacement efficiency of in situ fluid leads to decrease in the reservoir capacity (Jessen et al. 2001). Most of the coal seams around the world are faulted with very thin beds (1–5 m) and low permeability (1–5 md). The swelling of coal in coal bed seams during enhanced coal bed methane recovery leads to faulting to promote the leakage of carbon dioxide from coal seams.

The estimation of probability of leakage along faults or fractures has been studied by Zhang et al. (2009). The probability of leakage of carbon dioxide plumes into compartments through faults or fractures depends on geometric characteristics of conduit systems such as distribution and connectivity between storage reservoir and compartment and size and location of carbon dioxide and plume. Carbon dioxide leaks may exhibit three distinct behaviors like upward migration of the fluids through faults, lateral fluid movement through permeable layers and continued movement of carbon dioxide along the fault above the leakage pathways. Chang et al. (2009) developed a quasi-ID model for assessing the migration of buoyant fluid from reservoir along a conductive faults. These kind of simple models are valuable tools for operators, regulators and policy makers allowing them to make physics based risk assessment and thereby reduce the uncertainties associated with physical properties of storage mechanisms.

The presence of an intact confining layer is a necessary layer for several trapping mechanism. However, sedimentary basins contain geological discontinuities which are potential leakage pathways through confining layers. Hence it is important to assess the consequences of injected carbon dioxide, when it encounters a fault. A conductive fault can act as a major pathway for carbon dioxide plume due to its large capacity. Carbon dioxide can be trapped secondarily by shallow subsurface structures, dissolution and residual phase creation (Linderberg 1997). Migration of fluid attenuates upward flux as well spreads the influence of carbon dioxide across a wider area. The attenuation rate is sensitive to subsurface properties (Oldenberg and Unger 2003). Hence it is necessary to analyze the effect of conductive faults in net carbon dioxide storage based on geometric and petrophysical properties of formation of faults, of overlying permeable layer and on the boundary conditions (pressure in the storage formation and in overlying layers). One of the factors affecting the attenuation rate of carbon dioxide flowing in a fault is layer permeability (Chang et al. 2009).

5.5.2 Potential Risks and Adverse Effects of Leakage

Carbon dioxide accumulation in high concentrations causes adverse health, safety and environmental consequences. Slow carbon dioxide seepage into near subsurface harm flora and fauna and hence potentially disrupt local ecology and agriculture. Large surface release of carbon dioxide pose risk to humans either in form of immediate depth from asphyxiation or effects from prolonged exposure of high concentrations of carbon dioxide. The potential risks associated with injected carbon dioxide in underground geological reservoirs include displacement of saline ground water into potable aquifers, incitement of ground heave and inducement of seismic events. Although the probability of these risks is very low, managing carbon capture and storage injection for ensuring human and environmental safety is an important component of a successful carbon capture and sequestration program. Six main categories of risks associated with carbon dioxide leakage had been identified:

(1) Direct carbon dioxide leakage can contaminate groundwater or can catalyze other pollutants to contaminate water.
(2) Large volume of carbon dioxide injected underground and the resulting build up in pressure can induce seismicity risk.
(3) Carbon dioxide leakage to surface can pose risk to human health as it can act as asphyxiant at high concentration.
(4) Climatic risks associated with slow chronic, sudden or large releases of carbon dioxide to surface.
(5) Potential contamination of underground assets with carbon dioxide or displaced brines.
(6) General environment degradation caused by leakage of sequestered carbon dioxide.

However, the risks associated with carbon dioxide leakage are manageable, if the GCS sites are properly selected, operated and monitored. Hence storage verification and leakage detection are an integral part of GCS.

5.5.3 MVA and LCRM of GCS Projects

In general, MVA aims at (NETL 2009; Zhang and Surampalli 2013):

- Site performance assessment. This is to a) image and measure CO_2 in the reservoir (e.g., to make sure the CO_2 is effectively and permanently trapped in the deep rock formations), b) show if the site is currently preforming as expected, c) estimate inventory and predicate long-term site behaviors (e.g., enable site closure), and d) evaluate the interactions of CO_2 with formation solids and fluids for improved understanding of storage processes, model calibration, future expansion, design improvement;
- Regulatory compliance. This is to a) monitor the outer envelope of the storage complex for emissions accounting, b) collect information for regulatory compliance and carbon credit trading, and c) provide a technical basis to assist in legal dispute resulting from any impact of CCS; and
- Health, safety, and environmental (HSE) impact assessment. This is to a) detect potentially hazardous leakage and accumulations at or near surface, b) identify possible problems and impact on HSE, and c) collect information for designing remediation plans.

MVA of CO_2 sequestration in different geological formations for CO_2 storage is very challenging because for each setting, there are so many different layers that need monitoring, often, with different methods. For example, for on-shore storage systems (e.g., a CO_2-EOR system), monitoring and measurement are needed in a) CO_2 plume, b) primary seal, c) saline formation, d) secondary seal, e) groundwater aquifer, f) vadose zone, g) terrestrial ecosystem, and g) atmosphere, while for an off-shore storage system, it would need in a)–d), e) seabed sediments, f) water column and aquatic ecosystem, and g) atmosphere. As another example, the flux of CO_2 leaving a reservoir is extremely difficult to determine because they might be much smaller than the biological respiration rate and photosynthetic uptake rate of the ground cover. Currently, there exist several knowledge gaps with respect to MVA, such as: How redox conditions affect GCS and MVA? What are the CO_2 intrusion rates and composition in gas stream in different geo-formations? How microbial activities affect fate and transport of CO_2 in different GCS sites? Some details are described by Harvey et al. (2013).

The time course of the LCRM of CCS includes (Zhang and Surampalli 2013):

Development and quality CCS technology → Propose site → Prepare site → operate site → close site → post closure liability.

The LCRM can be classified as three phases:

- Pre-operation phase (about 1–2 years), including technology development, site selection, site characterization, and field design;
- Operation phase (about 10–50 years), including site construction, site preparation, injection, and monitoring; and
- Post-injection phase (about 100–1000 years), including site retirement program, and long-term monitoring (operation, seismic verification, HSE impact).

Table 5.3 shows potential risks associated with large-scale GCS. At each project decision point, the risk assessment needs to be reviewed, and the decision to proceed to the next phase will depend on the ability of the project partners to manage the assessed risks. It is recommended that contingency plants with mitigation strategies need to be established.

Table 5.3. Potential risks associated with large-scale injection of CO_2[a]

Phase	Associated risks	Qualification and mitigation strategy
Pre-operation	• Problems with licensing/permitting. • Poor conditions of the existing well bores. • Lower-than-expected injection rates.	• Revise injection rates, well members, and zonal isolation. • Test all wells located in the injection site and the vicinity for integrity and establish good conditions. • Determine new injection rates or add new wells/pools.
Operation	• Vertical CO_2 migration with significant rates. • Activation of the pre-existing faults/fractures. • Substantial damage to the formation/caprock. • Failure of the well bores. • Lower-than-expected injection rates. • Damage to adjacent fields/producing horizons.	• The monitoring program will allow for early warning regarding all associated risks and for the injection program to be reconfigured upon receiving of such warnings. • If wellbore failure, recomplete or shut it off. • Include additional wells/pools in the injection program.
Post-injec tion	• Leakage through pre-existing faults/or fractures. • Leakage through the wellbores. • Degradation of water quality (decreased pH, mobilization of contaminants, changes in TDS.	• Decrease formation pressure and treat with cement. • Test periodically all wells in the injection site. In case of leakage, wells will be recompleted and/or plugged.

[a] Adapted from NETL (2009) and Zhang and Surampalli (2013).

5.6 Future Trends and Summary

In order to stabilize the increasing GHG emissions it is always advisable to adopt a combination of mitigation strategies. OCS has been suggested as a scientifically and ecologically sound method for reducing atmospheric carbon dioxide emissions. On a global scale, OCS will help to lower the atmospheric carbon dioxide content, their rate of increase and in turn will reduce the detrimental effects of climate change and chance of catastrophic events. The physical capacity for OCS is large compared to fossil fuel resource, and the utilization of this capacity to its full range depends upon cost, equilibrium pCO_2 and environmental consequences. One of the main knowledge gaps for OCS is the environmental impacts that may pose to the marine biota due to the injection of carbon dioxide. Almost all the data available and predictions made are based on the models. Alterations in the biogeochemical cycles will have large consequences, which may be secondary, yet difficult to predict. Since oceans play a pivotal role in maintaining the ecosystem balance, any change in the oceanic environment should be dealt seriously. Hence detailed research is needed to develop techniques to monitor the carbon dioxide plumes, their biological and geochemical behavior in terms of long duration and on a large scale.

For GCS, the major limitations are that carbon dioxide may escape from formations used for geological storage, since carbon dioxide exists in a separate phase. Carbon dioxide can escape through pore systems in cap rocks where capillary entry pressure is exceeded in cap rock fractures or faults or through anthropogenic pathways. This in turn poses serious local health, safety and environmental hazards. Elevated gas phase carbon dioxide has direct effects on surface and subsurface shallow environments. It poses problems due to the effects caused by dissolved carbon dioxide on groundwater chemistry and due to displacement of fluids by injected carbon dioxide. Potential hazards to human health and safety arise from elevated carbon dioxide concentration in ambient air, either in a confined environment and caves or buildings. Hazards to terrestrial and marine ecosystems occurs when stored carbon dioxide and accompanying substances comes in contact with flora and fauna in surface, shallow surfaces and deep surfaces.

For TCS, a thorough understanding of the form of soil organic carbon sequestered and the contribution of above ground and below ground components for soil organic carbon is essential to understand the carbon sequestration potential of different ecosystems. Efforts should be taken for the development and restoration of prevailing ecosystems, such as forests, peat lands, degraded and dessert lands, mined lands so as to maintain or increase the carbon storage capacity and also for the creation of ecosystems that can store carbon in an increased rate. It is advisable to adopt an integrated approach, to reduce the GHG emissions, such as the use of low carbon fuels along with sequestration techniques for securing an energy efficient environment. Alteration of land management practices not only help in increasing the carbon sequestration potential, but also will promote the economic benefits of farmers. However, the risks and factors for the farmers, associated with the adoption of improved management practices should be taken into consideration.

A key regulation in the Underground Injection Control Program (UIC) program aimed to prevent leakages of injected fluids through wells is the Area of Review (AOR) requirement. Targeted research is needed on well integrity and isolation containment that could help regulators and decision makers to plan large scale injection projects. The impact of leakage that may if occur through seal are site specific and the consequences are more on ground water quality. Current regulation of underground injection primarily addresses the operational phase rather than long term monitoring and risk management issues (Wilson and Gerard 2007). Steps should be taken to close the site, monitor and verify the behavior of injected material underground, when the sequestration site reaches its storage capacity. Long term storage cost accounts to trivial percentage of CCS project (Herzog et al. 2005). Ensuring institutional and regulatory mechanisms to manage long term risks are one of the practical solutions for effective citing and implementation of sequestration projects (Schively 2007). The IPCC report on CCS (2005) points out that approximately 99% of injected carbon dioxide is likely to remain in appropriately selected geological reservoirs for over hundreds of years and probability of surface leakage appears low. To ensure that this technology is mature enough to address any problem, it is essential to identify potential risk for carbon capture and storage and developing mitigation strategies (Gerard and Wilson 2009).

Although many methods for carbon sequestration has been studied, it is clearly understood that relying on a single method for carbon sequestration will prove to ineffective in the long run to sequester carbon. Sequestration techniques differ in terms of their permanence, capacity, advantages, limitations, time period, cost factors and effectiveness. Carbon sequestration methodologies that offer practical and immediate solutions to remediate the atmosphere for a considerable long period of time should be considered. One of the cost effective methods for reducing greenhouse emissions is to develop and enhance the natural processes that will sequester more carbon. Proper understanding of the biological and ecological processes in unmanaged and managed terrestrial ecosystems will aid in the development of better strategies to more effective carbon sequestration. The social and ecological implications of carbon sequestration under different ecosystems should also be considered while developing new strategies. Research should be oriented to understand the genomics and biochemical pathways of the microorganisms that are important to the global carbon cycling. Moreover, fundamental research is needed to answer some of the following questions: 1) What are the physical, biological and chemical processes controlling carbon input, distribution, and longevity in an ecosystem? 2) How can these processes be exploited to enhance carbon sequestration? 3) How do carbon sequestration strategies relate to and influence other strategies to mitigate climate change? 4) What is the long-term potential and sustainability for terrestrial carbon sequestration to mitigate climate change at a global scale? It is imperative for us to answer these questions in the future.

5.7 Acknowledgment

Sincere thanks are due to the Natural Sciences and Engineering Research Council of Canada (Grant A 4984, Canada Research Chair) for their financial support. Views and opinions expressed in this article are those of the authors.

5.8 Abbreviations

atm	atmosphere
C	Carbon
cm	centimeter
GHG	Greenhouse Gas
GCS	Geological carbon sequestration
GT	Gigatonne (10^9 tonnes)
IEA	International Energy Agency
IPCC	Intergovernmental Planet on Climate Change
km	kilometer
LCRM	Life cycle risk management
md	millidarcies
Mha	Million hectare
mm	millimeter
MMT	Million metric tonnes (10^9 Kilograms)
MVA	Monitoring, verification and accounting
OCS	Ocean carbon sequestration
p CO_2	partial pressure of carbon dioxide
Pg	Picogram (10^{-12} grams)
Ppmv	Parts per million by volume
TCS	Terrestrial carbon sequestration
TDS	Total dissolved solids
Tg	Teragram (10^{12} grams)
yr	year

5.9 References

Adams, E.E., Caulfield, J.A., Herzog, H.J., and Auerbach, D.I. (1997). "Impacts of reduced pH from ocean CO_2 disposal: Sensitivity of zooplankton mortality to model parameters." *Waste Management,* 17(5/6), 375–380.

Alendal, G. and Drange, H. (2001). "Two-phase, near field modelling ofpurposefully released CO_2 in the ocean." *Journal of Geophysical Research-Oceans,* 106(C1), 1085–1096.

Ametistova, L., Twidell, J., and Briden, J. (2002). "The sequestration switch: removing industrial CO_2 by direct ocean absorption." *Science of the Total Environment*, 289, 213–223.

Anderson, D.M. (2004). "The growing problem of harmful algae." *Oceanus*, 43 (1), 1–5.

Angel, M. (1997). "Environmentally focused experiments: Pelagic studies." In *Proceedings of the Ocean Storage of Carbon Dioxide Workshop 4: Practical and Experimental Approaches*, IEA Greenhouse Gas R&D Programme, 59–70.

Archer, D.E., Kheshgi, H., and Maier-Reimer, E. (1998). "Dynamics of fossil fuel neutralization by Marine $CaCO_3$." *Global Biogeochemical Cycles*, 12(2), 259–276.

Aya, I., Yamane, K., and Shiozaki, K. (1999). *Proposal of self-sinking CO_2 sending system*. COSMOS. Elsevier Science Ltd, Pergamon, 269–274.

Aya, I., Yamane, K., and Yamada, N. (1995). "Simulation experiment of CO_2 storage in the basin of deep-ocean." *Energy Conversion and Management,* 36(6–9), 485–488.

Bachu, S. (2000). "Sequestration of CO_2 in geological media: Criteria and approach for site selection in response to climate change." *Energy Conversion and Management*, 41(9), 953–970.

Bachu, S. (2003). "Screening and ranking of sedimentary basins for sequestration of co_2 in geological media in response to climate change." *Environmental Geology*, 44, 277–289.

Bachu, S., and Adams, J.J. (2003). "Sequestration of CO_2 in geological media in response to climate change: capacity of deep saline aquifers to sequester CO_2 in solution." *Energy Conversion and Management*, 44, 3151–3175.

Bakker, D.C.E., Bozec, Y., Nightingale, P.D., Goldson, L.E., Messias, M.J., de Baar, H.J.W. , Liddicoat, M.I., Skjelvan, I., Strass, V., and Watson, A.J. (2005). "Iron and mixing affect biological carbon uptake in SOIREE and EisenEx, two Southern Ocean iron fertilisation experiments." *Deep-Sea Research*, Part I, 52, 1001–1019.

Benson, S.M., Gasperikova, E., and Hoversten, G.M. (2004). "Overview of monitoring techniques and protocols for geologic storage projects." IEA Greenhouse Gas R&D Programme Report No. PH4/29, 89.

Berg, G.M., Glibert, P.M., Jorgensen, N.O.G., Balode, M., and Purina, I. (2001). "Variability in inorganic and organic nitrogen uptake associated with riverine nutrient input in the Gulf of Riga, Baltic Sea." *Estuaries,* 24, 176–186.

Betts, P.A., Falloon, P.D., Goldewijk, K.K., and Ramankutty, N. (2007). "Biogeophysical effects of land use on climate: Model simulations of radiative forcing and large scale temperature change." *Agricultural and Forest Meteorology* 142, 216–223.

Bond, G.M., Stringer, J., Brandvold, D.K., Arzum, S.F., Medina, M.G., and Egeland, G. (2001). "Development of integrated system for biomimetic CO_2 sequestration using the enzyme carbonic anhydrase." *Energy & Fuels,* 15, 309–316.

Boyd, P.W., Watson, J.A., Law, C.A., Abraham, E.R., Trull, T., Murdoch, R., Bakker, D.C.E., Bowie, A.R., Buesseler, K.O., Chang, H,O., Charette, M., Croot, P.,

Downing, K., Frew, R., Gall, M., Hadfield, M., Hall, J., Harvey, M., Jameson, G., LaRoche, J., Liddicoat, M., Ling, R., Maldonado, M.T., McKay, R.M., Nodder, S., Pickmere, S., Pridmore, R., Rintoul, S., Safi, K., Sutton, P., Strzepek, R., Tanneberger, K., Turner, S., Waite, A., and Zeldis, J. (2000). "A mesoscale phytoplankton bloom in the polar Southern Ocean stimulated by iron fertilization." *Nature*, 407, 695–702.

Bradshaw, J., Boreham, and la Pedalina, F. (2005). "Storage retention time of CO_2 in sedimentary basins: Examples from petroleum systems." *Proceedings of the 7th International Conference on Greenhouse Gas Control Technologies (GHGT-7)*, September 5–9, 2004, Vancouver, Canada, v.I, 541–550.

Brewer, P.G., D.M. Glover, C. Goyet, and D.K. Shafer (1995). "The pH of the North-Atlantic Ocean–improvements to the global model for sound-absorption in seawater." *Journal of Geophysical Research-Oceans*, 100(C5), 8761–8776.

Brewer, P.G. (2004). "Dissolution rates of pure methane hydrate and carbon dioxide hydrate in under-saturated sea water at 1000 m depth." *Geochimica et Cosmochimica Acta,* 68(2), 285–292.

Brewer, P.G., Friederich, G., Peltzer, E.T. and Orr, F.M. (1999). "Direct experiments on the ocean disposal of fossil fuel CO_2." *Science*, 284, 943–945.

Bridgham S.D., Megonigal, J.P., Keller, J.K., Bliss, N.B., and Trettin, C. (2006). "The carbon balance of North American wetlands." *Wetlands*. 26(4), 889–916.

Buesseler, K.O., Doney, S.C., Karl, D.M., Boyd, P.W., and Caldeira, K. (2008) "Ocean iron fertilization–moving forward in a sea of uncertainty." *Science*, 319,162.

Burant, A., Lowry, G.V., and Karamalidis, A.K. (2013). "Partitioning behavior of organic contaminants in carbon storage environments: A critical review." *Environ. Sci. Technol.*, 47, 37–54.

Chang, K.W., Minkoff, S.E. and Bryant, S.L. (2009). "Simplified model for CO_2 leakage and its attenuation due to geological structures." *Green House Gas Control Technologies*, 3453–3460.

Chmura G.L., Anisfeld S.C., Cahoon D.R., and Lynch J.C. (2003) "Global carbon sequestration in tidal saline wetland sediments." *Global Biogeochemical Cycles* 17, 1111. doi: 10.1029/2002GB001917.

Chisholm, S.W., Falkowski, P.G., and Cullen, J.J. (2001). "Discrediting ocean fertilization." *Science*, 294, 309–310.

Coale, K.H., Johnson, K.S., Chavez, F.P., Buesseler, K.O., Barber, R.T., and Brzezinski, M.A. (2004). "Southern Ocean iron enrichment experiment: carbon cycling in high-and low-Si waters." *Science,* 304, 408–414.

Cole, K.H., Stegen, G.R., and Spencer, D. (1993). "The capacity of the deep oceans to absorb carbon-dioxide." *Proceedings of the International Energy Agency Carbon Dioxide Disposal Removal Symposium, Energy Conversion and Management*, special issue, 34(9–11), 991–998.

Converse, D.R., Hinton, S.M., Hieshima, G.B., Barnum, R.S. and Sowlay, M.R. (2003). "Process for stimulating microbial activity in a hydrocarbon bearing subterranean formation." Patent 6543535. Available: http://www.patentstorm.us/patents/6543535/description.html.

Damen, K., Faaij, A., and Turkenburg, W. (2003). "*Health, Safety, and Environmental Risks of Underground CO_2 Sequestration: Overview of Mechanisms and Current Knowledge*." Report NWS-E-2003-30, ISBN: 90-393-3578-8.

de Baar, H.J.W., Boyd, P.W, Coale, K.H., Landry, M.R., Tsuda, A., Assmy, P., Bakker, D.C.E., Bozec, Y., Barber, R.T., Brzezinski, M.A., Buesseler, K.O., Boyé, M., Croot, P.L., Gervais, F., Gorbunov, M.Y., Harrison, P.J., Hiscock, W.T., Laan, P., Lancelot, C., Law, C.S., Levasseur, M., Marchetti, A., Millero, F.J., Nishioka, J., Nojiri, Y., van Oijen, T., Riebesell, U., Rijkenberg, M.J.A., Saito, H., Takeda, S., Timmermans, K.R., Veldhuis, M.J.W., Waite, A.M., and Wong, C.-S. (2005). "Synthesis of eight in-situ iron fertilization in high nutrient low chlorophyll waters confirms the control by wind mixed layer depth of phytoplankton blooms." *Journal of Geophysical Research,* 110, C09S16. doi: 10.1029/2004JC002601.

Denman, K.L. (2008). "Climate change, ocean processes and ocean iron fertilization." *Marine Ecology Progress Series,* 364, 219–225.

Dixon, R.K. (1995). "Agroforestry systems: Sources and sinks of greenhouse gases?" *Agroforestry Systems*, 31, 99–116.

Dreux, R. (2006). "CO_2 R&D Program at Gaz de France: K12B & Altmark Cases." *CO_2 NET Annual Meeting,* Athens, Greece.

Duan, X., Xiaoke, W., Lu, F., and Zhiyun, O. (2008). "Primary evaluation of carbon sequestration potential of wetlands in China." *Acta Ecologica Sinica,* 28(2), 463–469.

Fertiq, B. (2004). "*Ocean gardening using iron fertilization.*" Available: http://www.csa.com/discoveryguides/oceangard/overview.php.

Forge, F. (2001). *Carbon sequestration by agicultural soils.* Science and Technology division. Parliamentary Research Branch, Government of Canada. Available: http://dsp-psd.pwgsc.gc.ca/Collection-R/LoPBdP/BP/prb0038-e.htm.

Freund, P. (2003). "*Capture and Geological Storage of Carbon Dioxide – A Status Report on the Technology.*" Report No. Coal R223 DTI/Pub URN 02/1384.

Friedmann, S.J., Upadhye, R., and Kong, F. (2009). "Prospects for underground coal gasification in carbon-constrained world." *Energy Procedia,* 1, 4551–4557.

Franzluebber, A.J., Hons, F.M. and Zuberer, D.A. (1994). "Long-term changes in soil carbon and nitrogen pools in wheat management systems." *Soil Science Society of American Journal,* 58, 1639–1645.

Freibauer, A., Rounsevell, M.D.A., Smith, P., and Verhagen, J. (2004). "Carbon sequestration in the agricultural soils of Europe." *Geoderma,* 122, 1–23.

Gerard, D. and Wilson, E.J. (2009). "Environmental bonds and the challenge of long-term carbon. sequestration." *Journal of Environmental Management*, 90, 1097–1105.

Glibert , P. (2008). "Ocean urea fertilization for carbon credits poses high ecological risks." *Marine Pollution Bulletin,* 56, 1049–1056.

Gnanadesikan, A., Sarmiento, J.L., and Slater, R.D. (2003). "Effects of patchy ocean fertilization on atmospheric carbon dioxide and biological production." *Global Biogeochemical Cycles,* 17, 1050.

Goldberg, D., and Slagle, A.L. (2009). "A global assessment of deep-sea basalt sites for carbon sequestration." *Energy Procedia, 1*, 3675–3682.

Gozalpour, F., Ren, S., and Tohidi, B. (2005). "CO_2 EOR and storage in oil reservoirs." *Oil & Gas Science and Technology- Rev. IFP*, 60(3), 537–546.

Graham, R.L., Wright, L.L. and Turhollow, A.F. (1992). "The potential for short-rotation woody crops to reduce CO_2 emissions." *Climatic Change,* 22, 223–233.

Harvey, O.R., Qafoku, N.P., Cantrell, K.J., Lee, G., Amonette, J.E., and Brown, C.F. (2013). Geochemical implications of gas leakage associated with geologic CO_2 storage–A qualitative review." *Environ. Sci. Technol.*, 47, 23–36.

Haugan, P.M. and Drange, H. (1992). "Sequestration of CO_2 in the deep ocean by shallow injection." *Nature,* 357, 318-320.

Haugan, P.M. and Alendal, G. (2005). "Turbulent diffusion and transport from a CO_2 lake in the deep ocean." *Journal of Geophysical Research-Oceans*, 110, C09S14, doi:10.1029/2004JC002583.

Heil, C.A., Revilla, M., Glibert, P.M., and Murasko, S. (2007). "Nutrient quality drives phytoplankton community composition on the West Florida shelf." *Limnology and Oceanography,* 52, 1067–1078.

Hepple, R.P., and Benson, S.M. (2003). "Implications of surface seepage on the effectiveness of geologic storage of carbon dioxide as a climate change mitigation strategy." *Proceedings of the Sixth International Greenhouse Gas Technologies Conference,* Gale, J., Kaya, Y., and Pergamon, I, (Ed.), Kyoto, Japan, (2002).

Herzog, H., Smekens, K., Dadhich, P., Dooley, J., Fujii, Y., Hohmeyer, O., Riahi, K., Akai, M., Hendriks, C., Lackner, K., Rana, A., Rubin, E., Schrattenholzer, L., and Senior, B. (2005). *"Costs and economic potential."* In: IPCC, Editor, IPCC Special Report on Carbon Dioxide Capture and Storage, IPCC, Geneva, (8), 8-37.Herzog, H., Caldeira, K., and Adams, E. (2001). "Carbon sequestration via direct injection." In: *Encyclopedia of Ocean Sciences,* Steele, J.H. Thorpe, S.A., and Turekian, K.K. (Ed.), vol. 1, London, UK, Academic Press, 408–414.

Herzog, H., Drake, E., and Adams, E. (1997). *"CO_2 capture, reuse, and storage technologies for mitigating global climate change." A White Paper Final Report,* Order No. DE-AF22- 96PC01257, U.S. Department of Energy, Washington, D.C.

Hoffman, D.L. (2009). "New Jelly pump rewrites carbon cycle." Available at <http:// theresilientearth.com/?q=content/> (accessed Feb. 2012).

Hoffman, D.L. (2010). "Ocean CO_2 storage revised." Available at <http://www. theresilientearth.com/?q=content/> (accessed Feb. 2012).

Hoppemma, M. (2004). "Weddell Sea is a globally significant contributor to deep-sea sequestration of natural carbon dioxide." *Deep Sea Research Part I: Oceanographic Research Papers,* 51(9), 1169–1177.

Horiuchi, K., Kojima, T., and Inaba, A. (1995). "Evaluation of fertilization of nutrients to the ocean as a measure for CO_2 problem." *Energy Conversion and Management*, 36, 915–918.

Houghton, J.T., Ding, Y., Griggs, D.J., Noguer, M., van der Linden, P.J., and Xiaosu, D. (Eds.) (2011). *Climate Change 2001: The Scientific Basis*, pp. 1–896, Cambridge Univ. Press, New York, 2001.

Huesemann, M.H., Skillman, A.D., and Crecelius, E.A. (2002). "The inhibition of marine nitrification by ocean disposal of carbon dioxide." *Marine Pollution Bulletin*, 44, 142–148.

Ide, S.T., Friedmann, S.J. and Herzog, H.J. (2006). "CO_2 leakage through existing wells: current technology and regulations." In: *Proceedings of 8th international conference on greenhouse gas control technologies, IEA Greenhouse Gas Programme,* Trondheim, Norway, June.

IPCC (Intergovernmental Panel on Climate Change) (2000). *Good Practice Guidance for Land Use, Land-Use Change and Forestry.* Special Report of the Intergovernmental Panel on Climate Change, Cambridge University Press, UK, 599.

IPCC (2005). *Special Report on Carbon Dioxide Capture and Storage.* Prepared by Working Group III of the IPCC, Cambridge University Press, Cambridge, U.K.

IEA GHG (2006). "*Updating the IEA GHG Global CO_2 Emissions Database: Developments since 2002.*" Technical Study Report Number 2006/7, United Kingdom.Jessen, K., Sam-Olibale, L.C., Kovscek, A.R., and Orr, F.M. (2001). "Increasing CO_2 storage in oil recovery." *First National Conference on Carbon Sequestration*, NETL, Washington, DC.

Jiao, N., Herndl, G.J., Hansell, D.A., Benner, R., Kattner, G., Wilhelm, S.W., Kirchman, D.L., Weinbauer, M.G., Luo, T., Chen, F., and Azam, F. (2010). "Microbial production of recalcitrant dissolved organic matter: Long-term carbon storage in the global ocean." *Nature Reviews Microbiology*, 8, 593–599.

Jin, X., and Gruber, N. (2003). "Offsetting the radiative benefit of ocean iron fertilization by enhancing ocean N_2O emissions." *Geophysical Research Letters,* 30(24), 2249.

Johannes, L., Gaunt, J., and Rondon, M. (2006). "Bio-char sequestration in terrestrial ecosystems–A review" *Mitigation and Adaptation Strategies for Global Change.* 11(2), 395–419.

Jun, Y.-S., Giammar, D.E., and Werth, C.J. (2013). "Impacts of geochemical reactions on geologic carbon sequestration." *Environ. Sci. Technol.*, 47, 3–8.

Kern, J.S. and Johnson, M.G. (1993). "Conservation tillage impacts on national soil and atmospheric carbon levels." *Soil Science Society of American Journal*, 57, 200–210.

Kheshgi, H.S. (1995). "Sequestering atmospheric carbon dioxide by increasing ocean alkalinity." *Energy*, 20(9), 915–922.

Klusman, R.W. (2003). "Evaluation of leakage potential from a carbon dioxide EOR / sequestration project." *Energy Conversion and Management*, 44, 1921–1940.

Korre, A., Shi, J., Imrie, C., and Durucan, S. (2009). "Modelling the uncertainty and risks associated with the design and life cycle of CO_2 storage in coalbed reservoirs." *Proceedings of 9th International Conference on Greenhouse Gas Control Technologies, Energy Procedia*, 1, 2525–2532.

Lewicki, J.L., Hilley, G.E., and Oldenburg, C.M. (2005). "An improved strategy to detect CO_2 leakage for verification of geologic carbon sequestration." *Geophysical Research Letters,* 32(19), L19403, DOI: 10.1029/2005GL024281.

Lindeberg, E. (1997). "Escape of CO_2 from Aquifers." *Energy Conversion Management*, 38, Suppl., 229–234.

Lucas, M.I., Seeyave, S., Sanders, R., Moore, C.M., and Williamson, R. (2007). "Nitrogen uptake responses to a naturally Fe-fertilised phytoplankton bloom during the 2004/2005 CROZEX study." *Deep-Sea Research II*, 54, 2138–2173.

Ma, X.H., Lü, X.G., and Yang, Q. (1996). "Carbon cycle of a marsh in Sanjiang Plain." *Scientia Geographica Sinica,* 16(4), 323–330.

Maltby, E., and P. Immirzi (1993). "Carbon dynamics in peatlands and other wetland soils: Regional and global perspectives." *Chemosphere*, 27(6), 999–1023.

Marchetti, C. (1977). "On geoengineering and the CO_2 problem." *Climate Change*, 1977, 1: 59–68.

Martin, J.H. (1990). "Glacial-interglacial CO_2 change: The iron hypothesis." *Paleoceanography*, 5, 1–13.

Matter, J.S., Broecker, W.S., Stute, M., Gislason, S.R., Oelkers, E.H., Stefansson, A., Wolff-Boenisch, D., Gunnlaugsson, E., Axelsson, G., and Bjornsson, G. (2009). "Permanent carbon dioxide storage into basalt: The CarbFix Pilot project, Iceland." *Energy Procedia,* 1, 3641–3436.

Matthiassen, O.M. (2003). "*CO_2 as Injection Gas for Enhanced Oil Recovery and Estimation of the Potential on the Norwegian Continental Shelf.*" Trondheim/Stavanger, Norwegian University of Science and Technology, Department of Petroleum Engineering and Applied Geophysics, Trondheim, Norway.

McCook, L.J., Jompa, J., and Diaz-Pulido, G. (2001). "Competition between corals and algae on coral reefs: a review of evidence and mechanisms." *Coral Reefs,* 19, 400–417.

McGrail, B.P., Reidel, S.P., and Schaef, H.T. (2003). "Use and features of basalt formations for geologic sequestration." In: *Proceedings of the 6th International Conference on Greenhouse Gas Control Technologies (GHGT-6)*, Gale, J., and Kaya, Y. (Ed.), 2002, Kyoto, Japan, Pergamon, vol. II, 1637–1641.

Mori, Y.H. (1998). "Clathrate hydrate formation at the interface between liquid CO_2 and water phases-a review of rival models characterizing 'hydrate films." *Energy Conversion Management,* 39, 1537–1557.

Moritis, G. (2002). "Enhanced oil recovery." *Oil and Gas Journal*, 100(15), 43–47.

Moulton, R., and Richards, K. (1990). *Costs of sequestering carbon through tree planting and forest management in the United States.* United States Forest Service Washington Office General Technical Report. GTR-WO-58.

Nair, R.P.K., and Nair, V.D. (2002). "Carbon sequestration in agroforestry systems." *In: Proceedings of the 17th World Congress on Soil Science,* Available: http://www.grida.no/climate/ipcc/land use/163.htm.

Nakashiki, N. (1997). "Lake-type storage concepts for CO_2 disposal option." *Waste Management*, 17 (5–6), 361–367.

NETL (National Energy Technology Laboratory). (2010). *Best Practices for: Terrestrial Sequestration of Carbon Dioxide.* The Energy Lab, NETL, U.S.DOE, Nov. 2010.

Nguyen, D.N. (2003). "Carbon Dioxide Geological Sequestration: Technical and Economic Reviews." Paper SPE 81199 presented at the SPE/EPA/DOE Exploration and Production Environmental Conference, San Antonio, Texas, 10–12 March.

Nordbotten J.M., Celia, M.A., Bachu, S., and Dahle, H.K. (2005). "Semi-analytical solution for CO_2 leakage through an abandoned well." *Environmental Science and Technology,* 39(2), 602–611.

Obernosterer, I., Christaki, U., Lefèvre, D., Catala, P., and van Wambeke, F. (2008). "Rapid bacterial mineralization of organic carbon produced during a phytoplankton bloom induced by natural iron fertilization in the Southern Ocean." *Deep Sea Research II*, 55, 777–789.

Oelbermann, M., Voroney, R.P., and Gordon, A.M. (2004). "Carbon sequestration in tropical and temperate agroforestry systems: A review with examples from Costa Rica and Southern Canada." *Agriculture, Ecosystems, and Environment,* 104 (3), 359–377.

Oldenburg, C.M., and Unger, A.J.A. (2003). "On leakage and seepage from geologic carbon sequestration sites: Unsaturated zone attenuation."*Vadose Zone Journal*, 2, 287–296.

Pacala, S.W. (2003). "Global constraints on reservoir leakage." Pergamon, I. (Ed.), *Proceedings of the Sixth International Greenhouse Gas Technologies Conference,* Kyoto, Japan, 267–272.

Palumbo, A.V., McCarthyb, J.F., Amonettec, J.E., Fishera, L.S., Wullschlegera, S.D. and Daniels, W.L. (2004). "Prospects for enhancing carbon sequestration and reclamation of degraded lands with fossil-fuel combustion by-products." *Advances in Environmental Research*, 8, 425–438.

Paustian, K., Elliott, E.T., Six, J., and Hunt, H.W. (2000). "Management options for reducing CO_2 emissions from agricultural soils." *Biogeochemistry,* 48, 147–163.

Pawar, R.J., Watson, T.L., and Gable, C.W. (2009). "Numerical simulation of CO_2 leakage through abandoned wells: Model for an abandoned site with observed gas migration in Alberta, Canada." *Green House Gas Control Technologies,* 9, 3625–3632.

Piessens, K., and Dusar, M. (2004). "Feasibility of CO_2 sequestration in abandoned coal mines in Belgium." *Geologica Belgica*, **7**–3/4.

Powell, H. (2008). "Will ocean iron fertilization work?" *Oceanus Magazine*, 46 (1), 10–13.

Prentice, C., Farquhar, G., Fasham, M., Goulden, M., Heimann, M., Jaramillo, V., Kheshgi, H., Quéré, C.L., Scholes, R., and Wallace, D. (2001). "The carbon cycle and atmospheric CO_2." J.T. Houghton (Ed.), *Climate Change 2001: The Scientific Basis: Contribution of WGI to the Third Assessment Report of the IPCC*. Cambridge University Press, New York, 183–237.

Ramanan, R., Kannan, K., Sivanesan, S.D., Mudliar, S., Kaur, S., Tripathi, A.K., and Chakrabarti, T. (2009). "Bio-sequestration of carbon dioxide using carbonic anhydrase enzyme purified from *Citrobacter freundii*." *World Journal of Microbiology and Biotechnology*, 25 (6), 981–987.

Rau, G.H., and Caldeira, K. (1999). "Enhanced carbonate dissolution: A means of sequestering waste CO_2 as clean bicarbonate." *Energy Conversion & Management*, 40, 1803–1813.

Raven, J.A., and Falkowski, P.G. (1999). "Oceanic sinks for atmospheric CO_2." *Plant, Cell and Environment*, 22, 741–755.

Rehder, G., Kirby, S.H., Durham, W.B., Stern, L.A., Peltzer, E.T., Pinkston, J., and Brewer, P.G. (2004). "Dissolution rates of pure methane hydrate and carbon dioxide hydrate in under-saturated sea water at 1000 m depth." *Geochimica et Cosmochimica Acta*, 68(2), 285–292.

Ricketts, E.R., Kennett, J.P., Hill, T.M., and Barry, J.P. (2009). "Effects of CO_2 hydrate emplacement on deep-sea foraminiferal assemblages: 3600m on the California Margin." *Marine Micropaleontology*, 72, 165–175.

Reeves, D.W. (1997). "The role of soil organic matter in maintaining soil quality in continuous cropping systems." *Soil and Tillage Research*, 43, 131–167.

Rudd, J.M., Harris, R., Kelly, C.A., and Hecky, R.E. (1993). "Are hydroelectric reservoirs significant sources of greenhouse gases?" *Ambio* 22, 246–248.

Rutqvist, J., and Tsang, C.-F. (2002). "A study of caprock hydromechanical changes associated with CO_2-injection into a brine formation." *Environmental Geology*, 42, 296–305.

Sabine, C.L., Feely, R.A., Gruber, N., Key, R.M., Lee, K., Bullister, J.L., Wanninkhof, R., Wong, C.S., Wallace, D.W.R., Tilbrook, B., Millero, F.J., Peng, T.H., Kozyr, A., Ono, T., and Rios, A.F. (2004). "The oceanic sink for anthropogenic CO_2." *Science*, 305, 367–371.

Saripalli, P., and McGrail, P. (2002). "Semi-analytical approaches to modeling deep well injection of CO_2 for geological sequestration." *Energy Conversion and Management*, 43, 1968–1974.

Schively, C. (2007). "Siting geologic sequestration: problem and prospects." In: *Carbon Capture and Sequestration Integrating Technology, Monitoring, Regulation,* E.J. Wilson and D. Gerard (Ed.), Blackwell Academic, Ames, I. A. 223–241.

Sheps, K.M., Max, M.D., Osegovic, J.P., and Tatro, S.R. (2009). "A case for deep-ocean CO_2 sequestration." *In: Proceedings of GHGT-9, Energy Proceedia 1*, 4961–4968.

Shi, J.Q., and Durucan, S. (2005). "CO_2 storage in unmineable coal seams." *Oil & Gas Science and Technology-Rev. IFP*, 60(3), 547–558.

Shindo, Y., Fujioka, Y., and Komiyama, H. (1995). "Dissolution and dispersion of CO_2 from a liquid CO_2 pool in the deep ocean." *International Journal of Chemical Kinetics*, 27(11), 1089–1095.

Sloan, E.D. (1998). *Clathrate Hydrates of Natural Gases.* Marcel Dekker Inc., New York, 705.

Smith, P., Goulding, K. W., Smith, K. A., Powlson, D. S., Smith, J. U., Falloon, P. D. and Coleman, K. (2001). "Enhancing the carbon sink in European agricultural soils: Including trace gas fluxes in estimates of carbon mitigation potential." *Nutrient Cycling in Agroecosystems*, 60, 237–252.

Song, Y., Chen, B., Nishio, M., and Akai, M. (2005). "The study on density change of carbon dioxide seawater solution at high pressure and low temperature." *Energy,* 30(11–12), 2298–2307.

Stevens, S., Kuuskraa, V., and Gale, J. (2000). "Sequestration of CO_2 in depleted oil and gas fields: Global capacity and barriers to overcome." *Fifth International Conference on Greenhouse Gas Control Technologies Cairns,* Australia.

Stevens, S.H., V.A. Kuuskra and J. Gale (2001). "Sequestration of CO_2 in depleted oil & gas fields: global capacity, costs and barriers." *Proceedings of the 5th International Conference on Greenhouse Gas Control Technologies (GHG-5),* (D.J. Williams, R.A. Durie, P. McMullan, C.A.J. Paulson and A.Y. Smith, eds.), CSIRO Publishing, Collingwood, VIC, Australia, 278-283.

Tan, X.L., and Zhang, Q.M. "Mangrove beaches's accretion rates and effects of relative sea-level rise on mangrove in China." *Marine Science Bulletin*, 16(4), 29–35.

Tarnocai, C. (1998). "The amount of organic carbon in various soil orders and ecological provinces in Canada." In *Soil Processes and the Carbon Cycle*, Lal, R., Kimble, J.M., Follett, R.F., and Stewart, B.A. (eds.), CRC Press, Boca Raton, FL, USA, p. 81–92.

Tarnocai, C., Kettles, I.M., and Lacelle, B. (2005). *Peatlands of Canada*. Ottawa. Agriculture and Agri-Food Canada, Research Branch, Ottawa, ON, Canada.

Teng, H., Yamasaki, A., and Shindo, Y. (1996). "The fate of liquid CO_2 disposed in the ocean." *International Energy,* 21(9), 765–774.

Teng, H., Yamasaki, A., and Shindo, Y. (1999). "The fate of CO_2 hydrate released in the ocean." *International Journal of Energy Research,* 23(4), 295–302.

Teng, F., and Tondeur, D. (2007). "Efficiency of carbon storage with leakage: Physical and economical approaches." *Energy,* 32, 540–548.

Tsouris, C., Brewer, P.G., Peltzer, E., Walz, P., Riestenberg, D., Liang, L., and West, O.R. (2004). "Hydrate composite particles for ocean carbon sequestration: Field verification." *Environment Science and Technology*, 38, 2470–2475.

Turner, S.M., Harvey, M.J., Law, C.S., Nightingale, P.D., and Liss, P.S. (2004). "Iron induced changes in oceanic sulfur biogeochemistry." *Geophysical Research Letters,* 31, L14307.

Verstraeten, W.W., Veroustraete, F., Coppinc, P.R. Feyenb, J.A., and Hansen, E.A. (1993). "Soil carbon sequestration beneath hybrid poplar plantations in the north central United States.: *Biomass and Bio-energy*, 5, 431–436.

Vleeshouwers, L.M., and Verhagen, A. (2002). "Carbon emission and sequestration by agricultural land use: A model study for Europe." *Global Change Biology*. 8(6), 519–530.

Watson, A.J., Bakker, D.C.E., Ridgwell, A.J., Boyd, P.W., and Law, C.S. (2000). "Effect of iron supply on Southern Ocean CO_2 uptake and implications for glacial atmospheric CO_2." *Nature*, 407, 730–733.

West, O.T., and Marland, G. (2002). "A synthesis of carbon sequestration, carbon emissions and net carbon flux in agriculture: comparing tillage practices in the United States." *Agriculture, Ecosystems and Environment*, 91, 217–232.

Wickett, M.E., Caldeira, K. and Duffy, P.B. (2003). "Effect of horizontal grid resolution on simulations of oceanic CFC-11 uptake and direct injection of anthropogenic CO_2." *Journal of Geophysical Research,* 108(C6), 3189–3208.

Wilson, M., and Monea, M. (2004). "IEA GHG Weyburn CO_2 Monitoring and Storage Project Summary Report 2000–2004." *Proceedings of the 7th International Conference on Greenhouse Gas Control Technologies, Volume III, Vancouver*, Available: http:// ieaagreen.org.uk/glossies/weyburn.pdf .

Wilson, E.J., and Gerard, D. (2007). "Geologic sequestration under current U.S. regulations." In: *Carbon Capture and Sequestration Integrating Technology, Monitoring, Regulation,* Wilson, E.J., and Gerard, D. (Ed.), Blackwell Publishing, Ames, I. A. 169–198.

Xie, X., and Economides, M.J. (2009). "The impact of carbon geological sequestration." *Journal of Natural Gas Science and Engineering,* 1, 103–111.

Zeng, N. (2008). "Carbon sequestration via wood burial." *Carbon Balance and Management*, 3:1, doi:10.1186/1750-0680-3-1.

Zhang, T.C., and Surampalli, R.Y. (2013). "Carbon capture and storage for mitigating climate changes." Chapter 20 in *Climate Change Modeling, Mitigation and Adaptation,* Surampalli, R., Zhang, T.C., Ojha, C.S.P., Gurjar, B.R., Tyagi, R.D., and Kao, C.M. (eds.), ASCE, Reston, Virginia, February 2013.

Zhang, Y., Oldenburg, C.M., Finsterle, S., Jordan, P., and Zhang, K. (2009). "Probability estimation of CO_2 leakage through faults at geologic carbon sequestration sites." *9th International Conference on Greenhouse Gas Technologies,* Washington, D.C. (2009). *Energy Procedia,* 1, 41–46.

Zhao, H.Y., Leng, X.T., and Wang, S.Z. (2002). "Distribution, accumulation of peat in the Changbaishan Mountains and climate change in Holocene." *Journal of Mountain Science*, 20(5), 513–518.

CHAPTER 6

Monitoring, Verification and Accounting of CO_2 Stored in Deep Geological Formations

Anushuya Ramakrishnan, Tian C. Zhang and Rao Y. Surampalli

6.1 Introduction

Since the pre-industrial era, the level of atmospheric CO_2 has shown significant increase. There has been a 104 parts per million (ppm) increase as per the current levels of 384 ppm (Tans 2008). Increased exploitation of fossil fuels for energy is the major cause for the observed rise in atmospheric CO_2 levels. It is anticipated that the increased energy use in this century would lead to a continued increase in atmospheric carbon emissions and elevated levels of CO_2 in the atmosphere unless drastic measures are implemented in energy production, usage and carbon management (Socolow et al. 2004). Important mitigation measures to reduce CO_2 emissions include a higher degree of energy conversion, higher energy efficiency and increasing renewable energy sources. According to the UN Frame Work Convention on Climate Change (UNFCCC), the primary goal of mitigation measures should be targeted at 80–90% decrease in power station emission profiles (IPCC 2005). Although CO_2 sinks occur as a part of carbon-cycle, they are not able to absorb the entire CO_2 emitted into the atmosphere each year. Therefore, it is imperative to develop long-term carbon storage (sequestration) technologies (e.g., terrestrial and geological, mineral, and ocean storage).

At present, underground storage or geological sequestration (geo-sequestration) of CO_2 is gaining rapid attention throughout the planet as it offers an attractive option for the mitigation of anthropogenic greenhouse gas (GHG) emission as it offers an opportunity to achieve significant reductions in atmospheric green-house gases used in alliance with other options such as energy efficiency and renewable energy sources (IPCC 2005; Frontier 2006). The main requirements for geological sequestration are a safe and secure underground disposal site which adequately stores CO_2 for a period of hundreds to thousands of years without polluting other important natural resources of potable ground water, coal or petroleum. However, due to the complexity of the systems and the related issues, storage of CO_2 in deep geological formations still may not be free of CO_2 leakage, and therefore, should be treated as a source for emissions, which has led to a higher degree of investment by public and private sectors to establish suitable technology and evaluate its safety and effectiveness.

Many papers, reports and proceedings have been published on CCS since the 1990s (e.g., Riemer et al. 1993; USDOE 1999; Herzog 2001; Anderson and Newell 2003; IPCC 2005; Dodds et al. 2006; Chadwick et al. 2007; IEA 2009; Lackner and Brennan 2009; CCCSRP 2010; Zhang and Surampalli 2013). Information on monitoring, verification and accounting (MVA) of CO_2 stored in deep geological formations and life cycle risk management of carbon capture and storage (CCS) have been addressed in many previous studies (e.g., NETL 2009a, b 2012). However, there still is a need to review the current status and issues related to MVA of CO_2 stored in deep geological formations, which is the focus of this chapter. This chapter starts with an introduction of the concepts of generic storage options for geological storage of CO_2, followed by descriptions on background and procedures of MVA of CO_2 stored in deep geological formations. The chapter further discusses monitoring plan design, key monitoring techniques, ecosystem stress monitoring, MVA data integration and analysis technologies as well as a few case studies. Finally, the chapter presents the current issues, future research needs and conclusions.

6.2 Generic Storage Options for Geological Storage of CO_2

Geological storage/disposal of CO_2 is a conventional technique that has been in practice for the last 100 years. Carbon dioxide storage in deep sub-surfaces has not only enabled oil production, but also has been used as a means of disposing waste gases (CO_2 and H_2S) from oil production. As shown in Fig. 6.1, three main options for geological sequestration towards the large scale disposal of CO_2 are:

- Disposal/storage in deep saline formations
- Disposal/storage in depleted or near depleted gas and oil reserves
- Disposal/storage in coal seams

Table 6.1 shows the storage capacity, concerns and needs for each option. A brief description on each of these options is as follows.

6.2.1 Disposal/Storage in Deep Saline Formation

For monetary and logistical reasons, CO_2 storage in hydrocarbon reservoirs or deep saline formations is efficiently achieved at depths greater than 800 m, where the ambient temperatures and pressure would facilitate the stored CO_2 to remain in a supercritical or liquid state (Funnell et al. 2009). These conditions guarantee an efficient storage of CO_2 and require a good cap rock seal above the reservoir formation to acquire the relatively buoyant CO_2. Trapping may occur by a number of mechanisms in saline formations (IPCC 2005; CO2CRC 2008):

- Physical trapping in structural and stratigraphic traps, much similar to the CO_2 trapping in oil and gas petroleum reservoirs;
- Adsorption trapping where CO_2 remains adsorbed onto the surface of a mineral such as coal;
- Hydrodynamic trapping, where immiscible CO_2 (in liquid or super-critical state) is contained by migration of formation waters;
- Residual trapping, where CO_2 becomes trapped in the pore-space by capillary forces;
- Solubility trapping, where CO_2 is soluble in formation waters; and
- Mineral trapping, where CO_2 is precipitated as a new mineral.

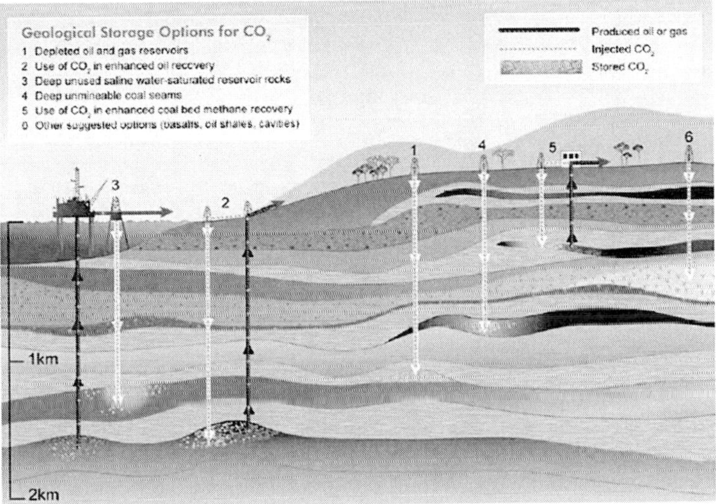

Figure 6.1. Options for geological storage of carbon dioxide (Source: CO2CRC 2008)

6.2.2 *Disposal/Storage in Depleted or Nearly Depleted Gas/Oil Reserves*

Depleted oil and gas reservoirs are specifically favorable for long-term storage of CO_2. Reservoirs with oil and gas production are promising traps with adequate reservoir and seal intervals capable of trapping oil and/or gas over periods of geological time–millions of years. There is an extensive characterization of their geological structure and petro-physical properties in search for petroleum. Availability of well and seismic data reduces the budget for implementation of sequestration projects, and most commonly, sophisticated numerical models of reservoir characteristics and behavior have already been developed by petroleum engineers. Most of the required infrastructure for injecting and storing CO_2 exists at oil/gas fields and most of the regulatory, compliance, permitting and public acceptance aspects of initiating a CO_2 storage project should be achieved with a significant ease. Moreover, the use of depleted oil/gas fields as storage sites should prove to be economical and offer better options than the new green-fields, associated with a larger capacity of deep saline formations.

Table 6.1. Capacity, concerns and needs of three options for geological storage of CO_2[a]

Geological storage options and worldwide CO_2 storage capacity (C in Gt CO_2):

- Saline formations contain brine in their pore volumes, commonly with salinities > 10,000 ppm. Capacity > 1000
- Depleted or nearly-depleted oil/gas fields have some combination of water and hydrocarbons in their pores. Examples include enhanced oil or gas recovery. Capacity = 675–900
- Unminable coalbeds (= CO_2 enhanced coalbed methane, CO_2-ECBM). Capacity = 3–200

Concerns and needs:

- Little is known about saline aquifers compared to oil fields; as the salinity of the water increases, less CO_2 can be dissolved into the solution. Larger uncertainty about the saline aquifers exists if the site appraisal study is limited
- Liquid CO_2 is nearly incompressible with a density of ~1000 kg/m^3; overpressuring and acidification of the reservoir may cause a) changes in the pore/mineral volume, and b) saline brines (or water) moving into freshwater aquifers or uplift; old oil wells may provide leak opportunities
- In CO_2-ECBM, the key reservoir screening criteria include laterally continuous and permeable coal seams, concentrated seam geometry, minimal faulting and reservoir compartmentalization, of which there is not much known
- New technologies are needed to ensure CO_2 stays in place forever
- Need comprehensive site appraisal studies to reduce harmful effects

[a] Adapted from Zhang and Surampalli (2013).

Under favorable geological conditions, CO_2 may be injected into oil or gas reservoirs to enhance the volume of gas or oil recovered. Typically in an enhanced oil recovery (EOR) projects the storage capacity of CO_2 is only 10% of that of the oil reservoir. However, EOR may provide early opportunities for commercialization of CCS and can form the basic step in developing a value chain for CO_2 storage. That is EOR projects may help in the development of infrastructure, policy and public acceptance for long-term storage projects in nearby sites. Presently, operators specifically focus on EOR, but in the future there may be an opportunity to focus on maximizing CO_2 storage rather than oil production.

6.2.3 Disposal/Storage in Coal Seams

Coal offers a good storage option, since CO_2 is specifically adsorbed onto the coal surface, substituting gases such as methane, and will remain trapped as long as temperature and pressure conditions within the coal remain stable. Storage might also be feasible in coal seams in association with coal-bed methane production (enhanced coal-bed methane or ECBM). A number of international projects are ongoing to research coal storage, but practical storage at industrial scales is yet to be fully tested. Storage volumes are judged to be relatively small in comparison to deep saline formations and depleted oil/gas fields (IPCC 2005).

6.3 MVA: Background and General Procedures

Guidance on different aspects of CSS is provided by the Intergovernmental Panel on Climate Change (IPCC) Special Report on Carbon Dioxide Capture and Storage (e.g., IPCC 2005). An important tool to an effective monitoring program is the implementation of protocols that can be verified. Protocols approved by the International Organization for Standardization (ISO) 140641 and 140652 for over 45 countries and the American National Standards Institute (ANSI 2007) laid the foundation for validation and verification of geo-sequestered CO_2. Each accredited project will develop a proposal that defines the overall framework, including site characterization, monitoring programs and verification. Independent bodies assess and verify stored CO_2 for its environmental and ecological safety. Application of ISO 14064 and 14065 standards (ISO 2006, 2007) for evaluation of the project ensures that there is a balance between the cost of a monitoring program and the effective implementation of it to accompany accuracy and transparency to guide the mechanism of efficient CCS. These standards are adopted by diverse industrial sectors and will also form a basis for various GHG programs, such as The Climate Registry, the California Climate Action Registry, the Chicago Climate Exchange (CCX) and Regional Greenhouse Gas Initiative (RGGI).

6.3.1 Regulatory Framework and Guidance of CCS–An Insight

Successful commercial scale CCS project implementation is dependent on a new regulatory framework that effectively deals with unresolved issues pertaining to the regulation of a large-scale industrial CCS program to guarantee an effective, economic and safe capture, transportation, sub-surface injection and long-term CCS and monitoring. The U.S. Environmental Protection Agency (EPA) proposed federal regulations under the Safe Drinking Water Act (SDWA) for underground injection of CO_2 (Federal Register 2008). The U.S. EPA is actively tracking the progress and results of national and international geo-sequestration projects. The U.S. Department of Energy (DOE) initiated an experimental field research on geo-sequestration in the U.S in association with the Recovery Community Services Program (RCSP). The RCSP program is an important tool for establishing an effective regulatory and legal environment for the safe, long-term underground injection and geo-sequestration of CO_2. Additionally, the information acquired from small and large-scale geologic sequestration projects will enable the accounting of stored CO_2 in an effort to support future GHG registries, incentives and other policy initiatives for the future.

6.3.2 Life Cycle Stages and Risk Management of CCS projects

Life Cycles Stages of CCS Projects. CCS will need to be regulated as an industrial process with regulations geared to each project stage: capture, transportation, site-selection and permitting, injection, site closure and long term stewardship (Fig. 6.2). Each of the elements exists and operates in isolation, yet they are not integrated into a single industrial process. The structure of future CCS industry could depend on relationships between CO_2 producers, CO_2 pipe-line operators and geological storage site operators. However, each CCS project will have four common stages: i) pre-operations phase, ii) operations phase, iii) post-operation phase (also called closure phase); and iv) post-closure phase. Descriptions about each stage are as follows:

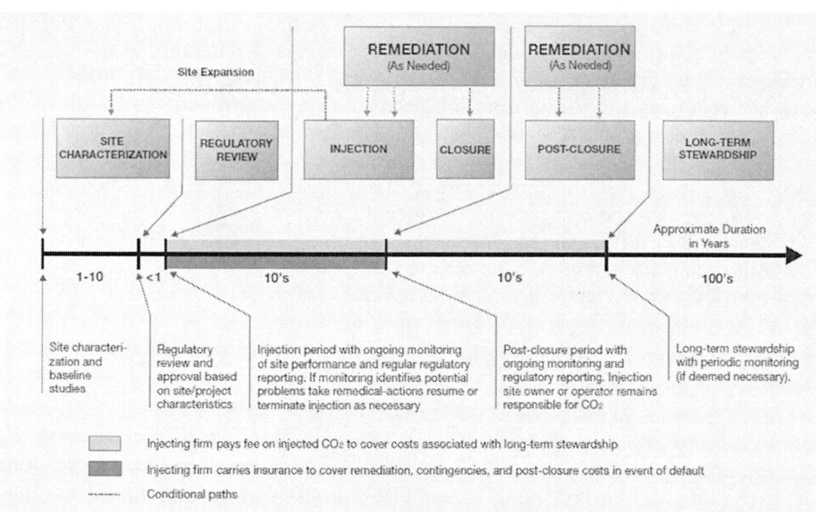

Figure 6.2. Life cycle stages of a CCS project (Rubin et al. 2007)

- *Stage 1–Pre-operations*: This stage is about 3–10 years, including technology development, site selection, site-characterization, field design, site construction, and site preparation. Careful site characterization is the most effective way to manage short and long-term risks of CCS. Establishment of generalized CCS siting guidelines that is customized to local geology is an important first regulatory step that can be planned immediately. Most of the countries focus on such efforts as the prime step in CCS, including Australia, US, Canada and throughout the EU. The site characterization phase will extend into site development. Installation of injection wells and monitoring systems will add detailed understanding to site geological features.

- *Stage 2–Operations*: This stage is about 10–50 yeas and mainly includes pipeline transport, injection and monitoring. Site operations focus on the pipeline transport, injection and monitoring. CCS projects are dependent on pipeline transport from source to sink. There are specific regulatory requirements on inventories including injection well design, allowable injection quantity, pressure and level of purity of CO_2 stream. Though current regulations cover most of these aspects, yet a face-lift of it is required to ensure that risks of the operations are adequately addressed. Moreover, each site will have its specific monitoring and verification requirements and will be adaptive as project progresses. High level monitoring requires efficient base line measurements before injection. Monitoring becomes important not only as regulatory requirement, but also to gain public acceptance.
- *Stage 3–Post-operation (closure phase)*: This stage is about 50–100 years and includes a site retirement program and long-term monitoring (operation, seismic verification, HSE impact). Closure requirements will center on operations including decommissioning, monitoring and verification and regulatory oversight throughout the project. All the stake holders involved in the project will be interested in meeting successful closure requirements at the end of the operation. After injection, the CCS operator needs to ensure that the stability of storage is established for a specific period. The duration of the post-closure liability period varies between several years to several decades across different projects.
- *Stage 4–Post-closure:* This stage is up to 10,000 years. CCS technique needs to ensure that CO_2 remains sequestered underground for more than hundred years and up to thousand years. There is a need for long-term monitoring to ensure that CO_2 storage is safe and is behaving as predicted. To guarantee HSE, regulations will need to specify the requirements pertaining to temporal and technical aspects that govern the ownership transfer.

Potential Risks and Life Cycle Risk Management of CCS Projects. Table 6.2 and Fig. 6.3 show potential risks associated with large-scale injection of CO_2 at different stages. Monitoring activities at these stages can be implemented effectively towards risk assessment at the storage site and development of mitigation strategies for handling possible problems at the site. Effective application of monitoring technologies ensure that the CCS are safe for human health and the environment and will play an important role in the establishment of relevant technical approaches for MVA (IRGC 2008). Table 6.3 lists potential monitoring objectives for different stages of a CCS project.

It is essential that MVA strategies for CCS projects be integrated with the multi-disciplinary team involved in design and operation of geo-sequestration projects. The site characterization and simulation phase will provide a thorough MVA system that helps with required data for validation of expected results, monitoring for leakage and ensuring that CO_2 remains in the subsurface. Inventory verification is an essential step in the national and international strategies to mitigate and control CO_2 emissions. Annual accounting is performed based on sector-specific methodologies. All the CCS projects will require an interactive risk assessment geared to identify and quantify potential risks to human health and the environment related to CCS and helps to ensure that these risks remain low throughout the life cycle of a CCS project.

Table 6.2. Potential risks associated with large-scale injection of CO_2[a]

Stage	Associated risks	Qualification and mitigation strategy
Pre-operation	• Problems with licensing/permitting. • Poor conditions of the existing well bores. • Lower-than-expected injection rates.	• Revise injection rates, well members, and zonal isolation. • Test all wells located in the injection site and the vicinity for integrity and establish good conditions. • Determine new injection rates or add new wells/pools.
Operation	• Vertical CO_2 migration with significant rates. • Activation of the pre-existing faults/fractures. • Substantial damage to the formation/caprock. • Failure of the well bores. • Lower-than-expected injection rates. • Damage to adjacent fields/producing horizons.	• The monitoring program will allow for early warning regarding all associated risks and for the injection program to be reconfigured upon receiving of such warnings. • If wellbore failure, recomplete or shut it off. • Include additional wells/pools in the injection program.
Post-operation	• Leakage through pre-existing faults/or fractures. • Leakage through the wellbores.	• Decrease formation pressure and treat with cement. • Test periodically all wells in the injection site. In case of leakage, wells will be recompleted and/or plugged.

[a]Adapted from Zhang and Surampalli (2013).

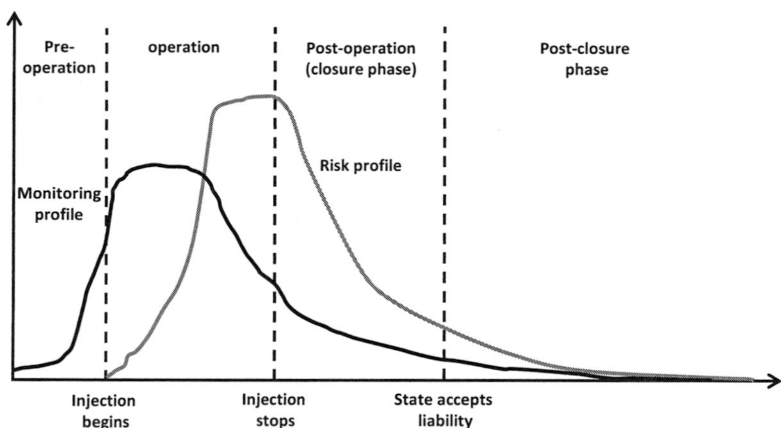

Figure 6.3. Risk and monitoring intensity profile of a CCS project at different stages (after Benson in WRI 2008). Note: in this chapter, the 4 stages are defined as: i) pre-operation, ii) operation, iii) post-operation; and iv) post-closure

Table 6.3. Potential monitoring objectives for different stages of CSS projects[a]

Stage (years)	Monitoring objectives
Pre-operation (3–10)	• Develop or update available geological models • Perform an environmental impact assessment • Develop predictive models of system behavior • Perform risk assessment with an uncertainty management plan used to support development of the monitoring program • Develop effective remediation strategies • Establish baseline data with which future site performance can be compared
Operation (10–50)	• Provide stake-holder assurance • Manage monitoring program to ensure that no CO_2 leaks to the shallow subsurface or surface • Verify the location and mass of stored CO_2 • Test accuracy of predictive models, and history match dynamic geological models • Meet local health, safety and environmental (HSE) performance criteria • Provide stakeholder assurance
Post-operation (50–100)	• Provide evidence that the system will behave as predicted by dynamic geological models so that the site can be closed • Manage monitoring program to ensure that no CO_2 leaks to the shallow subsurface or surface • Provide stakeholder assurance
Post-closure (Up to 10,000)	• Periodic monitoring if deemed necessary

[a]Benson in WRI (2008).

6.3.3 Objectives and General Procedures of MVA

Objectives of MVA. The main goal of carbon sequestration is to acquire an understanding of specific CCS options to result in an economic, effective and environmentally sound technology option that may aid in reducing CO_2 emissions. The overall goal of monitoring is to demonstrate to the regulatory bodies that the practice of carbon sequestration is safe and is an effective GHG mitigation technology and does not lead to significant adverse environmental impacts locally. The various objectives of MVA for carbon geo-sequestration are to (Litynski et al. 2008):

- Gain an understanding of different storage processes and evaluate their effectiveness;
- Assess the interactions of CO_2 with the components of the formation in different environmental compartments;
- Evaluate the EHS impacts that may occur in case of a leak to the atmosphere;
- Assess/monitor the sequence of remediation efforts in the event of a leak; and
- Provide scientific guidance for assisting legal disputes arising from the impacts of sequestration (ground water impacts, crop losses, seismic events, etc.).

General Procedures of MVA. Figure 6.4 shows the MVA flow chart in different stages of a CO_2 geological storage project. In general, the following steps are involved:

- Identification of sub-surface processes associated with the particular monitoring activity of interest;
- Selection of a chain of geophysical techniques that is relevant to specific sub-surface measurements;
- Performance of base-line measurements before CO_2 injection;
- Repetition of measurement at specific intervals during and after injection; and
- Interpretation of results that are focused on time-lapse changes (LBNL 2004).

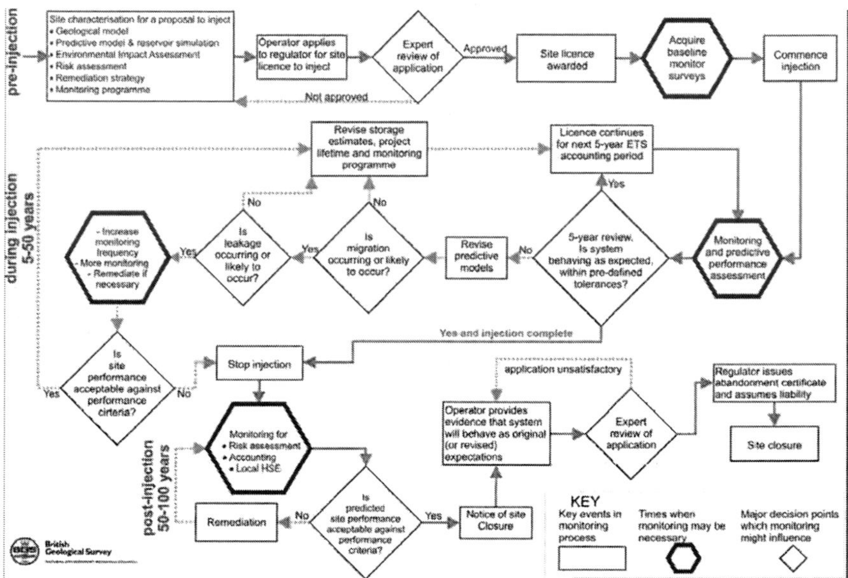

Figure 6.4. MVA flow chart in different stages of CO_2 geological storage projects (Pearce et al. 2006)

Monitoring Plan Design. The monitoring plan design forms the basis of a successful CO_2 injection project along with risk analysis and reservoir management. The main criteria for MVA plan are that it should be broad in scope and include CO_2 storage, conformance and containment, monitoring techniques for internal quality control and verification and accounting for regulators (DNV 2010a). A typical MVA plan will include components for meeting regulatory requirement, monitoring the CO_2 plume monitoring water/brine behavior, detecting potential release pathways, and quantifying releases (EC, 2011). A monitoring plan also outlines monitoring objectives, defines risk-based performance metrics and resources allocated for monitoring activities. In addition, a comprehensive plan should include reviews of monitoring tools' effectiveness, stakeholder communications, procedures for documenting monitoring activities, and processes used to evaluate monitoring performance.

MVA plans may change in scope as a project progresses from the pre-injection phase to the post-injection phase. In the pre-injection phase, project risks are identified, monitoring plans are developed to mitigate these risks, and baseline monitoring data is obtained. During the injection phase, monitoring activities are focused on containment and storage performance. Monitoring techniques may need to be adapted and evaluated to ensure that they continue to be effective for meeting MVA goals. In the post-injection phase, monitoring activities are focused on long-term storage integrity and managing containment risk.

Significance of MVA Protocols. Reliable and cost-effective monitoring would serve as an important tool to assess CSS as a safe and effective strategy and a dependable method for CO_2 control. Monitoring is an essential requirement for the permitting process for CO_2 injection, plume tracking, leak testing and verification. Further monitoring may be required for assessment of natural resources including ecosystems and groundwater to ensure that the exposure to CO_2 does not affect the health and safety of the local population. Also the regulators need to verify if the levels of CO_2 migration are within the pre-defined limits and that it meets the pre-injection levels predicted. MVA also informs the stakeholders (i.e., investors, public, site-operators and regulators) that CCS projects are being conducted safely and that the sites do not create a detrimental impact on the environment. MVA projects are especially important to gain stakeholder confidence in the primary stage of technology and to make sure that carbon credits gathered as a part of the emission trading remain in the ground (IPCC 2005).

6.4 Key Monitoring Techniques of MVA

6.4.1 General Descriptions

MVA of CO_2 sequestration in different geological formations for CO_2 storage is very challenging because for each setting, there are so many different layers that need monitoring, often with different methods. For example, for on-shore storage systems (e.g., a CO_2-EOR system), monitoring and measurements are needed in a) the CO_2 plume, b) the primary seal, c) saline formation, d) the secondary seal, e) groundwater aquifer, f) the vadose zone, g) the terrestrial ecosystem and g) the atmosphere. For an off-shore storage system, it would need in a)–d), e) seabed sediments, f) water column and aquatic ecosystem, and g) atmosphere. Each MVA program has to be designed for specific projects and sites, as there are wide variations between individual sites in terms of accessibility, perceived risks, total amount of CO_2 to be injected, the original site application (e.g., EOR or depleted oil and gas field), land use, geology, topography and technical needs (Benson et al. 2004; Pearce et al. 2005; Dodds et al. 2006; Benson 2007). Table 6.4 shows the examples with respect to this concern.

Table 6.4. Monitoring tools used in onshore and offshore global projects (open circles implies possible use of technique)[a]

Monitoring tool	**Gorgon (onshore)**	**Weyburn (onshore)**	**Sleipner (offshore)**	**In Salah (onshore)**	**Otway (onshore)**
Surface					
2D/3D seismic	•	•	•	•	•
Soil/sediment gas	•	•			•
Atmospheric			•	•	
Gravity				○	
Well-based					
Well head pressure & flow rates	•	•		•	•
Downhole pressure & temperature	•	•		•	•
CO_2 saturation logging	•	•		•	•
Casing/cement integrity	•	•		•	•
Passive					
VSP	•	•		•	•
Crosswell seismic				○	
Crosswell EM					
Geo-chemical sampling	•	•			•
Traces					•
Air-borne					
Spectral imaging				•	

[a]Bannister et al. (2009).

Currently, CO_2 MVA technologies can be broken down into four main categories (NETL 2012):

- Atmospheric monitoring tools: such as CO_2 detectors, eddy covariance, advanced leak detection system, laser systems and Light Detection and Ranging (LIDAR), tracers and isotopes (Campbell et al. 2009);
- Near-surface monitoring tools: such as ecosystem stress monitoring, tracers, groundwater monitoring, thermal hyperspectral imaging, synthetic aperture radar, color infrared transparency films, tiltmeter, flux accumulation chamber, induced polarization, spontaneous (self) potential, soil and vadose zone gas monitoring, shallow 2-D seismic;
- Subsurface monitoring tools: such as multi-component 3-D surface seismic time-lapse survey, vertical seismic profile, magnetotelluric sounding, electromagnetic resistivity, electromagnetic induction tomography, injection well logging (wireline logging), annulus pressure monitoring, pulsed neutron capture, electrical resistance tomography, acoustic logging, 2-D seismic survey, time-lapse gravity, density logging and optical logging. Cement bond long, Gamma

ray logging, microseismic survey, crosswell seismic survey, aqueous geochemistry, resistivity log; and

- MVA data integration and analysis technologies: such as intelligent monitoring networks and advanced data integration and analysis software.

The criteria of judging which methods are suitable for different settings are a) simple and cost effective (regarding explaining and implementing the method), b) defensible (sufficiently stringent to ensure that the method is of good QA/QC–quality assurance and quality control), and c) verifiable (the value obtained by the method can be assigned with confidence and certainty) (Zhang and Surampalli 2013).

In Section 6.4.2, we introduce tools and technologies for collecting CO_2 monitoring data in the atmosphere, the near-surface zone, and the subsurface. In Section 6.4.3, we discuss the techniques used for ecosystem stress monitoring. In Section 6.4.4, we briefly introduce the data integration and analysis technologies.

6.4.2 Key Monitoring Techniques for CO_2 MVA

This section presents techniques that are imperative for monitoring and verification of the location of geologically stored CO_2 in different underground reservoir systems. We present the techniques best suited to record the presence of CO_2 stored in different types of storage reservoirs (i.e., deep saline reservoirs, depleted oil and gas reservoirs, enhanced oil and gas recovery and coal seams). The key monitoring techniques for monitoring injected CO_2 applicable at the international level is given below (NETL 2009b):

- Surface geophysics
  - 3-D seismic reflection survey
  - Passive seismic array monitoring using down-hole seismometers
  - Vertical seismic profiling (VSP-combination of surface and borehole seismic monitoring)
  - Gravity surveys
  - Electomagnetic surveys (EM and MT)
- Earth deformation
  - GPS, surveying and use of tiltmeters
  - InSAR satellite interferometry
- Pressure and temperatures
  - Wellhead monitoring–pressure and flow rates
  - Down-hole pressure and temperature gauges (injection & monitoring wells)
- Chemistry
  - Down-hole water sampling–water chemistry
  - Down-hole water sampling–CO_2 tracers

- Down-hole
  - Well integrity logging
  - Saturation logging
  - Cross-hole seismic and EM
  - Well gravimetry
- Environmental (assurance)
  - Remote sensing (hyperspectral imaging)
  - Atmospheric gas analyses
  - Soil gas surveys
  - Ecosystem studies
  - Shallow groundwater measurements
  - Marine high-resolution imaging (side-scan sonar, bubble surveys, high-resolution acoustics)

Details of some of these techniques are described as follows.

Geophysical Detection of Subsurface CO_2. Various geophysical techniques including 3-D and 4-D seismic reflection surveys (Angerer et al. 2001; Chadwick et al. 2005; Arts et al. 2008), gravity surveys (Eiken et al. 2000) and electromagnetic measurements (Hoverston and Gasperikova 2005) have been employed for monitoring plumes of CO_2 to verify that they are not found beyond their geographical limits. The key technologies are summarized in Table 6.5. Many of these technologies have already been tested for geologic storage of oil and gas industry as well as investigation of hazardous waste disposal sites. Most of these techniques have the ability to identify and link changes observed in physical measurements to changes in the properties of the reservoir (Hoversten et al. 2002; Benson et al. 2004). Repeated measurements of storage sites over weeks, months and/or years would be required to record changes by many geophysical techniques. Time lapse measurements that are recorded will help in identifying saturation of fluids with CO_2 based on comparative analysis (Chadwick et al. 2008). Some geophysical techniques that serve as useful tools for tracking and migration of CO_2 stored in sub-surface are discussed as follows.

Seismic. Among the geophysical techniques currently employed, seismic methods are the most matured and highly developed. These techniques are based on the principle of seismic wave migration in rocks saturated with CO_2 and deployment of receiver arrays (active source seismic) or the use of seismic recorders (passive source) for the monitoring and verification of CO_2. Active source seismic techniques, including 3D-seismic reflection surveys have been employed with great success and have generated excellent images of migrating plumes of CO_2. However, 4-D time lapse seismic reflection data are helpful in monitoring the distribution of CO_2 within the reservoir (Chadwick et al. 2005). Advanced seismic techniques like AVO (amplitude versus offset) studies and multi-component seismic may be employed as instruments for sensitive discrimination of saturated CO_2 in the plume (Brown et al. 2007).

Time-lapse 3D seismic reflection surveys have been employed both offshore and onshore. They have found higher applicability in marine environments where there is an enhancement of the penetration of sub-surface CO_2 into sub-sea rock formations by seismic waves. Hence higher quality data is attained and CO_2 accumulations to the extent of 1000 tons or more at the depths of 1–2 km could be detected (Myer et al. 2002; Arts et al. 2004; Chadwick et al. 2005). In 3D-Seismic reflection surveys, the source and receivers are arranged in strings along the ground or sea surface (close to it). Sources and receivers could also be placed in monitoring wells to the depths of up to several kilometers. Two important down-hole seismic techniques that have highest applicability are the vertical seismic profiling (VSP) and cross-well seismic profiling. In VSP a source is used at the surface and down-hole receivers to detect changes in seismic reflectivity arising from the presence of CO_2. In cross-well seismic, down-hole sources and receivers are employed to measure the change in bulk seismic properties such as P-wave velocity through the use of tomographic techniques (Majer et al. 2006).

Table 6.5. Key geophysical technologies for CO_2 verification[a]

Technique	Capabilities	Detection limits	Where applicable
3-D seismic	Images seismic reflectivity, for a volume beneath a 3D surface array. Can be used in a 4D sense, by repeated surveys, permanent arrays may be utilized	Limited by the wavelength of the seismic waves, depth of target and the acoustic velocity of the sediments.	Both onshore and offshore. Costs are significantly cheaper offshore.
Borehole EM	Measures changes in formation resistivity using semi-permanent down-hole transmitters and receivers	Could be valuable for detecting fine- scale changes in CO_2 saturation, both within and outside the storage reservoir.	Primarily onshore.
Cross-well Electrical Resistance Tomography (ERT)	Involves the measurement of resistivity in the subsurface between wells. Pilot studies indicate that cross-hole ERT can detect resistivity changes due to CO_2 injection above a certain threshold.	Depending on electrode configuration, CO_2 accumulations > ~30 m thick could potentially be imaged with a borehole separation of 200 m. Further research Is needed to refine models and optimize electrode setups.	Primarily onshore.
Cross-well EM	Utilizes time-variant source field to derive information about subsurface electrical structure.	Potentially ~5 m resolution, but dependent on the separation of transmitters and receivers that are placed in adjacent monitoring wells.	Primarily onshore.
Gravimetry	Useful to detect density changes in a volume and to track the migration of the CO_2. Repeat surveys useful for improving vertical resolution.	Site specific	Primarily onshore, although offshore work has been carried out using seafloor plinths and ROV'.
Passive-seismic	Used to record micro-fracturing occurring in the vicinity of the seismometers, which may indicate movement on natural fractures in the vicinity of the CO_2 plume	5-10 m accuracy obtainable, depending on whether borehole sensors are used, and the design of the sensor array.	Primarily onshore; seafloor seismic recording is routine in North Sea oil fields, using OBS's and seafloor ocean-bottom cables.
Satellite Interferometry	Repeated radar surveys detect changes in elevation potentially caused by CO_2 injection.	InSAR can detect millimetre-scale changes in elevation.	Onshore in regions of limited vegetation
Seafloor EM	Induced electrical and magnetic fields are created by towed (marine) electromagnetic sources, which are detected by a series of seabed receivers. These data can determine subsurface electrical profiles that may be influenced by the presence of highly resistive CO_2.	EM methods are likely to be most suitable for monitoring storage in saline formations, where CO2 is displacing more conductive formation waters. The technique could be sensitive to thin resistive anomalies at depths between tens of meters, and several km.	Offshore, involving repeated surveys using seafloor EM instruments, recording from a ship-source.
VSP	Can form a high-resolution image of seismic reflectivity, in the vicinity of the CO2 plume. Can involve multi-component recording, offering potential for pressure /saturation discrimination and anisotropy characterization. Also potential to image leakage from primary container.	Has a high seismic resolution, but usually only in 2D. Often site specific.	Both onshore and offshore

[a]McCurty et al. (2009).

Passive seismic techniques enable the recording of micro-earthquakes that result from the movement of fractures or for detecting tremor and passive signals arising from fluid movement through rock mass. Increase in seismicity gets triggered by the pressure and stress distribution related to the injection and migration of CO_2 (Maxwell and

Urbancic 2001; Maxwell et al. 2004). The resulting seismicity that arises from CO_2 movement and rock fracturing are employed to track the movement of CO_2. Micro-seismicity also enables the location as well as the size of rock fractures that produce the deformation of rock masses.

Gravity. Gravity recorders are based on the principle of density changes in reservoir (Eiken et al. 2000; Westrich et al. 2001; Wilson and Monea 2004, Preston et al. 2005; Arts et al. 2008). Limitations of gravity-based techniques include the limited distance between plume and gravity meters and the density contrast between the injected CO_2 and the surrounding material. When injected CO_2 displaces brine-rich water in the plume, changes in the density of reservoir would be recorded (Gasperikova and Hoverston 2006). This technique is also dependent on the geometry of plume, where vertically elongated plumes would give much higher peak gravity than the thin widely spread plumes.

Gravity surveys are primarily performed onshore. Main advantages of gravity techniques are that they offer lower spatial resolution and cost than seismic approaches and enhance the monitoring of sub-surface mass changes, thus enabling estimates of the amount of CO_2 that dissolves in the plume.

Electrical. Electrical methods involve the measurement of resistivity changes to be mapped within the imaged rock volume. The resistance of CO_2 can be easily identified in rock formations saturated with conductive brine-rich water. Electrical methods include cross-well electromagnetic (EM), borehole EM and electrical resistance tomography (ERT). They provide information on the spatial distribution and pore-space saturation of injected CO_2. However, EM techniques are sensitive to the type, amount and interconnectivity of fluid (liquid or gas) contained within rock pore space (Hoversten et al. 2002; Gasperikova and Hoversten 2006), while integrated changes in resistivity are used to provide a valid measurement of the total injected CO_2 volume (Christensen et al. 2006). Electrical techniques are employed as down-hole instruments as they are used in time-lapse mode and as they are sensitive to thin resistive anomalies at depths between tens of meters and several kilometers.

Cross-well EM and ERT have been tested and employed for onshore sites although marine EM via electromagnetic sources towed by ships and seabed receivers is still in development. Cross-well EM requires transmitters and receivers in adjacent monitoring wells and has the capability to resolve CO_2 layers as thin as five meters. Resolution of this technique is dependent on the separation of transmitters and receivers and their location relative to the plume. To improve the resolution of cross-well EM, it has to be run in conjunction with cross-hole seismic. ERT is based on the measurement of resistivity in the sub-surface between wells. With an optimal electrode configuration and a borehole separation of 200 m, cross-well ERT achieves the capability to detect

CO_2 accumulations more than about 30 m thick. Hence, ERT requires two or more closely spaced monitoring wells, which could be costly. Research and development efforts are required for redefinition of current models and optimization of electrode setups that are essential for testing of the technique at full-scale storage sites.

Earth Deformation. Earth deformation techniques have the capability for indirect mapping and location of sub-surface CO_2. These techniques have been frequently employed to indirectly map the location and migration of sub-surface CO_2. These techniques are based on the increase in pressure that results from the injection of CO_2 and ultimate rock deformation and changes in the altitude of the ground surface above the plume in low permeability reservoirs (e.g., < 10–20 mD). The earth deformation techniques, including global positioning system (GPS) surveying, tiltmeters and InSAR satellite interferometry, are efficient to record even a millimeter scale uplift or subsidence at the ground surface.

InSAR data proves to be the least expensive option among the three deformation techniques mentioned, and it has been employed to record uplift/subsidence of up to 4–5 mm/yr over the last four years in response to the injection of CO_2 (Onuma and Ohikawa 2008). Increasing rates of ground deformation are accompanied by increases in reservoir pressure (e.g., > 2 MPa or 300 psi) for CO_2 plumes at the depth of 1–2 km. However, suitable measures taken to mitigate the pressure increases may result in lower ground deformation that is far below the resolution for InSAR and GPS techniques. Tiltmeters are based on the principle of recording tilts of nanoradians too small to be observed by GPS and InSAR and are still useful for tracking ground deformation accompanied by lower pressure changes (e.g. < 2 MPa). GPS, InSAR and tiltmeters have the ability to locate the plume approximately. While tiltmeters and GPS can produce continuous records of CO_2 migration, InSAR can provide weekly or monthly time-lapse data. Among all these techniques, InSAR is extremely cost-effective as data are collected remotely. However, selection of a suitable technique is based on its sensitivity to resolve ground deformation arising from CO_2 injection. More research and feasibility studies are required to evaluate these techniques for their utility and implementation.

Direct Measurement of CO_2. Direct measurement techniques are used in conjunction with geophysical techniques and are primarily focused on measuring and recording the chemistry, isotopes, concentrations, pressure and temperature of CO_2 formation water, and other subsurface hydrocarbons using well, soil and atmospheric data sources (Table 6.6). Direct measurement of atmospheric gas, soil gas and down-hole water/gas chemistry/temperature/pressure enable the evaluation of the behavior of the injected sub-surface CO_2 plume. Moreover, the data will also identify and determine the rate of movement from the storage container into overlying reservoirs and, in extreme cases, leakage to the surface. Pressure and temperature measurements at the wellhead and within reservoir rocks are used to monitor changes in the reservoir conditions induced by injection and to ensure that the well integrity is maintained and there is not too much of rock mass fracturing (Wright and Majek 1998). Direct measurement techniques primarily focus on onshore storage sites; with additional offshore monitoring option. The following techniques are described in this section:

Table 6.6. Direct measurement technologies for CO_2 monitoring[a]

Technique	Capabilities	Detection limits	Where applicable
Pressure / temperature	PT conditions typically measured at wellhead and down hole locations in well. Can utilize electronic, fiber optic or capillary tube in-situ systems, or run PLT logs in well. Preferable to use continuous telemetered systems esp. wellhead.	Dependent on system chosen. Should be able to acquire high accuracy with most specialist equipment.	Onshore and offshore. Offshore access may increase costs.
Downhole logging	Measures rock-fluid properties in the reservoir immediately surrounding the well, with very high (cm) resolution. Routinely used in the oil-gas industry. New saturation logging tools recently developed applicable for CCS.	Very high resolution, for a large range of rock properties, including resistivity, density, sonic velocity, nuclear magnetic resonance (NMR), borehole gravity (each with the potential to discriminate different lithologies/ properties).	Onshore and offshore. Offshore access more costly–requires an offshore platform.
Well fluid chemical sampling	Well head or down hole sampling techniques are commonly deployed in wells either as permanent or temporary completions. Depending on the parameter being measured various field & lab techniques are commonly available to measure a host of geochemical attributes of fluid & gas samples	Depending on the parameter being measures, concentrations can be obtained to ppb. Cost of sample analysis is often related to the sensitivity and accuracy of the measurements required.	These techniques can be deployed when there is access to a well
Groundwater and Surface water gas analysis	Well-developed gas content measurement techniques but care must be taken to account for rapid degassing of CO_2 from the water.	Background levels likely to be in low ppm range.	Onshore and offshore. Should be used in combination with flux measurements as provides another route for CO_2 leaks.
Soil gas analysis	Measures CO_2 and other gas levels in soil using probes, or from wells. Sampling usually on a grid using a portable IR laser detector or into gas-tight canisters for lab analysis.	NDIR detectors can resolve changes in CO_2 concentration down to at least ± 1–2 ppm. Small variations in CO_2 concentration can be detected. Stable isotopes can indicate origin of CO_2.	Onshore. Useful for detailed measurements especially around detected low flux leakage points.
Soil gas flux	Gas in accumulation chamber is analyzed (e.g., by non-dispersive infrared (NDIR)) and then returned to chamber to monitor build-up over time. Detects fluxes through the soil.	Easily capable of detecting fluxes of 0.04 g CO_2/ m^2-day = 14.6 t/km^2-yr (Klusman 2003). Need to differentiate genuine underground leak against varying biogenic background.	Onshore. Powerful tool when used with analysis of other gases and stable and radiogenic carbon isotope analysis to identify the CO_2 source.
Eddy covariance	Equipment mounted on a platform or tower. Gas analysis data is integrated with wind speed and direction to define upwind footprint and calculate CO_2 flux.	Realistic flux detectable in biologically active area with hourly measurements = 4.4 x 10^{-7} kg/m^2-s = 13870 t/km^2-yr (Miles et al. 2005).	Mainly used onshore. Proven cheap technology. Can survey large areas to determine fluxes and detect leaks. Once a leak is detected, it is likely to require detailed survey of footprint to pinpoint it.
Long open path infrared laser gas analysis	Measures absorption by CO_2 in air infrared laser of a specific part of the infrared gas analysis spectrum along the path of a laser beam, and thus CO_2 levels in air near ground level	Needs development but estimate potential at ±3% of ambient (ca conc.11 ppm) or to cover several km^2 with one device relatively cheap ($1000s).	Onshore. Requires detailed soil gas survey to pinpoint leak.
Portable personal safety-oriented hand-held NDIRs	Measures CO_2 levels in air which could also be useful for pinpointing high concentration leaks detected by wider search methods.	Resolution of small handheld devices for personal protection is typically ca. 100 ppm.	Can be used onshore and on offshore infrastructure. A proven technology. Small hand-held devices for personal protection are <$1000 per unit.
Airborne infrared laser gas analysis	Helicopter or aero plane mounted open or closed path infrared laser gas detectors have potential to take measurements of CO_2 in air every ~10 meters.	Brantley and Koepenick (1995) quote a ±1 ppm above ambient detection limit for the airborne closed path technique. Less information is available on the open path technique, maybe < ±1%.	Onshore. Proven technology for detecting CH_4 leaks from pipelines and large CO_2 leaks. Could detect CO_2 leaks from infrastructure or from underground.
Satellite or airborne hyperspectral imaging	Detects changes in the plant health that could be due to CO_2 seepage. Can also detect faults that may be pathways for gases. Uses parts of visible and IR spectrum.	Spatial resolution of images 1–3 meters. Not calibrated in terms of flux or volume fraction of CO_2 in air or soil gas, but may give indications of areas that should be sampled in detail.	Onshore.

[a] Modified from IPCC (2006).

- Injection well flow rates pressures and temperatures using thermocouples, pressure transducers, borehole logs, including casing integrity logs, temperature logs and radiotracer;

- Fluid/gas chemistry and pH sampling from wells (e.g., U-tube (Freifeld et al. 2005) and down hole pH sensors) and surface samples using geochemical tracers (Stalker et al. 2006); and
- Atmospheric and soil gas chemistry and CO_2 concentrations using eddy towers, soil gas meters, Fourier Transform Infrared Spectroscopy (FTIRS), Perflourocarbon (PFC) and noble gas tracers.

Flow Rates, Pressures and Temperatures. Application of down-hole instruments to measure CO_2 flow rates together with injection pressures and temperatures (both at the wellhead and within the reservoir) is a usual practice, and it can be implemented by employing commercially available sensors (Wright and Majek 1998). These properties enable the verification of the volumes of CO_2 being injected into storage reservoirs that may be applied towards the calculation of carbon credits in Emissions Trading Schemes and for input into fluid flow modeling. They are also commercially used to check the integrity of injection and monitoring wells and to ensure that the reservoir does not reach pressures that would induce slip on existing fractures or create new fractures. Also, it becomes important to monitor temperature of the reservoir as it is an important indicator of the changes in properties of the reservoir (e.g., making it amenable to fracturing) and the state of the CO_2.

Measurement of CO_2 flow rates, pressures and temperatures are imperative for the safe maintenance and effective operation of storage sites. Commercially these measurements are carried out at a number of sites, such as Frio, Texas (Hovorka et al. 2006), Weyburn, Canada (Wilson and Monea 2004) and Otway, Australia (Urosevic et al. 2008). Most of these sites employ techniques that measure pressure in conjunction with temperature using gauges that measure well bore annulus pressure as well as employ orifice differential flow meters (Benson 2007) suitable for operating in remote environments. Fiber-optic cables that supply continuous (e.g., every 15 seconds) measurements from well bores several kilometers below the ground surface are used to connect the instrument to the surface. The important characteristic of sub-surface environments is that pressures generally increase and temperature decreases with rising CO_2 injection rates. There are several factors governing the relationships between injection rates and pressure/temperature conditions, including reservoir permeability and depth, and the conditions in the reservoir prior to injection (e.g., type of fluid present, pressures and temperatures). Hence, the three types of measurements are required to warrant well and reservoir conditions within acceptable limits. Effective management of reservoir can be achieved by decreasing CO_2 injection rates and/or by introducing a pressure relief well(s) to remove water from the reservoir interval.

Location of pressure relief and monitoring wells need to be carefully selected based on the outcome of the site assessment of containment security issues identified and risked. Their cost needs to be balanced against their application towards monitoring and costs of employing alternative pressure management techniques (e.g., decreasing the rate of CO_2 injection). Though offshore monitoring wells are more expensive than local wells, they could be economically viable if there is a need to utilize long reach access wells (located along the coast) to reduce costs.

Fluid Chemistry. In order to track CO_2 migration and understand how the CO_2 is reacting with saline formation water and/or the host rock, geochemical monitoring of groundwater is highly desirable in both shallow and deep subsurface in monitoring wells (Gunter et al. 1998). Fluid and gas reactions may result in removal (dissolution) of host rock near the well bore and/or chemical trapping (precipitation) of carbon as minerals in the reservoir. To enable efficient tracking of fluid chemistry changes in reservoir, fluid sampling in monitoring wells should be carried out before, during, and after injection.

Standard analytical techniques are employed for measuring major ions (e.g., Ca, Mg, Si and SO_4), stable isotopes and gases (e.g., CH_4 and CO_2). Analytical techniques are also widely available for measurement of pH, alkalinity and evaluation of CO_2 flux from carbon dissolved in groundwater (Evans et al. 2002). Great attention needs to be paid to sampling fluids at pressure as depressurization may result in the loss of gas that may significantly change the water chemistry during the analysis of CO_2 dissolved in water. Application of U-tube at Frio, Texas (Hovorka et al. 2006) and Otway, Australia (Stalker et al. 2006) are described to avoid depressurization of low temperature fluid samples (Freifield et al. 2005).

It is imperative to characterize the aquifer fluids adjacent to storage containers chemically and isotopically prior to the commencement of injection due to the presence of many potential sources of CO_2 residing in fluids. Use of tracers will be crucial for a unique distinction and identification of injected CO_2 from other sources of CO_2 in water samples. Both natural (e.g., isotopes of carbon, oxygen or hydrogen) and introduced (e.g., SF_6 and perfluorocarbons) tracers may be employed during the injection of CO_2 (Benson 2007). Along with chemical tracers, tracking fluid pH may provide an empirical evidence of the arrival of the injected CO_2. Tracers and pH changes will be used to track the movement of the injected CO_2 in the storage container, through the seal and, in extreme cases, into the soil and atmosphere. Analytical results for tracers and pH changes enable the confirmation of the arrival of the CO_2 plume at monitoring wells. Sometimes, it is possible to constrain the lateral and vertical migration of the CO_2 plume when samples are taken at multiple locations in the well bore. Also, the application of multiple tracers at different migration rates may enable the description of flow properties of fluids in sub-surface.

Soil Gas. Soil gas flux measurements record flow rates or rates of change of gas concentrations when injected CO_2 leaks from the storage container and reaches the ground surface to accumulate in the soil profile (Oldenburg and Unger 2003). A variety of instruments, including soil probes (e.g., Fig. 6.4) and flux accumulation chambers placed on the ground surface can be used for soil gas flux measurements. Additionally, soil gas chemistry can be determined by the insertion of sample canister (e.g., pipes) into soil and/or pumping soil gas into these canisters. Flux meters, which usually take several minutes for each measurement, are capable of detecting CO_2 fluxes of 0.04 g/m^2-day as compared to gas chemistry samples that must be analyzed in the laboratory and are significantly more time-consuming than the earlier (Klusman 2003). Both flux meters and soil gas chemistry measurements are capable of detecting CO_2 concentrations down to at least ± 1–2 ppm, leaving small footprints.

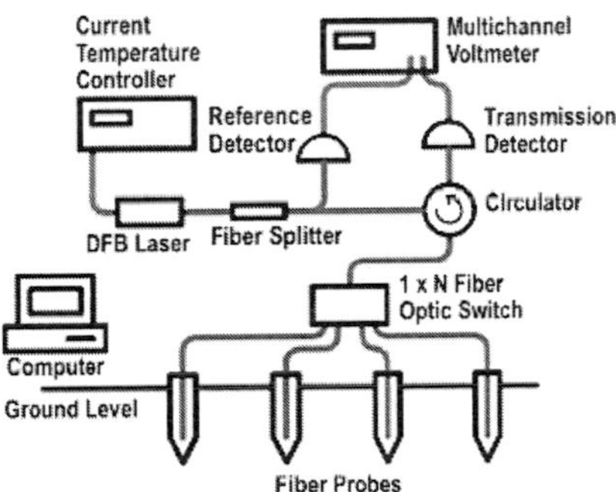

Figure 6.5. Schematic of fiber-optic sensor arrays for soil-CO_2 detection over a large area using solid-state IR sensors developed at Montana State University (NETL 2012).

It is important to differentiate between natural CO_2 in the soil, which may be produced from a number of ecosystems and geological sources and the injected CO_2 that leaked into the soil. There are many ways to distinguish between the different sources of CO_2, including (a) pre-injection baseline surveys to distinguish natural CO_2 in the soil from the injected CO_2 leaked into the soil, (b) recognition of the isotopic signatures of natural CO_2 sources, and (c) using chemical tracers to fingerprint the anthropogenic CO_2. Leakage of CO_2 in volcanic regions may have a profound influence on ecosystems close to (e.g., < 10 m) emission sources (Vodnik et al. 2006; Beaubien et al. 2008). Additional monitoring for recording the impact on ecosystems at the surface and within aquifers is encouraged.

Atmospheric Gas. Atmospheric gas measurements are carried out to obtain continuous and precise measurements of CO_2 concentrations. Point measurements (using closed path analyzers or samplers) or line integrated measurements (using open path analyzers) should be carried out both up and downwind of the storage area. Closed path analysis techniques include NDIR, cavity ring down spectroscopy (CRDS), gas chromatography (GC), Mass Spectrometer (MS) and tuneable diode laser spectrometer (TDLS). All these techniques along with Fourier transform infrared spectroscopy (FTIR) can be employed for the open path analysis upon connection to line path inlets. Open path analyzers that employ infrared lasers have the advantage of detecting leakage over a wider area than point source measurements.

Accurate measurement and recording of the location and rate of CO_2 leaks will require that the background signal is accounted towards the measurements. Monitoring of the background carbon composition and flux prior to injection is imperative in order to achieve this goal. Background concentration should account of CO_2 generated by the ecosystem and anthropogenic activities (e.g., oil and gas infrastructure, urban centers and agriculture). The use of tracers will enhance detection and quantification of CO_2 leakage into the atmosphere. Tracers can be used to detect background atmospheric carbon or from sub-surface natural sources or may be injected along with CO_2. Variations result when the injected gas is isotopically different from the background sources, thus enhancing the application of isotopes of CO_2 and CH_4 as tracers (Leuning et al. 2008). As these carbon isotopes are natural, they migrate almost identically to the injected CO_2; hence their use is highly encouraged. The injected tracers gain the capability to tag the injected CO_2 with a specific chemical signature that clearly distinguishes it from background CO_2 sources. However, application of tracers may be governed by environmental concerns and regulations; hence the use of injected tracers is limited. As an example, sulphur hexafluoride (SF_6) is relatively cheap and easily measured at ppt levels, but like many similar compounds is a powerful greenhouse gas. New spectroscopic measurement technologies are emerging to deal with these demands for CO_2, CH_4 and their isotopes.

Offshore monitoring of atmospheric gas is more challenging. Leakage from the seafloor is likely to be significantly modified by the overlying water, with signals being significantly modified and dispersed to prevent them from reaching the atmosphere. Despite these modifications, preliminary calculations suggest that chemical signals resulting from 1000 tCO_2/year leakage be significant against the background levels in the ocean. Sea-floor monitoring is less developed in spite of improved capabilities to detect and monitor sub-surface changes including pH. Sea-floor deployment and data transmission back to surface are extra difficulties. Sea-floor MVA capability for measuring gas chemistry and fluxes still need lot of research and development.

6.4.3 Techniques for Ecosystem Stress Monitoring

Plants are susceptible to stress caused by elevated levels of CO_2 in the soil, and measurements of vegetative stress can be used as an independent indicator of possible CO_2 release from the subsurface. Vegetative stress can be measured by aerial photography, satellite imagery, and spectral imagery. Initial surveys are required to establish baseline conditions, including seasonal changes that take place at a particular site, as well as natural variations in temperature, humidity, and light and nutrient availability at the site. Once the baseline is established, anomalous vegetative stress may be observed.

Hyperspectral imaging collects and processes radiation across a broad portion of the electromagnetic (EM) spectrum, typically including wavelengths from 400 to 900 nanometers. This includes the high absorbance region in the visible spectrum associated with chlorophyll absorbance, and high reflectance in the near-IR region that is typical of spongy leaf tissues. Spectral imaging has the ability to detect changes in light reflectance and absorption that occur in vegetation that is struggling. Multispectral imaging may be simpler and less costly, and it affords continuous daytime operation in both clear and cloudy weather (Rouse et al. 2010). Whereas hyperspectral imaging collects a continuous spectrum of wavelengths, multispectral imaging collects discrete spectral bands. Spectral imaging sensors may be airborne, satellite-mounted, or handheld.

Pickles and Cover (2005) proposed the use of satellite- or airborne-based spectral imaging to assess vegetative stress over a large area. Remote sensing techniques were tested in central Italy in 2005 to detect CO_2 emanating from natural seeps at the Latera caldera (Bateson et al. 2008). Hyperspectral imaging, multispectral imaging, LIDAR (Light Detection and Ranging), orthophoto, and high-resolution photographic data were all acquired during two airborne surveys over an area with known CO_2 gas venting. These imaging methods were successful in locating some, but not all of the major gas vents. The researchers concluded that different remote sensing techniques work best in different conditions, depending on the amount of vegetation and steepness of topography at the site, and depending on the season and time of day during which the data are collected. In all cases, complementary soil-gas geochemical data were required to interpret the remote sensing results in terms of CO_2 concentrations and flux rates.

Researchers at the MGSC Phase II Sugar Creek site in Kentucky tested several monitoring techniques, including aerial hyperspectral imaging, during a real, short-duration CO_2 release from a buried pipeline (Wimmer et al. 2010). DOE's Core R&D Program carried out a controlled release experiment at the Naval Petroleum Reserve Site #3 in Wyoming in 2006. Aerial hyperspectral imagery was acquired using Moderate Resolution Imaging Spectroradiometer/Advanced Spaceborne Thermal Emission Reflection Radiometer (MASTER) technology. Data analysis demonstrated that MASTER could identify major CO_2 and CH4 surface seeps.

In summary, sustained releases of significant CO_2 flux are detectable with hyperspectral and multispectral imaging techniques, and the vegetative stress indicators have been found to be proportional to soil CO_2 levels and proximity to the CO_2 release.

6.4.5 *MVA Data Integration and Analysis*

Throughout this chapter, numerous tools and technologies used to collect CO_2 monitoring data in the atmosphere, the near-surface zone, and the subsurface have been discussed. There are also a number of cross-cutting technologies being developed to better integrate and analyze the wide variety of monitoring data that are acquired. These data integration and analysis technologies include computer-based intelligent monitoring networks and advanced data integration and analysis software tools. Intelligent monitoring networks are automated, computer-based systems that gather field information from injection and monitoring equipment, evaluate GS conditions, and recommend appropriate actions. Systematic data collection, analysis, and modeling are key components of these systems. Intelligent monitoring networks are designed to show that site performance meets pre-defined objectives and to ensure release of CO_2 is promptly identified and mitigated.

An intelligent monitoring network may combine data from CO_2 monitoring wells, surface monitoring sensors, subsurface monitoring tools, and injection equipment. The data are compiled in real time in a database that is updated continuously. The intelligent monitoring network may also compare field data to available models and historical field data. Measurements that lie outside normal operational limits or historical trends are flagged as potential risks. In some cases, an intelligent monitoring network may determine the cause of an anomaly and proceed to rectify the problem. If a CO_2 transport line registers an increase in pressure, for example, the monitoring network may decrease the flow rate or utilize a bypass line. The system may also recommend action items based on analysis of the field data. For example, if a surface sensor shows increased levels of CO_2, the system may recommend further investigation in the vicinity of the sensor and specify potential release pathways present in the area. This information will aid field operators in promptly locating and identifying a release.

The selection of sensors and methods employed in a monitoring network is site-specific and requires testing, planning, and scheduling within the project plan. Conditions that may affect the selection of monitoring network components include site access, surface geography, type and complexity of storage formation, and size of the monitoring area. Project developers may perform a risk assessment of the site in order to determine the appropriate techniques required to monitor and mitigate risks. Smart-well technology may be utilized to provide real-time well data to a monitoring network. Smart wells contain permanent, downhole sensors and flow equipment that allow for continuous monitoring and regulation of fluid flow, formation pressure, and formation temperature in the injection formation

6.5 Two Case Studies

6.5.1 SECARB Phase III "Early Test" at Cranfield Field

The SECARB Phase III "early test" is underway at Cranfield Field, approximately 10 miles east of Natchez, Mississippi. The Validation Phase test is focused on the Denbury Onshore, LLC CO2-EOR project in the depleted oil reservoir, and the Development Phase test is focused on the downdip water leg on the east side of the same reservoir. At Cranfield, the lower Tuscaloosa D-E sandstone, a 60- to 80-foot thick injection zone, is in a broad four-way structural closure at a depth greater than 10,000 feet. Complexly incised channels form a regionally continuous sandstone flow unit with lateral variability in permeability over short distances. Reservoir-scale vertical compartmentalization has isolated oil charge to the lower part of the lower Tuscaloosa Formation at Cranfield. The middle marine Tuscaloosa forms the lowest regional confining zone.

Monitoring Plan, Results, and Lessons Learned. The monitoring plan was targeted to the research goals of the RCSP Development Program: (1) evaluation of protocols to demonstrate that it is probable to retain 99 percent of CO_2, and (2) predict storage capacities within ±30 percent. Observations were linked through a large number of models, allowing the significance of the measurement to be assessed. Some monitoring data were collected at points distributed across the study area and at a wide range of time intervals; other data sets were collected in focused study areas or during intensive sampling campaigns (Hovorka and others 2009; 2011).

The SECARB early test at Cranfield was highly leveraged by participation of groups that brought non-SECARB-funded expertise to the project. For example, the project hosted experiments funded by the National Risk Assessment Program (NRAP); the Research Institute of Innovative Technology for the Earth (RITE); the DOE-funded SIM-SEQ; Stanford, Princeton, and CCP rock-physics analyses; American Water Works Association (AWWA) funded controlled release; analyses by ORNL; University of Tennessee-funded biological sampling; BP test of wellbore gravity; and Scottish Carbon Capture and Storage (CCS) Centre-funded noble gas sampling.

Three Findings Relevant to Future Monitoring. These are (1) high-frequency pressure data contains information about reservoir response; however, all the events have to be recorded at the same frequency (minutes to hours); (2) low-cost, easily repaired wellhead tubing pressure gauges have value if calibrated to density of fluid in tubing; and (3) doubt remains that injection zone mass balance or pressure monitoring would be sufficient to detect release, mostly because of large uncertainties about boundary conditions.

Four-dimensional seismic and time-lapse vertical seismic profiling (VSP) data were collected to explore the uncertainty of downdip, off-structure, and out-of-injection-zone migration of CO_2. Injected CO_2 was successfully detected in the injection zone, subtracting the pre-injection 3-D survey from the 2010 repeat survey; however, noise is high. Resolution of these methods is limited in terms of their ability to detect thin, saturated zones; heterogeneous reservoir zones; and complex fluids. No above-zone migration of CO_2 has been detected.

Other Findings Relevant to Future Monitoring. Above Zone Monitoring Interval (AZMI) pressure monitoring shows promise as a sensitive release detection method. In future installations, it is recommended that baseline hydrologic characterization of the AZMI interval, as well as well construction, is invested more heavily to ensure that AZMI pressure gauge is well-connected to the formation and isolated from well construction.

6.5.2 BSCSP Kevin Dome Phase III Development Test

BSCSP is in the early stages of conducting a large-scale storage test at Kevin Dome in North Central Montana. The Dome is an ~700-mi^2 feature extending from Shelby, Montana, to just south of the Canadian border. It contains naturally occurring CO_2 in Devonian Duperow (dolostone), which was likely generated via geochemical reactions caused by a sweep of hot fluids initiated by igneous intrusions that formed the Sweetgrass Hills to the southeast of the dome. The CO_2 resides in a 100-foot-thick porous section in the middle Duperow and in a thinner porous section in the lower Duperow. Estimated CO_2 in place is ~0.6 GT, or 10 TCF, equivalent to Jackson Dome. The CO_2 is estimated to have an areal extent of ~540 mi^2 and does not fill the dome to its spill point. The Kevin Dome project plans to drill and core producing wells, produce the natural CO_2, pipe it laterally 6 to 8 miles, and re-inject into the Duperow porosity zone in the brine. leg. The primary seal is the upper Duperow (~200 feet of tight carbonate with inter-bedded anhydrites), and the secondary seal is the Potlach Anhydrite (175 feet), with multiple tertiary seals that have contained oil and gas in shallower horizons. This project will combine studies of natural reservoir storage capacity and carbonate geochemistry with studies of engineered injection and storage. While the injection is into a saline formation, the project provides valuable information concerning the use of structural features for CO_2 warehousing in a regional CCUS hub concept.

Monitoring Plan, Results, and Lessons Learned. Descriptions about some of the current status of the project are as follows:

Monitoring Wells. Three to four monitoring wells are planned. One will be placed more distal to the injector, updip, with an estimated breakthrough of ~750,000 tonnes CO_2 injected. The remaining monitoring wells will be placed symmetrically about the injector at the appropriate crosswell seismic distance. At least two wells will be used for geochemical fluid sampling and tracer studies using U-tubes.

Seismi. The planned geophysical program is designed to use the highest resolution, greatest sensitivity method applicable to image the current plume dimensions. Resolution and areal extent are addressed by use of both borehole and surface seismic methods.

Surface Seismic. As mentioned previously, a 58-mi^2, 3-D, nine-component survey is underway. This survey serves multiple purposes: (1) it will be used for hazard identification and avoidance; (2) it will provide data to the static geologic model in the site characterization phase; (3) it will provide a test of potential for multi-component seismic detection of CO_2 without time-lapse (spatially, because it is being shot across the gas-brine interface); and (4) it will serve as a baseline for subsequent surveys used for time-lapse monitoring of the plume.

Vertical Seismic Profiling (VSP). Vecta's vibroseis trucks will be used with downhole, multi-component receivers in the monitoring wells to perform 3-D and 4-D nine-component VSP. Crude preliminary simulations indicate the CO_2 plume can be imaged for three to four years via this technique. VSP is intermediate in resolution and areal coverage to crosswell and surface seismic.

Geochemical Monitoring. Up to four U-tubes will be deployed in monitoring wells to collect fluid samples. In addition to pH, alkalinity, cation and anion analysis, rare earth elements will be analyzed and tracers (including phase partitioning tracers) will be used to study geochemistry. One of the U-tubes will likely be used to monitor above injection zone fluids.

Assurance Monitoring. Soil flux chambers, EC towers, DIAL, and hyperspectral imaging will all be used in the Assurance Monitoring Program. Additionally, drinking water and surface water analysis will be performed in the vicinity of the injection.

6.6 Current Issues and Future Research Needs

Currently, many problems exist, such as detection limits and precision levels of different methods have not been completely established; strategies for different locations have not been fully established. The procedures for detecting, locating and then quantifying leakage have not been developed. Sensitivity analysis of different methods is still in their infancy. Current underground storage accounting is at best qualitative. For example, seismic data can show where CO_2 exists qualitatively but not quantitatively. Similarly, it is difficult to use chemical samples to verify storage, since CO_2 can take on many different forms and will mingle with carbon resources that are at the site prior to injection (Lankner and Brennan 2009). Several methods have been proposed to improve accounting of stored carbon, such as using C-14 as a tracer for a) monitoring fluxes from geologic sequestration (Bachelor et al. 2008), b) facilitating measurement via sampling (Landcar and Brennan 2009), c) optical techniques with path lengths of ~1 km, and d) computer simulation and model development.

In the future, improvement is needed for a) direct emission measurements from existing CO_2-EOR projects, b) controlled release experiments for demonstrating the ability to detect, locate and quantify emissions in various settings, c) best practices and procedures that can be used to respond to any detected changes, d) approaches to distinguish natural ecosystem fluxes, and other anthropogenic emissions from geological storage reservoir emissions, and e) improve detection of small secondary accumulations of CO_2.

6.7 Conclusions

MVA features are common to all the sites whether they are onshore or offshore. They are required to track the sub-surface location of the injected plume and to detect any CO_2 movement into the shallow subsurface or leakage at the ground surface (or sea bed). Baseline surveys will be required for most of the techniques used to establish site conditions prior to injection. Chemical tracers (either natural or injected) can be used to enable positive identification of injected CO_2. All sites will need a risk management plan which outlines remediation measurements or adjustments to the MVA program throughout the project life to reduce the risk associated with unexpected migration or CO_2 movement from the primary reservoir.

Numerous geophysical techniques may help define the location of injected CO_2 plumes in the sub-surface. Of these techniques, time-lapse 3-D seismic reflection surveys show the greatest promise for offshore sites This technique will probably have greater utility offshore than onshore because in the former case it is often of a higher quality and both easier and cheaper to acquire. This technique will likely be most useful for tracking CO_2 plume migration in saline reservoirs at depths of up to 2 km.

Direct observations of the reservoir interval, CO_2 plume, and gas in sea water above the storage container are desirable for all future storage projects. As usual, the following should be monitored/measured: i) fluid chemistry and pressure (and temperature) recording at the wellhead; ii) the optimal locations of these wells will be determined using CO_2 reservoir flow simulations for the storage system, iii) water/gas chemistry samples must be taken without being depressurized (e.g., using U-tube device), and iv) tracers must be used to confidently identify the injected CO_2 in fluid samples. Offshore monitoring of atmospheric gas presents additional challenges and may not be feasible because the signals are modified and dispersed to the point that they may not reach the atmosphere (this is particularly so with increasing water depth. Further development of these sea-floor water-gas chemistry and flux monitoring systems (including the design and implementation of their sea floor deployment and data transmission back to surface) is required before fully operational systems will be available for offshore storage areas.

6.8 List of Acronyms and Abbreviations

ANSI	American National Standards Institute
AZMI	Above Zone Monitoring Interval
CCS	Carbon Capture and Storage
CCCSRP	California Carbon Capture and Storage Review Panel
CCX	Chicago Climate Exchange
CRDS	Cavity Ring Down Spectroscopy
EM	ElectroMagnetic
EOR	Enhanced Oil Recovery
ERT	Electrical Resistance Tomography
FTIRS	Fourier Transform Infrared Spectroscopy
GC	Gas Chromatography
GHG	Greenhouse Gas
GPS	Global Positioning System
IPCC	Intergovernmental Panel on Climate Change
ISO	International Organization for Standardization
LIDAR	Light Detection and Ranging
NETL	National Energy Technology Laboratory
MASTER	Moderate Resolution Imaging Spectroradiometer/Advanced Spaceborne
TERR	Thermal Emission Reflection Radiometer
MS	Mass Spectrometer
MVA	Monitoring, Verification and Accounting
NETL	National Energy Technology Laboratory
PFC	Perflourocarbon
PPM	Parts Per Million
RCGI	Regional Greenhouse Gas Initiative
RCSP	Recovery Community Services Program
SDWA	Safe Drinking Water Act
SF6	Sulphur hexaflouride
TDLS	Tuneable Diode Laser Spectrometer
U.S. DOE	U.S. Department of Energy
U.S. EPA	U.S. Environmental Protection Agency
UNFCC	UN Frame Work Convention on Climate Change
VSP	Vertical seismic profiling
WRI	World Resources Institute

6.9 References

Anderson, S., and Newell, R. (2003). "Prospects for Carbon Capture and Storage Technologies." Discussion Paper published by Resources for the Future, Washington, DC. Available at <http://rff.org/rff/Documents/RFF-DP-02-68.pdf> (accessed Feb. 2012).

Angerer, E., Crampin, S., Li, X.Y., and Davis, T.L. (2001). "Processing and modeling time-lapse effects of over-pressured fluid-injection in a fractured reservoir." *Geophysical Journal International*, 149, 267–280.

Arts, R., Chadwick, A., Eiken, O., Thibeau, S., and Nooner, S. (2008). "Ten years' experience of monitoring CO_2 injection in the Utsira Sand at Sleipner, offshore Norway." *First Break*, 26. 65–72.

Arts, R., Eiken, O., Chadwick, A., Zweigel, P., van der Meer, B., and Kirby, G., (2004). "Seismic monitoring at the Sleipner underground CO_2 storage site (North Sea)." *Geological Society Special Publication*, 233. 181–191.

ANSI (2007). *New Standard Offers Tool to Address Climate Change*. American National Standards Institute, April 17, 2007. Available at <http://www.ansi.org/news_publications/news_story.aspx?menuid=7&articleid=1472> (Accessed June 8, 2008).

Bachelor, P.P., McIntyre, J.I., Amonette, J.E., Hayes, J.C., Milbrath, B.D., and Saripalli, P. (2008). "Potential method for measurement of CO_2 leakage from underground sequestration fields using radioactive tracers." *J. Radioanal. Nucl. Chem.*, 277(1), 85–89.

Bannister, S., Nicol, A., Funnel, R., Etheridge, D., Christenson, B., Underschultz, J., Caldwell, G., Stagpoole, V., Kepic, A., and Stalker, L. (2009). *Opportunities for underground geological storage of CO_2 in New Zealand*. Report CCS-08/11. GNS Science Report 2009/64.

Bateson, L., Vellico, M., Beaubien, S.E., Pearce, J.M., Annunziatellis, A., Ciotoli, G., Coren, F., Lombardi, S., and Marsh, S. (2008). "The application of remote sensing techniques to monitor CO_2 storage sites for surface leakage: method development and testing at Latera (Italy) where naturally-produced CO_2 is leaking to the atmosphere." *International Journal of Greenhouse Gas Control*, 2, 388–400.

Benson, S.M. (2007). "Monitoring geological storage of carbon dioxide." In: Wilson, E.J., and Gerard, D. (eds). *Carbon Capture and Sequestration: Integrating Technology, Monitoring and Regulation*. Blackwell Publishing, USA, 73–101.

Benson, S.M., Hoversten, E., and Gasperikova, E. (2004). *Overview of monitoring techniques and protocols for geologic storage projects*. IEA Greenhouse Gas R&D Program Report PH4/35. 108 p.

Brantley, S.L., and Koepenick, K.W. (1995). "Measured carbon dioxide emissions from

Oldoinyo Lengai and the skewed distribution of passive volcanic fluxes." *Geology,* 23(10), 933–936.

Brown, S., Bussod, G., and Hagin, P. (2007). "AVO monitoring of CO_2 sequestration: A benchtop-modeling study." *The Leading Edge*. 26(12), 1576–1583.

Beaubien, S.E., Ciotoli, G., Coombs, P., Dictor, M.C., Kruüger, M., Lombardi, S., Pearce, J.M., and West, J.M. (2008). "The impact of a naturally occurring CO_2 gas vent on the shallow ecosystem and soil chemistry of a Mediterranean pasture (Latera, Italy)." *Int. J. Greenhouse Gas Control*, 2, 373–387.

CCCSRP (California Carbon Capture and Storage Review Panel) (2010). *Findings and Recommendations by the California Carbon Capture and Storage Review Panel.* CCCSRP, Dec. 2010. Available at <www.climatechange.ca.gov/carbon_capture_review_panel/.../2011-01-14...> (accessed Feb. 2012).

Chadwick, A., Noy, D., Arts, R., and Eiken, O. (2008). "Latest time-lapse seismic data from Sleipner yield new insights into CO_2 plume development." *Proceedings of the Greenhouse Gas Control Technologies Conference (GHGT-9)*, Washington D.C., USA.

Chadwick, A., Arts, R., Bernstone, C., May, F., Tibeau, S., and Zweigel, P. (eds) (2007). *Best Practise for the Storage of CO_2 in Saline Aquifers–Observations and guidelines from the SACS and CO2STORE projects.* Nottingham, British Geological Survey, pp 273.

Chadwick, A., Arts, R., and Eiken, O. (2005). "4D seismic quantification of a growing CO_2 plume at Sleipner, North Sea." In: Dore, A.G., and Vining, B.A. (eds), *Petroleum Geology: North West Europe and Global Perspectives–Proceedings of the 6th Petroleum Geology Conference. Geological Society*, London. 1385–1399.

Christensen, N.B., Sherlock, D., and Dodds, K. (2006). "Monitoring CO_2 injection with cross-hole electrical resistivity tomography." *Exploration Geophysics*, 37, 44–49.

Cooperative Research Centre for Greenhouse Gas Technologies (CO2CRC). *A review of existing best practice manuals for carbon dioxide storage and regulation.* Available at <http://cdn.globalccsinstitute.com/sites/default/files/publications/14591/> (accessed May 2013).

DNV (2010a). *CO2QUALSTORE Guideline for selection and qualification of sites and projects for geological storage of CO_2.* Det Norske Veritas, DNV Report No.: 2009-1425.

Dodds, K., Sherlock, D., Urosevic, M., Etheridge, D., de Vries, D., and Sharma, S. (2006). "Developing a Monitoring and Verification Scheme for a Pilot Project, Otway Basin, Australia." *Proceedings of the Greenhouse Gas Control Technologies Conference (GHGT-8)*, Trondeim, Norway, June 2006. ISBN: 0-08-046407-6, Elsevier.

European Commission (EC) (2011). *Implementation of Directive 2009/31/EC on the Geological Storage of Carbon Dioxide, Guidance Document 2: Characterisation of the Storage Complex, CO_2 Stream Composition, Monitoring and Corrective*

Measures. Available at: http://ec.europa.eu/clima/policies/lowcarbon/ccs/implementation/docs/gd2_en.pdf. (accessed March 2013).

Eiken, O., Brevik, I., Arts, R., Lindeberg, E., and Fagervik, K., (2000). *Seismic monitoring of CO_2 injected into a marine aquifer.* 70th Annual Meeting. Society of Exploration Geophysics. SEG Expanded Abstracts 19, 1623–1626. DOI:10.1190/1.1815725.

Evans, W.C., Sorey, M.L., Cook, A.C., Kennedy, B.M., Shuster, D.L., Colvard, E.M., White, L. D., and Huebner, M. A., (2002). "Tracing and quantifying magmatic carbon discharge in cold groundwaters: Lessons learned from Mammoth Mountain, USA." *Journal of Volcanology and Geothermal Research*, 114(3–4) 291–312.

Federal Register (2008). *Part II Environmental Protection Agency, 40 CFR Parts 144 and 146, Federal Requirements Under the Underground Injection Control (UIC) Program for Carbon Dioxide (CO_2) GS (GS) Wells; Proposed Rules*, July 25, 2008, p. 43492-43541, http://edocket.access.gpo.gov/2008/pdf/E8-16626.pdf (accessed March 2013).

Freifield, B.M., Trautz, R.C., Kharaka, Y.K., Phelps, T.J., Myer, L.R., Hovorka, S.D., and Collins, D.J. (2005) "The U-tube: A novel system for acquiring borehole fluid samples from a deep geologic CO_2 sequestration experiment." *Journal of Geophysical Research*, 110, B10203. DOI: 10.1029/2005JB003735.

Frontier, W. (2006). *Safe storage of CO_2: experience from the natural gas storage industry*. IEA Greenhouse Gas R & D Program, Technical Study, Report Number: 2006/2. January 2006. 123 p.

Funnell, R., King, P., Edbrooke, S., Bland, K., and Field, B. (2009). *Opportunities for underground geological storage of* CO_2 *in New Zealand* - Report CCS-08/1 – Waikatoand onshore Taranaki overview. GNS Science Report 2008/53.

Gasperikova, E., and Hoversten, G.M., (2006). "A feasibility study of nonseismic geophysical methods for monitoring geologic CO_2 sequestration." *The Leading Edge*, 25(10), 1282–1288. DOI:10.1190/1.2360621.

Gunter, W.D., Chalaturnyk, R.J., and Scott, J.D., (1998). "Monitoring of aquifer disposal of CO_2: Experience from underground gas storage and enhanced oil recovery." In: Eliasson, B., Riemer, P., and Wokaun, A., (eds). *Proceedings of the Greenhouse Gas Control Technologies Conference (GHGT-4)*, Interlaken, Switzerland, August-September 1998. Elsevier. 151–156.

Herzog, H.J. (2001). "What future for carbon capture and sequestration?" *Environ. Sci Technol.*, 35(7), 148A–153A.

Hovorka, S.D., Benson, S.M., Doughty, C., Freifeld, B.M., Sakurai, S., Daley, T.M., Kharaka, Y.K., Holtz, M.H., Trautz, R.C., Nance, H.S., Myer, L.R., and Knauss, K.G. (2006). "Measuring permanence of CO_2 storage in saline formations: the Frio experiment." *Environmental Geosciences*, 13(2), 105–121.

Hovorka, S.D., Meckel, T.A., Trevino, R.H., Lu, J., Nicot, J.P., Choi, J. -W., Freeman, D., Cook, P., Daley, T.M., Ajo-Franklin, J.B., Freifeld, B.M., Doughty, C.,

Carrigan, C.R., La Brecque, D., Kharaka, Y.K., Thordsen, J.J., Phelps, T.J., Yang, C., Romanak, C.D., Zhang, T., Holt, R.M., Lindler, J.S., and Butsch, R.J. (2011). "Monitoring a large volume CO_2 injection: Year two results from SECARB project at Denbury's Cranfield, Mississippi, USA." *Energy Procedia,* 4, 3478–3485.

Hovorka, S.D., Meckel, T.A., Treviño, R.H., Nicot, J.–P., Choi, J.–W., Yang, C., Paine, J., Romanak, K., Lu, J., Zeng, H., and Kordi, M. (2009) *Southeast Partnership early test update—Cranfield field, MS: GCCC Digital Publication #09-05.*

Hoversten, G.M., Gritto, R., Daley, T.M., Majer, E.L., and Myer, L., (2002). "Crosswell seismic and electromagnetic monitoring of CO_2 sequestration." In: Gale, J and Kaya, I (eds). *Sixth International Conference on Greenhouse Gas Control Technologies (GHGT-6)*, Kyoto, Japan, October 2002. Elsevier. 371–376.

Hoversten, G.M., and Gasperikova, E., (2005). "Non-seismic geophysical approaches to monitoring." In: Thomas, D.C. and Benson, S.M. (ed.) *Carbon dioxide Capture for Storage in Deep Geologic Formations - Results from the CO_2 Capture Project. Volume 2: Geological Storage of Carbon Dioxide with Monitoring and Verification.* Elsevier, Oxford, 1071–1112. IEA (International Energy Agency) (2009). *Technology Roadmap: Carbon Capture and Storage.* An IEA report, Available at <http://www.iea.org/papers/2009/CCS_Roadmap.pdf> (accessed Feb. 2012).

Intergovernmental Panel on Climate Change (IPCC) (2005). *IPCC Special Report of Carbon Dioxide Capture and Storage.* Prepared by Working Group III of the Intergovernmental Panel on Climate Change, Metz, B., Davidson, O., de Coninck, H.C., Loos, M., and Meyer, L.A. (eds.). Cambridge University Press, New York.

Intergovernmental Panel on Climate Change (IPCC) (2006). IPCC *Guidelines for National Greenhouse Gas Inventories, Prepared by the National Greenhouse Gas Inventories Programme*, Eggleston, H.S., Buendia, L., Miwa, K., Ngara, T., and Tanabe, K. (eds). Published: IGES, Japan, Volume 2 Energy, Chapter 5 Carbon Dioxide transport, injection and geological storage.

International Risk Governance Council (2008). *Policy Brief: Regulation of Carbon Capture and Storage.* Available at http://www.irgc.org/IMG/pdf/Policy_Brief_CCS.pdf, ISO, 2006. New ISO 14064 standards provide tools for assessing and supporting greenhouse gas reduction and emissions trading, March 3, 2006, Accessed June 8, 2008. http://www.iso.org/iso/pressrelease.htm?refid=Ref994.

ISO, 2007. *ISO 14065 standard–new tool for international efforts to address greenhouse gas emissions, International Organization for Standardization*, April 17, 2007, Accessed June 8, 2008. http://www.iso.org/iso/pressrelease.htm? refid=Ref1054.

Kessels, J.R., and Matheson, T.W. (2002). "Technical and policy issues associated with sequestration of carbon dioxide emissions." In: *2002 New Zealand Petroleum Conference proceedings*, 24–27 February 2002, Carlton Hotel, Auckland, New

Zealand. Wellington: Publicity Unit, Crown Minerals, Ministry of Economic Development. 128–132.

Klusman, R.W. (2003). "Rate measurements and detection of gas micro-seepage to the atmosphere from an enhanced oil recovery/sequestration project, Rangely, Colorado, USA." *Applied Geochemistry* 18(12), 1825–1838.

Lackner, K.S., and Brennan, S. (2009). "Envisioning carbon capture and storage: expanded possibilities due to air capture, leakage insurance, and C-14 monitoring." *Climatic Change*, 96, 357–378.

Lawrence Berkeley National Laboratory (LBNL) (2004) *GEO-SEQ Best Practices Manual, Geologic Carbon Dioxide Sequestration: Site Evaluation to Implementation*. U.S. Department of Energy, Office of Fossil Energy, September 30, 2004.

Litynski, J.T., Plasynski, S., McIlvried, H.G., Mahoney, C., and Srivastava, R.D. (2008). "The United States Department of Energy's Regional Carbon Sequestration Partnerships Program Validation Phase." *Environ. International,* 34, 127–138.

Leuning, R., Etheridge, D., Luhar, A., and Dunse, B. (2008). Atmospheric monitoring and verification technologies for CO_2 geosequestration. International *Journal of Greenhouse Gas Control,* 2(3), 401–414.

Majer, E.L., Daley, T.M., Korneev, V., Cox, D., and Peterson, J.E. (2006). "Cost-effective imaging of CO_2 injection with borehole seismic methods." *The Leading Edge*., 25(10), 1290–1302.

McCurdy, M., Clemens, A., Matheson, T., and Rumsey, B. (2009). *Opportunities for underground geological storage of CO_2 in New Zealand.* Report CCS-08/9-Technical reviews on carbon, capture, transport and injection technologies. GNS Science Report, 2009/62.

Maxwell, S.C., and Urbancic, T. (2001). "The role of passive microseismic monitoring in the instrumented oil field." *The Leading Edge,* 20(6), 636–639.

Maxwell, S.C., White, D.J., and Fabroil, H. (2004). "Passive seismic imaging of CO_2 sequestration at Weyburn." SEG abstract, 74th Annual Meeting, 2004.

Miles, N.L., Davis, K.J., and Wyngaard, J.C. (2005). "Detecting leaks from below ground CO_2 reservoirs using eddy covariance." In: Benson, S.M. (ed.) *Carbon Dioxide Capture for Storage in Deep Geologic Formations*, Volume 2, 1031-1044. Elsevier Ltd, Oxford.

Myer, L.R., Hoversten, G.M., and Gasperikova, E. (2002) "Sensitivity and Cost of Monitoring Geologic Sequestration Using Geophysics." In: Gale, J., and Kaya, I. (eds). *Sixth International Conference on Greenhouse Gas Control Technologies (GHGT-6)*, Kyoto, Japan, October 2002. Elsevier.

National Energy Technology Laboratory (NETL) (2009a). *Monitoring and Verification, andaccounting of CO_2 stored in deep geological formations*. DOE/NETL-311/081508, Availablehttp://fossil.energy.gov/news/techlines/2009/09016\DOE_ReleasesMVA_Report.html.

NETL (2009b). http://www.co2captureandstorage.info/co2tool_v2.2.1/co2tool_panel.php.

NETL (2012). *Best Practices for Monitoring, Verification, and Accounting of CO_2 Stored in Deep Geologic Formations – 2012 Update.* DOE/NETL-2012/1568, Oct. 2012.

Oldenburg, C.M., and Unger, A.J. (2003). "On leakage and seepage from geologic carbon sequestration sites: unsaturated zone attenuation." *Vadose Zone Journal* 2, 287–296.

Onuma, T., and Ohkawa, S. (2008). *Detection of surface deformation related with CO_2 injection Conference (GHGT-9),* Washington DC, United States of America.

Pearce, J.M., Chadwick, R.A., Bentham, M., Holloway, S., and Kirby, G.A. (2005). *Monitoring technologies for the geological storage of CO_2.* UK Department of Trade and Industry Technology Status Review Report. 98 p. No DTI/Pub URN 05/1033.

Pearce, J.M., Chadwick, R.A., Kirby, G., and Holloway, S. (2006). "The objectives and design of generic monitoring protocols for CO_2 storage." *Proceedings of the Greenhouse Gas Control Technologies Conference (GHGT-8),* Trondeim, Norway, June 2006. ISBN: 0-08-046407-6, Elsevier. Published on CD.

Pickles, W.L., and Cover, W.A. (2005) "Hyperspectral geobotanical remote sensing for CO_2 storage monitoring." In *Carbon Dioxide Capture for Storage in Deep Geologic Formations—Results from the* CO_2 *Capture Project,* Thomas, D., and Benson, S. (eds.), Amsterdam, Elsevier. 2, 1045–1070.

Preston, C., Monea, M., Jazrawi, W., Brown, K., Whittaker, S., White, D., Law, D., Chalaturnyk, R., and Rostron, B. (2005). "IEA GHG Weyburn CO_2 monitoring and storage project." *Fuel Processing Technology,* 86(14 –15), 1547–1568.

Riemer, P., Audus, P., and Smith, A. (1993). *Carbon Dioxide Capture from Power Stations. Cheltenham, United Kingdom*: International Energy Agency Greenhouse Gas R&D Program.

Rouse, J.H., Shaw, J.A., Lawrence, R.L., Lewicki, J.L., Dobeck, L.M., Repasky, K.S., and Spangler, L.H. (2010). "Multi-spectral imaging of vegetation for detecting CO_2 leaking from underground." *Environmental Earth Sciences*, 60(2), 313–323.

Rubin, E.S., McCoy, S.T., and Apt, J. (2007). "Regulatory and Policy Needs for Geological Sequestration of Carbon Dioxide." http://www.irgc.org/Expert-contributions-and-workshop.html. (accessed August 29, 2012).

Socolow, R., Hotinski, R., Greenblatt, J.B., and Pacala, P. (2004). "Solving the climate problem – Technologies available to curb CO_2 emissions." *Environment* 46 (10), 8–19.

Stalker, L., Boreham, C., and Perkins, E., (2006). *A review of tracers in monitoring CO_2 breakthrough: Properties, uses, case studies and novel tracers.* A report for the CO2CRC *Report No: RPT06-0070. 32pp.*

Tans, P. (2008). *Trends in Atmospheric Carbon Dioxide – Mauna Loa.* National Oceanic and Atmospheric Administration, US Department of Commerce. Accessed May, 2008. http://www.esrl.noaa.gov/gmd/ccgg/trends/.

Urosevic, M., Kepic, A., Sherlock, D., Daley, T., Freifeld, B., Sharma, S., and Dodds, K. (2008). "Application of geophysical monitoring within the Otway Project, S.E. Australia." SEG Las Vegas 2008 Annual Meeting.

USDOE (U.S. Department of Energy) (1999). *Carbon Sequestration Research and Development.* Washington, DC: Office of Science and Office of Fossil Energy.

Vodnik, D., Kastelec, D., Pfanz, H., Maþek, I., and Turk, B. (2006). "Small-scale spatial variation in soil CO_2 concentration in a natural carbon dioxide spring and some related plant responses." *Geoderma*, 133, 309–319.

Westrich, H., Lorenz, J., Cooper, S., Colon, C. J., Warpinski, N., Zhang, D., Bradley, C., Lichtner, P., Pawar, R., Stubbs, B., Grigg, R., Svec, R., and Byrer, C. (2001). "Sequestration of CO_2 in a depleted oil field: an overview." *Proceedings of the First National Symposium on Carbon Sequestration.* U.S. National Energy Technology Laboratory. Washington DC.

Wilson, M., and Monea, M. (eds.) (2004). *IEA GHG Weyburn CO_2 Monitoring and Storage Project Summary Report 2000–2004:* Petroleum Technology Research Center, Regina, Saskatchewan, 273 pp.

Wimmer, B.T., Krapac, I.G., Locke, R., and Iranmanesh, A. (2010). "Applying monitoring, verification, and accounting techniques to a real-world, enhanced oil recovery operational CO_2 leak." In *Proceedings of the Greenhouse Gas Control Technologies Conference,* Amsterdam, The Netherlands, September 19–23, 2010.

World Resources Institute (WRI) (2008). *CCS Guidelines: Guidelines for Carbon Dioxide Capture, Transport and Storage.* Washington DC: WRI.

Wright, G., and Majek, A. (1998). "Chromatograph, RTU monitors of CO_2 injection." *Oil and Gas Journal*, 20 July 1998.

CHAPTER 7

Carbon Reuses for a Sustainable Future

M. Verma, F. Pélletier, S.K. Brar, S. Godbout, R.D. Tyagi and R.Y. Surampalli

7.1 Introduction

Carbon dioxide (CO_2) emission constitutes around 60% of the global carbon emissions. The recovery of CO_2 can contribute to the mitigation of carbon emission related problems. At present, two main options are explored: i) entrapment in deep geological cavities in oceans or land such as depleted petroleum wells; and ii) utilization by fixation or recycling in the form of chemicals. Between these two, the second option is getting increasing attention (Agnolucci et al. 2009). Over the last several decades, CO_2 was mainly used to manufacture urea, urea-melamine resins, animal feed additive, and other organic chemicals, such as alkylene carbonates (solvent), β-oxynaphthoic acid (raw materials of dyes), salicylic acid and its derivatives (pharmaceuticals, food preservatives, etc.) in small quantities (Darensbourg et al. 2004; Du et al. 2005; Liu et al. 2006; Halloran 2007; Liu et al. 2009). However, the global CO_2 utilization never exceeded 0.2 billion tons per year (Omae 2006). On the other hand, the recent CO_2 emission trend shows about 8 billion tons per year increase which is now difficult to check in comparison to about 0.45 billion tons per year for the last two centuries. These factors have driven the research and development efforts of environmentally benign techniques to transform and reuse CO_2 as valuable product, fuelled by deteriorating effects of global warming, environmental risks, and frequent fuel crises.

The global CO_2 reuse market currently amounts to approximately 80 million tonnes/year, of which about 50 million tonnes per year are used for enhanced oil recovery at a price of $15–19/tonne. Potentially, the global supply of anthropogenic CO_2 is about 500 million tonnes of low-cost (< $20/tonne) high concentration CO_2; at a much higher cost ($50–100/tonne), around 18,000 million tonnes per year could also be captured for CO_2 reuse (PB and GCCSI 2011). Advances of CO_2 reuse technologies depends on future carbon restrictions and prices and their interaction with other CCS technologies (Zhang and Surampalli 2013).

CO_2 has several commercial applications such as industrial chemical feedstock, fire extinction, carbonated drinks, among many others, and most of the uses end up in releasing CO_2 in the atmosphere (Kannan 2009; Yeh and Sperling 2010). Nevertheless, CO_2 can replace use of steam, CO, and many carbon requiring chemical processes to produce feedstock chemicals for fuels, such as synthesis gas, methanol, and CO. There are numerous organics which can be synthesized using CO_2 such as formic acid, formic acid esters, formamides, other hydrogenation products, carbonic acid esters, carbamic acid esters (urethanes), lactones, carboxylic acids, polycarbonate (bisphenol-based engineering polymer) and aliphatic polycarbonates, but only a few are industrially feasible so far (Omae 2006). At present, syntheses of urea and its derivatives, organic carbonates, increasingly utilize CO_2 as the carbon building unit in place of highly toxic alternatives such as phosgene ($COCl_2$). CO_2 is also heavily consumed by the electrochemical Kolbe-Schmitt process for the production of salicylic acid. CO_2 can also react with hydrogen, alcohols, acetals, epoxides, amines, carbon-carbon unsaturated compounds, and oxetanes under favorable reaction conditions, namely, metal catalysts, high temperature and pressure among others (Yin and Moss 1999; Wang et al. 2005; Yasuda et al. 2005; van Schilt et al. 2006; Wang et al. 2006; van Alphen et al. 2010). Currently, the reactions are mainly carried out in supercritical CO_2 to avoid use of solvents hazardous to the environment. Additionally, there are other beneficial uses of CO_2 such as i) extractant, ii) food/products, iii) enhanced and fuel recovery, iv) inerting agent, iv) fire suppression, vi) refrigerant, and vii) others (fertilizer, secondary chemicals, dry ice pellets used for sand blasting, added to medical O_2 as a respiratory stimulant, which have been also covered in small sections.

In the present scenario, CO_2 can be an economic, nontoxic, and easily available carbon feedstock if the technical challenges are matched. In addition, utilization of CO_2 for production of chemicals may or may not be fail-safe strategy to check or mitigate carbon emission in the environment (Yeh and Sperling 2010). In order to address this issue with CO_2 use, the conversion technology must be considered on the basis of life-cycle assessment (LCA) before endorsing carbon emission mitigation potential of such option. As CO_2 is a thermodynamically stable compound relative to other naturally occurring carbon sources, its reduction requires high energy substances or energy intensive electroreductive processes (Bennaceur and Gielen 2006). Thus, the reuse of CO_2 could be a viable option in terms of CO_2 mitigation and climate change if more research is carried out on CO_2 conversion techniques.

After going through different major reuse technologies of CO_2, it is desirable to understand the different reuse options, how the technology is evolved, and what are the challenges should these technologies be adopted, which are the focuses of this chapter.

7.2 CO_2 Reuse as Fuel

The CO_2 emission from power plants can be mitigated significantly via ''recycling'' of the CO_2 into fossil fuel that could reduce the overall use of fossil fuels (Jiang et al., 2010; Olah et al., 2011). Nevertheless, conversion of CO_2 back to hydrocarbon fuels requires ≥ 80% of the energy equivalent of a typical coal (Omae 2006; Hamilton et al. 2009; Aresta et al. 2010; Seyfang 2010). Furthermore, if processing losses are to be considered, the net gain of energy can easily become negative. Many researchers have reported that unless the energy comes from non-fossil sources, net additional CO_2 is generated. Therefore, if non-fossil energy is economically feasible, in most cases it should be used to substitute coal and other forms of fossil fuels (Herzog et al. 1997; Herzog and Golomb 2004). Thambimuthu et al. (2002) proposed a non-fossil fuel based energy cycle for CO_2 utilization (Fig. 7.1). In this approach, the CO_2 captured from the flue gas stream of a power plant or industrial process is transformed into a hydrocarbon fuel or chemicals such as (CO, HCOOH, CH_2O, CH_3OH, CH_4) using non-fossil energy input (e.g., solar, wind, or nuclear energy). Thus, CO_2 can be transformed to compounds, useful as "energy vector" such as hydrogen carrier. The use of CO_2 as fuel compounds can improve the validity and interest in ''hydrogen based'' economy for future as it would contribute mitigate two major problems of hydrogen based fuel applications, i.e., storage and transportation. For example, storage and transportation of hydrogen in the form of formic acid can substantially reduce both cost and risks.

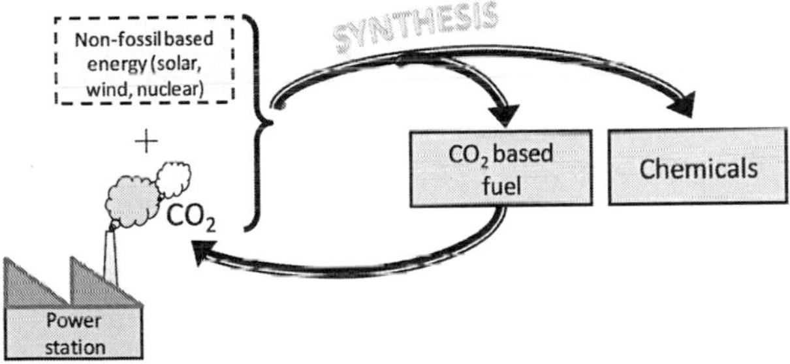

Figure 7.1 An energy cycle for CO_2 based secondary fuel (adopted from Thambimuthu et al. 2002)

However, the existing technologies of converting CO_2 into fuel compounds is far from optimal due to various factors: i) the amount of CO_2 used for fuel might not be substantial enough to lower its global concentration; ii) the cost of the alternative non-fossil source of energy for CO_2 reduction in terms of capital investment and/or operating expenses could be discouraging; and iii) the lower rate of CO_2 conversion can increase production cost and affect process economy (Andrew et al. 2010). On the positive side, CO_2 reuse as fuel presents an option to decelerate the growth of fossil fuel consumption by utilization of the carbon fixed as a fuel (followed by the continuous recycling of CO_2) or with permanent fixation of carbon while producing products with longer half-life such as char/activated carbon (Zevenhoven et al. 2006). At present, CO_2 reuse as fuel is possible via chemical, biological (facilitated via photosynthetic processes) and photochemical processes, as shown in Figure 7.1 and discussed in the subsequent sections.

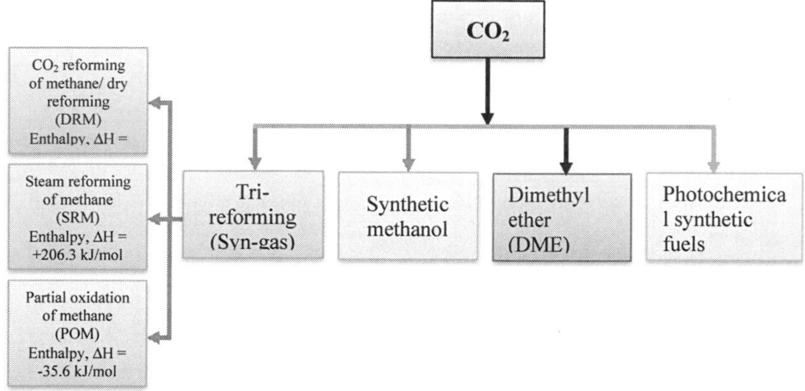

Figure 7.2 Different options for CO_2 reuse as fuels

7.2.1 Chemical Conversion of CO_2 to Fuel

The chemical conversion of reactants to products is possible for thermodynamically favourable reactions. During such conversion process, the reactants move from a higher to a lower energy state under suitable reaction conditions (e.g., if activation energy is provided), in other words, from a less stable form towards chemically more stable form. The energy state (chemical stability) of a compound is measured in terms of Gibbs free energy of formation ($\Delta G°$). Table 7.1 shows $\Delta G°$ of some common chemical compounds including CO_2. Thus, lower or more negative Gibbs free energy value of compounds indicates higher chemical stability. The formation of a more stable compound (higher negative magnitude of $\Delta G°$) results in a net release of energy (exothermic reaction). Thus, combustion of fossil fuels results in production of energy and CO_2 as a very stable end product, which can be interpreted from Table 7.1. As CO_2 is chemically more stable than hydrocarbons, the reactions for CO_2 conversion require positive change in enthalpy (ΔH), i.e., substantial input of energy (endothermic reaction) where effective reaction conditions and often active catalysts are necessary for chemical conversion of CO_2 (Stangland et al. 2000; Song 2006). Therefore, if fossil fuels such as coal have to be used for chemical synthesis of CO_2 into fuels, the CO_2 emissions of the entire process should be lower than the net CO_2 consumption. Alternatively, a carbon free source such as solar, wind, geothermal, or nuclear energy has to be used (Omae 2006; Aresta et al. 2010).

Syn-gas by Tri-reforming. A novel process has been pioneered by Song (2006), centred on the unique advantages of directly utilizing flue gas, rather than pre-separated and purified CO_2 from flue gases, for the production of hydrogen-rich syngas from methane reforming of CO_2 (also known as 'dry reforming'). The overall process, defined as 'tri-reforming', combines the processes of CH_4/CO_2 reforming, steam reforming of CH_4, and partial oxidation and complete oxidation of CH_4. The reactions involved are shown below, together with the corresponding enthalpies.

$$CH_4 + CO_2 \Leftrightarrow 2CO + 2H_2 \qquad \Delta H_{298K} = +247.3 \text{ kJ/mol} \qquad (7.1)$$

$$CH_4 + H_2O \Leftrightarrow CO + 3H_2 \qquad \Delta H_{298K} = +206.3 \text{ kJ/mol} \qquad (7.2)$$

$$CH_4 + \tfrac{1}{2} O_2 \Leftrightarrow CO + 2H_2 \qquad \Delta H_{298K} = +35.6 \text{ kJ/mol} \qquad (7.3)$$

Synthesis gas (equimolar mixture of CO and H_2) can be produced via Eq. 7.1. It was found that the combination of CO_2 and H_2O can produce syngas with the desired H_2/CO ratios for methanol and dimethyl ether (DME) synthesis and higher-carbon Fischer–Tropsch synthesis of fuels.

Synthetic Methanol and Dimethyl Ether (DME). Olah et al. (2011) have advanced the process for synthetic methanol economy. Currently, methanol synthesis is one of the most promising processes for the utilization of CO_2 as synthetic fuels. The heat of reaction of hydrogen with atmospheric CO_2 under proper conditions can provide economical solutions to methanol production and mitigate the substantial rise of CO_2 concentration in the atmosphere (Omae 2006; Ma et al. 2009; Aresta et al. 2010). Methanol can be produced from a mixture of $CO/CO_2/H_2$ via various reaction pathways (Eqs. 7.4–7.6). Similarly, DME can also be produced by direct catalytic hydrogenation of CO_2.

$$CO + 2H_2 \Leftrightarrow CH_3OH \qquad \Delta H_{298K} = -94.08 \text{ kJ/mol} \qquad (7.4)$$

$$CO_2 + 3H_2 \Leftrightarrow CH_3OH + H_2O \qquad \Delta H_{298K} = -52.81 \text{ kJ/mol} \qquad (7.5)$$

$$CO_2 + H_2 \Leftrightarrow CO + H_2O \qquad \Delta H_{298K} = +41.27 \text{ kJ/mol} \qquad (7.6)$$

However, the conversion of CO_2 to methanol is limited by the thermodynamic equilibrium constraints and catalyst deactivations. At relatively lower temperatures, the higher conversion towards the right side of the reversible exothermic reactions (Eqs. 7.4

and 7.5) is possible, but this must be compensated by the use of a large amount of catalyst (Rahimpour 2007). The common catalysts for CO_2 hydrogenation contain Cu and Zn as the principal ingredients along with different modifiers (Zr, Ga, Si, Al, B, Cr, Ce, V, Ti, etc.) (Toyir et al. 2001; Ma et al. 2009). The crucial understanding of the characteristics, and reaction mechanisms of CO_2 hydrogenation catalysts are still lacking. Therefore, CO_2 hydrogenation catalysts are still marginally exploited in industrial applications (Ma et al. 2009).

Table 7.1. Gibbs free energy of formation, ΔG° for CO_2 and other chemicals (Thambimuthu et al. 2002)

Compound – molecular formula (phase)	ΔG°_{298} (at 298 K) (kJ/mole)
Hydrocarbon fuels – CxHy (l and g)	Higher positive value
Acetylene – C_2H_2 (g)	+209
Benzene – C_6H_6 (g)	+130
Ethylene – C_2H_4 (g)	+68
Propylene – C_3H_6 (g)	+62
Methane – CH_4 (g)	-51
Carbon mono-oxide – CO (g)	-137
Methanol – CH_3OH (g)	-162
Ethanol – C_2H_5OH (g)	-168
Urea – NH_2CONH_2 (s)	-197
Water – H_2O (l)	-228
Steam – H_2O (g)	-237
Acetic Acid – CH_3COOH (l)	-374
Carbon Dioxide – CO_2 (g)	-394
Dimethyl Carbonate – $(CH_3)_2CO_3$ (s)	-492
Silicon dioxide – SiO_2 (s)	-805
Magnesium Carbonate – $MgCO_3$ (s)	-1012
Calcium Carbonate – $CaCO_3$ (s)	-1129

As CO_2 and CH_4 are naturally abundant and can also be economically synthesized (biomethanation) or captured (CO_2 from flue gases), their conversion to higher value energy feedstock is of great interest (Toyir et al. 2001; Rahimpour 2007). For example, sulphur free diesel from synthesis gas via Fischer-Tropsch synthesis (Eq. 7.7), and methanol (Eq. 7.8) are commercially important energy feedstock with many other applications in synthesis of industrial chemicals (Herzog et al. 1997; Herzog and Golomb 2004). The hydrogen extracted from synthesis gas has tremendous potential for use in fuel cells, which has been widely regarded as a fuel-efficient means of powering automobiles (Conte 2009; Cormos et al. 2010).

$$nCO + (2n + 1)H_2 \Leftrightarrow C_nH_{(2n+2)} + nH_2O \quad (7.7)$$

$$CO + 2H_2 \Leftrightarrow CH_3OH \quad (7.8)$$

Photochemical Production of Synthetic Fuels. Inoue et al. (1979) invented the photocatalytic reduction of CO_2 in aqueous solutions to produce a mixture of formaldehyde, formic acid, methanol and methane using various wide-band-gap semiconductors. Afterwards, several researchers have worked on the photochemical production of fuels by CO_2 reduction using a variety of photocatalysts (Halmann 1993; Hwang et al. 2005; Indrakanti et al. 2009; Olah et al. 2010). The characteristics of potential photocatalysts are determined by the redox potentials of the rate-limiting steps of water oxidation and CO_2 reduction:

$$2H_2O(l) \rightarrow O_2(g) + 4H^+(aq) + 4e^- \quad E^0_{ox} = -1.23\ V \qquad (7.9)$$

$$CO_2(aq) + e^- \rightarrow CO_2^-(aq) \quad E^0_{red} = -1.65\ V \qquad (7.10)$$

The redox potential values (E^0) represent a minimum threshold for the energy of photo-excited electrons to reduce CO_2 and also the energy levels of the conduction and valence bands of a photocatalyst.

7.2.2 *Biological Conversion of CO_2 to Fuel*

Biological conversion of CO_2 to fuel is possible via microalgae systems (Omae 2006; Aresta et al. 2010). Microalgae can be used to capture CO_2 emitted from fossil fuel based power plants, ethanol plants, concrete plants, and in fact almost any industry with CO_2 emissions. In comparison to plant species (e.g., oil crops), microalgae are of particular interest because of their rapid growth rates similar to microorganisms (more than 10 times that of the plants) and potential for significantly higher efficiency of photosynthesis process with respect to land use or foot print (Herzog et al. 1997). In general, these microscopic plants could be cultivated in large open ponds, purged with flue gas or pure CO_2 (captured from power plants) as small bubbles (Fig. 7.3). The algae biomass generated is separated from the liquid phase using mechanical, chemical, gravity and/or, combination processes, and the algal oil is extracted by disruption of algae cells via chemical or physical methods. The oil extracted from algal biomass can be processed using transesterification to produce biodiesel for energy and transportation (Hickman et al. 2010). In addition, non-photosynthetic routes for biological fixation of CO_2 into valuable industrial products and fuels are seeing a transition from concept to reality. Electrofuel processes are being developed for a range of microorganisms and energy sources (e.g. hydrogen, formate, electricity) to produce a variety of target molecules (e.g. alcohols, terpenes, alkenes), albeit the yields are low, but efforts are underway to build-up on their optimization processes. In this particular field, these days the focus is more on the biochemistry of hydrogenases and carbonic anhydrases, and the state of genetic systems for current and prospective electrofuel-producing microorganisms for enhanced yields which has been extensively reviewed by Hawkins et al. (2013). In the last decade, there has been significant growth of algae harvest, as well as total production volume of phycocolloids. It should be emphasized, however, that the significant drawback in all of the technologies is the big share of cost for algae cultivation. The economic feasibility of seaweed biomass as feedstock for obtaining low value products (e. g. biomass for energy) still seems to be a far-fetched dream despite given advantages as there are still questions on its stratus as carbon neutral. Some drawbacks include: seasonality/vulnerability to potential impacts (natural/ anthropogenic); over-exploitation (in mechanical harvesting); mechanical impacts on the ecosystem; effects on marine biodiversity; and limited culture techniques (e.g., Gelidium corneum).

Moreover, technologies are also available for electricity generation from co-firing, combined heat and power (CHP), pyrolysis, gasification, and anaerobic digestion of algal biomass (Omae 2006). In co-firing and CHP, the heat generated is used as a prime power source. On the other hand, biofuels, e.g., biodiesel, ethanol, bio-butanol, gasoline, jet fuel, and biogas can be used as an alternative renewable fuel source in the internal combustion (IC) engine for heavy machinery or transportation.

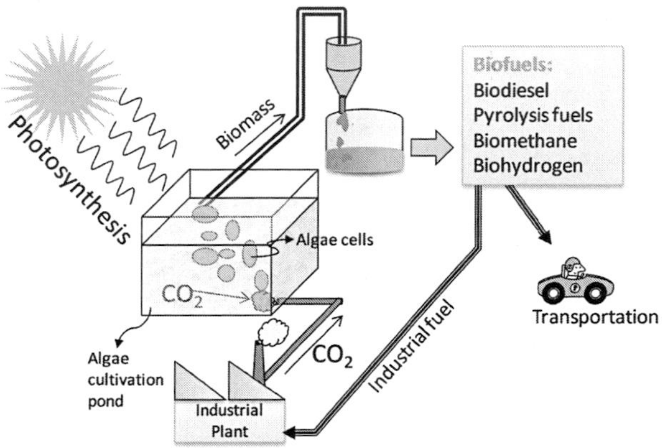

Figure 7.3 Algae cultivation for CO_2 conversion to fuels

Algae are present in varying environments throughout the planet and are capable of producing more biofuel per acre than any terrestrial plant species. The present condition of volatile oil supplies, political, and economic environments have influenced the federal government and many state governments to allocate subsidies for biofuels, which can significantly offset their production and capital costs. Moreover, utilization of CO_2 for biofuels production can further attract industrial investments to exploit benefits of carbon credit, where applicable.

On the other side is the trend in developing photocatalytic reactors which can convert CO_2 into novel organic compounds. The photocatalytic route for reducing CO_2

to hydrocarbon fuels and/or valuable chemicals by solar energy is attracting great interest. However, this promising prospect was limited by the low quantum efficiency and selectivity of photocatalytic materials, which has been now taken over by the nanocatalysts to enhance the reaction rate. In the actual scenario, the overall conversion of the CO_2 into hydrocarbons is energy intensive as compared to the energy obtained from the final product, the fuel. The length of the hydrocarbon chain that forms from this reaction is determined by the type of catalyst used and the reactor process conditions. The current technology can make methane using Eq. 7.11:

$$\text{Solar Energy} + x\,CO_2 + (x+1)\,H_2O \rightarrow C_xH_{2x+2} + (1.5x + 0.5)\,O_2 \tag{7.11}$$

where x=1, with an efficiency of 10.2% with reference to the amount of incident solar energy compared to the change in enthalpy of formation of the methane produced. The reaction above is a multi-step process in which both the CO_2 and H_2O must be split into their constituent atoms so as to re-form hydrocarbon products (Roy et al. 2010). One current approach is to split these molecules using a Zn/ZnO electrode that uses solar power to provide the driving force (Louitzenhiser et al. 2010).

However, the inputs (e.g., waste CO_2, water and sunlight) to this reaction are nearly costless to the power plant. The economic feasibility of this process will ultimately depend on the cost of raw hydrocarbon resources balanced against the cost of the capital equipment for the reverse combustion reaction and the potential savings from reduced CO_2 emissions. Likewise, for further development of solar fuels, it is necessary to reduce the cost of the best performing routes, to determine in more detail the material impacts of those routes and to ensure that carbon capture becomes a common practice and hence a reality towards sustainability.

At present, biological conversion of CO_2 to fuel is in its infancy, and most of the processes are under laboratory or pilot scale investigation. This is mainly due to the technical challenges such as: impurities in algal oil in comparison to oil obtained from oil-crops; separation of algae biomass from water; CO_2 mixing, variation in sun intensity; requirements of additional nutrients; bioaccumulation of metals and toxic compounds in algal biomass; and extraction of oil from algae cell. Therefore, further research and development efforts are needed for algal-based CO_2 utilization for fuels.

7.3 Carbon Reuse as Plastics

Using CO_2 for the manufacturing of plastics has great potential to mitigate global warming as plastics, in general, have very long shelf life (Omae 2006; Zevenhoven et al. 2006; Zhang et al. 2006). Some of the industrial applications already in use are polycarbonate formation without the utilization of phosgene, an alternating epoxide-CO_2

copolymerization, condensation with benzenedimethanol, and an alternating diynes-CO_2 copolymerization.

7.3.1 CO_2 Reuse for Bisphenol-Based Engineering Polymers

In Japan, use of CO_2 has already replaced the consumption of toxic substances such as phosgene and methylene chloride long back by major chemical plants (Aresta et al. 2010). For example, since 2002, Asahi Chemical Industry, Japan has reduced about 8,650 tons/year of CO_2 for the production of around 50,000 tons/year of polycarbonates as a novel environmentally benign process. The CO_2 polymerization process mentioned above consists of four steps:

(1) pre-polymerization between diphenylcarbonate and bisphenol A to produce a clear amorphous prepolymer;

(2) crystallization of the molten prepolymer to a porous, white, opaque material using acetone as solvent;

(3) the crystallized prepolymer is heated upto 210–220 °C under a flow of heated nitrogen, or under mild vacuum conditions (e.g., 67 Pa) to produce a solid polymer; and

(4) finally, the ''self-mixing melt polymerization'' process utilizes gravity instead of a conventional twin-screw type reactor to polymerize 6,200 Da prepolymers to 11,700 Da polymers.

The utilization of CO_2 for the production of engineering polymers (polycarbonates) has many advantages: (1) the use of relatively toxic compounds, phosgene and dichloromethane is replaced with raw materials such as CO_2, ethylene oxide and bisphenol A; (2) the products, polycarbonate and ethylene glycol, are of high quality due to the absence of halide constituents; (3) high yield and high selectivity of intermediate products, ethylene carbonate, dimethylcarbonate, methylphenylcarbonate, and diphenylcarbonate. Moreover, the intermediates and two of the raw materials (i.e., methanol and phenol) are completely recycled; and (4) a substantial decrease in CO_2 emission (Omae 2006). It was estimated that if this method of polycarbonate production is applied worldwide, the global decrease in CO_2 emission could exceed 450,000 tons per year.

7.3.2 CO_2 Reuse for Aliphatic and Other Polymers

In the past, many aliphatic polymers have been reported to be manufactured using CO_2 with oxetane, epoxides in the presence of organotin complexes or zinc-based catalysts under mild polymerization conditions (Inoue et al. 1969; Baba et al. 1987). The polymerization reactions were found to be highly susceptible to the catalyst types in terms of % polymerization and molecular weight. For example, for three different catalysts, namely, organotin-phosphine complex (Bu_2SnI_2-PBu_3), organotin (Bu_2SnI_2), and Bu_3SnI-hexamethylenephosphoric triamide (HMPA), the yields were 89, 98, and 100%, respectively (Eq. 7.12) with varying molecular weights.

$$\text{oxetane} + CO_2 \xrightarrow{\text{cat.}} -(CH_2CH_2CH_2\text{-O-}\underset{\underset{O}{\|}}{C}\text{-O})_n- \quad (7.12)$$

cat. (Bu_3SnI-HMPA) → cyclic carbonate (1,3-dioxan-2-one): Yield = 100%, Mol. Wt. = 102

cat.	Yield (%)	Mol. Wt.
Bu_2SnI_2	89	2100
Bu_2SnI_2-PBu_3	98	4250

Furthermore, Inoue et al. (1969) reported use of Zn catalyst for alternative copolymerization of CO_2 with epoxide under mild conditions (Eq. 7.13). However, the preferred catalysts systems are the reaction mixtures of diethylzinc with an equimolar amount of a compound, such as water (which has two active hydrogens), a primary example of the preferred catalyst system, which forms many kinds of oligomers. These oligomers are alternating copolymers with a resorcinol group at the terminal position. The mechanism of this alternating copolymerization is suggested as follows: the zinc alkoxide produced by the reaction between dialkyl zinc and diol, nucleophilically attacks CO_2 to give a zinc carbonate; and subsequently, zinc carbonate reacts with epoxide to reproduce zinc alkoxide. Thus, the repetitive nature of these two reactions results into the alternating copolymerization products. Many active zinc catalysts such as zinc phenoxides, bulky fluorophenoxides, zinc diimines, and zinc bis-Schiff bases have been studied recently. The production of a copolymer with narrow molecular weight distribution (Mw/Mn = 1.07–1.17) and high molecular weight (Mw = about 420,000) at a high turn-over-number (1441 g/g cat) were reported (Inoue et al., 1969). These alternating copolymers have been found to be biodegradable with high oxygen permeability. Therefore, research is also underway to explore the potential of these polymers for application in sustained-release drug delivery systems.

$$\text{epoxide (R, R')} + CO_2 \xrightarrow[Et_2Zn/H_2O]{\text{30–50 atm, RT}} -(CHR\text{-}CHR'\text{-O-}\underset{\underset{O}{\|}}{C}\text{-O})_n- \quad (7.13)$$

CO_2 has also been polymerized with several others compounds such as, ethyleneimine (aziridine), epithioxide (three-membered ring thioether), a vinyl ether, benzenedimethanol, and an aromatic diamine (Lu et al. 2004). Moreover, polycondensation of CO_2 with diamines results in a high yield of polyureas, which have application in widely used elastomers such as spandex (Muradov and Veziroglu 2008). Other polymer-mediated fixation of CO_2 includes transformation of oxirane groups of methacrylate derivatives into corresponding cyclic carbonate groups quantitatively using a lithium salt as a catalyst.

7.3.3 Role of Catalysts in Reuse of CO_2

There are numerous efficient industrial catalysts for CO_2 utilization reactions, consisting of both transition metal compounds and main-group metal compounds (Paddock et al. 2004). Catalysts for manufacturing formic acid, formic acid methyl ester and formamide include transition metal-based, namely, ruthenium phosphine complexes, heteropolytungstate, and heteropolymolybdate (Aouissi et al. 2010). In contrast, the catalysts of main-group metal compounds are used for the synthesis of dimethyl carbonate, ethyl carbamate, diphenyl carbonate, and the alternating copolymerization of CO_2 and epoxide (Darensbourg and Holtcamp 1996; Bai et al. 2009). Some newer economical catalysts such as $AlCl_3$ have also been reported for the carbonylation of aromatic compounds under mild conditions (Muradov and Veziroglu 2008). Nevertheless, research in catalyst systems are expected to produce better catalysts because, though thermodynamically stable, CO_2 is able to react with various kinds of metal compounds.

Zn-Y based catalyst systems (e.g., $Y(CF_3CO_2)_3$, $Zn(Et)_2$ and m-hydroxybenzoic acid) were used by Hsu and Tan (2002) for an alternating poly-propylene carbonate with a 100% carbonate content. The poly-propylene carbonate was effectively generated from the copolymerization of CO_2 and propylene oxide in 1,3-dioxolane. The yield and molecular weight of the resultant poly-propylene carbonate were higher than that reported in the literature under milder temperature (60 °C), and pressure (≈27 atm). Sun and Zhai (2007) have used a catalyst system consisting of n-Bu_4NBr, α_2-(n-$Bu_4N)_9P_2W_{17}O_{61}(Co^{2+}\cdot$ Br) and polyethylene glycol (Mol. Wt. 400) for the coupling reaction between CO_2 and propylene oxide. The authors reported that the yield and selectivity were 98% and 100%, respectively at 120 °C for 1 h; however, the catalytic activity slowly diminished with recycling.

Furthermore, high turn-over-numbers are also expected in the supercritical CO_2 solvent with many kinds of inducer compounds. One of the novel methods would be the use of ionic liquids to immobilize or slow down the molecules of CO_2, during the fixation and transformation process (Zevenhoven et al. 2006). Ionic liquids are a novel concept of reaction media composed entirely of ions. Table 7.2 presents typical cation/anion combinations comprising the main types of ionic liquids (Zhang et al. 2006). Several authors have investigated the use of ionic liquids for CO_2 fixation and transformation (Tominaga et al. 2010; Zhang et al. 2006; Palgunadi et al. 2004). The ionic liquids in combination with -NH_2 groups have shown an absorption capacity of

CO_2 as high as 7–8 wt% under ambient conditions. Ionic liquids have also resulted in high activity, high yield for the reactions of CO_2 and epoxides, propargyl alcohols and amines (Zhang et al. 2006).

7.4 CO_2 Reuse towards Low Carbon Economy

"Low carbon economy" refers to minimization of fossil-based energy uses, implementation of renewable/non-fossil energy sources, and adoption of energy policies to curtail/reduce greenhouse gas (GHG) emission. Nowadays, the modernization and economic potential of countries are measured by the effectiveness of the policies adopted to protect and restore the environment with emphasis on subduing GHG emissions (Nader 2009). Thus, the current international community is pragmatically moving toward the low carbon economy concept. Meanwhile, low carbon economy requires integration of novel technologies into existing domestic and industrial infrastructure. Therefore, this approach is demanding in terms of infrastructure developments and energy policies amendments. Low carbon economy also encircles non-fossil based fuels such as biofuels for fixation and storage of CO_2. In order to make the low carbon economy approach feasible, the global regime needs to price the use of carbon. Many G-20 as well as emerging nations have started already taking initiatives to emerge as low carbon economies. Economies such as, European Union, Australia, Japan, UAE, and others are gradually implementing renewable energy and sustainability technologies in order to generate the skilled man-power, institutions and intellectual capital necessary for a low carbon future (Ockwell et al. 2008). For example, Masdar City, UAE could be a good example of a carbon neutral, zero waste urban development, despite having abundant and cheaper fossil fuel (Nader 2009). The Masdar City supports a world-scale carbon capture and storage (CCS) project, which could be a techno-economic feasibility model for the rest. The city uses scaled-up applications of existing renewable technologies, integrating them into effective systems and encouraging innovation.

7.4.1 Decarbonization Options of Fossil Fuels

There are several conventional fossil decarbonization options, such as, post-combustion, precombustion and oxyfuel combustion, which are instrumental in low carbon economy approach (Richels et al. 2008; Peace and Juliani 2009). Currently, post-combustion decarbonization process is widely used option due to research advancements and higher compatibility to existing infrastructure. Fig. 7.4 shows the general scheme of CO_2 mitigation for low carbon approach. All of these technologies involve carbon capture and storage (CCS), which is an energy intensive process involving several costly steps: CO_2 separation, pressurization, liquefaction/solidification, and transportation to the final disposal site (injection of liquid CO_2 into geologic formations/exhausted petroleum reserves, or in aquifers), or as industrial feedstock for synthesis of chemicals or hydrocarbon fuels (Omae 2006). Technological options for CO_2 capture from diluted streams consists of chemical solvents (e.g., amines, potassium carbonate, aqueous ammonia); physical solvents (e.g., glycol, methanol, ionic liquids); chemical sorbents (e.g., amine-modified sorbents, metalorganic frameworks); membranes (e.g., polymer, ceramic); enzymatic processes; novel methods (e.g., CO_2 hydrates) (Aresta et al. 2010). These technologies are discussed in detail in other chapters of this book.

Table 7.2. Cation/anion compounds comprising the main types of ionic liquids (adapted from Zhang et al. 2006)

Cations	Anions
H_3C—N(imidazolium)N^+—R	I^-, Br^-, Cl^-
	BF_4^-, PF_6^-
	$ZnCl_3^-$, $CuCl_2^-$, $SnCl_3^-$
(pyridinium) N^+—R	$N(SF_3SO_2)_2^-$
	$N(C_2F_5SO_2)_2^-$
(pyrrolidinium) N^+, R_1, R_2	$N(FSO_2)_2^-$
	$CF_3CO_2^-$
	$CF_3SO_3^-$
(oxazolium) R_1, R_2, O, N^+—R_3, R_4	$Al_2Cl_7^-$, $Al_3Cl_{10}^-$
	$Fe_2Cl_7^-$
	$Sb_2F_{11}^-$

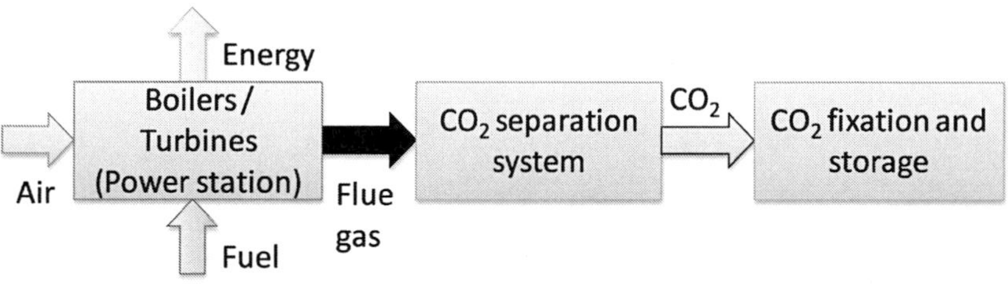

Figure 7.4. Energy production intended for low carbon economy

7.4.2 *Risks of CCS*

Despite diverse technological advancements in CCS, the major challenges for global implementation of CCS are: (i) significant investment cost for infrastructure; (ii) lack of comprehensive understanding of the risk factors associated with the long-term ecological consequences of CCS via existing technologies; and (iii) risk factors of the reservoir options (Liu and Gallagher 2010). CO_2 storage in deep geological formations and ocean is actively debated in the literature, due to seemingly plausible risks of adverse impact on the aquatic environment (i.e., ocean acidification, an effect on marine life and eco-balance, etc.) (Zevenhoven et al. 2006). Moreover, on a long-term basis these solutions are temporary. For example, disposal/storage in deep geological formations seems to be a less expensive and risky option than ocean sequestration, however, any accidental leakage of CO_2 can lead to leaching of harmful trace elements in freshwater aquifers by lowering of pH and can also adversely affect soil chemistry. It was estimated that even 1% leak of captured CO_2 could nullify the sequestration effort in a century and could be catastrophic for humans and animals (Liu and Gallagher 2010). The CO_2 leakage happened in the Lake Nahos in Africa asphyxiated about three thousand people (Muradov and Veziroglu 2008). Thus, the low carbon economy concept has immense challenges due to lack of reliability of existing CCS technologies.

7.4.3 *Production of Carbon Free or Carbon Neutral Fuels*

Another prospective decarbonisation strategy for low carbon economy could be the production of hydrogen from fossil fuels (e.g., natural gas, petroleum or coal) coupled with CO_2 sequestration. In this approach, hydrogen is produced from fossil fuel and the CO_2 formed as a reaction by-product is captured and sequestered. The energy conversion efficiency for coal and natural gas as a feed for the production of hydrogen are 50–60% and 70–75%, respectively (Muradov and Veziroglu 2008). The commendable features of this approach are that the energy infrastructure could be based on a range of carbonaceous fuels, either biomass or fossil based, without the pollution load of CO_2 in the atmosphere. Similarly, the technologies under active research and development such as steam methane reforming can play a major role in driving low carbon economies. Strategies for achieving the goal of low carbon economy at the utilization level of fuels/energy are also important (Damm and Fedorov 2008). Use of electric vehicles, hydrogen powered vehicles, and substitution of petroleum fuels with carbon neutral biofuels are few examples discussed in the following.

Electric Vehicles. Electric vehicles are attractive mode to eliminate direct carbon emissions by the transportation sector. The electricity consumed by the electric vehicles can be generated at a large-scale centralized location from renewable/ alternative energy sources or from fossil fuels coupled with CO_2 sequestration. This electric energy will be used by the vehicle to produce mechanical energy with no direct

CO_2 emissions. Fig. 7.5 shows the general scheme for low carbon economy at the utilization level of fuels/energy. The energy efficiency of electric vehicles can be better than internal combustion engines. Currently, however, energy density and the charging time of batteries are limiting factors (Damm and Fedorov 2008). Use of rare earth metals (e.g., lithium) in batteries are also a matter of concern as they may hamper the cost competitiveness of electric vehicles.

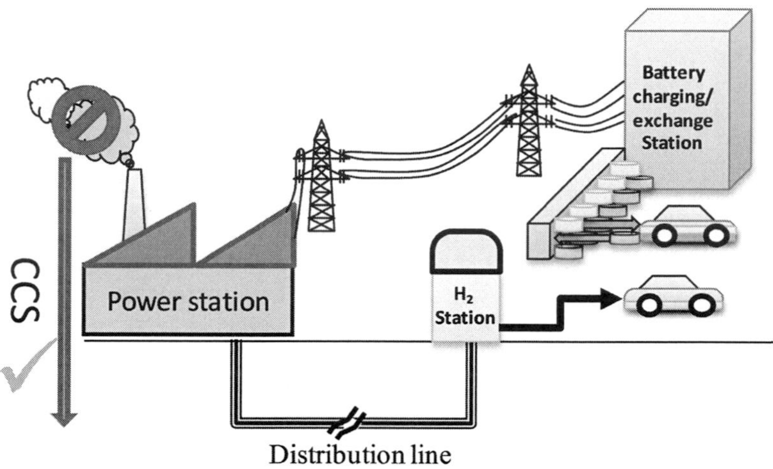

Figure 7.5. Strategies for achieving the goal of low carbon economy at the utilization level of fuels/energy

Hydrogen Fueled Vehicles. Use of hydrogen in automobiles can provide zero carbon emission in two ways: fuel cell technology and internal combustion engine. Similar to electricity, the pure hydrogen can be produced at a central location (using renewable energy sources or from fossil fuels coupled with CO_2 sequestration) and distributed through infrastructure (refuelling stations, pipelines, trucking, etc.) tailored for hydrogen. The hydrogen can be burned in an internal combustion engine or electrochemically converted to electricity via a fuel cell system (Fig. 7.5). Analysis of several feasible scenarios suggests that the broad scale use of hydrogen-fuelled vehicles can stabilize the atmospheric levels of CO_2 (Dutton and Page 2007; Conte 2009; Cormos et al. 2010). Nevertheless, multiple techno-economic barriers such as refuelling, hydrogen storage, infrastructure investment and safety must be addressed.

Carbon Neutral Biofuels. It is an alternative pathway that employs the use of modified internal combustion engine. Biofuels can be carbon neutral or even negative, if they are produced from biomass using renewable/alternative non-fossil energy sources.

Carbon neutral biofuels have huge potential in achieving the goal of low carbon economy as in most cases; carbon neutral biofuels can simply replace fossil fuels without major changes to the infrastructure (Omae 2006; Aresta et al. 2010). Nevertheless, the implementation of carbon neutral biofuels will require major initiatives and intensives from the governments.

7.5 Conclusions

The planet earth cannot restore the imbalance caused by anthropogenic activities such as growing release of CO_2 in the atmosphere via use of fossil fuels. If the carbon emission is not mitigated or neutralized in time, the adverse effects of global warming due to higher level of GHGs can be severe. Fortunately, the current momentum of research/development and initiations taken by international community is a positive note in the direction of restoration of the environment. Several novel as well as well-established technologies for CCS are gradually becoming mainstream pathways for fighting climate change.

Utilization of CO_2 for the production of synthetic fuels, chemical feedstock, polymers, and polycarbonates are some exemplary steps. CO_2 has been successfully transformed into methanol, synthesis gas, and synthetic hydrocarbon fuels to compete with petroleum products. On the other hand, CO_2 also has great potential as a feedstock for industrially important chemicals such as formic acid, formic acid esters, formamides, other hydrogenation products, carbonic acid esters, carbamic acid esters (urethanes), lactones, carboxylic acids. Developments in polymer science have paved way for fixation of CO_2 as commercial polymeric materials. The thermodynamic stability of CO_2 requires efficient metal catalysts to achieve economically feasible conversion processes. The recent literature is full of novel catalyst systems for CO_2 transformation, thereby, proving the potential of future use of CO_2.

Low carbon economy is based on CCS and sustainable energy production technologies. International community is in unison of need for low carbon global economy and taking all possible steps to achieve this goal. Therefore, nowadays, low carbon uses are also considered to measure modernization and development of any nation. However, risks associated with carbon capture and sequestration in deep ocean, geological formations are significant and pose challenge to the implementation of low carbon economy on a global basis. Utilization of diversified alternative energy resources such as biofuels, solar, wind, tidal, nuclear, geothermal could provide more feasible ways of mitigating carbon emission as well as energy security for the future generations.

7.6 References

Agnolucci, P., Ekins, P., Iacopini, G., Anderson, K., Bows, A., Mander, S., and Shackley, S. (2009). "Different scenarios for achieving radical reduction in carbon emissions: A decomposition analysis." *Ecological Economics*, 68(6), 1652–1666.

Andrew, J., Kaidonis, M.A., and Andrew, B. (2010). "Carbon tax: Challenging neoliberal solutions to climate change." *Critical Perspectives on Accounting*, 21(7), 611–618.

Aouissi, A., Al-Deyab, S.S., Al-Owais, A., and Al-Amro, A. (2010). "Reactivity of Heteropolytungstate and Heteropolymolybdate Metal Transition Salts in the Synthesis of Dimethyl Carbonate from Methanol and CO2." *International Journal of Molecular Sciences*, 11(7), 2770–2779.

Aresta, M., Quaranta, E., and Tommasi, I. (2010). "Prospects for the utilization of carbon dioxide." *Energy Conversion and Management*, 33(5–8), 495–504.

Baba, A., Kashiwagi, H., and Matsuda, H. (1987). ''Reaction of carbon dioxide with oxetane catalyzed by organotin halide complexes: control of reaction by ligands.'' *Organometallics*, 6, 137–140.

Bai, D., Jing, H., Liu, Q., Zhu, Q., and Zhao, X. (2009). "Titanocene dichloride-Lewis base: An efficient catalytic system for coupling of epoxides and carbon dioxide." *Catalysis Communications*, 11(3), 155–157.

Bennaceur, K., and Gielen, D. (2010) "Energy technology modelling of major carbon abatement options." *International Journal of Greenhouse Gas Control*, 4(2), 309–315.

Conte, M. (2009). "ENERGY | Hydrogen Economy." *Encyclopedia of Electrochemical Power Sources*, Elsevier, Amsterdam, 232–254.

Cormos, C.-C., Starr, F., and Tzimas, E. (2010). "Use of lower grade coals in IGCC plants with carbon capture for the co-production of hydrogen and electricity." *International Journal of Hydrogen Energy*, 35(2), 556–567.

Damm, D.L., and Fedorov, A.G. (2008). "Conceptual study of distributed CO_2 capture and the sustainable carbon economy." *Energy Conversion and Management*, 49(6), 1674–1683.

Darensbourg, D.J., and Holtcamp, M.W. (1996). "Catalysts for the reactions of epoxides and carbon dioxide." *Coordination Chemistry Reviews*, 153, 155–174.

Darensbourg, D.J., Rodgers, J.L., Mackiewicz, R.M., and Phelps, A.L. (2004). "Probing the mechanistic aspects of the chromium salen catalyzed carbon dioxide/epoxide copolymerization process using in situ ATR/FTIR." *Catalysis Today*, 98(4), 485–492.

Du, Y., Kong, D.-L., Wang, H.-Y., Cai, F., Tian, J.-S., Wang, J.-Q., and He, L.-N. (2005). "Sn-catalyzed synthesis of propylene carbonate from propylene glycol and CO_2 under supercritical conditions." *Journal of Molecular Catalysis A: Chemical*, 241(1–2), 233–237.

Dutton, A.G., and Page, M. (2007). "The THESIS model: An assessment tool for transport and energy provision in the hydrogen economy." *International Journal of Hydrogen Energy*, 32(12), 1638–1654.

Halloran, J.W. (2007). "Carbon-neutral economy with fossil fuel-based hydrogen energy and carbon materials." *Energy Policy*, 35(10), 4839–4846.

Halmann, M.M. (1993). *Chemical fixation of carbon dioxide: Methods for recycling CO_2 into useful products.* Boca Raton, FL: CRC Press.

Hamilton, M.R., Herzog, H.J., and Parsons, J.E. (2009). "Cost and U.S. public policy for new coal power plants with carbon capture and sequestration." *Energy Procedia*, 1(1), 4487–4494.

Hawkins, A.S., McTernan, P.M., Lian, H., Kelly, R.M., and Adams, M.W.W. (2013). "Biological conversion of carbon dioxide and hydrogen into liquid fuels and industrial chemicals." *Current Opinion in Biotechnology*, 24(3), 376–384.

Herzog, H., and Golomb, D. (2004). "Carbon Capture and Storage from Fossil Fuel Use." *Contribution to Encyclopedia of Energy.* Available online: http://sequestration.mit .edu/pdf/enclyclopedia_of_energy_article.pdf (accessed July 2013).

Herzog, H., Drake, E., and Adams, E. (1997). *CO_2 Capture, Reuse, and Storage Technologies for Mitigating Global Climate Change.* A White Paper Final Report. Energy Laboratory. Massachusetts Institute of Technology Available on line: http://sequestration.mit.edu/pdf/WhitePaper.pdf.

Hickman, R., Ashiru, O., and Banister, D. (2010). "Transport and climate change: Simulating the options for carbon reduction in London." *Transport Policy*, 17(2), 110–125.

Hsu, T.-J., and Tan, C.-S. (2002). "Block copolymerization of carbon dioxide with cyclohexene oxide and 4-vinyl-1-cyclohexene-1,2-epoxide in based poly(propylene carbonate) by yttrium-metal coordination catalyst." *Polymer*, 43(16), 4535–4543.

Hwang, J.S., Chang, J.S., Park, S.E., Ikeue, K., and Anpo, M. (2005). "Photoreduction of carbondioxide on surface functionalized nanoporous catalysts." *Top. Catal.* 35, 311–319.

Indrakanti, V.P., Kubicki, J.D., and Schobert, H.H. (2009). "Photoinduced activation of CO2 on Ti-based heterogeneous catalysts: Current state, chemical physics-based insights and outlook." *Energy Environ. Sci.*, 2, 745–758.

Inoue, S., Koinuma, H., and Tsuruta, T. (1969). "Copolymerization of carbon dioxide and epoxide." *J. Polym. Sci. Polym. Lett.*, 7, 287–292.

Inoue, T., Fujishima, A., Konishi, S., and Honda, K. (1979). Photoelectrocatalytic reduction of carbon dioxide in aqueous suspensions of semiconductor powders. *Nature*, 277, 637–638.

Jiang, Z., Xiao, T., Kuznetsov, V.L., and Edwards, P.P. (2010). "Turning carbon dioxide into fuel." *Phil.Trans. R. Soci. A*, 368, 3343–3364.

Kannan, R. (2009). "Uncertainties in key low carbon power generation technologies–Implication for UK decarbonisation targets." *Applied Energy*, 86(10), 1873–1886.

Liu, B., Zhao, X., Guo, H., Gao, Y., Yang, M., and Wang, X. (2009). "Alternating copolymerization of carbon dioxide and propylene oxide by single-component cobalt salen complexes with various axial group." *Polymer*, 50(21), 5071–5075.

Liu, H., and Gallagher, K.S. (2010). "Catalyzing strategic transformation to a low-carbon economy: A CCS roadmap for China." *Energy Policy*, 38(1), 59–74.

Liu, Y., Huang, K., Peng, D., and Wu, H. (2006). "Synthesis, characterization and hydrolysis of an aliphatic polycarbonate by terpolymerization of carbon dioxide, propylene oxide and maleic anhydride." *Polymer*, 47(26), 8453–8461.

Louitzenhiser, P.G., Meier A., and Steinfeld, A. (2010)."Review of the two-step H_2O/CO_2-splitting solar thermochemical cycle based on Zn/ZnO redox reactions." *Materials*, 3, 4922–4938.

Lu, X.-B., Zhang, Y.-J., Jin, K., Luo, L.-M., and Wang, H. (2004). "Highly active electrophile-nucleophile catalyst system for the cycloaddition of CO_2 to epoxides at ambient temperature." *Journal of Catalysis*, 227(2), 537–541.

Ma, J., Sun, N., Zhang, X., Zhao, N., Xiao, F., Wei, W., and Sun, Y. (2009). "A short review of catalysis for CO_2 conversion." *Catalysis Today*, 148(3–4), 221–231.

Muradov, N.Z., and Veziroglu, T.N. (2008). "Green" path from fossil-based to hydrogen economy: An overview of carbon-neutral technologies." *International Journal of Hydrogen Energy*, 33(23), 6804–6839.

Nader, S. (2009). "Paths to a low-carbon economy–The Masdar example." *Energy Procedia*, 1(1), 3951–3958.

Ockwell, D.G., Watson, J., MacKerron, G., Pal, P., and Yamin, F. (2008). "Key policy considerations for facilitating low carbon technology transfer to developing countries." *Energy Policy*, 36(11), 4104–4115.

Olah, G.A., Prakash, G.K.S., and Goeppert, A. (2011). "Anthropogenic chemical carbon cycle for a sustainable future." *Journal of the American Chemical Society*, 133 (33), 12881–12898.

Omae, I. (2006). "Aspects of carbon dioxide utilization." *Catalysis Today*, 115(1–4), 33–52.

Paddock, R.L., Hiyama, Y., McKay, J.M., and Nguyen, S.T. (2004). "Co(III) porphyrin/DMAP: an efficient catalyst system for the synthesis of cyclic carbonates from CO_2 and epoxides." *Tetrahedron Letters*, 45(9), 2023–2026.

Palgunadi, J., Kwon, O.S., Lee, H., Bae, J.Y., Ahn, B.S., Min, N.-Y., and Kim, H.S. (2004). "Ionic liquid-derived zinc tetrahalide complexes: structure and application to the coupling reactions of alkylene oxides and CO_2." *Catalysis Today*, 98(4), 511–514.

Parsons Brinckerhoff (PB) and Global CCS Institute (GCCSI) (2011). "Accelerating the uptake of CCS: Industrial use of captured carbon dioxide." Available at http://www.globalccsinstitute.com/publications/accelerating-uptake-ccs-industrial-use-captured-carbon-dioxide (accessed July 2013).

Peace, J., and Juliani, T. (2009). "The coming carbon market and its impact on the American economy." *Policy and Society*, 27(4), 305–316.

Rahimpour, M.R. (2007). "A two-stage catalyst bed concept for conversion of carbon dioxide into methanol." *Fuel Processing Technology*, 89, 556–566.

Richels, R.G., and Blanford, G.J. (2008). "The value of technological advance in decarbonizing the U.S. economy." *Energy Economics*, 30(6), 2930–2946.

Roy, C.C., Varghsese, O.K., Paolose, M., and Grimes, C.A. (2010). "Toward solar fuels: photocatalytic conversion of carbon dioxide to hydrocarbons." ACS *Nano*, 4, 1259–1278.

Seyfang, G. (2010). "Community action for sustainable housing: Building a low-carbon future." *Energy Policy*, 38(12), 7624–7633.

Song, C. (2006). "Global challenges and strategies for control, conversion and utilization of CO_2 for sustainable development involving energy, catalysis, adsorption and chemical processing." *Catalysis Today*, 115, 2–32.

Stangland, E.E., Stavens, K.B., Andres, R.P., and Delgass, W.N. (2000). "Characterization of gold-titania catalysts via oxidation of propylene to propylene oxide." *Journal of Catalysis*, 191(2), 332–347.

Sun, D., and Zhai, H. (2007). "Polyoxometalate as co-catalyst of tetrabutylammonium bromide in polyethylene glycol (PEG) for coupling reaction of CO_2 and propylene oxide or ethylene oxide." *Catalysis Communications*, 8(7), 1027–1030.

Thambimuthu, K., Davison, J., and Gupta, M. (2002). "CO_2 capture and reuse." *Proceedings of the IPCC Workshop on Carbon Dioxide Capture and Storage*. 18–21 November 2002, Regina, Canada.

Tominaga, Y., Shimomura, T., and Nakamura, M. (2010). "Alternating copolymers of carbon dioxide with glycidyl ethers for novel ion-conductive polymer electrolytes." *Polymer*, 51(19), 4295–4298.

Toyir, J., Ramírez de la Piscina, P., Fierro, J.L.G., and Homs, N. (2001). "Highly effective conversion of CO_2 to methanol over supported and promoted copper-based catalysts: influence of support and promoter." *Applied Catalysis B: Environmental*, 29, 207–215.

van Alphen, K., Noothout, P.M., Hekkert, M.P., and Turkenburg, W.C. (2010). "Evaluating the development of carbon capture and storage technologies in the United States." *Renewable and Sustainable Energy Reviews*, 14(3), 971–986.

van Schilt, M., Kemmere, M., and Keurentjes, J. (2006). "Process development for the catalytic conversion of cyclohexene oxide and carbon dioxide into poly(cyclohexene carbonate)." *Catalysis Today*, 115(1–4), 162–169.

Wang, J.T., Zhu, Q., Lu, X.L., and Meng, Y.Z. (2005). "ZnGA-MMT catalyzed the copolymerization of carbon dioxide with propylene oxide." *European Polymer Journal*, 41(5), 1108–1114.

Wang, J.-Q., Kong, D.-L., Chen, J.-Y., Cai, F., and He, L.-N. (2006). "Synthesis of cyclic carbonates from epoxides and carbon dioxide over silica-supported quaternary ammonium salts under supercritical conditions." *Journal of Molecular Catalysis A: Chemical*, 249(1–2), 143–148.

Yasuda, H., He, L.-N., Sakakura, T., and Hu, C. (2005). "Efficient synthesis of cyclic carbonate from carbon dioxide catalyzed by polyoxometalate: the remarkable effects of metal substitution." *Journal of Catalysis*, 233(1), 119–122.

Yeh, S., and Sperling, D. (2010). "Low carbon fuel standards: Implementation scenarios and challenges." *Energy Policy*, 38(11), 6955–6965.

Yin, X., Moss, J.R. (1999). ''Recent developments in the activation of carbon dioxide by metal complexes.'' *Coordinat. Chem. Rev.*, 181, 27–59.

Zevenhoven, R., Eloneva, S., and Teir, S. (2006). "Chemical fixation of CO_2 in carbonates: Routes to valuable products and long-term storage." *Catalysis Today*, 115(1–4), 73–79.

Zhang, S., Chen, Y., Li, F., Lu, X., Dai, W., and Mori, R. (2006). "Fixation and conversion of CO_2 using ionic liquids." *Catalysis Today*, 115(1–4), 61–69.

Zhang, T.C., and Surampalli, R.Y. (2013). "Carbon capture and storage for mitigating climate changes." Chapter 20 in *Climate Change Modeling, Mitigation and Adaptation,* Surampalli, R., Zhang, T.C., Ojha, C.S.P., Gurjar, B.R., Tyagi, R.D., and Kao, C.M. (eds.), ASCE, Reston, Virginia, February 2013.

CHAPTER 8

Carbon Dioxide Capture Technology for Coal-powered Electricity Industry

C. M. Kao, Z. H. Yang, R. Y. Surampalli, and Tian C. Zhang

8.1 Introduction

Many different sources contribute significant amounts of carbon into the atmosphere, including combustion, industrial processes, respiration and decay, and volcanic activities This has caused the buildup of greenhouse gases (GHGs) in the atmosphere and also resulted in the global climate change. It is well known that carbon dioxide (CO_2) is the main GHG, contributing to the greenhouse warming effect by 81% (VGB 2004), and fossil-fuel-burning power plants are the single-largest contributor to CO_2 emission. Therefore, innovative technologies for capturing CO_2 from these fossil-fuel-burning power plants and storing it in geologic formations, that is, carbon capture and storage (CCS), have been developed. CCS is a process consisting of the separation and capture of CO_2 from industrial and energy-related sources, transport to a storage location and long-term isolation from the atmosphere (IPCC 2005; Plasynski et al. 2009; Wang et al. 2011; Padurean et al. 2012; Zhang and Surampulli 2012). Currently, CCS has become one of the major methods to reduce the CO_2 emissions to the atmosphere and mitigate the deterioration of global warming. To establish significant CCS capabilities, more efforts are necessary to make the CCS technologies more applicable, practical, efficient, and cost effective.

The focus of this chapter is on CO_2 capture technology for the coal-fired power plants as coal-fired plants emit significantly more CO_2 than natural gas plants. This topic has been addressed and researched/reviewed since the early 1940s (e.g., Tepe and Dodge 1943; Spector and Dodge 1946; Riemer et al. 1993; USDOE 1999; Herzog 2001; Anderson and Newell 2003; VGB 2004; IPCC 2005; IEA 2009; Lackner and Brennan 2009; CCCSRP 2010; ITF 2010). However, information related to the topic is overwhelming and difficult to digest. Therefore, this chapter will serve as an introductory guideline to the topic. Specifically, the chapter will discuss the basic principles of CO_2 capture technologies, the major approaches and alternatives to capturing CO_2, the current issues and future perspectives. The chapter also provides a list of references for the audiences who need detailed reviews of CO_2 capture technologies and cutting-edge research.

8.2 CO_2 Capture Technologies

8.2.1 General Means for CO_2 Capture

CO_2 capture technologies themselves are not new. In the 1940s, chemical solvents (e.g., monoethanolamine (MEA)-based solvents) were developed to remove acid gases (e.g., CO_2 and H_2S) from impure natural gas to boost the heating value of natural gas. The same or similar solvents were used to recover CO_2 from their flue gases for application in the foods-processing and chemicals industries by power plants. On the other hand, the feasibility of capturing CO_2 from ambient air was evaluated in the 1940s (Tepe and Dodge 1943; Spector and Dodge 1946). Carbon dioxide removal technology has been developed and used as a standard process in gas production industry (e.g., natural gas, hydrogen gas). Carbon dioxide needs to be removed before the product can be used and sold. Therefore, many CO_2 removal systems have been constructed and operated in these gas production plants (Plasynski et al. 2009).

Carbon capture technologies can be categorized as a) physical/chemical and biological technologies (Table 8.1) and b) technologies for carbon capture from concentrated point sources and mobile/distributed point- or non-point sources (Table 8.2). On-site capture is the most viable approach for large sources and initially offers the most cost-effective avenue to sequestration. The remainder of this chapter will mainly cover CO_2 capture technologies for the coal-fired power plants, that is, the physical and chemical (sorption-based) technologies for capturing CO_2 from post-, pre- or oxy-combustion processes from coal-fired plants. Those interested in carbon capture via biological technologies or from mobile/diluted point- or non-point sources are directed to elsewhere (e.g., Zhang and Surampalli 2012).

8.2.2 CO_2 Capture Technologies for Coal-fired Power Plants

The choice of a suitable technology for CO_2 capture depends on power plant technology as it dictates the characteristics of the flue gas stream. In a fossil-fuel power station, coal, natural gas or petroleum (oil) can be used to produce electricity. Table 8.3 shows the variety of power plant fuels and technologies that affect the choice of CO_2 capture systems. As shown in Table 8.4 (VGB 2004), of the fossil-fuel plants, for a fixed amount of fuel feed, coal-fired plants emit about twice as much CO_2 as natural gas plants, and thus, will be the focus for us to introduce CO_2 capture technology.

Table 8.1. Physical/chemical and biological technologies for CO_2 capture[a]

Methods	Description
1) Physical/chemical	• <u>Cryogenic distillation</u>: widely used for other gas separation. However, high energy cost is involved; thus, not considered as a practical means for CO_2 capture • <u>Membrane separation/purification</u>: extensively used for CO_2 separation from relatively concentrated sources. A relatively new technology that has not been optimized for large-scale applications. Comments: i) not suitable for post-combustion CO_2 capture due to the low CO_2 concentration (5-15%) in the off-gases; and ii) maybe suitable for pre-combustion (or new power generation plants') CO_2 capture due to relatively high CO_2 concentration in the off-gases • <u>Sorption</u>: o Sorption with liquid sorbents: A benchmark, mature, widely-used technology, including physical sorption (e.g., using methanol or poly(ethylene glycol) dimethyl ether as sorbing phases) or chemical sorption (e.g., using amines solutions or fluids with basic character o Sorption with solid sorbents: Operated via weak physisorption processes or strong chemi-sorption interactions. Now being widely considered as an alternative, potentially less-energy-intensive technology. Examples of major solid sorbents include: zeolites, activated carbons, calcium oxides, hydrotalcites, organic-inorganic hybrids, and metal-organic frameworks (see Table 8.4)
2) Biological	• <u>Tree and organisms</u>: Capture CO_2 via photosynthesis (e.g., reforestation or avoiding deforestation); cost range 0.03-8\$/t-CO_2, one-time reduction (i.e., once the forest mature, no capture; release CO_2 when decomposed). Comments/problems/concerns: i) develop dedicated biofuel and biosequestration crops (e.g., switchgrass); ii) enhance photosynthetic efficiency by modifying Rubisco genes in plants to increase enzyme activities; iii) choose crops that produce large numbers of phytoliths (microscopic spherical shells of silicon) to store carbon for thousand years • <u>Ocean flora</u>: Adding key nutrients to a limited area of ocean to culture plankton/algae for capturing CO_2; utilize biological/microbial carbon pump (e.g., jelly pump) for CO_2 storage. Comments/problems/concerns: i) large-scale tests done but with limited success; ii) limited by the area of suitable ocean surface; iii) may have problems to alter the ocean's chemistry; and d) mechanisms not fully known • <u>Biomass-fueled power plant, bio-oil and biochar:</u> i) growing biomass to capture CO_2 and later captured from the flue gas. Cost range = 41\$/t-CO_2; ii) by pyrolyzing biomass, about 50% of its carbon becomes charcoal, which can persist in the soil for centuries. Placing biochar in soils also improves water quality, increases soil fertility, raises agricultural productivity and reduce pressure on old growth forests; pyrolysis can be cost-effective for a combination of sequestration and energy production when the cost of a CO_2 ton reaches \$37 (in 2010, it is \$16.82/ton on the European. Climate Exchange). • <u>Sustainable practices:</u> i) Farming practices (e.g., no-till, residue mulching, cover cropping, crop rotation) and conversion to pastureland with good grazing management would enhance carbon sequestration in soil; ii) Peat bogs inter ~25% of the carbon stored in land plants and soils. However, flooded forests, peat bogs, and biochar amended soils can be CO_2 sources

[a]Major references: Choi et al. (2009); Greenleaf and SenGupta (2009); Liu et al. (2009); Plasynski et al. (2009); Pfeiffer et al. (2011); de Richter (2012); and Zhang and Surampulli (2012).

In all coal-fired power plants, the major component of flue gas is nitrogen, which enters originally with the air feed. If there were no nitrogen, CO_2 capture from flue gas would be greatly simplified (VGB 2004). Therefore, for the existing coal-fired combustion plants, there are two main options for CO_2 capture: removal of nitrogen from a) flue gases or b) air before combustion to obtain a gas stream ready for transport or geo-sequestration. Therefore, there are three major options in CO_2 capture, i.e., post-, pro-, and oxy-combustion capture (Fig. 8.1) as described below.

8.2.2.1 Post-Combustion Technologies

Post-combustion CO_2 capture refers to the capture of CO_2 from the flue gas stream of a PC power plant. Conventional coal plants use the heat from burning coals to produce steam, which drives turbines to generate electric power. Most coal-powered plants are conventional subcritical or supercritical PC plants (Esber 2006). Post-combustion capture involves the CO_2 removal from the flue gas produced by burning a fossil fuel. Post-combustion carbon capture has a great potential to reduce greenhouse gas emissions, because it can be retrofitted to the existing units that generate two-thirds of the CO_2 emissions in the power sector (Plasynski et al. 2009). Post-combustion capture offers some advantages as the existing combustion technologies can still be used without significant changes. This makes post-combustion capture easier to implement as a retrofit option to existing plants. The advantage comes at the expense of the efficiency of the power generation process (Davison 2007; Wang et al. 2011). The post-combustion CO_2 capture system will reduce the plant's overall thermal efficiency by 24%, about one-third (8%) is due to compression, with the rest (16%) attributable to separation (Herzog et al. 2009).

Table 8.2. CO_2 capture technologies from difference sources[a]

Methods	Description
1) Concentrated sources	• Post-combustion: CO_2 is removed after combustion of the fossil fuel as in power plants. The technology is well understood and is used in other industrial applications. Comments/ problems/concerns: i) it would reduce energy efficiency by 10–40%; ii) the thermodynamic driving force for capture CO_2 is low; and iii) compatible with the power plants, flexible, and a leading candidate for gas-fired power plants • Pre-combustion: The CO_2 is recovered from some process stream before the fuel is burned. Widely applied in fertilizer, chemical, gaseous fuel plants. Comments/problems/concerns: i) the partial pressure of CO_2 is much higher than in a typical flue gas, and a cheaper CO2 capture process can be used as a result; ii) in the US, only two IGCC plants are in operation in the power industry and both were built as demonstration plants; and iii) the ultimate commercial success of IGCC to provide coal-fired electricity remains uncertain • Oxy-combustion: The fuel is burned in oxygen, resulting in an almost pure CO_2 stream that can be transported. The oxy-fuel plant can eliminate all air pollutants (i.e., zero emission). Comments/Problems/concerns: i) may add ~7₵/kWh to the production cost of electricity; ii) the need for a cryogenic oxygen plant and flue gas recycle is costly; and iii) chemical looping combustion method (using a metal oxide as a solid oxygen carrier) is a promising emerging technology
2) Mobile/diffused point- or non-point source	• Work is still in its infancy. Capture costs are higher than from point sources. May be feasible for carbon capture from distributed sources such as automobiles and aircraft. Examples: i) an anionic exchange resin as a solid sorbent that absorbs CO_2 when dry and releases it when wet; ii) ion-exchange fibers to sequester CO_2 into an aqueous Ca or Mg alkalinity while concurrently softening hard water; and iii) the "Air Capture" system captures ~ 80% of CO_2 from the air

[a]Major references: Osborne and Beerling (2006); Stolarooff (2004); VGB (2006); Beerling (2008); Greenleaf and SenGupta (2009); and Zhang and Surampulli (2012).

Table 8.3. Technology options for fossil-fuel based power generation

Fuel	Oxidant	Technology
• Coal o Combustion-based o Gasification-based • Gas o Direct combustion o Gas reforming	• Air • Pure oxygen	• Simple cycle o Pulverized coal (PC) o Gas turbines • Combined cycle o GTCC o IGCC o Others

GTCC = gas turbine combined cycle; IGCC = integrated gasification combined cycle.

A number of separation technologies could be employed with post-combustion capture. These include: (a) adsorption; (b) physical absorption; (c) chemical absorption; (d) cryogenics separation; and (e) membranes (IPCC 2005; Wang et al. 2011). Post-combustion capture of flue gas CO_2 is a feasible option for CO_2 emission control. Although the current systems use chemical solvent absorption/regeneration to obtain CO_2 capture, higher additional capital costs and energy consumption are needed. A significant amount of energy is lost during the processes of solvent regeneration and CO_2 transport and storage.

Table 8.4. CO_2 from fossil fuel combustion

Fuel	T CO_2/TJ	Fuel	T CO_2/TJ
Lignite	101.2	Residual fuel oil	77.4
Sub-bituminous coal	96.1	Diesel/gas oil	74.1
Coking coal	94.6	Natural gas	56.1

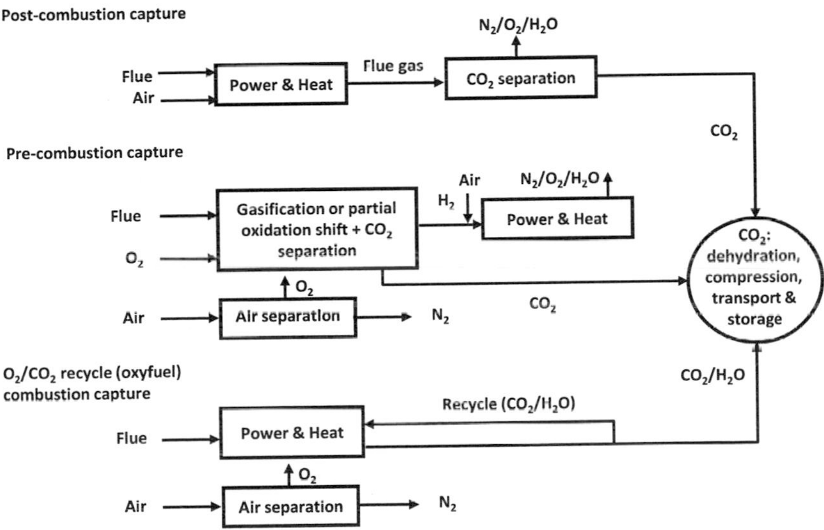

Figure 8.1. Overview of CO_2 capture from power plants (adapted from VGB 2004)

Considerable research has been conducted to obtain technological solutions to reduce the cost of CO_2 capture of post-combustion systems currently in use. Several alternative solvents have been investigated for their applicability, and they all have their advantages and disadvantages, but none seems to be clearly superior to MEA. Thus, deeper integrations of a chemical absorption system into the overall plant could yield significant improvements without requiring overall innovations. Some innovative developments, such as the stimulus-responsive separation aids and structured fluids, the lithium zirconate wheel, cryogenic CO_2 frosting process, and membrane systems have been investigated (IPCC 2005; Wang et al. 2011). These novel designs have unique characters and advantages including reducing heat of regeneration requirements and reduced energy requirements (IPCC 2005; Wang et al. 2011). However, it is still not clear that which design is able to achieve overall or superior improvements (including the reduction of the capture cost) than the other systems. It is very difficult to compare processes or technologies that have been studied by different groups because they all have different analytical bases and design considerations, and their results are usually obtained from laboratory or pilot-scale studies without further support by full or field-scale data. Thus, more efforts are needed to analyze these technologies carefully. Reduction of capital cost, energy consumption, and operational cost are still the top priority issues in designing the future post-combustion CO_2 capture systems. The government policies cannot be ignored to commercialize the developed post-combustion CO_2 capture systems.

8.2.2.2 Pre-combustion Technologies

In the pre-combustion CO_2 capture process, CO_2 is recovered from the stream of different processes before the fuel is burned (Kanniche and Bouallou 2007; GTC 2011). The partial pressure of CO_2 can be increased if the concentration and pressure of the CO_2 stream are increased, and this can reduce the CO_2 capture cost. Thus, development of combustion technologies, which can produce concentrated CO_2 streams at a higher pressure, is a way for operational cost reduction (Ordorica-Garcia et al. 2006; Pennline et al. 2008; Plasynski et al. 2009).

The integrated gasification combined cycle (IGCC) process mainly contains a gasifier island, a cold-gas conditioning section, and a combined cycle (Spliethoff 2010). After preparation, coal is converted under reducing conditions to a raw gas composed of hydrogen and carbon monoxide. Pure oxygen from an air separation unit is used as an oxidizing agent. In the gasification process, many species that are critical for the environment are released into the gas phase. Different species such as ammonia, chlorine, sulfur, and CO_2 are removed in a low-temperature gas treatment section. The remaining fuel gas is diluted using waste N_2 from the air separation unit and utilized in the combined cycle section (Spliethoff 2010).

Because of the favorably high concentration of the component in the raw gas stream, IGCC has become one of the effective technologies for CO_2 capture. The advantages of using IGCC include the following: high thermal efficiency, low byproducts (e.g., NOx, SOx, solids) emissions, and the capability of processing lower grade coals. Thus, it has been proved competitive against conventional coal power plants.

However, the drawback for IGCC plants is the production of significant amounts of CO_2, which are released into the atmosphere. Thus, it is very important to design more efficient power plants, and in the meantime, adopt integrated CO_2 capture technologies (Cormos 2009; Spliethoff 2010; Padurean et al. 2012).

Research and development of IGCC plant technology began in the 1970s but, even so, coal-based IGCC plants are still not completely commercialized (GTC 2011). The studies published so far include conceptual designs, flow sheet modeling and cost estimation based on different technology selections and assumptions (Haslbeck 2002; IEA 2010). However, there are no generally available process models that can be easily used or modified to systematically study the performance and cost of CO_2 capture and storage options from IGCC systems (Rubin et al. 2004; Chen and Rubin 2009; Strube and Manfrida 2011; Padurean et al. 2012).

Selexol® and Purisol® Technology. The Selexol® technology uses dimethyl ethers of polyethylene glycol ($CH_3O(C_2H_4O)_nCH_3$) as the solvent to physically absorb H_2S and CO_2 from syngas (Padurean et al. 2010; 2011). The main factors affecting the efficiency of Selexol® technology include the following: requirements for the level of H_2S and CO_2 selectivity, the efficiency of sulfur removal, the efficiency for CO_2 removal, and the efficiency of gas dehydration. When the Selexol® technology is used for H_2S removal, it is removed from a product stream. As for the CO_2 removal, it is usually conducted as a separate product stream.

In the Selexol® process, syngas enters the first absorber unit at approximately 32 bar and 40 °C, and the H_2S-free gas obtained at the top of the absorption column. Significant amounts of CO_2 are sent to the second absorber for its removal. The rich H_2S solution leaving the bottom of the first absorber is partially decompressed using a hydraulic turbine, and then it is regenerated in a stripper, through the indirect application of thermal energy via condensing low-pressure steam in a boiler (Padurean et al. 2010; 2011). After the desulfurization unit, the CO_2 is removed in the second absorption column, with regenerated solvent resulting from H_2S and CO_2 desorption. The purified gas obtained at the top of the second absorber, and the CO_2 rich solution discharges from the bottom of the absorber, which is regenerated by gradually reducing the pressure. The CO_2 stream is dried, compressed, and cooled for further transportation and storage (Kohl and Nielsen 2005; Kanniche et al. 2010). The Purisol® technology uses N-Methyl-2-pyrolidone as the solvent to absorb H_2S and CO_2 from gas streams. The flow scheme used for this solvent is similar to the one used for Selexol® solvent and the process can be operated either at ambient temperature or to refrigeration down to about -15 °C (Kohl and Nielsen 2005).

Rectisol® Technology. The Rectisol® technology uses refrigerated methanol as the solvent to purify syngas generated from the gasification of oils and coals. This process operates at a low temperature (between -40□ and -60□) and is more complex compared to other physical solvent processes to obtain higher H_2S removal rates. The Rectisol® process is designed to remove high levels of sulfur and CO_2 from a syngas (Korens et al. 2002; Kohl and Nielsen 2005). The cooled syngas is first discharged into the first absorption column where Rectisol® is preloaded with CO_2, for H_2S removal. The H_2S solution is then discharged from the bottom of the absorber. It is regenerated first by flashing at medium pressure to recover the useful gases, and then by heating to boiling temperature and stripping with methanol vapor. The desulfurized gas reenters in the CO_2 absorption column where CO_2 is removed. The CO_2 solution, leaving the absorber, is regenerated in a flash regeneration unit (Korens et al. 2002; Kohl and Nielsen 2005).

8.2.2.3 Oxy-Fuel Technologies

Oxy-fired coal combustion technology has been used as a way to lower down the CO_2 capture cost. Some pilot-scale demonstration plants of oxy-firing systems have been designed and operated. However, field-scale or large-scale oxy-fired coal power plants have not been deployed. Oxy-fired technology is able to improve CO_2 capture from post-combustion flue gases. There is currently no incentive to use oxy-firing for the coal power production if CO_2 capture is not operated concurrently (Esber 2006).

In this system, N_2 and O_2 are first separated prior to combustion, and thus, the flue gas is made up primarily of CO_2 and water vapor (USDOE 2002). The purification and following transport and storage for CO_2 is made easier and less expensive than for an air-fired system. The flue gas recycle step is necessary to control flame temperature and stability and to ensure proper heat flux in the boiler (Buhre et al. 2005). Most oxy-fired systems are designed with a supercritical steam cycles with an overall plant thermal efficiency of around 30% (Deutch and Moniz 2006; Esber 2006).

There are several features that differentiate oxy-fired systems from conventional pulverized coal (PC) power plants. The O_2 content of the gas fed to the boiler of the oxy-fired systems is higher, typically around 30%, compared to the air gas (with 21% of O_2), which is fed to a conventional boiler (Buhre et al. 2005). The high CO_2/H_2O gas in the furnace has a higher gas emissivity, which allows the same radioactive heat transfer with a smaller volume of boiler gases than a PC system (Buhre et al. 2005). The volume of flue gas is reduced by 80% prior to CO_2 purification and compression (Buhre et al. 2005; Esber 2006).

The technology used for the oxy-fired system is available in the market, and thus, the system can be built and operated without a technology breakthrough. However, large scale systems have not been developed because there is no incentive to build a CO_2 capture system in place only for the CO_2 capture purpose (Deutch and Moniz 2006).

While the oxy-fuel combustion and O_2-enhanced combustion have been used in some industries, the concept of oxy-fuel combustion in the context of providing a CO_2-rich flue gas for EOR was proposed in 1982 (Deutch and Moniz 2006). In the early 1990s, reduction of CO_2 emissions has been assigned as a priority in the power generation industry. Many research institutes and government agencies (e.g., Argonne National Laboratory, International Flame Research Foundation, the Japanese New Energy and Industrial Technology Development Organization) conducted different pilot-scale studies regarding the oxy-fuel combustion technologies during the 1990s (Deutch and Moniz 2006; Esber 2006).

There are two different techniques that have been developed for the purpose of CO_2 emission reduction in electric power plants using fossil fuels as energy sources: 1) efficiency improvement and 2) CO_2 separation from fuel or flue gases. The efficiency-improving measure can further reduce CO_2 emissions from fossil-fired power plants by approximately 25% (Buhre et al. 2005; Esber 2006). Development of the separation methods concentrates on power plant processes that use coals as the energy source. The CO_2 emissions are particularly high in the case of power generation from coal, which means that separation of CO_2 will enable greater reduction in CO_2 emissions. However, significant efficiency losses might be observed depending on the separation method and the power plant type.

In the case of the steam power cycle with a coal-fired boiler that is used most frequently for electric power generation, the CO_2 could soon be scrubbed out of the flue gas using a scrubber with monoethanolamine (MEA) or similar solvents (Aroonwilas and Veawab 2006; Kather et al. 2008). The low CO_2 concentration in the wet flue gas implies that this involves considerable energy input. As the largest portion of the remaining flue gas components consists of the atmospheric N_2 introduced into the process with the combustion air, holding back this N_2 before combustion ensures a significant increase in the CO_2 concentration. This is the reason why the concept underlying the oxy-fuel process is to extract N_2 from the combustion air before combustion via an air separation unit, which indicates that pure O_2 is fed to the combustion process. Currently, oxy-fuel technology is undergoing rapid development towards commercialization (Wall and Wu 2009). A comprehensive overview on recent developments in pilot- or large-scale plants or demonstration projects for the oxy-fuel CCS technology has been reported (Wall and Yu 2009). These projects have been conducted to characterize the combustion performance of coals under oxy-fuel combustion conditions.

Compared to conventional air-fired combustion, oxy-fuel combustion affects the PC combustion process and heat transfer processes due to the variation in the oxidant and consequently in the furnace gas environment. Many projects have been conducted that covered many scientific and engineering fundamental issues, such as: the recycled flue gas ratio, ignition, flame stability, heat transfer, combustion characteristics, and pollutant formation and reduction (Aroonwilas and Veawab 2006; Kather et al. 2008). Carbon dioxide has different characteristics from N_2, which influence the heat transfer and combustion reaction kinetics. Because CO_2 gas has a higher density than N_2, and the CO_2-H_2O mixture has a higher specific heat capacity as well as different radiation and absorption characteristics, the flue gas recycle ratio and the oxygen concentration during oxy-fuel combustion are the main parameters that determine the optimum firing conditions.

8.2.2.4 Carbon Dioxide Drying and Compression

As shown in Fig. 8.1, before transporting CO_2, we need to dry, condition and compress the captured CO_2. Captured CO_2 may contain impurities like water vapor, H_2S, N_2, methane (CH_4), O_2, mercury, and hydrocarbons that may require specific handling or treatment. The sources of impurities in the CO_2 stream include: 1) fuel (e.g., H_2O, CO, SOx, NOx, H_2S, HCl, HF, H_2, CH_4, heavy metals, hydrocarbons, particulates are from the fuel); 2) air or oxidant used for combustion of the fuel (e.g., O_2, N_2 and Ar); and 3) the CO_2 capture and CO_2 clean-up process (e.g., NH_3 and solvents). Transport and storage related aspects of CO_2 quality mainly include: 1) corrosion of pipeline and injection well; and 2) hydrate formation and plugging of pipeline and injection well. For example, CO_2 reacts with water to form carbonic acid, which is corrosive. Thus, it is desirable to reduce the water level to be lower than 50 ppm (Tohidi 2008; WRI 2008). In addition, hydrates (solid ice-like crystals) can form and plug the pipeline under the certain thermodynamic conditions. Thus, limiting the content of hydrate forming components (e.g., H_2O, H_2S, CO, CH_4) is needed.

The commonly used techniques for CO_2 dehydration include the following: glycol dehydration (used as liquid desiccant), molecular sieves (used as solid adsorbent), and refrigeration. In the CO_2 drying process, the liquid desiccant contactor-regenerator process using tri-ethylene-glycol (TEG) is the most common applied method (Carrol 2002). This process contains two columns (absorber and desorber) and the complementary units (e.g., reboiler, condenser, pump, heat exchanger). Gaseous CO_2 containing water flows into the bottom of the absorber tower and the liquid TEG from the stripper enters the top of the tower. The gas and the liquid flows counter currently within the tower, and water is absorbed in the liquid. The dried CO_2 stream leaves from the top of the absorber tower, and the TEG stream rich in water leaves at the bottom. The thermal TEG regeneration is conducted in a stripper, and the dried TEG stream being cooled and pump back to the absorber (Carrol 2002).

Unlike glycol dehydration, which is an absorption process, dehydration with molecular sieves is an adsorption process. Molecular sieves are usually used when dry gas is required (such as a cryogenic process). In the molecular sieve process, the wet gas enters a bed of adsorbent material. The water in the gas adsorbs onto the bed and a dry gas is produced. Once the bed becomes saturated with water, this must be regenerated (Carrol 2002).

After drying, CO_2 stream is sent to a multistage compression unit where it is compressed to more than 120 bar and then sent via pipelines to the geological storage, like saline aquifers or used for Enhanced Oil Recovery (EOR) (Kohl and Nielsen 2005; Yang et al. 2008). Table 8.5 presents the quality specification of captured CO_2 considering the need of transport and storage conditions.

Table 8.5 Carbon dioxide stream specifications (Padurean et al. 2012)

Compound	Limit concentration
H_2O	< 200 ppm
CO_2	> 95 vol.%
H_2S	< 200 ppm
CO	< 2000 ppm
O_2	< 10 ppm (EOR)
Non-condensable gases (CH_4, N_2, Ar)	< 4 vol.%
SO_x	< 50 ppm
NO_x	< 50 ppm

8.3 Principles of Sorption-based CO_2 Capture Technologies

Table 8.6 shows that one can classify sorbents according to different criteria. Here, we present information on the basis of physical/chemical processes. Solvents used in these processes can be broadly grouped into two categories: physical solvents and chemical solvents (Plasynski et al. 2009; Padurean et al. 2012). When sorbents are used in the aforementioned commercially available processes for CO_2 capture, all of the processes are similar in concept, that is, a two-vessel process with liquid, solid or liquid-impregnated solid sorbents. In the first vessel, the CO_2-containing gas contacts a lean solvent (e.g., monoethanolamine (MEA)-based), and the CO_2 is absorbed there. The CO_2-rich solvent is regenerated by low pressure and high temperature in the second vessel and then returned to the first vessel (VGB 2004). The principles of sorbent-based physical and chemical processes are described below.

Physical Sorption Processes. The CO_2 is absorbed onto the sorbents mainly through physical mechanisms without chemical processes. Depending on the sorbents selected, the amount of CO_2 absorbed can be determined via sorption experiments. Table 8.7 shows properties of two CO_2 physisorbents (Choi et al. 2009).

Table 8.6. Classification of different sorbents for CO_2 capture[a]

- **Physical or chemical sorbents:**
  - o Physical sorbents: zeolites, activated carbons, selexol (a mixture of dimethyl ethers of polyethylene glycol), rectisol (chilled methanol) and propylene carbonate; and Purisol
  - o Chemical sorbents: Amines (monoethanol, diethanol, and methyl diethanol amine); NaOH/ $Ca(OH)_2/NH_3$; novel liquid sorbents (e.g., CO_2 hydrates, liquid crystals, ionic liquids); solid and/or liquid-impreganted sorbents
- **Liquid or solid sorbents:**
  - o Liquid sorbents: i) Amines (monoethanol, diethanol, and methyl diethanol amine); ii) NaOH/ $Ca(OH)_2/NH_3$; and iii) novel liquid sorbents (e.g., CO_2 hydrates, liquid crystals, ionic liquids)
  - o Solid sorbents: i) metal-organic frameworks (MOFs) (e.g., zeolitic imidazolate frameworks, ZIFs); ii) ion exchange resins/ion exchange fibers/functionalized fibrous matrices; iii) ceramics/steel slag and waste concrete; iv) metal-based sorbents (alkalinemetal oxides (Na_2O, K_2O) and alkaline earth metal oxides (CaO, MaO)
  - o Liquid-impreganted sorbents: Liquid-impreganted clay
- **Inorganic or organic sorbents:**
  - o Inorganic sorbents: zeolites, activated carbons, limestone/dolomite/CaO/$Ca_{12}Al_{14}O_{33}$, metal-based sorbents (e.g., calcium or magnesium oxides, lithium zirconates), hydrotalcite-like compounds (e.g., Mg-Al-CO_3, $[Ca_2Al(OH)_6]_2CO_3 \cdot mH_2O$)
  - o Organic or organic-inorganic hybrid sorbents: amines physically sorbed on oxide supports (the molecular basket), amine covalently tethered to oxide supports, amines supported on solid organic materials (e.g., carbon, polymer, resins, etc.)

[a]Major references: Herzog (1999); VGB (2004); Choi et al. (2009); Plasynski et al. (2009); Zhang and Surampalli (2012).

Table 8.7. Properties of two inorganic physisorbents (adapted from Choi et al. 2009)

Name	CO_2 sorption capacity (mmol/g)	H_2O effect on CO_2 capacity	Regenerability
Zeolite	0.09–4.9, usually, natural zeolites: 0.09–1.25, and synthetic: around 1–2.5	Either favorable or detrimental, depending on the concentration of CO_2 in the feed	Good, recovering their fresh sorption capacity without significant degradation after numerous cycles
Activated carbons	0.057–3.5, usually, GAC-MEA, GAC-NH_3: ~0.06, and others: around 1–3.3	Negatively effect. Water competes with CO_2 for sorption on the activated carbons	Excellent. One of the best CO_2 sorbent materials in terms of regenerability.

Factors affecting the absorption efficiency include temperature and partial pressures. Plasynski et al. (2009) reported that the CO_2 absorption efficiency is proportional to CO_2 partial pressure. Solvent regeneration is endothermic and requires energy consumption, and the energy required is usually supplied by steam extracted from the turbines (MGSC 2004; Plasynski et al. 2009). The main drawback for physical solvents is that the absorption efficiency and capacity is significantly affected by the temperature, and low temperature is required to maintain a high absorption efficiency and capacity. Furthermore, improvement of the regeneration system is another key issue to enhance the applicability of the physical solvent-based CO_2 capture technology.

Liquid Sorbent-based Chemical CO_2 Capture Technology. The current benchmark technology being considered for post-combustion CO_2 capture is sorption by amines solution. In this case, amines react with CO_2 to form water soluble compounds for CO_2 removal:

$$2RNHNH_2 + H_2O + CO_2 \leftrightarrow (RHN_3)_2CO_3 \quad (8.1)$$

This reaction is reversible, so the CO_2 gas can be released by heating in a separate stripping column. The interaction of CO_2 with amines is governed by several mechanisms. Primary and secondary amines can react directly with CO_2 to produce carbamates through the formation of zwitterionic intermediates. Tertiary amines catalyze the formation of bicarbonate (Choi et al. 2009). Amines, such as monoethanolamine (MEA), diethanolamine (DEA), methyldiethanolamine (MDEA), or diisopropanolamine (DIPA) may be used. For example, CO_2 and MEA react to form a protonated amine and a bicarbonate anion in solution as per the following equation:

$$C_2H_4OHNH_2 + H_2O + CO_2 \leftrightarrow C_2H_4OHNH_3^+ + HCO_3^- \quad (8.2)$$

The remaining flue gases are washed to remove any residual MEA, and exhausted to the atmosphere. This technology can capture about 75–90% of the CO_2 and get a nearly pure (> 99%) CO_2 product stream (Rao and Rubin 2002).

When amines are used, lower CO_2 partial pressure is used for CO_2 capture. The major factors affecting the absorption efficiency include amine concentration and chemical equilibrium constant. Amines are usually limited to concentrations of about 30% because of corrosion problems at higher concentrations. Furthermore, the presence of contaminants such as O2, SO2, NO_X, HCl, hydrocarbons, particles, and Hg in the flue gas reduces the absorption capacity of amines significantly, degrades the conventional amines rapidly, and results in operational problems such as foaming and corrosion. The formation of stable compounds with impurities in the flue gas can cause the gradual loss of effectiveness. Therefore, significant operational cost is usually required to obtain higher CO_2 removal efficiency (Polasek and Bullin 1985; MGSC 2004; Plasynski et al. 2009). In addition, the process is associated with high energy requirements for the CO_2 regeneration step. More efforts on the increase in capture capacity, system stability, and cost effectiveness are necessities in the future studies (Plasynski et al. 2009; IEA 2010).

Solid Sorbent-based Chemical CO_2 Capture Technology. As shown in Table 8.6, CO_2 can also be captured with solid sorbents. Table 8-8 shows properties of several metal-based chemisorbents, and some of the reactions of these metal-based sorbents with CO_2 are as follows:

$$MO(s) + CO_2(g) \leftrightarrow MCO_3(s) \text{ with } M = Ca, Mg, Sr, Ba, etc. \quad (8.3)$$
$$M_2CO_3(s) + H_2O(g) + CO_2(g) \leftrightarrow 2MHCO_3(s) \text{ with } M = K \text{ and } Na \quad (8.4)$$
$$Li_2ZrO_3 + CO_2(g) \leftrightarrow Li_2CO_3(s) + ZrO_2(s), \Delta H\ (298\ K) = -160\ kg/mol \quad (8.5)$$

$$Li_4SiO_4 + CO_2(g) \leftrightarrow Li_2CO_3(s) + Li_2SiO_3(s),\ \Delta H\ (298\ K) = -142\ kg/mol \quad (8.6)$$

For example, the reaction pair below is an effective technique for the in-situ removal of CO_2 in many high-temperature applications:

$$CaO(s) + CO_2(g) \leftrightarrow CaCO_3(s),\ \text{exothermic} \quad (8.7)$$
$$CaO_3(s) \leftrightarrow CaO(s) + CO_2(g),\ \text{endothermic} \quad (8.8)$$

The CO_2 sorption capacity can be as high as ~12 mmol/g of CaO at temperature > 850 K with a cost of ~\$0.0015/mol of CO_2 being sorbed (Abanades et al. 2007). In general, however, these metal-based sorbents are much slower than physisorbents such as zeolites or activated carbons, and thus, need significant improvement, particularly the adsorption kinetics.

Table 8.8. Properties of metal-based chemisorbents (adapted from Choi et al. 2009)

Name	CO_2 sorption capacity (mmol/g)	CO_2 sorption kinetics	Regenerability
CaO, $Ca(OH)_2$, $CaCO_3$	2.3–10.7, CaO: 2.3–4.5, $Ca(OH)_2$: 6.5–10.7; $CaCO_3$: 1.5–10.5	Much slower than physisorbents (e.g., zeolites & activated carbons)	Reduced by 75% of the initial values after several cycles
MgO, Li_2ZrO_3, Li_4SiO_4	0.043–6.14, MgO: 0.6–2.36, Li_2ZrO_3: 0.23–6.14; Li_4SiO_4: 6.14	Slow sorption kinetics (> several days to achieve equilibrium)	Not being able fully regenerated after several cycles

Choi et al. (2009) described detailed information about solid sorbents for CO_2 capture and the criteria to evaluate these sorbents, such as fast sorption and desorption kinetics, large sorption capacity, infinite regenerability and stability, and a wide yet tunable range of operating conditions. Recently, hydrotalcite-like compounds, amines adsorbed on oxide supports (the molecular basket) or solid organic materials, and metal-organic frameworks are attracting most people's attention. For example, hydrotalcite-like compounds, also known as mixed-metal layered hydroxides or layered double hydroxides (LDHs) have been studies for CO_2 capture even though their capacities are typically < 1.0 mmol/g, lower than other chemisorbents. This is because the presence of water molecules is favorably influence the CO_2 sorption capacities of these compounds because of the formation of bicarbonates (Eq. 8.4); in addition, hydrotalcite-like compounds have sufficient regenerability and hydrothermal stability (Choi et al. 2009). Different approaches have been envisioned to enhance chemical absorption of CO_2 by amines, such as: i) use of solid amines or polyamines directly as absorbents; ii) amines or polyamines chemically bonded to the surface of a solid; or iii) amines or polyamines deposited (physical adsorption) on a solid support such as silica or alumina (Olah et al. 2011). Moreover, materials with pore structure (e.g., the ordered mesoporous silicas, OMS) or crystalline solid (e.g., metal-organic frameworks, MOFs) with or without functionalization with amino groups have been studied for the removal of CO_2. Recently, it has been discovered that MOF 177, composed of zinc clusters joined by 3,5-benzenetribenzoate units, has a surface area of 4500 m^2/g and a CO_2

storage capacity of 33.4 mmol CO_2 per gram of MOF at a pressure of 30 atm (Olah et al. 2011). However, MOFs had much more limited absorption capacity at lower pressure and with gas mixtures.

8.4 Major Issues and Future Perspectives

Briefly, we would consider the following two major issues in this section: a) implementation of CO_2 capture technologies, and b) other options and concepts related to CO_2 capture technologies for the electricity industry. We will address several sub-issues associated with issue a), including cost assessment, water consumption, and cost-effective retrofits. For issue b), several options to enhance power plant efficiency and decrease the cost of CO_2 capture will be discussed.

Implementation of CO_2 Capture Technologies. Currently, CO_2 removal technologies are not completely implemented on coal-based power plants. This is due to the fact that these technologies have not been demonstrated at the scale necessary to establish confidence for power plant application. The CO_2 capture capacities used in industrial processes are usually smaller than the capacity required for the purposes of GHG emission mitigation at a power plant, and thus, there are uncertainties associated with process scale-up (Kuuskraa 2007).

Estimating exactly what the cost of CO_2 capture for the power generation industry is extremely difficult. Many factors affect CCS costs, such as a) choice of power plant and CCS technology, b) process design and operating variables, c) economic and financial parameters, d) choice of system boundaries (e.g., one facility vs. multi-plant system; GHG gases considered (CO_2 only vs. all GHGs); power plant only vs. partial or complete fuel cycle), and e) time frame of interest (e.g., first-of-a-kind plant vs. n^{th} plant; current technology vs. future systems; consideration of technological "learning") (Rubin 2011). Since it is almost impossible to summarize the different methods for cost estimation, one can only provide some general ranges. For example, the cost (in US$/tonne of CO_2) is ~ $35 for post-combustion CO_2 capture by chemical sorption from a coal-fired power plant, ~$40 for a natural gas-fired power station, and ~$7 for a newly built IGCC plant with shift conversion of the fuel gas (Hunt et al. 2010). These costs should always be considered together with other steps of a CCS alternative. Figure 8.2 shows DOE's estimation of the range of CCS costs for various types of power plants (IFT 2012).

The technical and economic evaluations are not the only parameters used in determining the selection of a system for CO_2 capture. Other parameters can become the factors affecting the system design and method selection. For example, CO_2 capture requires large quantities of water because of the cooling water requirements of capture and compression (Ciferno et al. 2010). DOE studies investigated the water withdrawal and consumption for the subcritical pulverized coal fired power plants (PC), supercritical PC, oxy-combustion, and IGCC configurations (USDOE 2010a and b). Their results indicate that there is an 80 to 90% increase in water usage for the subcritical and supercritical PC plants. More moderate increases in water usage (35 to 60%) for IGCC and oxy-combustion plants were observed when constant net power outputs were maintained (USDOE 2010a and b). Implementation of CCS to retrofit plants will be a challenge as the size and space required for CO_2 capture process facilities are greater than the size and space for conventional air pollution controls. The energy required for solvent regeneration is another retrofit issue. Diversion of power plant steam requires careful integration of the steam cycle and the CO_2 capture technology. Retrofits could also face challenges associated with proximity to a geologic sequestration site and/or a CO_2 pipeline (IFT 2010).

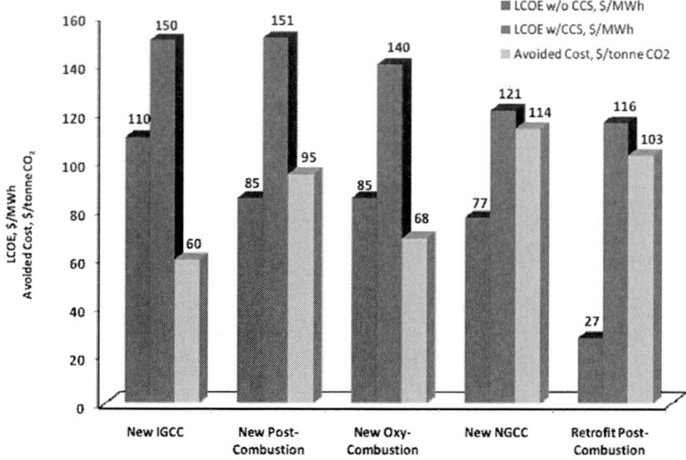

Figure 8.2. DOE's estimation of the range of CCS costs for various types of power plants (adapted from IFT 2010). LCOE = Levelized Cost of Electricity, a cost of generating electricity for a particular system. It is an economic assessment of the cost of the energy-generating system including all the costs over its lifetime: initial investment, operations and maintenance, cost of fuel, cost of capital

Other Options and Concepts. Table 8.9 shows some concepts that are less well developed but combine several novel technologies for CO_2 capture. Other processes, such as fuel cells are not included here. Many of these second-generation technologies need a considerable development effort for scale up and commercialization even though the basic principles are well understood.

Table 8.9. Other novel technologies related to CO_2 capture and storage (VGB 2004)

Option	Basic principle	R&D needs
Chemical looping combustion (CLC)	An indirect combustion method where fuel reacted with an O_2 carrier (M_yO_x) that transfers O_2 from the air to the fuel and then transports the chemical energy released by the fuel. The metal oxide O_2 carrier circulates between a reactor containing air and another containing fuel. CO_2 can be captured with a much lower energy penalty than for other capture concepts. The CO_2 is dehydrated and compressed for storage or use	• O_2 carrier (reactivity, improved thermal efficiency and suitability) • Demonstration of CLC for obtaining engineering data, system design info and cost estimation • Development of solid fuel CLC system • Development of highly reactive particles
ALSTOM CO_2 wheel	A ljungström-like wheel to use Li_4SiO_4 as the sorbent to absorb ~500 times its own volume of CO_2 with an achievable capture rate of 63% of the CO_2 on the cold side. The captured CO_2 can be desorbed into a gas stream	• An efficient integration is necessary to make the overall process viable • Heat recovery and lower associated energy penalties
High temperature carbonation process	Calcined limestone is circulated between a high temperature calciner and a high temperature region of the furnace where CO_2 is absorbed, similar to a moving bed heat exchanger	• Sorbents and the reactor system suitable for the process • Process integration with the power plant
ZECA (Zero Emission Coal Alliance) process	The process includes an exothermic gasification section where H_2 is used to produce a CH_4 rich intermediate gas. The CH_4 is reformed using water and a CaO-based sorbent. The sorbent supplies the energy needed to drive the reforming reaction and also removes the generated CO_2 by producing $CaCO_3$.	• Integration of different steps to solve energy requirements and demonstration of sufficient reaction or conversion rates • Materials for high temperature requirement

8.5 Conclusions

There are three major approaches that can be used for CO_2 capture for the coal-fired power plants: a) pre-combustion systems, designed to separate CO_2 and H_2 in the high-pressure syngas produced at Integrated Gasification Combined Cycle (IGCC) power plants; b) post-combustion systems, designed to separate CO_2 from the flue gas produced by fossil-fuel combustion in air; and c) oxy-combustion, using high-purity oxygen (O_2), rather than air, to combust coal and therefore produces a highly concentrated CO_2 stream. Each of these approaches significantly increases the cost of electricity due to the increased capital and operating costs and decreased electricity output (or energy penalty). The energy penalty occurs because the CO_2 capture process uses some of the energy produced from the plant. Many novel, emerging processes and physical, chemical and hybrid sorbents have been investigated. In general, choosing the most promising sorbent and the CO_2 capture technology may not be possible due to the fact that multiple parameters would affect the overall process performance and economics. An effective integration and scale-up aspects need to be thoroughly examined for these new concepts and second-generation technologies.

8.6 References

Abanades, J.C., Grasa, G., Alonso, M., Rodriguez, N., Anthony, E.J., and Romeo, L.M. (2007). "Cost structure of a postcombustion CO_2 capture system using CaO." *Environ. Sci. Technol.* 2007, 41, 5523–5527.

Anderson, S., and Newell, R. (2003). "Prospects for Carbon Capture and Storage Technologies." Discussion paper published by Resources for the Future, Washington, DC. Available at <http://rff.org/rff/Documents/RFF-DP-02-68.pdf> (accessed Feb. 2012).

Aroonwilas, A, and Veawab, A., (2007). "Integration of CO_2 capture unit using single- and blended-amines into supercritical coal-fired power plants: implications for emission and energy management." *Int. J. Greenhouse Gas Control*, 1, 143–150.

Buhre, B.J.P., Elliot, L.K., Sheng, C.D., Gupta, R.P., and Wall, T.F. (2005). "Oxy-fuel Combustion Technology for Coal-fired Power Generation." *Progress in Energy and Combustion Science*, 31, 283–307.

Carrol, J. (2002). Natural Gas Hydrates-A Guide for Engineers. Elsevier.

CCCSRP (California Carbon Capture and Storage Review Panel) (2010). *Findings and Recommendations by the California Carbon Capture and Storage Review Panel.* CCCSRP, Dec. 2010. Available at <www.climatechange.ca.gov/carbon_capture_review_panel/.../2011-01-14...> (accessed Feb. 2012).

Chen, C., and Rubin, E.S. (2009). "CO_2 control technology effects on IGSS plant performance and cost." *Energy Policy*, 37, 915–924.

Choi, S., Drese, J. H., Jones, C. W. (2009). "Adsorbent materials for carbon dioxide capture from large anthropogenic point sources." *ChemSuschem*, 2, 796–854.

Ciferno, J. P., Munson, R. K., Murphy, J. T., and LaShier, B. S. (2010). "Determining carbon capture and sequestration's water demands." *POWER*, 154, 71–76.

Cormos, A.M., Gaspar, J., and Padurean, A. (2009). "Modelling and simulation of carbon dioxide absorption in monoethanolamine in packed absorption columns." *Studia Universitatis Babes-Bolyai, Chemia,* 37–48.

Davison, J. (2007). "Performance and costs of power plants with capture and storage of CO_2." *Energy*, 32, 1163–1176.

Deutch and Moniz. (2006). "The Future of Coal." *MIT Laboratory for Energy and the Environment, Cambridge, MA.*

Esber, S. (2006). "Carbon Dioxide Capture Technology for the Coal-Powered Electricity Industry: A Systematic Prioritization of Research Needs." Massachusetts Institute of Technology.

Gasification Technologies Council (GTC). (2011). "September. gasification: redefining clean energy." *GTC*, http://www.gasification.org.

Haslbeck, J.L. (2002). *Evaluation of Fossil Fuel Power Plants with CO_2 Recovery.* Parsons Infrastructure & Technology Group Inc.

Herzog, H. J., Meldon, J., and Hatton, A. (2009). *Advance Post-Combustion CO_2 Capture.* Technical report prepared for the clean air task force. <http://web.mit.edu/mitei/docs/reports/herzog-meldon-hatton.pdf> (accessed Feb. 2012).

Hunt, A.J., Sin, E.H.K., Marriott, R., and Clark, J.H. (2010). "Generation, capture, ad utilization of industrial carbon dioxide." *ChemSusChem*, 3, 306–322.

IEA (International Energy Agency) (2009). *Technology Roadmap: Carbon Capture and*

Storage. An IEA report, Available at <http://www.iea.org/papers/ 2009/CCS_Roadmap.pdf> (accessed Feb. 2012).

IEA (International Energy Agency) (2010). "World Energy Outlook 2010." Available at <http://www.worldenergyoutlook.org/> (accessed Jan. 2013).

IPCC (International Panel on Climate Change) (2005). *IPCC Special Report on Carbon Dioxide Capture and Storage.* Cambridge University Press, Cambridge.

ITF (Interagency Task Force) (2010). *Report of the Interagency Task Force on Carbon Capture and Storage.* Available at <http://www.fe.doe.gov/programs/sequestration/ccstf/CCSTaskForceReport2010.pdf> (accessed Jan. 2013).

Kanniche, M., and Bouallou, C. (2007). "CO_2 capture study in advanced integrated gasification combined cycle." *Appl. Thermal Engineering*, 27, 2693–2702.

Kanniche, M., Gros-Bonnivard, R., Jaud, P., Valle-Marcos, J., Amann, J-M, and Bouallou, C., (2010). "Pre-combustion, post-combustion and oxy-combustion in thermal power plant for CO_2 capture." *Appl. Therm. Eng.*, 30, 53-62.

Kather, A., Rafailidis, S., Hermsdorf, C., Klostermann, M., Maschmann, A., Mieske, K., Oexmann, J., Pfaff, I., Rohloff, K., and Wilken, J. (2008). *Research & Development Needs for Clean Coal Deployment.* IEA Clean Coal Centre Report Nr. CCC/130, London.

Kohl, A. and Nielsen, R. (2005). "Gas purification." *5th ed. Gulf Publishing Company*, Houston.

Korens, N., Simbeck, D.R., and Wilhelm, D.J. (2002). *December. Process Screening Analysis of Alternative Gas Treating and Sulfur Removal for Gasification.* Prepared for U.S. Department of Energy by SFA Pacific, Inc., Revised Final Report.

Kuuskraa, V. A. (2007). *A Program to Accelerate the Deployment of CO_2 Capture and Storage: Rationale, Objectives, and Costs.* Coal Initiative Reports, White Paper Series. Arlington, VA, Pew Center on Global Climate Change.

Lackner, K.S., and Brennan, S. (2009). "Envisioning carbon capture and storage: expanded possibilities due to air capture, leakage insurance, and C-14 monitoring." *Climatic Change*, 96, 357–378.

MGSC (Midwest Geological Sequestration Consortium). (2004). *Carbon Dioxide Capture and Transportation Options in the Illinois Basin.* Technical Report. Contract: *DE-FE26-03NT41994*, National Energy Technology Center.

Olah, G.A., Surya Prakash, G.K., and Goeppert, A. (2011). "Anthropogenic chemical carbon cycle for a sustainable future." *JACS*, 133, 12881–12898.

Ordorica-Garcia, G., Douglas, P., Croiset, E., and Zheng, L. (2006). "Technoeconomic evaluation of IGCC power plants for CO_2 avoidance. Energy Convers." *Manage.*, 47, 2250–2259.

Padurean, A., Cormos, C.C., Cormos, A.M., and Agachi, P.S. (2010). "Technical assessment of CO_2 capture using alkanolamines solutions." *Studia Universitatis Babes-Bolyai, Chemia,* LV, 55–64.

Padurean, A., Cormos, C.C., Cormos, A.M., and Agachi, P.S. (2011). "Multicriterial analysis of post-combustion carbon dioxide using alkanolamines." *Int. J. Greenhouse Gas Control.*, 5, 676–685.

Padurean, A., Cormos, C. C. and Agachi, P. S. (2012). "Pre-combustion carbon dioxide capture by gas-liquid absorption for Integrated Gasification Combined Cycle power plants." *INT. J. GREEN GAS CON.*, 7, 1–11.

Pennline, H.W., Luebke, D. R., Jones, K. L.,Myers,C.R., Morsi, B. I.,Heintz,Y. J., and Ilconich, J. B. (2008). "Progress in carbon dioxide capture and separation research for gasification-based power generation point sources." *Fuel Process. Technol.*, 89, 897–907.

Plasynski, S. I., Litynski, J. T., McIlvried, H. G., and Srivastava, R. D. (2009). "Progress and new development in carbon capture and storage." *Critical Reviews in Plant Science.*, 28, 123–138.

Polasek, J., and Bullin, J. (1985). "Process considerations in selecting amines." *In: Acid and Sour Gas Treating Processes, Newman, S.A.*, (ed.), 190–211.

Rao, A.B., and Rubin, E.S. (2002), "A technical, economic, and environmental assessment of amine-based CO_2 capture technology for power plant greenhouse gas control." *Environmental Science & Technology*, 36(20), (2002), 4467–75.

Riemer, P., Audus, P., and Smith, A. (1993). *Carbon Dioxide Capture from Power Stations.* Cheltenham, United Kingdom: International Energy Agency Greenhouse Gas R&D Program.

Rubin, E.S., Rao, A.B., and Chen, C. (2004). "Comparative assessments of fossil fuel power plants with CO_2 capture and storage." *In: Proceedings of 7th International Conference on Greenhouse Gas Control Technologies.*

Rubin, E.S. (2011). "Methods and measures for CCS Costs." Presentation given at CCS Cost Workshop, Paris, France. March 22, 2011. *Proceedings from a CCS Cost Workshop* by Carnegie Mellon University CMU), Electric Power Research Institute (EPRI), Global CCS Institute, IEA Greenhouse Gas R&D Programme (IEAGHG), International Energy Agency (IEA), Massachusetts Institute of Technology (MIT), Vattenfall. Available at <http://www.global ccsinstitute.com/publications/proceedings-ccs-cost-workshop> (accessed Oct. 2012).

Spector, N.A., and Dodge, B.F. (1946). "Removal of carbon dioxide from atmospheric air." *Trans. Am. Inst. Chem. Eng.*, 42, 827–848.

Splietpoff, H. (2010). *Power Generation from Solid Fuels, Power Systems*. Springer.

Strube, R., and Manfrida, G. (2011). "CO_2 capture in coal-fired power plants-impact on plant performance." *Int. J. Greenhouse Gas Control*, 5, 710–726.

Tepe, J.B., and Dodge, B.F. (1943). "Absorption of carbon dioxide by sodium hydroxide solutions in a packed column." *Trans. Am. Inst. Chem. Eng.*, 39, 255–276.

Tohidi, B. (2008). "December. Risk of hydrate formation in low water content CO_2 and rich CO_2 systems." *Hydrafact for Progressive Energy*, Carried out Under the EC Dynamis Project.

USDOE (U.S. Department of Energy) (1999). *Carbon Sequestration Research and*

Development. Washington, DC: Office of Science and Office of Fossil Energy.

USDOE (2002). *Advanced Fossil Power Systems Comparison Study.* National Energy Technology Laboratory, USDOE, Morgantown, WV.

USDOE (2010a). *Cost and Performance Baseline for Fossil Energy Plants. Volume 1: Bituminous Coal and Natural Gas to Electricity.* U.S. Department of Energy and National Energy Technology Laboratory.

USDOE (2010b). *Advanced Oxycombustion 2015+ Bituminous Coal Fossil Energy Plants, Draft Final Report (Revision 2).* U.S. Department of Energy, National Energy Technology Laboratory.

U.S. Energy information administration (US EIA) (2010a). *Annual Energy Outlook 2010 with Projections to 2035*, Energy Information Administration.

VGB (2004) *CO_2 Capture and Storage – VGB Report on the State of the Art.* VGB PowerTech e.V., Essen/Germany, August 2004. Available at http://www.vgb. org /en/CO2cace.html. (accessed Oct. 2012).

Wall, T.F., and Yu, J. (2009). "Coal-fired oxyfuel technology status and progress to deployment." *In: 34th International Conference on Coal Utilisation and Fuel Systems, Clearwater.*

Wang, M., Lawal, A., Stephenson, P., Sidders, J., and Ramshaw, C. (2011). "Psot-combustion CO_2 capture with chemical absorption: A state-of-the-art review." *Chemical Engineering Research and Design*, 89, 1609–1624.

WRI (World Resources Institute). (2008). *CCS Guidelines: Guidelines for Carbon Dioxide Capture, Transport, and Storage.* Published by WRI, 10 G St., NE, Suite 800, Washington, DC.

Yang, H., Xu, Z., Fan, M., Gupta, R., Slimane, R., Bland, A., Wright, I. (2008). "Progress in carbon dioxide separation and capture." *J. Environ. Sci.*, 20, 14 27.

Zhang, T.C., and Surampalli, R.Y. (2013). "Carbon capture and storage for mitigating climate changes." Chapter 20 in *Climate Change Modeling, Mitigation and Adaptation,* Surampalli, R., Zhang, T.C., Ojha, C.S.P., Gurjar, B.R., Tyagi, R.D., and Kao, C.M. (eds.), ASCE, Reston, Virginia, February 2013.

CHAPTER 9

CO_2 Scrubbing Processes and Applications

Wenbiao Jin, Guobin Shan, Tian C. Zhang and Rao Y. Surampalli

9.1 Introduction

CO_2 separation and capture from point and nonpoint sources is regarded as one of the grand challenges for the 21st century (Yu et al. 2008; Jacobson 2009). Conventional technologies for large-scale capture of CO_2 from flue gases by the use of amine absorbers ("scrubbers") have been commercially available for over 50 years (Rochelle 2009). The Intergovernmental Panel on Climate Change (IPCC) estimates that CO_2 emissions to the atmosphere could be reduced by 80–90% for a power plant equipped with carbon capture and storage technology (IPCC 2005). CO_2 scrubbing is the most promising technology due to its wild conditions, low costs, easier regeneration and faster loading (Aaron and Tsouris 2005; Davision and Thambinuthu 2009). Absorption with liquid amines and adsorption with solid adsorbents have been proposed as possible retrofits to existing pulverized coal (PC) power plants that are integral to modern power generation infrastructure (Steeneveldt et al. 2006).

This chapter starts with process overview to post-combustion CO_2 scrubbing technologies, followed by discussing advantages and disadvantages, scrubber materials, and applications of CO_2 scrubbing processes. CO_2 scrubbing materials are divided into liquid and solid. In additional, current status and future perspectives of CO_2 scrubbing technologies are reviewed and discussed.

9.2 Process Overview

Post-combustion CO_2 scrubbing technologies are based on CO_2 absorption from flue gas streams by a liquid solvent or solid matrix, predominantly primary alkanolamine (e.g., monoethanolamine, MEA) (Yu et al. 2008). The process mainly includes two steps: the first step is to use a liquid solvent or solid matrix to absorb/enrich CO_2 that results in a CO_2-rich solution or matrix. Optimal conditions for the first step are low-temperature and high pressure. The second step is to recover the CO_2 from the CO_2-rich solution or matrix via a desorption process. Usually, low pressure and high temperature favor the second step resulting in regeneration of the solvent.

In the process of CO_2 absorption (Fig. 9.1), a solvent is used that dissolves CO_2, but not O_2, N_2, or any other components of a flue gas stream. The CO_2-rich solution is then pumped to a regeneration operation unit, where the CO_2 is stripped from the solution and the solvent recycled for a new batch of flue gas. Prior to CO_2 removal the CO_2 containing stream is cooled, and other impurities are removed as much as possible. Typically, the process involves the passage of an aqueous amine solution (typically 25-30 wt.%) down the top of an absorption tower, while a gaseous stream of flue gas containing CO_2 is introduced at the bottom. At a temperature of approximately 40 °C, the reaction of CO_2 with the amine occurs through a zwitterion mechanism to form carbamate. The CO_2 rich solvent from the bottom of the absorber is passed into another column (stripper) where it is heated with steam to reverse the CO_2 absorption reactions. CO_2 released in the stripper is further treated (dehydration, compression) for transport and storage. The CO_2 free solvent is recycled to the absorption column (Wilson et al. 1992).

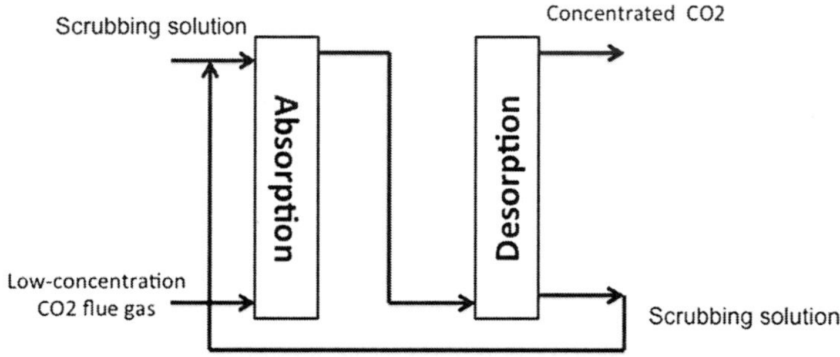

Figure 9.1. Schematic diagram of absorption/desorption scrubbing for CO_2 capture

Solid sorbents work in a similar way to amine scrubbing in that CO_2 is selectively taken out from the flue gas into another solid phase. The sorbent is then separated and regenerated, releasing relatively pure CO_2 and regenerating sorbent. The main difference from amine scrubbing is that the sorbent is a solid instead of a liquid.

An alternative to chemical absorption is the use of physical solvents in which the solvent selectively binds CO_2 at high partial pressures and low temperatures. Due to physical absorption, the physical solvents interact less strongly with CO_2. Thus, the advantage of such solvents is the lower heat consumption in the solvent regeneration step, as the stripping process can be driven by heat or a pressure reduction (i.e., "flash distillation") (Abu-Khader 2006). Main physical solvents include chilled (-40 °C) methanol (Rectisol process), a mix of dimethylether of polyethylene glycol (Selexol process), propylene carbonate (Fluor process) and n-methyl-2pyrollidone (NMP-purisol). Majority of physical solvents are organic solvents with high boiling points and low vapor pressures. Except methanol, most of these solvents can be used at ambient temperatures without appreciable vaporization losses. Physical solvents are suitable for CO_2 capture from high-pressure streams.

9.3 Advantage and Disadvantage

The CO_2 scrubbing technology is considered as the preferable method for separating CO_2 since they are as efficient in application as they have been in pilot testing. There are many advantages as follows. The conditions for the process of CO_2 scrubbing are relatively easy to meet for absorption and regeneration, causing a relatively low cost. Because absorption is a well-established process, much is known about it, promising new solvents are already being developed and pilot-tested by separate companies. Hindered amines are developed to defend against degradation. For example, KS-1, KS-2, and KS-3 solvents, developed and pilot-tested by Mitsubishi Heavy Industries Research (MHIR 2001), exhibit higher CO_2 loading per unit solvent, lower regeneration conditions, and almost no corrosion, degradation, or amine loss. When the CO_2-rich solution is sent into the regeneration operation unit, the solvent can be recycled and reused, thus reducing the cost of material. Continuous monitoring and automation minimizing human efforts in absorption/desorption processes have been industrially realized that thus minimizes labor cost (Chapel et al. 1999). The recovered CO_2 from the process of CO_2 scrubbing is of high purity, typically, higher than 95%.

Though the process of CO_2 scrubbing does have strong advantages, the total cost including the cost of new solvent as well as operating and maintenance (O&M) costs is relatively high, about \$40–\$70 per ton CO_2 separated, as reported by Chakma and Tontiwachwuthikul (1999). Specially, the loss of amine reagents and transfer of water into the gas stream during the desorption stage, degradation of amine reagents to form corrosive byproducts, and high energy consumption during regeneration, as well as insufficient carbon dioxide/hydrogen sulfide capture capacity, result in an increase in cost (DuPart et al. 1993; Kittel et al. 2009). For example, approximately 3.5 lbm of solvent are lost for each ton of CO_2 separated due to the degradation (Aaron and Tsouris 2005). The regeneration step may be 70% of the total operating costs of the capture process (4 GJ/ton of CO_2) (Rao and Rubin 2002) due to the degradation and loss of solvents. Additionally, different conditions require different solvents. For low-partial pressures of CO_2 (< 15 vol. %), liquid solvents like MEA are preferable. For high-partial pressures of CO_2 (> 15 vol. %), solid solvents, such as lithium hydroxide and lithium zirconate, are better because they can absorb more CO_2 and are more easily regenerated (Audus 1997).

9.4 CO_2 Scrubbing Materials

CO_2 scrubbing materials can be categorized into liquid solvents and solid sorbents, and each of them are briefly described below (Duke et al. 2010).

9.4.1 Liquid

Absorption processes involving CO_2 capture by liquid media are widely established (Choi et al. 2009). The liquid media are often aqueous amine (e.g., MEA) solutions or other fluids with basic character, such as chilled ammonia that chemically absorb the acid gases (Danckwerts and Sharma 1966; Danckwerts 1979).

9.4.1.1 Amines

Primary, secondary and tertiary amines are generally used as organic chemical solvents. Typical amines include monoethanolamine (MEA), diethanolamine (DEA), diglycol-amine (DGA), N-methyldiethanolamine (MDEA), and 2-amino-2-methyl-1-propanol (AMP), that have been widely practiced for several years for CO_2 capture from gas streams in natural gas, refinery off-gases and synthesis gas processing. Alkanolamine-based solvent was reported to be less corrosive at higher concentrations and have greater resistance to oxygen than other CO_2 stripping solvents (Bastin et al. 2008). The properties and behavior of amine solutions and amine-CO_2 chemistry have been empirically and theoretically documented over time to provide engineers with a useful database for process plant design (Branan 2002).

Organic amines are the majority of solvents that act as CO_2 scrubbers by chemisorptive formation of N-C bonded carbamate species with typical bonding energies being 100 kJ mol^{-1} (Arstad et al. 2007). Regeneration of the amine requires cleavage of this covalent bond by heating (at 100 to 150 °C) to release CO_2. CO_2 reacts with aqueous solutions of amines that reach equilibrium of carbamate, bicarbonate, and carbonate (Figure 9.2). Chemical mechanisms have been proposed to describe the reaction process (Sartori et al. 1983; Blauwhoff et al., 1984; Maddox et al. 1987; Crooks and Donnellan 1990). For example, the initial absorption reaction is the formation of the carbamate, which can then hydrolyze to produce the bicarbonate and, under suitable pH, the carbonate species. The degree of hydrolysis of carbamate is dependent on amine concentration, solution pH, and the chemical stability of the carbamate. At low temperatures the equilibrium favors the formation of carbamate and bicarbonate, whilst heating the equilibrium favors the formation of amine and carbon dioxide (Figure 9.2).

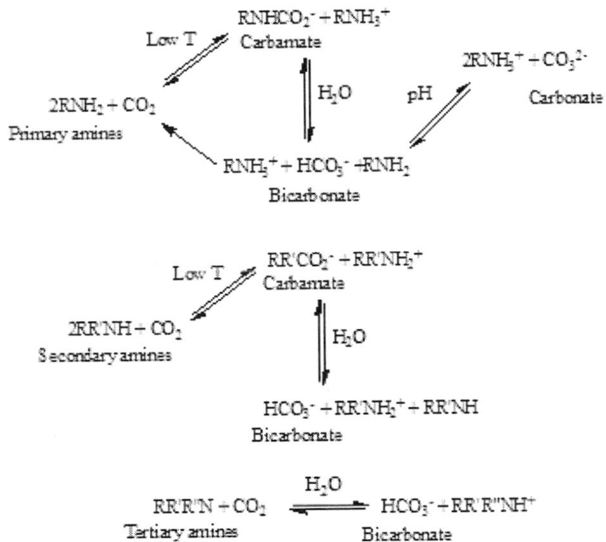

Figure 9.2. General reaction of pathway among CO_2 and primary, secondary, and tertiary amines (adapted form Hook 1997; Vaidya and Kenig 2007)

The CO_2 loading capacity for primary and secondary amines is in the range 0.5-1 mol of CO_2 per mol of amine, since a fraction of the carbamate species is hydrolyzed to form hydrogen carbonates (Fig. 9.2). The reaction of CO_2 with tertiary amines such as N-methyldiethanolamine (MDEA) occurs with a higher loading capacity of 1 mol of CO_2 per mol of amine, with a relatively lower reactivity towards CO_2 compared with the primary amines. The carbamation reaction cannot proceed for tertiary amines, which result in a base-catalyzed hydration of CO_2 to form hydrogen carbonate. MDEA is commonly employed for natural gas treatment and exhibits lower solvent degradation rates in addition to a low energy penalty for regeneration of the solvent in the stripper. In practice, the addition of small amounts of primary and secondary amines enhances the CO_2 absorption rates for tertiary amines.

Aqueous ethylenediamine (EDA, 12 M) exhibits 0.48 mol CO_2 per equivalent of EDA, and can be used up to 120 °C in a stripper without significant thermal degradation. EDA does not result in excessive foaming. The capacity of 12 M EDA is 0.72 mol CO_2/kg (H_2O + EDA) for P = 0.5 to 5 kPa at 40 °C, which is about double that of MEA. The apparent heat of CO_2 desorption in EDA solution is 84 kJ·mol^{-1} CO_2, greater than most of other amine systems (Zhou et al. 2010).

9.4.1.2 Sterically hindered amines

The instability of the conventional primary, secondary, and tertiary amines under scrubbing conditions has been widely reported, with the degradation of the amine resulting in reduced scrubbing efficiency, the production of ammonia, increased viscosity, and excessive foaming (Niswander et al. 1993). To overcome such limitations and to improve the efficiency of the CO_2 cycling process, substituents at the carbon adjacent to the amine group can be introduced to amine-based solvents (sterically hindered amines). Sterically hindered amines using for CO_2 scrubbing are mainly 2-amino-2-methyl-1-propanol (AMP), 1,8-pmethanediamine (MDA) and 2-piperidine ethanol (PE), and the proprietary solvents marketed by Mitsubishi Heavy Industries, KS-1, KS-2 and KS-3 (White et al. 2003). Sterically hindered amines have more favourable reaction stoichiometry such that the theoretical capacity is higher than MEA, and lower heat of absorption/regeneration.

Chakraborty et al. (1986) have demonstrated that the introduction of substituents at the carbon adjacent to the amine group results in the carbamate instability and subsequently the increase in bicarbonate, allowing greater CO_2 loading of the amine solutions. Lee and Kitchin (2012) have evaluated the reactivity of three groups of functionalized amines: alkylamines, alkanolamines, and trifluoroalkylamines, and studied the two main CO_2 capture pathways (i.e., the formation of bicarbonate and carbamate species) using the density functional theory. Electron withdrawing groups tend to destabilize CO_2 reaction products, whereas electron-donating groups tend to stabilize CO_2 reaction products. Hydrogen bonding stabilizes CO_2 reaction products. Electronic structure descriptors based on electronegativity were found to describe trends in the bicarbonate formation energy. A chemical correlation was observed between the carbamate formation energy and the carbamic acid formation energy. The local softness on the reacting N in the amine was found to partially explain trends of carbamic acid formation energy. Methyl groups substituted adjacent to the amine were found to increase solution absorption capacities but with an overall reduction in absorption rate. Two methyl groups substituted R to the amine reduce the stability of the subsequent carbamate, fully shifting the CO_2 absorption equilibrium to bicarbonate (Hook 1997). Another advantage of the substituted amines is that reaction energies of reactions of CO_2 with the substituted amines can be tuned by adjusting both the nature and the placement of functional groups (Mindrup and Schneider 2010).

Studies on the thermodynamic capacity and absorption/desorption rates of CO_2–amine reactions have shown that the steric hindrance and basicity of the amine are major factors controlling the efficiency of CO_2 capture reactions (Mimura et al. 1998; Yeh et al. 2008). Sterically hindered amines such as 2-amino-2- methyl-1-propanol (AMP) containing bulkier substituents have been identified as the most promising absorption solvents due to the lower stability of their carbamates (carbamate stability constant: AMP, 0.1 < DEA, 2.0 < MEA, 12.5 at 303 K) (Duke et al. 2010). The sterically hindered amines allow CO_2 loadings well in excess of 0.5 mol equivalents with higher regeneration rates (and therefore lower regeneration costs) compared with the conventional alkanolamines (e.g., the CO_2 regeneration rate ratio for AMP/MEA is 1.83 (Dibenedetto et al. 2002). Sterically hindered amines are amines in which a bulky alkyl group is attached on the amino group. Impact of alkyl chain length between amino groups has been investigated (Lepaumier et al. 2008). For example, four tertiary polyamines (DMP, TMEDA, TMPDA, and PMDPTA) are more stable than MEA while two tertiary polyamines (TMBDA and PMDETA) are much less stable than MEA. If the molecule can give easily five- or six-membered rings, amine will highly degrade even if it has only tertiary amine functions. However, if ring closure is not favorable (three-, four-, or more than six-membered rings formation), a better stability of amine is expected.

In summary, sterically hindered amines have higher adsorption abilities, better thermal stability, and lower circulation rate over conventional amines. For example, only 1 mol of the sterically hindered amine, instead of 2 mol of alkanolamine, is required to react with 1 mol of CO_2.

9.4.1.3 Aqueous Ammonia

CO_2 capture from flue gas streams can be achieved by reacting the CO_2 with ammonia gas and/or water vapor in a gas-phase reactor or by bubbling raw flue gas through an aqueous ammonia solution. Possible reaction mechanisms between ammonia and CO_2 have been proposed as follows (Brooks and Audrieth 1946; Hatch and Pigford 1952; Brooks 1953; Kuchervavvi and Gorlovskii 1970; Koutinas et al. 1983).

$$CO_2(g) + 2NH_3(g) \rightleftharpoons NH_2COONH_4(s) \overset{H_2O}{\rightleftharpoons} (NH_4)_2CO_3(s) \tag{9.1}$$

$$CO_2(g) + 2NH_3(g) \rightleftharpoons CO(NH_2)_2(s) + H_2O(g) \tag{9.2}$$

$$CO_2(g) + 2NH_3(aq) \rightleftharpoons NH_4^+(aq) + NH_2COO^-(aq) \tag{9.3}$$

$$CO_2(g) + 2NH_3(g \text{ or } aq) + H_2O(g \text{ or } aq) \rightleftharpoons (NH_4)_2CO_3(s) \tag{9.4}$$

$$CO_2(g) + NH_3(g \text{ or } aq) + H_2O(g \text{ or } aq) \rightleftharpoons NH_4HCO_3(s) \tag{9.5}$$

At room temperature and atmospheric pressure, in the dry condition (reaction equation (1)), ammonium carbamate (NH_2COONH_4) is formed by the reaction of carbon dioxide and ammonia; under moist air, the hydration product of ammonium carbonate ($(NH_4)_2CO_3$) is produced (Zhou et al. 2010). While under high pressure and with temperatures greater than 140 °C, the CO_2-NH_3 reaction is directed to the formation of urea ($CO(NH_2)_2$) (reaction equation (2)). Reaction equations (3)-(5) also possibly occur. The formation of ammonium (NH_4^+) and carbamate (NH_2COO^-) ions is very fast, and reaction equation (3) is irreversible (White et al. 2003). On the other hand, reaction equations (4)-(5) are reversible, with ammonium carbonate ($(NH_4)_2CO_3$) or bicarbonate (NH_4HCO_3) as the products (Chakraborty et al. 1986; Niswander et al. 1993). The forward reactions are dominant at room temperature while the backward reactions occur at temperatures of around 38-60 °C (Grayson 1978; Pelkie et al. 1992).

The NH_4HCO_3 and the $(NH_4)_2CO_3$ form stable solids while the N_2 and other gases that were in the flue gas stream continue through for release or treatment. The solid products are to be used as soil fertilizers, instead of being regenerated to recover the CO_2. The overall CO_2 removal efficiencies could be above 95% by aqueous ammonia scrubbing under proper operation conditions (Bai and Yeh 1997). The absorption capacity of ammonia was around 0.9 kg of CO_2/kg of ammonia, which is higher than that by a MEA solution.

You et al. (2008) have demonstrated that aqueous ammonia can be modified by the use of additives including amines and hydroxyl groups to improve the performance of CO_2 scrubbing. For example, the removal efficiency of CO_2 scrubbing using aqueous ammonia (10 wt %) has been improved by adding 2-amino-2-methyl-1-propanol (AMP), 2-amino-2-methyl-1,3-propandiol (AMPD), 2-amino-2-ethyl-1,3-propandiol (AEPD), and tri(hydroxymethyl) aminomethane (THAM). The improvement is attributed to the interactions among ammonia and additives or absorbents and CO_2 via hydrogen bonding that is verified by Fourier transform infrared spectroscopy (FTIR) spectra and computational calculation.

9.4.1.4 Ionic liquids

Ionic liquids (ILs) have emerged in the last few years as promising acid gas absorbents (Karadas et al. 2010), and have also been considered as an important class of physical solvents which are selective for CO_2 absorption (Smiglak et al. 2007; Radosz et al. 2008). Ionic liquids consist of large organic cations and smaller inorganic anions and are typically viscous liquids near room temperature. In addition to their extremely low vapor pressures, they are non-flammable, environmentally benign, and can exhibit exceptional thermal stability. Typically, the interaction between ionic liquids and CO_2 is often based on physisorption, with heats of adsorption of around -11 $kJ \cdot mol^{-1}$ (Cadena et al. 2004). In view of this low heat of reaction, the benefit for CO_2 capture is the minimal energy required for solvent regeneration. The capacity is directly proportional to the partial pressure of CO_2 and improves at pressures above 1-2 bar.

Shiflett et al. (2010) have demonstrated that the use of an ionic liquid, 1-butyl-3-methylimidazolium acetate can reduce the energy losses by 16% compared to a monoethanolamine process. Furthermore, engineering design estimates indicate that the investment for the ionic liquid process will be 11% lower than the amine-based process and provide a 12% reduction in equipment footprint. After being optimized, the ionic liquid technology may reduce even the energy and cost required for CO_2 capture. Brennecke et al. (1999, 2001) have reported a 0.0881% increase in mass of ionic liquid, 1-hexyl-3-methyl imidazolium hexafluorophosphate ([6-mim]PF_6), upon exposure to CO_2 at 1 atm, that confirmed by the FTIR spectrum of the gas-treated ionic liquid. The FTIR spectrum has peaks characteristic of dissolved CO_2 at 2380 and 2400 cm^{-1}. When 1.2896 g of pure1-butylimidazole modified with 2-bromopropylamine hydrobromide is exposed to a stream of bone dry CO_2 for 3 h at 1 atm and room temperature, a total mass gain of 0.0948 g (7.4%) is observed, a vastly greater increase than that observed for conventional ionic liquid like [6-mim] PF (Bates et al. 2002). An alkanolamine-based ionic liquid, *N*-methyl-diethanolammonium tetrafluoroborate ([MDEA][BF_4]), has also been developed for CO_2 scrubbing, and its optimal performance for CO_2 capture was found at 45 °C, 1.50 MPa, probably due to a synergistic action of the reaction and the transport (Zhang et al. 2010). Possible reactions of some ionic liquids with CO_2 are shown in Figure 9.3.

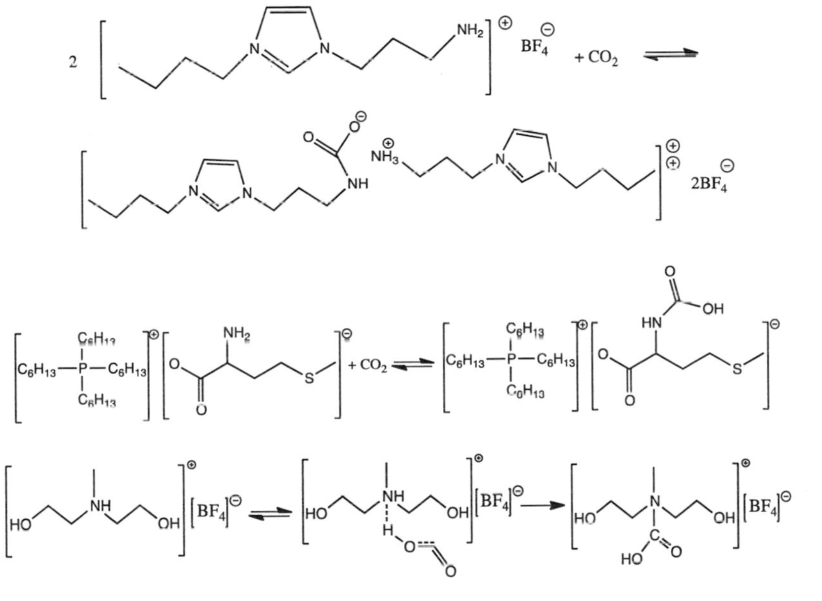

Figure 9.3. Reaction equations of ionic liquids with CO_2

There are some limitations of ILs using as CO_2 absorbents. Firstly, the knowledge about their toxicity is lack, and various ILs have been found to be combustible, which are barriers to recognize them as green solvents. Secondly, the cleaning of ILs involves washing with water or volatile organic compounds, leading to another waste stream. Thirdly, the drawback of ILs is their high viscosities. For example, the viscosity of [bmim][BF_4] (79.5 cP) is reported to be much higher than that of pure monoethanolamine (25 cP) and 30% aqueous MEA solution (2 cP) (Sanchez et al. 2007). The high viscosity results in the slow rate of absorption that is limited by the slow diffusion of the ILs. Moreover, it would increase the pumping and related operating costs, such as reduced mass-transfer rates and poor heat transfer. Fourthly, a certain corrosion of ILs toward some metals and alloys has been reported, especially at high temperatures with the presence of certain impurities (e.g., halides) (Tolstoguzov et al. 2009). Additionally, the high prices of most ILs hinder their extension to large-scale applications.

9.4.1.5 Other Solvents

Inorganic solvents using in CO_2 absorption include potassium, sodium carbonate, aqueous ammonia and aqueous alkali hydroxide solution (Keith et al. 2006). Among these, potassium carbonate is the major one. Typically, the potassium carbonate system uses an aqueous solution of about 20-40 wt% of the potassium salt. The absorption of CO_2 is operated at temperatures (typically 70-120 ^{o}C) close to the atmospheric boiling point of the solvent. This feature can eliminate the use of the heat exchangers used to cool the solvent flow between the regenerator and absorption column. The drawback of inorganic solvents is that they may release Na, K and V in the product gas that could result in deposition, erosion and corrosion in gas turbines and fuel cells.

9.4.2 Solid

The CO_2 can be chemically absorbed by solid sorbents and then subsequently released in a second step to produce a concentrated stream of CO_2 at which point the sorbents are regenerated and recycled. Sorption on solid using pressure and/or temperature swing approaches is an emerging alternative that has advantages such as reduced energy for regeneration, greater capacity, selectivity, ease of handling, etc. Criteria for selecting CO_2 sorbent material typically include adsorption capacity and selectivity for CO_2, and adsorption/desorption kinetics (Samanta et al. 2012). For example, CaO, Li_2ZrO_3, K_2O, and Na_2CO_3 can be used as CO_2 sorbents (Abanades et al. 2004a). Whereas CO_2 molecules dissolve into a liquid in absorption, CO_2 adsorption involves van der Waals, static electric interaction, hydrogen bonding or covalent bonding interactions between the gas molecules and the surface of a solid (Choi et al. 2009). Furthermore, solid sorbents impregnated with amines facilitate the adsorption of CO_2 through the formation of carbamate species, and thus improves their capability of CO_2 capture (Harlick and Sayari 2006). Solid sorbent chemical absorption for post-combustion capture (PCC) is also a mature process in technological terms, but no pilot trials yet. The carbonation reactions are relatively slow, and the process kinetics needs to be considered at high temperature, e.g., 450 ~ 700 °C, which is energy intensive. The spent absorbent is then removed and may be regenerated by higher temperature calcination (> 900 °C) to release relatively pure CO_2 and regenerated absorbent. Spent sorbent may be sold as a mineral carbonate product.

9.4.2.1 Metal-based Solid

The acidity of CO_2 facilitates adsorption on the basic sites of some metal-based solid materials, especially those with a low charge/radius ratio, which present more strongly basic sites (Audus 1997; Ye et al. 2012). Generally, metal-based solid for CO_2 capture can be categorized into two large groups: metal oxides and metal-layered hydroxides.

Metal oxides mainly include alkaline metal oxides and alkaline earth metal oxides, on which CO_2 molecules can adsorb forming mono- or multidentate species (Yong et al. 2002). Many metal oxides display CO_2 adsorption properties including calcium oxides, magnesium oxides, lithium oxides (Mosqueda et al. 2006), sodium oxides (López-Ortiz et al. 2004; Alcerreca-Corte 2008), potassium oxides (Li et al. 20001), rubidium oxides (Doskocil et al. 1997), cesium oxides (Tai et al. 2004), barium oxides (Tutuianu et al. 2006), iron oxides (Ismail et al. 1997), tantalum oxides (Dobrova et al. 2006), copper oxides (Pohl and Otto 1998), chromium oxides (Funk et al. 2007), and aluminum oxides (Horiuchi et al. 1998; Yong et al. 2000; Casarin et al. 2003).

At high temperatures, calcium carbonates liberate CO_2 and generate calcium oxides; while calcium oxides can adsorb CO_2 to yield $CaCO_3$, as described below (Wang et al. 2007):

Carbonation:

$$CaO(s) + CO_2(g) \rightarrow CaCO3(s), \text{ exothermic} \tag{9.6}$$

Calcination (decomposition):

$$CaCO_3(S) \rightarrow CaO(s) + CO_2(g) \rightarrow CaCO3(s), \text{ endothermic} \tag{9.7}$$

The maximum theoretical amount of adsorbed CO_2 is 17.8 mmol/g. Practically, calcium oxides offer a large capacity for CO_2 uptake, up to 13.4 mmol CO_2 per gram of adsorbent when being operated at high temperatures, ~1000 K, and are also thought to be advantageous compared to other adsorbents on account of the low cost and wide availability of precursors such as limestones or dolomites (Oliveira et al. 2008). The corresponding hydroxides of calcium oxides can also be used for CO_2 scrubbing. CO_2 is absorbed by an alkaline NaOH solution to produce dissolved sodium carbonate, and then the carbonate ion is removed from the solution by reacting with $Ca(OH)_2$, which results in the precipitation of calcite ($CaCO_3$).

Magnesium oxides have also been investigated as adsorbents for CO_2 separation because of its lower energy requirement for regeneration compared to calcium oxides (Lee et al. 2008a). CO_2 adsorbed by magnesium oxides can be recovered by 1 h of regeneration under vacuum at 973 K, whereas ca. 4 h of heating is required to remove CO_2 from calcium oxides under similar conditions (Beruto et al. 1987; Philipp and Fujimoto 1992). At given adsorption conditions, the typical capacity of magnesium oxide corresponds to less than a half of that of calcium oxide, and substantially decreases from 0.64 to 0.13 mmol/g when the adsorption temperature was increased from 273 to 773 K (Philipp and Fujimoto 1992).

Metal layered hydroxides, also call "hydrotalcites, or layered double hydroxides," are a class of anionic clays represented by the general formula $[(M_{1-x}^{2+} \cdot M_x^{3+} (OH)_2]^{x+} (A_{x/m}^{m-} \cdot nH_2O)^{x-}$, where M^{2+}=Mg^{2+}, Ni^{2+}, Zn^{2+}, Cu^{2+}, Mn^{2+}, or others, M^{3+} =Al^{3+}, Fe^{3+}, Cr^{3+}, or others, and A^{m-} =CO_3^{2-}, SO_4^{2-}, NO_3^-, Cl^-, OH^-, or others (Yong et al. 2002). Usually, metal-layered hydroxides consist of positively charged brucite ($Mg(OH)_2$) layers in which trivalent cations partially substitute for divalent cations located at the center of octahedral sites in the hydroxide layer (Yong et al. 2002). The excess positive charge is compensated by species such as CO_3^{2-} anions located in the interlayer region, resulting in a charge-balanced framework (Yong et al. 2002; Oliveira et al. 2008). For example, amorphous Mg-Al mixed oxides, derived from Mg-Al-CO_3 layered double hydroxide, have a CO_2 sorption capacity of 0.49 mmol/g at 200 °C, and regeneration restored the oxide to 98% of its initial CO_2 sorption after several cycles of CO_2 adsorption (Ram Reddy et al. 2006). Thus, the Mg-Al mixed oxide can be an excellent candidate for CO_2 capture from flue gases at high temperatures (up to 200 °C).

Besides, sodium hydroxide, potassium hydroxide, and lithium hydroxide are able to capture CO_2 by chemically reacting with it. For example, CO_2 is firstly absorbed by an alkaline NaOH solution to produce dissolved sodium carbonate; the carbonate ion is then removed from the solution by reaction with calcium hydroxide ($Ca(OH)_2$), which results in the precipitation of calcite ($CaCO_3$); the calcite is subsequently filtered from solution and thermally decomposed to produce gaseous CO_2 and lime (CaO), and the lime is finally hydrated to regenerate $Ca(OH)_2$. These reactions are shown below.

$$2NaOH(aq) + CO_2(g) \rightarrow Na_2CO_3(aq) + H_2O(l) \text{ (strongly exothermic, } \Delta H° = -109.4 \text{ kJ/mol)} \quad (9.8)$$

$$Na_2CO_3(aq) + Ca(OH)_2(s) \rightarrow 2NaOH(aq) + CaCO_3(s) \text{ (mildly exothermic, } \Delta H° = -5.3 \text{ kJ/mol)} \quad (9.9)$$

$$CaCO_3(s) \rightarrow CaO(s) + CO_2(g) \text{ (endothermic, } \Delta H° = +179.2 \text{ kJ/mol)} \quad (9.10)$$

$$CaO(s) + H_2O(l) \rightarrow Ca(OH)_2(s) \text{ (exothermic, } \Delta H° = -64.5 \text{ kJ/mol)} \quad (9.11)$$

Similarly, lithium hydroxide is used to remove CO_2 as shown in the reactions below:

$$2LiOH(s) + 2H_2O(g) \rightarrow 2LiOH \cdot H_2O(s) \quad (9.12)$$

$$2LiOH \cdot H_2O(s) + CO_2(g) \rightarrow Li_2CO_3(s) + 3H_2O(g) \tag{9.13}$$

9.4.2.2 Carbonaceous Materials

Adsorption studies on activated carbon, charcoal, and virgin coal have focused on high pressure CO_2 capture applications. Activated carbons are well-known adsorbent materials, and have been used for CO_2 capture (Siriwardane et al. 2001). Adsorption of CO_2 is reversible on activated carbon. The equilibrium adsorption capacities of activated carbon at 300 psi and 25 °C are about 8.5 mol of CO_2/kg of the sorbent. CO_2 can be separated from gas mixtures containing both CO_2/N_2 and $CO_2/H_2/He$ utilizing activated carbon. The adsorption capacities of activated carbons rapidly decrease with slight temperature increases, e.g., the decrease in CO_2 capacity from 3.2 mmol of CO_2 per grain of activated carbons to 1.5 mmol CO_2 per gram of adsorbent with only a temperature increase from 288 K to 328 K; the decrease in CO_2 uptake from 4 mmol CO_2 per gram of adsorbent to 1 mmol CO_2 per gram of adsorbent under wet conditions.

Other carbon materials like carbon molecular sieves (Jayaraman et al. 2002; Rutherford and Coons, 2003; Bae and Lee 2005) and carbon nano-tubes (Matranga and Bockrath 2005) have also been emerged as adsorbents. CO_2 can be selectively captured by carbon molecular sieves, such as Bergbau–Forschung, Takeda 3A, in CH_4/CO_2 mixture (Jayaraman et al. 2002). Temperature is an important factor, and a higher temperature is favorable for CH_4/CO_2 separation on molecular sieve where diffusion is slow. At 25°C, increasing the temperature decreases the equilibrium adsorption amount, but increases the diffusivities. The rate of diffusion (rather than equilibrium) is the dominating factor in kinetics-based separations. FITR has been used to demonstrate hydrogen-bonded and physisorbed CO_2 in single-walled carbon nanotubes in which surface hydroxyl groups are produced by acid purification steps of carbon nanotubes synthesis (Matranga and Bockrath 2005).

To improve its CO_2 adsorption property, carbon structural material can be modified. For example, activated carbons with basic surface groups are more resistant to the aging effect in humid atmospheres, and these sites also can increase the adsorption capacity of the activated carbons (Menéndez et al. 1996). The modification has been carried out either by removing the oxygen-containing groups with heat treatment at temperatures above 973 K in inert atmosphere, or by replacing the surface groups with other constituents such as amines (Menéndez et al. 1996). Incorporation of alternative functional groups on carbon surface has been achieved by impregnation or high temperature heat treatment using appropriate chemical agents (Plaza et al. 2007; Pevida et al. 2008). High-temperature ammonia treatment is one of the representative methods for carbon surface modification, whereby the chemical fixation of amines is carried out at temperatures of 673–1173 K by substitutional exclusion of oxygen-containing groups (Krishnankutty and Vannice 1995). Aliphatic amine-modified nanocarbon materials include a nanocarbon support, such as C60, nano-graphite, grapheme, and an aliphatic amine, such as polyethyleneimine (PEI) (US Patent 2012). Carbon nanotubes (CNTs) impregnated with tetraethylenepentamine (TEPA) exhibit an enhanced adsorption behavior toward CO_2; 2.0 vol % of CNTs−TEPA reached a sorption capacity of 2.97 $mmol \cdot g^{-1}$ at 298 K in a fixed-bed column (Ye et al. 2012). The CO_2 adsorption capacity increases with increasing temperature and can be 3.56 $mmol \cdot g^{-1}$ at 313 K. The adsorption capacity is also affected by moisture and can be up to 3.87 $mmol \cdot g^{-1}$ at 2.0% H_2O. After being modified with 3-aminopropyltriethoxysilane, carbon nanotubes (CNTs) and granular activated carbon (GAC) are improved concerning the physicochemical properties and adsorption behaviors of CO_2 from gas streams. The CO_2 sorption capacity increases from 69.2 to 96.3 $mg \cdot g^{-1}$ for CNTs and from 72.9 to 79.5 $mg \cdot g^{-1}$ for GACs, after being modified with the 3-aminopropyltriethoxysilane (Lu et al. 2008). N-doped porous carbon produced via chemical activation of polypyrrole functionalized graphene sheets shows selective adsorption of CO_2 (4.3 $mmol \cdot g^{-1}$) over N_2 (0.27 $mmol \cdot g^{-1}$) at 298 K (Chandra et al. 2012). Copper oxide-decorated porous carbons enhance adsorption capacity of carbon dioxide molecules due to electron-donor feature of copper oxide (Kim et al. 2010).

9.4.2.3 Silicon-containing Solid

There are many silicon-containing solid materials that are used for CO_2 adsorption, such as silica, zeolites, and molecular sieves. Zeolites, a typical class of porous crystalline aluminosilicates built of a periodic array of TO_4 tetrahedra (T = Si or Al), have been widely used in separation applications mainly because of their unique ability of molecular sieving (Walton et al. 2006; Zukal et al. 2009). The presence of aluminum atoms in these silicate-based molecular sieve materials introduces negative framework charges that are compensated with exchangeable cations in the pore space (usually alkali cations), and these structural characteristics of zeolites enable them to adsorb a wide variety of gas molecules, including acidic gas molecules such as CO_2. Naturally occurring zeolites, such as X and Y Faujasite systems (Maurin et al. 2005), and synthetic zeolites including 5A, 13X and MCM-41 (Harlick and Sayari 2006), have been used to remove CO_2 from low-pressure flue gas. The CO_2 adsorption capacity of zeolite is significantly high at ambient temperature or high pressure (Siriwardane 2005). The CO_2 adsorption capacity substantially decreases even with only minor increases in operating temperature or the presence of a small amount of moisture. For example, zeolite, CaX, decreases its CO_2 absorption capacity from 2.5 mmol CO_2 per gram of absorbent to 0.1 mmol CO_2 per gram of absorbent with a H_2O concentration increase from 1 wt% to 16 wt%.

Amines modification of these materials has also contributed an improvement of CO_2 adsorption ability. For example, aminopropyl-grafted pore-expanded MCM-41 silica (MONO-PE-MCM-41) with a mean pore size of 7.2 nm has been developed for

CO_2 scrubbing (Serna-Guerrero et al. 2008). The impregnation of polyethylenimine into MCM-41 mesoporous molecular sieves leads to a 24-fold enhancement in the CO_2 absorption capacity of the solid support using a pressure swing adsorption approach (Xu et al. 2008). Compared with conventional MCM-41, triamine surface-modified PE-MCM-41 shows a significant increase in CO_2-adsorbent interaction after the amine functionalization, consistent with the high CO_2 uptake in the very low range of CO_2 concentration (Belmabkhout and Sayari 2009). Mesoporous silica modified with polyaziridine exhibits reversible CO_2 binding (with a capacity of 2 mmol CO_2/g adsorbent) and multi-cycle stability under simulated flue gas conditions using a temperature swing adsorption (TSA) approach (Hicks et al. 2008). SBA-15 grafted with monoamino, diamino, and triamino ethoxysilanes, shows their capacities of 0.52, 0.87, and 1.10 mmol $CO_2 \cdot g^{-1}$ adsorbent, respectively (Hiyoshi et al. 2004). The adsorption of CO_2 on SBA-16 functionalized with N-(2-aminoethyl)-3-aminopropyltrimethoxysilane apparently increases, and the maximum adsorption capacity of CO_2 at 333 K was 0.727 $mmol \cdot g^{-1}$ (Wei et al. 2008). Triamine-grafted pore-expanded mesoporous silica (TRI-PE-MCM-41) exhibited high CO_2 and H_2S adsorption capacity as well as high selectivity toward acid gases versus CH_4, and exhibits extremely high CO_2 selectivity over methane, regardless of the CO_2 concentration and the occurrence of moisture (Belmabkhout et al. 2009). Silica supported amines have been using for CO_2 capture by Tsuda et al. (1992). The surface silanol density of mesoporous silica SBA-15 increases from 3.4 to 8.5 $OH \cdot nm^{-2}$, consequently the grafted amine loading is increased from 2.2 to 3.2 $mmol \cdot g^{-1}$, and thus the CO_2 adsorption capacity is increased from 1.05 to 1.6 $mmol \cdot g^{-1}$ at conditions relevant to CO_2 capture (0.15 bar and 25 °C), or a 52% increase. Hyperbranched aminosilica (HAS) adsorbents are obtained by covalently binding mesoporous silica SBA-15 support with aminopolymers and capture CO_2 reversibly in a temperature swing process (Drese et al. 2009). Incorporation of triptycene into benzimidazole-linked nanoporous organic polymer networks leads to the highest CO_2 uptake (5.12 mmol g^{-1}, 273 K and 1 bar) and results in high CO_2/N_2 (63) and CO_2/CH_4 (8.4) selectivities (Rabbani et al. 2012).

9.4.2.4 Metal-organic Frameworks

Metal–organic frameworks (MOFs) present a class of porous materials that offer these advantages for using as CO2 adsorbents: ordered structures, high thermal stability, adjustable chemical functionality, extra-high porosity, and the availability of hundreds of crystalline, and well-characterized porous structures (Thallapally et al. 2008; Wang et al. 2008). In general, these materials consist of three-dimensional organic–inorganic hybrid networks formed by multiple metal–ligand bonds. In these networks, there are many different metal–ligand combinations. The first MOF was MOF-5, synthesized by Yasghi research group (Li et al. 1999) MOF-5, with the formula $[Zn_4O(BDC)_3] \cdot (DMF)_8(C_6H_5Cl)$(BDC=1,4-benzenedicarboxylate, DMF=N,N'-dimethylformamide), consisted of tetranuclear supertetrahedral clusters linked by bidentate BDC ligands into an octahedral arrangement, exhibits stable porosity in the absence of solvent or guest ions in the framework galleries. A pore volume of ca. 1 $mL \cdot g^{-1}$ and a surface area of 2900 $m^2 \cdot g^{-1}$ were estimated for MOF-5 by N_2 physisorption. These relatively large values could make MOF to be a good candidate for CO_2 adsorption. Later, many MOF-5 derivatives were synthesized by using different dicarboxylate ligands (Eddaoudi et al. 2002). Zn(BDC)-(bpy)0.5 (MOF-508b, bpy=4,4'-bipyridine) with one-dimensional micropores of 4 Å diameter, has been considered for the separation of CO_2 from N_2 and CH_4 (Bastin et al. 2008). MOFs have similar behavior as most solid adsorbents that the CO_2 capacities of MOFs decrease with increasing adsorption temperature. In general, the heats of adsorption for the interaction of CO_2 with MOFs are low and comparable to those of physical adsorbents such as zeolites. For example, heats of adsorption for MIL-53 (Al, Cr) at pressures from 1 to 4 bars are in the range of -30 to -45 $kJ \cdot mol^{-1}$ (Bourrelly et al. 2005). Remarkably, at 30 bar, MOF-177 exhibits a CO_2 sorption capacity of 35 mmol of CO_2 per gram of sorbent material and has a better CO_2 capacity than the benchmark materials (zeolite 13X (7.4 $mmol \cdot g^{-1}$) and activated carbon (25 $mmol \cdot g^{-1}$)) (Millward and Yaghi 2005). Chemical-functionalized frameworks carboxylic acids, amines, hydroxyl, and methyl groups can outperform more widely studied amine sorbents in CO_2 capture and separation application (Dawson 2011). For example, The incorporation of N,N'-dimethylethylenediamine (mmen) into $H_3[(Cu_4Cl)_3(BTTri)_8$ (CuBTTri; H_3BTTri = 1,3,5-tri(1H-1,2,3-triazol-4-yl)benzene), a water-stable, triazolate-bridged framework, drastically enhance CO_2 adsorption. At 25 °C under a 0.15 bar CO_2/0.75 bar N_2 mixture, mmen-CuBTTri adsorbs 2.38 mmol CO_2 g^{-1} (9.5 wt%) with a selectivity of 327 (McDonald et al. 2011). Amine-functionalized MOFs have been tested as adsorbents for carbon dioxide, and show the highest CO_2 adsorption capacities, the best adsorbing around 14 wt% CO_2 at 1.0 atm CO_2 pressure (Arstad et al. 2008). Currently, most MOFs have relatively low adsorption capacities at low CO_2 partial pressures.

9.4.3 Others

Solid amine sorbents are the amine functionality chemically bonded to solid porous supports such as silica gels, fly ash carbon, molecular sieves, activated carbon, and polymers. Lee et al. () have investigated the absorption properties of CO_2 in different solid amine absorbents, such as absorption capacity, absorption/desorption rate, cyclic capacity, cycle decay effect, and temperature of reaction in the absorber (Lee et al. 2008b). Polymeric amine adsorbents have been investigated for CO_2 capture that can be prepared by incorporating polymers or oligomers with high amine content into polymer supports via impregnation or covalent bonding, and by copolymerization with amine-containing monomers or monomers that are easily derivatized to create amines. For example, polymeric amine adsorbents include, polyethyleneimine-bonded poly(methylmethacrylate) (Satyapal et al. 2001), oligomeric ethyleneimines (E-100) or tetraethylenepentamine modified poly(methylmethacrylate) (Schladt et al. 2007), polyethyleneimine-bonded poly(acrylonitrile) (Yang, et al. 2012), and aminated polystyrene copolymer (Diaf et al. 1994). Amine-containing solid organic resins can also potentially be CO_2 adsorbents (Drage et al. 2007).

The carbonate-based system is another CO_2 capture system and is based on the ability of a soluble carbonate to react with CO_2 to form bicarbonate, which reverts to a carbonate when heating the bicarbonate to release CO_2. A K_2CO_3/piperazine (PZ) system (5 molar K; 2.5 molar PZ) has an absorption rate 10–30% faster than a 30% solution of MEA, and furthermore oxygen is less soluble in the K_2CO_3/PZ solvents. A major advantage of carbonates over amine-based systems is the significantly lower energy required for regeneration. Analysis has indicated that the energy requirement is approximately 5–15% lower compared with MEA (Figueroa et al. 2008).

Zeolitic imidazolate frameworks (ZIFs) have been designed and prepared because the Si–O–Si preferred angle in zeolites (145°) is coincident with that of the bridging angle in the M-Im-M fragment (where M is Zn or Co and Im is imidazolate). The ZIFs have been investigated for CO_2 capture (Phan et al. 2010).

9.5 Current Status of CO_2 Scrubbing Technology

9.5.1 History of CO_2 Scrubbing Technology

Chemical absorption (scrubbing) is a well-known chemical engineering process, and mostly based on the chemical fixation of CO_2 with aqueous solutions of amines. Amine scrubbing is one of widely used chemical absorption technologies of CO_2 scrubbing from natural gas and has been established for over 70 years in the chemical and petroleum industries (Rochelle et al. 2009). The basic process, patented in 1930 (Bottoms et al. 1930), is that CO_2 is absorbed from a fuel at around ambient temperature into an aqueous solution of amine with low volatility, and then the amine is regenerated by stripping with water vapor at 100° to 120°C, and the water is condensed from the stripper vapor, leaving pure CO_2 that can be compressed to 100 to 150 bar for geologic sequestration. Mono-ethanolamine (MEA) is by far the most common absorbent that is used in amine scrubbing due to its higher reactivity than secondary amines. The idea of separating CO_2 from flue gas streams was started in the 1970s as a potentially economic source of CO_2 for enhanced oil recovery (EOR) operations. Several commercial CO_2 capture plants were built in the U.S. in the period of 1970–1980s (Kaplan 1982; Pauley et al. 1982). The first commercial CO_2 sequestration facility built in Norway in September 1996 (DOE 1999). Most of these plants captured CO_2 using processes that were based on chemical absorption with MEA solvent. Fluor Daniel Inc., Dow Chemical Co., Kerr-McGee Chemical Corp., Mitsubishi/Kansai Electric Power Company (KEPCO) and ABB Lummus Crest Inc. were the major developers of MEA-based technology of

CO_2 scrubbing. Typically, 75~90% of the CO_2 was captured by this technology with a purity of the CO_2 product stream being > 99%. 150 tons·d^{-1} of CO_2 is captured in the U.S. Warrior Run coal fired power station. Commercial processes that are based on physical absorption exist as well, using methanol or poly(ethylene glycol) dimethyl ether as absorbing phases (Sircar 2006). The choice of a suitable technology depends on the characteristics of the flue gas stream and thus the power plant technology. Generally, the scrubbing technique must be adapted to the conditions of a plant. Industrially produced CO_2 has been captured for enhanced oil recovery at the Rangely field in Colorado, USA since 1986 and at the Weyburn field in Saskatchewan, Canada since 2000. In 2009, the pilot CO_2-scrubbing plant at Niederaussem was built to capture CO_2 from the flue gas of a conventional power plant using the CO_2-scrubbing process. The factors affecting the choice of separation technology are CO_2 partial pressure of flue gas, sensitivity to other impurities, CO_2 recovery, capital and operating costs and by-products. For separating CO_2 from flue gas with very low CO_2 partial pressure, absorption processes using chemical solvents offer distinct advantages over the alternatives: lower energy use and costs, higher capture efficiency and better selectivity.

9.5.2 Application of CO_2 Scrubbing Technology in Industrial Plants

9.5.2.1 Amines Scrubbing Processes

Amine-based solvent chemical absorption is a mature and widely used technology in CO_2 scrubbing (Chapel et al. 1999; Abanades et al. 2004a; Iijima 2004). Most industrial processes use standard packed or plate columns with an 8–20 wt% amine aqueous solution (Astarita 1983). The conditions are relatively easy to meet for absorption and regeneration, causing the energy penalty to be fairly low. By considering the column height, amine plants could reduce CO_2 to 50 ppm (Jou et al. 1995) or even 10 ppm (Dodge 1972). Currently, the largest scrubber in Trona California captures 800 tons·d^{-1} of CO_2, cf ~6900 tons·d^{-1}, which would be required for a 350 MW power station unit (IEA 2007). Typically, the process operates at about 40 °C in the scrubbing column and approximately 120 °C in the regeneration column.

An effective, economical, and traditional solvent that can be used for CO_2 absorption is monoethanolamine (MEA). For the CO_2–MEA system, the rate of absorption is determined by the rate of reaction among the compounds (Dabckwerts and McNeil 1967), and the diffusivity of CO_2 in air and the diffusivity and solubility in solution are affected by the operating conditions in the scrubber, including temperature, pressure, and solution chemistry. MEA concentration in industry has been limited to around 15wt% MEA because of plant corrosion, which increases with increased MEA concentration. The addition of corrosion inhibitors allows the use of higher MEA concentrations, up to 30 per cent (Diaf et al. 1994; DeMontigny et al. 2001). Absorption took place at approximately 50 °C for the MEA, regardless of which packing was used.

Regeneration took place at 120 °C. This process is often used in current applications at most plants with MEA scrubbers (Table 9.1).

Four coal-fired plants with power outputs of 6 to 30MW separate CO_2 from flue gas using 20% monoethanolamine (MEA). Most of plants use 30% aqueous MEA (developed by Fluor) on gases with substantial O_2 content, including a gas-fired turbinewith a flue gas rate equivalent to that of a 40-MW coal-fired power plant that produces flue gas with 15% O_2. Kerr-McGee/ABB Lummus Global has licensed four units that use 15-20 wt% MEA to recover CO_2 from coal-fired flue gas (Steeneveldt et al. 2006). The plant capacities vary between 180 and 720 tons·d^{-1}. Some corrosion inhibitors in conjunction with a quantitative oxygen and NOX removal system allow the MEA concentration to be raised to 25–30 weight percent. The Bridgeport CO_2 plant (Table 9.1), which used an early Amine GuardTM process, is one example. There are now over 500 Amine Guard units worldwide including the UCARSOLTM family of formulated amines, licensed by UOP. Ucarsol plants using atmospheric pressure absorption are offered; however, not on oxygen-containing gas streams.

Table 9.1. Some industrial plants for CO_2 capture[a]

Operator	Tech Supplier	Scale (ton·d^{-1} CO_2)	CO_2 use	Plant location	Status
CO_2 technology	DOW MEA	1200	EOR	Lubbock, Texas	Shut (1982-1984)
Mitchell Energy	Inhibited MEA	493	EOR	Bridgeport, Texas	Shut (1991-1999)
BOC	Fluor	350	Foods	Bellingham, Massachusetts	Operating, 1991-
Sua Pan	Kerr-McGee/ABB Lummus	300	Soda ash production	Sua Pan, Botswana	Operating, 1991-
AES	Kerr-McGee/ABB Lummus	200	Foods	Shady Point, Oklahoma	Operating, 1991-
AES	ABB Lummus	150	Foods	Warrior Run	Operating, 2000-
Sumitomo Chemicals	Fluor and MHI	165	Foods	Chiba, Japan	Operating, 1994-
Luzhou Natural Gas Chemicals	Fluor	160	Urea production	Luzhou, China (Fertilizer plant)	Operating, 1998-
Indo Gulf Fertilizer Co.	Fluor	150	Ammonia production	Jagdishpur, India (Fertilizer plant)	Operating, 1988-
Petronas Fertilizer Co.	MHI	145	Ammonia and urea production	Petronas Fertilizer Co., Malaysia	Operating, 1999-
Prosint AGA	Fluor	90	Foods	Rio de Janeiro, Brazil	Operating, 1997-
Liquid Air	Fluor	60		Altona and Botany, Australia	Operating, 1985-
CO_2CRC, Loy Yang Power, CSIRO, International Power		12	PCC demo	Loy Yang Power Station, Victoria Australia	Operating, 2008-
CSIRO, China HuaNeng Group		8	PCC demo	HuaNeng Beijing Cogeneration Powerplant, China	Operating, 2008-

[a]Herzog (1999); Reddy (2003); Bolland (2004). IEA (2005); Grad (2009); Duke et al. (2010).

In 1982, N-ReN Southwest plant was designed to recover a maximum of 104 tons·d^{-1} of CO_2, using 18-20% monoethanolamine from boiler flue gas and primary reformer exhaust gas of two ammonia plants (Bourrelly et al. 2005). MEA as a CO_2 scrubber can be simply constructed in a compact packed bubble column (PBC), in which one can use a high MEA concentration in aqueous solution and operates at atmospheric temperature and pressure. Typically, the main reactor of the PBC is made of a stainless steel vessel. The CO_2 scrubbing performance of a PBC with 50 wt% MEA solution was found to scrub a 20 m^3·h^{-1} flue gas to 10 ppm (Wallace and Krumdieck 2005). The ability to scrub to a given CO_2 concentration depends on the MEA strength in solution and the residence time of the gas in the PBC, the packing size and type, and the degree of turbulent mixing generated by the high gas flow rate.

The limits on the SO_2 and NO_2 concentration in the flue gas being treated for CO_2 removal by MEA scrubbing are recommended to be in the range from 10 to 50 ppmv. Amine tolerance levels are reported to be 90 ppm O_2, 10 ppm SO_2 and 20 ppm NO_x (at 6% excess oxygen) (IEA 2004). Fly ash and soot removal is also important to prevent both foaming in the absorber and further reactions with the solvent.

GAS/SPEC FT-1 technology uses alkanolamine solvent to remove CO_2 from low-pressure streams containing low levels of CO_2 and oxygen (Bastin et al. 2008). Kansai Electric Power Company (KEPCO) and Mitsubishi Heavy Industries (MHI) have developed a proprietary hindered amine called KS-1 as an MEA replacement for flue gas applications. KS-1 has a lower circulation rate (due to its higher lean to rich CO_2 loading differential), lower regeneration temperature (110 ^{o}C), and 10–15% lower heat of reaction with CO_2. It is non-corrosive to carbon steel at 130 ^{o}C in the presence of oxygen. KS-1 has been applied in a commercial gas scrubbing operation for Petronas Fertilizer Kedah Sdn Bhd's fertilizer plant in Gurun Kedah in Malaysia to produce a pure CO_2 stream for urea production (Suda et al. 1992). Energy usage is about 4.2 MJ/kg-CO_2 for conventional MEA solvent (Chapel et al. 1999) and ~3.3 MJ/kg-CO_2 for KS-1 (Iijima et al. 1999). Another sterically hindered amine, AMP, (2-amino-1-methyl-1-propanol) may have similar properties to KS-1 (White et al. 2003).

Piperazine has also been used as a solvent for absorption/stripping systems for the removal of CO_2 from the flue gas in coal-fired power plants. The CO_2 absorption rate of aqueous piperazine is more than double that of 7 M MEA and the amine volatility at 40 ^{o}C ranges from 11 to 21 ppm. Oxidation of aqueous piperazine is appreciable in the presence of copper (4 mM), but negligible in the presence of chromium (0.6 mM), nickel (0.25 mM), iron (0.25 mM), and vanadium (0.1 mM). Initial system modeling indicates that 8 M piperazine will use 10-20% less energy than 7 M MEA. The fast mass transfer and low degradation rates suggest that concentrated, aqueous PZ have the potential to be a preferred solvent for CO_2 capture (Freeman et al. 2010). Piperazine has also been used for improving the activated potassium carbonate process (Cullinane and Rochelle 2003).

The main disadvantages with MEA and other amine solvents are corrosion in the presence of O_2 and other impurities, high solvent degradation rates from reaction with SO_x and NO_x and the large amounts of energy required for regeneration. As much as 80% of the total energy consumption in an alkanolamine absorption process is required during solvent regeneration (White et al. 2003). These factors result in large equipment, high solvent consumption and large energy losses. Although the MEA process is a promising system for the control of CO_2 emissions from massive discharging plants, it is an expensive option since the cost of CO_2 separation may range from US$40 to 70 per ton of CO_2 removed (Chakma 1995). The challenge is thus to couple efficient CO_2 capture with facile release in a sorbent material. Improved strategies for CO_2 scrubbing technologies include the use of liquids with lower heats of adsorption, increasing the concentration of the adsorbent molecules and improving the mass transfer and reaction kinetics.

Hybrid absorption processes mean a combination technology of chemical and physical absorption. Currently, main hybrid absorption processes for removal of CO_2 and sulphur compounds from flue gas are the Shell Sufinol process and Amisol process developed by Lurgi (Collot Anne-Gaelle Collot 2003). The Shell Sufinol process combines a physical solvent, sufolan (tetrahydroehiophene dioxide), and a chemical solvent, DIPA (di-isopropanolamine) or MDEA. In this process, the sufinol unit tolerates a much higher acid gas loading before becoming corrosive. The Amisol process is based on a mixture of methanol and either MEA or MDEA as the chemical absorbent.

9.5.2.2 Chilled Ammonia Processes

The chilled-ammonia process for CO_2 capture involves the reversible formation of ammonium hydrogen carbonate, with the forward reaction to capture CO_2 as solid NH_4HCO_3 occurring at temperatures below 20 °C (Johnson 2008). In the regeneration stage, the CO_2-rich aqueous ammonium carbonate solution is heated to about 80 °C to redissolve the solids. The maximum CO_2 removal efficiency by NH_3 absorbent can reach 99%, and the CO_2 loading capacity is up to 1.2 kg $CO_2 \cdot kg^{-1}$ NH_3 (Yeh et al. 2002); whilst the maximum CO_2 removal efficiency and loading capacity by MEA absorbent are 94% and 0.40 kg $CO_2 \cdot kg^{-1}$ MEA (Sircar 2006). At pH of 11.0, when the total ammonium carbonate concentration is 0.1 M, CO_2 removal efficiency is observed to be 100% from an initial 12% CO_2 in flue gas (Huang and Zhang 2002). Ammonia can be employed to capture all three major acid gases (SO_2, NO_x, CO_2) in addition to any Hg, HCl and HF, which may exist in the flue gas of coal combustors (Huang and Zhang 2002; Yel et al. 2002). Unlike the MEA process, ammonia is not expected to have absorbent degradation problems that are caused by sulphur dioxide and oxygen in flue gas and to cause equipment corrosion, and thus, could potentially reduce the energy requirements for CO_2 capture. Powerspan Corp. announced the start-up of a pilot plant for testing the application of ammonia for simultaneous reduction of SO_2, NO_x and mercury that includes CO_2 removal in 2006.

Yeh et al. (2005) have compared CO_2 capture capacity and energy requirement of aqueous ammonia solution and MEA in a semi-batch reactor. The results showed that CO_2 carrying capacity is 0.07 g CO_2 per g of ammonia solution (8 wt.%) as compared with 0.036 g CO_2 per g MEA solution (20 wt.%). The energy requirement for liquid mass circulation of ammonia solution is approximately 50% of MEA solution for equal weight of CO_2 carried. Besides, ammonium bicarbonate required the least thermal energy among the ammonium compounds for CO_2 regeneration.

However, the disadvantages of ammonia absorption include the highly volatile nature of ammonia and lacks in the regeneration of ammonia from its carbonate salts (Reddy et al. 2003). Released ammonia will react with the remaining ammonium bicarbonate to form ammonium carbonate, which results in the resin's inability to completely regenerate ammonia.

9.5.2.3 Solid Sorbents

Adsorption process for gas separation via selective adsorption on solid media is well-known (Yang 1997). These adsorbents can operate via weak physisorption processes or strong chemisorptions interactions. Solid adsorbents are typically employed in cyclic, multimodule processes of adsorption and desorption, with desorption induced by a pressure or temperature swing. Reactor systems for this type of processes include fixed bed reactors, moving bed reactors and fluidized bed reactors. Alkaline fuel cell (AFC) technology (McLean et al. 2002), pressure swing adsorption (PSA) (Ko et al. 2003) and temperature swing adsorption (TSA) are potential technologies that could be applicable for removal of CO_2 from flue gas. Molecular sieves (e.g., 13X, 4A) and activated carbons are the main sorbents that could be used in the PSA process (Siriwardane et al. 2001). The adsorption capacity of zeolite 13X for CO_2 was higher than zeolite 4A and activated carbon in the PSA process. Over 99%-purity of CO_2 can be obtained by zeolite 13X at higher recoveries and higher productivities than activated carbon (Chue et al. 1995). But at higher pressures (> 25 psi) activated carbon exhibited significantly higher CO_2 capacities than that were found for molecular sieves. Both zeolite 13X and activated carbon can be utilized for separation of CO_2 from gas mixtures.

The main challenges for solid sorbents on a large scale application are the costs of solids handling and dust elimination equipment, the cyclic absorption capacity and mechanical strength of the absorbent (Abanades et al. 2004a & b).

9.5.3 Cost Information

The energy requirements and technical and economic analyses by means of amine scrubbing integration into a commercial power plant have been reported (Romeo et al. 2008). Generally, the cost of CO_2 separation and compression dominates 50–80% of total specific CO_2 costs across the value chain of CO_2 capture and storage from power station flue gas. The cost of CO_2 capture is the sum of several terms, including the large capital costs associated with the capture plant, the operating and maintenance costs, the cost of additional fuel resources needed to compensate for the efficiency penalty introduced by the capture plant, and the cost of CO_2 compression (Rao 2001; Rao and Rubin 2002). The total capital requirement (TCR) of a system is the sum of direct equipment costs, plus various indirect costs that are calculated as fractions of the total process facilities cost (PFC). The absorber capital cost depends mainly on the flue gas flow rate. The cost of the regenerator section and the CO_2 compressor scale mainly with the mass flow rate of CO_2 captured. In operating and maintenance costs, major variable cost items include the cost of sorbent and the costs of CO_2 transport and storage. The cost of sorbent per kilogram of CO_2 removed ($C_{sorbent}$) can be estimated using the following equation with Fig. 9.4 showing the CO_2 scrubbing process.

$$C_{sorbent} = \left(\frac{F_0}{F_R}\right)\left(\frac{F_R}{F_{CO_2}}\right)\frac{bM_S}{M_{CO_2}}C_S = \left(\frac{F_0}{F_R}\right)\frac{C_S^k}{M_{CO_2}} \tag{9.14}$$

where F_0 = modified molar flow of absorbent; F_R = molar flow of sorbent flowing in the capture-regeneration loop, F_{CO2} = molar flow of CO_2; b = constant (= 2); M_s = molecular weight of scrubber; M_{CO2} = molecular weight of CO_2; C_s = concentration of scrubber; C_s^k – $M_sC_sb(F_R/F_{CO2})$. A modified flow of sorbent (F_0) is required to compensate for the natural decay of activity and/or sorbent losses during many sorption-desorption cycles. For example, in a typical MEA-based system, b=2, M_{MEA}=0.061, C_s=\$1.25/kg of MEA, and the modified flow is 1.5 kg of MEA/ton of CO_2. Correspondingly, F_0/F_R=0.000 152 and F_R/F_{CO2}=3.57. Thus, C_{MEA} is ~\$2/ton of CO_2 captured (Abanades et al. 2004a).

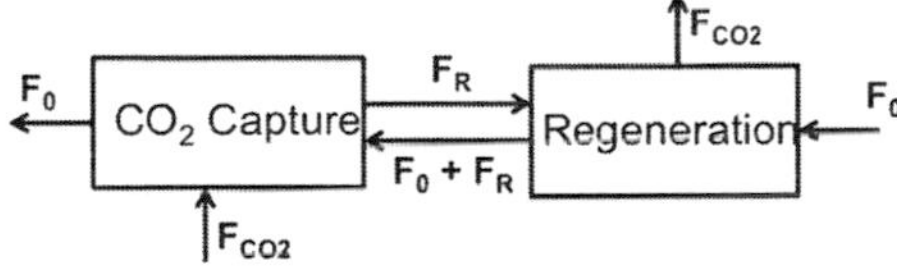

Figure 9.4. The general scheme for CO_2 scrubbing processes

Fixed costs include the costs of maintenance and labor. The cost of electricity (COE) for the overall power plant can be calculated by dividing the total annualized plant cost (\$·year^{-1}) by the net electricity generated (kWh·year^{-1}). Two key parameters are the levelized fixed charge factor (used to amortize capital expenses) and the plant capacity factor. The fixed charge factor is based on the plant lifetime and after-tax discount rate (or interest rate, or rate of return), while the capacity factor reflects the average annual hours of plant operation. Cost of CO_2 avoided needs to be considered. Since the purpose of adding a capture unit is to reduce the CO_2 emissions per net kWh delivered, the cost of CO_2 avoidance (relative to a reference plant with no CO_2 control) is the economic indicator most widely used. It can be calculated as

$$Cost\ of\ CO_2\ avoided\ (\$/tonne) = \frac{(\$/kWh)_{capture} - (\$/kWh)_{reference}}{(tonne\ CO_2/kWh)_{reference} - (tonne\ CO_2/kWh)_{capture}} \quad (9.15)$$

In order to minimize the overall cost of CO_2 capture, strict operational requirements are necessary for the desulfurization unit in a power plant in order to keep concentrations of SO_2 in the flue gas below 10 ppmv so as to minimize the absorbent modification that otherwise deactivates rapidly because of reactions with SO_2 and other pollutants in the flue gases (Chapel et al. 1999).

In addition, the cost of CO_2 capture can decrease by optimizing process design. For example, the efficiency penalty for CO_2 capture for pulverized coal (PC) plant with flue gas scrubbing using an amine solvent can be reduced to 20% from 28% with improved thermodynamic integration and lower-energy solvent, giving a 10% reduction in electricity costs, from \$63.5/MWh to \$57.4/MWh, and a 25% reduction in the cost of CO_2 avoided, from \$45/ton to \$34/ton (Gibbins and Cran 2004). Reductions in the average cost of electricity of 6–7% were estimated using solvent storage, giving a cost of electricity of \$56.73/MWh and \$33/ton CO_2 avoided for an integrated plant with MEA as the solvent.

9.5.4 Current Research

9.5.4.1 Development of CO_2-scrubbing Solvents

A large research effort is being directed at improved solvents to improve the CO_2 loading, reduce the energy requirement for solvent circulation and regeneration and overcome solvent degradation (Aresta et al. 2003; Zheng et al. 2003). To evaluate the potential of absorbents in CO_2 scrubbing technology, solvent's physical properties like the thermodynamic and the kinetics of CO_2 absorption (Porcheron et al. 2011), and the rate of solvent degradation, are main factors to be considered in designing efficient solvents (Bonenfant et al. 2003; Puxty et al. 2009). So far, amine technology is still the most widely used process in industry for CO_2 capture. Thus, developing new amine-based solvents that are fast and of high capability to absorption, resistant to degradation and not corrosive to the equipment is necessary. Currently, there exist some developed amine-based solvents, such as the combination of MEA with additives developed by

Fluor-Daniel Ecoamine, KS-1 and KS-3 developed by Mitsubishi, CORAL developed by TNO (future), PSR developed by University of Regina, Amine Blends developed by Praxair, and CANSOLV developed by CANSOLV. In addition, the capture process efficiency can be substantially improved by careful design of a mixture of solvents (Melien 2005).

Acetamidoxime showed the highest CO_2 capacity (2.71 $mmol \cdot g^{-1}$) when compared to terephthalamidoxime (two amidoximes per molecule) and tetraquinoamidoxime (four amidoximes per molecule). Highly porous, amidoxime rich solids based on clays, mesoporous materials and metal–organic frameworks (Fig. 9.6) are currently considered as valuable alternatives to MEA (Zulfiqar et al. 2011).

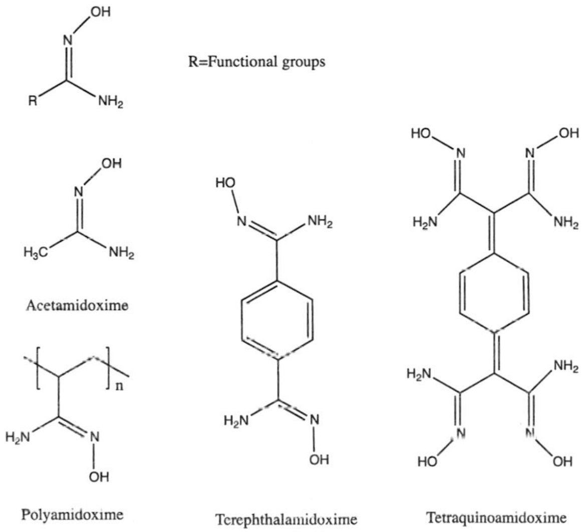

Figure 9.6. Acetamidoxime and its derivatives that could be used for CO_2 scrubbing

Amino acid salt solutions have been developed as competitive absorption liquid to alkanol amine solutions to separate CO_2 from CH_4 and can be used as competitive absorption liquids for energy-effective removal of CO_2 (Simons et al. 2010). Amino acids, such as sarcosine, have the same functionality as alkanol amines, but they exhibit a better oxidative stability and resistance to degradation. Adding a salt functionality can significantly reduce the liquid loss due to evaporation at elevated temperatures in the absorber. Aminosilicone solvents have been explored for the capture of CO_2 (Perry et al. 2010). Hydroxyether was used as a co-solvent to enhance physisorption of CO_2.

Regeneration of the capture solvent system was demonstrated over 6 cycles, and absorption isotherms indicate a 25–50% increase in dynamic CO_2 capacity over 30% MEA.

Ionic liquid is a new type of physical absorbents while often suffers from low rates of absorption. To overcome these shortcomings and increase the capacity of simple ionic liquids, "task specific ionic liquids" have been being developed (Bates et al. 2001). The introduction of functional groups such as amines into ionic liquids, has allowed higher rates of absorption at pressures relevant to flue streams (ca. 1 bar). Extremely high CO_2/N_2 selectivity in polymerized ionic liquids has been exhibited with enhanced CO_2 solubility compared with monomeric ionic liquid (Tang et al. 2005).

9.5.4.2 Development of Novel Sorbents

Novel concepts for CO_2 capture require the understanding of the chemical reactivity of the gas molecules and the selectivity of a separation process at a molecular level. The selectivity is considered as a combination of adsorption and diffusion selectivity. The introduction of a functional group that specifically binds one species, and improves on the adsorption selectivity, will simultaneously decrease the diffusion of these molecules. This inverse relationship between the adsorption and diffusion selectivity has been investigated in meso- and microporous materials including zeolites, carbon nanotubes, carbon molecular sieves, and metal–organic frameworks (Krishna 2009). Thus, novel CO_2 sorbents should be designed in which one can independently tune the diffusion and adsorption selectivity at the molecular level. In this regard, the latest developments in CO_2 capture have been considering about micro-crystal porous solids or metal–organic frameworks. For example, 'supramolecular chemistry' based porous metallosupramolecular networks are constructed mainly by coordination bonds between metal ions and ligands, together with other intermolecular interactions (Suh et al. 2008). Three-dimensional coordination polymers incorporating flexible pillars exhibit highly selective adsorption of CO_2 over N_2, H_2, and CH_4, thermal stability up to 300 °C, as well as air and water stability, and allow efficient CO_2 capture and storage (Choi and Suh 2009). Metal–organic frameworks (MOFs), constructed by metal-containing nodes connected by organic bridges, are such a new type of porous materials (Li et al. 2009).

9.5.4.3 Process Design

Although improved amines can save regeneration operating costs, they may have slower reaction kinetics and thus require longer gas/liquid contact time in the absorber. The design of improved contacting equipment has been investigated to overcome these problems with packed columns (flooding, channeling, entrainment and foaming), such as improved packing (Aroonwilas et al. 2003; Kvamsdal et al. 2005), the use of contacting membranes (Feron et al. 2002; Søybe Grønvold et al. 2005). For example, compared to structured packing, polymeric contactor shows significantly higher mass transfer coefficients (deMontigny et al. 2005).

9.6 Future Perspectives

Carbon dioxide scrubbing from large point sources such as power plants is an important technology to reduce anthropogenic CO_2 emissions; however, conventional CO_2 capture using amine scrubbers will increase the energy requirements of a plant by 25-40% (IPCC 2005; Haszeldine 2009). EPRI report estimates that the utilization of an amine scrubbing system could result in an increase in the electricity cost of \$0.06 kWh, or an "avoided cost of capture" of \$57-60/ton CO_2 (EPRI 2008). Thus, the existing technologies of CO_2 capture are energy intensive and are not cost-effective for carbon emissions reduction. Additionally, flue gas for CO_2 capture from fossil-fueled based thermal power plants are the large volumetric flow rates at essentially atmospheric pressure with large amounts of CO_2 at low partial pressures and in the temperature range of 100–150 °C. The presence of SO_x, NO_x, and significant oxygen partial pressure in the flue gas add to additional problems for implementation of the amine absorption process for CO_2 capture from flue gas streams. To overcome these limitations, the future direction should improve CO_2 capture processes and materials. There exists a serious need for research on innovative new materials and concept in order to reduce the time to commercialization and lower the overall cost of CO_2 capture.

Firstly, advanced amine solvents, solid sorbents, ionic liquids and metal-organic frameworks representing new materials will be designed and prepared (Chen et al. 2002; MCCI 2008). Conventional amine solvents can be degraded by high temperature (> 120 °C) in oxidizing environments and contaminants (SO_x and NO_x) that generally need to be less than 10 ppmv to minimize loss. New solvents for PCC flue gas should be required to be tolerant of SO_2, NO_x and high temperature, and to overcome slow absorption rate and small solvent capacity. Research direction of new solid absorbents requires high regeneration capacity (> 1000 cycles) and an increased amount of CO_2 absorbed through the carbonation reactions. MOFs can be closely integrated with hydrophobic polymers to produce block co-polymers, which prohibit the permeation of water, and have been paying particular attention on progress for CO_2 capture (D'Alessandro et al. 2010). MOFs capture CO_2 by physical adsorption except in cases where amines are incorporated into the structure or, possibly, when open metal coordination sites are generated. Serious advantages over fixed-bed adsorption methods are also expected for the application of metal–organic frameworks to gas separations if reliable methods can be developed for integrating these free-flowing powder materials into membranes (Aaron and Tsouris 2005). CO_2-sorbent interactions are critical for the design of better carbon-capture systems. The combination of appropriate pore size, strongly interacting amine functional groups, and the cooperative binding of CO_2 guest molecules are responsible for the low-pressure binding and large uptake of CO_2 in this sorbent material. MOFs functionalized with amine groups are current research direction and will be an important research area in the future (Vaidhyanathan et al. 2010). This prospect has prompted research on many amine-functionalized MOFs, and these studies have demonstrated that amines can enhance CO_2 uptake (An et al. 2010). Isostructural lanthanide coordination polymers with the empirical formula $[Ln_2(PDA)_3(H_2O)]\cdot 2H_2O$ surprisingly showed the adsorption to CO_2 (Pan et al. 2003). Selective metallic and ceramic membranes offer a promising new technology, and the conditions are the most easily attainable since the ideal operating pressure is atmospheric, and temperature can reach up to 350 °C (Dyer et al. 2000). In addition, due to the nature of the membranes, the energy required for operation is relatively small.

Secondly, with respect to new materials, the key scientific challenges are the development of molecular-level control as well as modern characterization and computational methods that will support, guide and provide further refinement to the most promising structures. Characterization of these new materials at the molecular level is essential. To accelerate the process, high-throughput characterization should be employed in cases where high-throughput materials synthesis is possible.

Thirdly, CCS will complement other crucial strategies, such as improving energy efficiency, switching to less carbon-intensive fuels such as natural gas and phasing in the use of renewable energy resources (e.g., solar energy, wind, and biomass).

Additionally, CO_2 scrubbing technology has been mainly used to capture CO_2 from flue gas in mobile/diffused point- and concentrated-sources (e.g., power plants, aircraft and home furnaces). Clearly, the low concentration of CO_2 in non-point sources (e.g., air, 0.04 %) presents a significantly higher thermodynamic barrier to capture compared with post-combustion technologies, while the expense of moving large volumes of air through an absorbing material presents a further challenge (Keith et al. 2006). The cost of CO_2 capture from low concentration point sources varies with a number of factors, but will certainly be higher than the recovery of CO_2 from point sources.

9.7 Conclusions

This chapter describes and discusses mechanism, materials, applications, cost information and current status of CO_2 scrubbing processes for CO_2 capture. Critical factors for the process are the properties of CO_2 scrubbing materials that can be categorized into liquid solvents and solid sorbents. Improved materials will give the maximum separation efficiency that will have the greatest potential for lowering the overall cost of capture systems in near-term.

Organic amines and aqueous ammonia are the majority of solvents that act as CO_2 scrubbers by chemisorptive formation of N-C bonded carbamate species. Developing new solvents are required to be resistant to degradation and not corrosive to the equipment and easier regeneration and faster loading.

Solid sorbents are promising alternatives, mainly including zeolites, carbon materials, silica, polymers, and MOFs. The interaction between CO_2 with sorbents includes chemisorption and physisorption. CO_2 adsorption on sorbents is strongly influenced by the temperature, pressure, and the presence of moisture. The CO_2 adsorption capacities of these physisorbents decrease significantly at high temperatures. The presence of water vapor in flue gas may negatively affect the capacity of sorbents and reduces the availability of the active surface area. In addition, other contaminants in flue gas, such as SO_x and NO_x, also have a detrimental impact on the CO_2 adsorption capacity. In contrast to sorbents based on physisorption, chemisorbents hold great potential for CO_2 capture from flue gas. Alkali carbonates such as sodium and potassium carbonates have reached pilot-scale trials with simulated and coal combustion flue gas. However, these sorbents-based systems still have challenges, such as high heat of reaction and long-term stability.

Ionic liquids and MOFs, new scrubber candidates for post-combustion CO_2 capture, are expected to have very high adsorption capacity but require substantial research efforts to be suitable under flue gas conditions. Research on functionalizing solid supports with amine functional groups for CO_2 capture has reached various stages of development, for example, amine-impregnated and grafted silica and amine-functionalized polymer sorbents.

9.8 References

Aaron, D., and Tsouris, C. (2005). "Separation of CO_2 from flue gas: A review." *Separation Science and Technology*, 40, 321–348.

Abanades, J.C., Rubin, E.S., and Anthony, E.J. (2004a). "Sorbent cost and performance in CO_2 capture systems." *Ind. Eng. Chem. Res.*, 43, 3462–3466.

Abanades, J.C., Anthony, E.J., Alvarez, D., Lu, D.Y., and Salvador, C. (2004b). "Capture of CO_2 from combustion gases in a fluidised bed of CaO." *AIChE J*, 50(7), 1614–1622.

Abu-Khader, M.M. (2006). "Recent progress in CO_2 capture/sequestration: A review." *Energy Sources*, Part A, 28, 1261–1279.

Alcerreca-Corte, I., Fregoso-Israel, E., and Pfeiffer, H. (2008). "CO_2 absorption on Na_2ZrO_3: A kinetic analysis of the chemisorption and diffusion processes." *J. Phys. Chem. C*, 112, 6520–6525.

An, J., Geib, S.J., and Rosi, N.L. (2010). "High and selective CO_2 uptake in a cobalt adeninate metal-organic framework exhibiting pyrimidine- and amino-decorated pores." *J. Am. Chem. Soc.*, 132(1), 38–39.

Aresta, M.A., and Dibenedetto, A. (2003). "New amines for the reversible absorption of carbon dioxide from gas mixtures." *Greenhouse Gas Control Technologies, Proceedings of the 6th International Conference on Greenhouse Gas Control Technologies (GHGT-6)*, 1–4 Oct. 2002, Kyoto, Japan, Gale, J., and Kaya, Y. (eds). Elsevier Science Ltd, Oxford, UK, 1599–1602.

Aroonwilas, A., Chakma, A., Tontiwachwuthikul, P., and Veawab, A. (2003). "Mathematical modeling of mass-transfer and hydrodynamics in CO_2 absorbers packed with structured packings." *Chem Eng Sci.*, 58, 4037–4053.

Arstad, B., Blom, R., and Swang, O. (2007). "CO_2 absorption in aqueous solutions of alkanolamines: Mechanistic insight from quantum chemical calculations." *J. Phys. Chem. A*, 111, 1222–1228.

Arstad, B., Fjellvåg, H., Kongshaug, K.O., Swang, O., and Blom, R. (2008). "Amine functionalised metal organic frameworks (MOFs) as adsorbents for carbon dioxide." *Adsorption*, 14, 755–762.

Astarita, G. (1983). *Gas treating with chemical solvents*, Wiley, New York.

Audus, H. (1997). "Greenhouse gas mitigation technology: An overview of the CO_2 capture and sequestration studies and further activities of the IEA greenhouse gas R&D programme." *Energy*, 22 (2/3), 217–221.

Bae, Y.S., and Lee, C.H. (2005). "Sorption kinetics of eight gases on a carbon molecular sieve at elevated pressure." *Carbon*, 43, 95–107.

Bai, H., and Yeh, A.C. (1997). "Removal of CO_2 greenhouse gas by ammonia scrubbing." *Ind. Eng. Chem. Res.*, 36, 2490–2493.

Bastin, L., Barcia, P.S., Hurtado, E.J., Silva, J.A.C., Rodrigues, A.E., and Chen, B. (2008). "A microporous metal−Organic Framework for Separation of CO2/N2 and CO_2/CH_4 by Fixed-Bed Adsorption" *J. Phys. Chem. C*, 112, 1575–1581.

Bates E.D., Mayton, R.D., Ntai, I., Davis. J.H. (2001). "CO_2 capture by a task-specific ionic liquid." *J. Am. Chem. Soc.*, 124(6) 926–927.

Belmabkhout, Y., and Sayari, A. (2009). "Effect of pore expansion and amine functionalization of mesoporous silica on CO_2 adsorption over a wide range of conditions." *Adsorption*, 15, 318–328.

Belmabkhout, Y., Weireld, G.D., and Sayari. A. (2009). "Amine-bearing mesoporous silica for CO_2 and H_2S removal from natural gas and biogas." *Langmuir*, 25(23), 13275–13278.

Beruto, D., Botter, R., and Searcy, A.W. (1987). "Thermodynamics of two, two-dimensional phases formed by carbon dioxide chemisorption on magnesium oxide." *J. Phys. Chem.*, 91(13), 3578–3581.

Blanchard, L.A., Hancu, D., Beckman, E.J., and Brennecke, J.F. (1999). "Green processing using ionic liquids and CO_2." *Nature*, 399, 28–31.

Blanchard, L.A., Gu, Z., and Brennecke, J.F.J. (2001). "High-pressure phase behaviour of ionic liquids/CO_2 systems." *J. Phys. Chem. B*, 105, 2437–2444.

Blauwhoff, P.M.M., Versteeg, G.F., van Swaaij, W.P.M. (1984). "A study on the reactions between CO_2 and alkanolamines in aqueous solutions." *Chem. Eng. Sci.*, 39(2), 207–225.

Bolland, O. (2004). "CO_2 capture technologies–an overview." In: *Proceedings of the second Trondheim conference on CO_2 capture, transport and storage*, 2004, 1–29.

Bonenfant, D., Mimeault, M., and Hausler, R. (2003). "Determination of the structural features of distinct amines important for the absorption of CO_2 and regeneration in aqueous solution." *Ind. Eng. Chem. Res.*, 42, 3179–3184.

Bottoms, R.R. (Girdler Corp.) (1930). "Separating acid gases," U.S. Patent 1783901, 1930.

Bourrelly, S., Llewellyn, P.L., Serre, C., Millange, F., Loiseau, T., and Fercy, G. (2005). *J. Am. Chem. Soc.*, 127, 13519.

Branan, C. (2002). *Rules of thumb for chemical engineers: A manual of quick, accurate solutions to everyday process engineering problems*, 3rd ed., Gulf Professional Pub., Amsterdam; New York.

Brooks, L.A., and Audrieth, L.F. (1946). "Ammonium carbamate." *Inorg. Synth.*, 2, 85–86.

Brooks, R. (1953). "Manufacture of *Ammonium Bicarbonate*." British Patent 742,386, 1953.

Cadena, C., Anthony, J.L., Shah, J.K., Morrow, T.I., Brennecke, J.F., Maginn, E.J. (2004). "Why is CO_2 so soluble in imidazolium-based ionic liquids?" *J. Am. Chem. Soc.* 2004, 126, 5300–5308.

Casarin, M., Falcomer, D., Glisenti, A., and Vittadini, A. (2003). "Experimental and theoretical study of the interaction of CO_2 with α-Al_2O_3." *Inorg. Chem.*, 42,436–445.

Chakraborty, A.K., Astarita, G., and Bischoff, K.B. (1986). "CO_2 absorption in aqueous solutions of hindered amines. *Chem. Eng. Sci.*, 41 (4), 997–1003.

Chakma, A., and Tontiwachwuthikul, P. (1999) "CO_2 separation from combustion gas streams by chemical reactive solvents." Online Library: Combustion Canada.

Chakma, A. (1995). "Separation of CO_2 and SO_2 from flue gas streams by liquid membranes." *Energy Convers. Manage.*, 36, 405–410.

Chandra, V. Yu, S.U., Kim, S.H., Yoon, Y.S., Kim, D.Y., Kwon, A.H., Meyyappan M., and Kim, K.S. (2012). "Highly selective CO_2 capture on N-doped carbon produced by chemical activation of polypyrrole functionalized graphene sheets." *Chem. Commun.*, 48, 735–737.

Chapel, D., Ernest, J., and Mariz, C. (1999) "Recovery of CO_2 from flue gases: commercial trends." In: Canadian Society of Chemical Engineers annual meeting, 1999, Saskatchewan, Canada, October 4–6, 1–16.

Choi, H.S., and Suh, M.P. (2009). "Highly selective CO_2 capture in flexible 3D coordination polymer networks." *Angewandte Chemie Int. Ed.*, 48, 6865–6869.

Choi, S, Drese, J.H., and Jones, C.W. (2009). "Adsorbent materials for carbon dioxide capture from large anthropogenic point sources." *ChemSusChem*, 2, 796–854.

Chue, K.T., Kim, J.N., Yoo, Y.J., Cho, S.H., and Yang, R.T. (1995). "Comparison of activated carbon and zeolite 13X for CO_2 recovery from flue gas by pressure swing adsorption." *Ind. Eng. Chem. Res.*, 34 (2), 591–598.

Collot Anne-Gaelle Collot (2003). *Draft-Prospects for hydrogen from coal*, IEA Coal Research, The Clean Coal Centre, UK, Aug. 2003.

Crooks, J.E., and Donnellan, J.P. (1990). "Kinetics of the reaction between carbon dioxide and tertiary amines." *J. Org. Chem.*, 55, 1372–1374.

Cullinane, J.T., and Rochelle, G.T. (2003). "Carbon dioxide absorption with aqueous potassium carbonate promoted by piperazine." In *Greenhouse Gas Control Technologies*, Vol. II, Gale, J., and Kaya, Y. (eds), 1603–1606 (Elsevier Science, UK).

Danckwerts, P.V. (1979). "The reaction of CO_2 with ethanolamines." *Chem. Eng. Sci.*, 34, 443–446.

Danckwerts, P.V., and Sharma, M.M. (1966). "Absorption of carbon dioxide into solutions of alkalis and amines." *Trans. Inst. Chem. Eng.*, 44, 244–277.

Danckwerts, P.V., and McNeil, K.M. (1966). "The absorption of carbon dioxide into aqueous amine solutions and the effects of catalysis." *Trans. Institut. Chem. Eng.*, 45, T32–T49.

Davison, J., and Thambimuthu, K. (2009). "An overview of technologies and costs of carbon dioxide capture in power generation." *Proc. Inst. Mech. Eng.*, Part A, 223, 201–212.

Dawson, R., Adams, D.J., and Cooper. A.I. (2011). "Chemical tuning of CO_2 sorption in robust nanoporous organic polymers." *Chem. Sci.*, 2, 1173–1177.

D'Alessandro, D.M., Smit, B., and Long, J.R. (2010). "Carbon dioxide capture: prospects for new materials." *Angew. Chem. Int. Ed.*, 49, 6058–6082.

DeMontigny, D., Tontiwachwuthikul, P., and Chakma, A. (2001). "Parametric studies of carbon dioxide absorption into highly concentrated monoethanolamine solutions." *Can. J. Chem. Eng.*, 79, 137–142

deMontigny, D., Tontiwachwuthikul, P., and Chakma, A. (2005). "Comparing the absorption performance of packed columns and membrane contactors." *Ind. Eng. Chem. Res.*, 44(15), 5726–5732.

Diaf, A., Garcia, J.L., and Beckman, E.J. (1994). "Thermally reversible polymeric sorbents for acid gases: CO_2, SO_2 and NO_x." *J. Appl. Polym. Sci.*, 53, 857–875.

Dibenedetto, A., Aresta, M., Fragale, C., and Narracci, M (2002). "Reaction of silyl-mono and diamines with carbon dioxide: Evidence of formation of inter- and intra-molecular ammonium carbamates and their conversion into organic carbamates of industrial interest by trans-esterification of carbonates under carbon dioxide catalysis." *Green Chem.*, 4, 439–443.

Dobrova, E.P., Bratchikova, I.G., and Mikhalenko, I.I (2006). "Adsorption of carbon fioxid on tantalum oxide coated with palladium chloride." *Russ. J. Phys. Chem.*, 80(9), 1528–1531.

Dodge, B.F. (1972). "Removal of impurities from gases to be processed at low temperatures." *Adv. Cryogen. Eng.*, 17, 37–55.

DOE (U.S. Department of Energy). (1999). *Carbon sequestration: research and development*, A U.S. Department of Energy Report; Office of Science, Office of Fossil Energy, U.S. Department of Energy, 1999.

Doskocil, E.J., Bordawekar, S.V., and Davis, R.J. (1997). "Catalysis by Solid Bases." *J. Catal.*, 169, 327-337.

Drage, T.C., Arenillas, A., Smith, K.M., Pevida, C., Piippo, S., and Snape, C.E. (2007). "Preparation of carbon dioxide adsorbents from the chemical activation of urea–formaldehyde and melamine–formaldehyde resins." *Fuel*, 86, 22–31.

Drese, J.H., Choi, S., Lively, R.P., Koros, W.J., Fauth, D.J., Gray, M.L., and Jones, C.W. (2009). "Synthesis–structure–property relationships for hyperbranched aminosilica CO_2 adsorbents." *Adv. Funct. Mater.*, 19, 3821–3832.

Duke, M.C., Ladewig, B., Smart, S., Rudolph, V., and Costa, J.C.D. (2010). "Assessment of postcombustion carbon capture technologies for power generation." *Front. Chem. Eng. China*, 4(2), 184–195.

DuPart, M.S., Bacon, T.R., and Edwards, D.J. (1993). "Understanding corrosion in alkanolamine gas treating plants." *Hydrocarbon Process*, 72, 89–94.

Dyer, P.N., Richards, R.E., Russeka, S.L., and Taylor, D.M. (2000). "Ion transport membrane technology for oxygen separation and syngas production." *Solid State Ionics*, 134, 21–33.

Eddaoudi, M., Kim, J., Rosi, N., Vodak, D., Wachter, J., O'Keeffe, M., Yaghi, O.M. (2002). "Systematic design of pore size and functionality in isoreticular MOFs and their application in methane storage." *Science*, 295, 469–472.

EPRI (Electric Power Research Institute). (2008). *Program on Technology Innovation: Post-combustion CO2 Capture Technology Development*, Electric Power Research Institute, Palo Alto, 2008.

Feron, P.H.M., and Jansen, A.E. (2002). "CO_2 separation with polyolefin membrane contactors and dedicated absorption liquids: Performances and prospects." *Separation and Purification Technology*, 27(3), 231–242.

Figueroa, J.D. Fout, T., Plasynski, S., McIlvried, H., and Srivastava, R.D. (2008). "Advances in CO_2 capture technology–The U.S. Department of Energy's carbon sequestration program, *International Journal of Greenhouse Gas Control*, 2 (2008) 9–20.

Freeman, S.A., Dugas, R., van Wagener, D.H., Nguyen, T., and Rochelle, G.T. (2010). "Carbon dioxide capture with concentrated, aqueous piperazine." *International Journal of Greenhouse Gas Control*, 4, 119–124.

Funk, S., Nurkic, T., Hokkanen, B., and Burghaus, U. (2007). "CO_2 adsorption on Cr(110) and Cr_2O_3(0001)/Cr(110)." *Appl. Surf. Sci.*, 253, 7108–7114.

Gibbins, J.R., and Crane, R.I. (2004). "Scope for reductions in the cost of CO_2 capture using flue gas scrubbing with amine solvents." *Proceedings of the Institution of Mechanical Engineers, Part A: Journal of Power and Energy,* 218, 231–239.

Grayson, M. (1978). *Kirk-Othmer Encyclopedia of Chemical Technology*, 3rd ed., John Wiley & Sons: New York, 1978, Vol. 2.

Grad, P. (2009). "Trials of carbon capture." *Engineers Australia*, 81, 45–47.

Harlick, P.J.E., and Sayari, A. (2006). "Applications of pore-expanded mesoporous silica. 3. triamine silane grafting for enhanced CO_2 adsorption." *Ind. Eng. Chem. Res.*, 45, 3248–3255.

Haszeldine, R.S. (2009). "Carbon *Capture and Storage*: How green can black be?" *Science*, 325, 1647–1652.

Hatch, T.F., and Pigford, R.L. (1962). "Simultaneous absorption of carbon dioxide and ammonia in water." *Ind. Eng. Chem. Fundam.*, 1(3), 209–214.

Herzog, H.J. (1999). "An introduction to CO_2 separation and capture technologies." Available at <http://sequestration.mit.edu/pdf/introduction_to_capture.pdf> (accessed Oct. 2011).

Hicks, J.C., Drese, J.H., Fauth, D.j., Gray, M.L., Qi, G., and Jones, C.W. (2008). "Designing adsorbents for CO_2 capture from flue gas-hyperbranched aminosilicas capable of capturing CO2 reversibly." *J. Am. Chem. Soc.*, 130, 2902–2903.

Hiyoshi, N., Yogo, D.K., and Yashima, T. (2004). "Adsorption of carbon dioxide on modified SBA-15 in the presence of water vapor." *Chem. Lett.*, 33, 510–511.

Hook, R.J. (1997). "An investigation of some sterically hindered amines as potential carbon dioxide scrubbing compounds." *Ind. Eng. Chem. Res.*, 36, 1779–1790.

Horiuchi, T., Hidaka, H., Fukui, T., Kubo, Y., Horio, M., Suzuki, K., and Mori, T. (1998). "Effect of added basic metal oxides on CO_2 adsorption on alumina at elevated temperatures." *Appl. Catal. A*, 167, 195–202.

Huang, H., and Chang, S.-G. (2002), "Method to regenerate ammonia for the capture of carbon dioxide." *Energy & Fuels*, 16, 904–910.

IEA (International Energy Agency). (2004). *Prospects for CO_2 Capture and Storage*, ISBN 92-64-10881-5.

IEA. (2005). *CO_2 capture and storage—R&D projects database.* In: *International energy agency greenhouse gas R&D programme*, Cheltenham UK, 2005.

IEA. (2007). *ERM—Carbon dioxide capture and storage in the clean development mechanism.* In: *International energy agency greenhouse gas R&D programme 2007/TR2*, Cheltenham UK, 2007, A10.

Iijima, M. (2004). "Flue gas CO_2 capture (CO_2 capture technology of KS-1)." In: Global Climate & Energy Project, Stanford USA, 2004, 1–28.

IPCC (The Intergovernmental Panel on Climate Change) (2005). *IPCC Special Report on Carbon Dioxide Capture and Storage*, Cambridge University Press, Cambridge, 2005.

Ismail, H.M., Cadenhead, D.A., Zaki, M.I. (1997). "Surface reactivity of iron oxide pigmentary powders toward atmospheric components: XPS, FESEM, and gravimetry of CO and CO_2 adsorption." *J. Colloid Interface Sci.*, 194, 482–488.

Jacobson, M.Z. (2009). "Review of solutions to global warming, air pollution, and energy security." *Energy Environ. Sci.*, 2, 148–173.

Jayaraman, A., Chiao, A.S., Padin, J., Yang, R.T., and Munson, C.L. (2002). "Kinetic separation of methane/carbon dioxide by molecular sieve carbons." *Sep. Sci. Technol.*, 37, 2505–2528.

Johnson, J. (2008). "Mixed impact of cutting CO_2." *Chem. Eng. News*, 86(12), 13.

Jou, F.Y., Mather, A.E., and Otto, F.D. (1995). "The solubility of CO_2 in a 30-mass-percent monoethanolamine solution." *Can. J. Chem. Eng.*, 73, 140–147.

Kaplan, L.J. (1982). "Cost-saving process recovers CO_2 from power-plant flue gas." *Chem Eng.*, 89(24), 30–31.

Karadas, F., Atilhan, M., Aparicio, S. (2010). "Review on the use of ionic liquids (ILs) as alternative fluids for CO_2 capture and natural gas sweetening," *Energy Fuels*, 24, 5817–5828.

Harlick, P.J.E., and Sayari, A. (2006). "Applications of pore-expanded mesoporous silica. 3. Triamine silane grafting for enhanced CO_2 adsorption." *Ind. Eng. Chem. Res.*, 45, 3248–3255.

Keith, D.W., Minh, H.-D., and Stolaroff, J.K. (2006). "Climate strategy with CO_2 capture from the air." *Clim. Change*, 74, 17–45.

Kim, B.J., Cho, K.S., and Park, S.J. (2010). "Copper oxide-decorated porous carbons for carbon dioxide adsorption behaviors." *J. Colloid Interface Sci.*, 342, 575–578.

Kittel, J., Idem, R., Gelowitz, D., Tontiwachwuthikul, P., Parrain, G., and Bonneau, A. (2009). "Corrosion in MEA units for CO_2 capture: Pilot plant studies." *Energy Procedia*, 1, 791–797.

Ko, D., Siriwardane, R., and Biegler, L.T. (2003). "Optimization of a pressure-swing adsorption process using zeolite 13X for CO_2 sequestration." *Ind. Eng. Chem. Res.*, 42(2), 339–348.

Koutinas, A.A., Yianoulis, P., and Lycourghiotis, A. (1983). "Industrial scale modelling of the thermochemical energy storage system based on $CO_2 + 2NH_3 \leftrightarrow NH_2COONH_4$ equilibrium." *Energy Convers. Manage.*, 23, 55–63.

Krishna, R. (2009). "Describing the diffusion of guest molecules inside porous structures." *J. Phys. Chem. C*, 113, 19756–19781.

Krishnankutty, N., and Vannice, M.A. (1995). "Effect of Pretreatment on Surface Area, Porosity, and Adsorption Properties of a Carbon Black." *Chem. Mater.*, 7, 754–763.

Kvamsdal, H., Mejdell, T., Steincke, F., Weydahl, T., Aspelund, A., Hoff, K.A., Skouras, S., and Barrio, M. (2005). *Tjeldbergodden power/methanol CO_2 reduction efforts, SP2 CO_2 capture and transport.* SINTEF Technical Report, Sintef Energy Research Trondheim, ISBN 82-594-2762-1.

Lee, S.C., Chae, H.J., Lee, S.J., Choi, B.Y., Yi, C.K., Lee, J.B., Ryu, C.K., and Kim, J.C. (2008a). "Development of regenerable MgO-based sorbent promoted with K_2CO_3 for CO_2 capture at low temperatures." *Environ. Sci. Technol.*, 42, 2736–2741.

Lee, S., Filburn, T.P., Gray, M., Park, J.W., and Song, H.J. (2008b) Screening test of solid amine sorbents for CO_2 capture" *Ind. Eng. Chem. Res., 47,* 7419–7423.

Lee, A.S., and Kitchin, J.R. (2012). "Chemical and molecular descriptors for the reactivity of amines with CO_2." *Ind. Eng. Chem. Res*., 2012, 51, 13609–13618.

Lepaumier, H., Martin, S., Picq, D., Delfort, B., and Carrette, P.-L. (2010). "New amines for CO_2 capture. III. Effect of alkyl chain length between amine functions on polyamines degradation." *Ind. Eng. Chem. Res.*, 49, 4553–4560.

Li, H.S., Zhong, S.H., Wang, J.W., and Xiao, X.F. (2001). "Effect of K_2O on adsorption and reaction of CO_2 and CH_3OH over Cu-Ni/ZrO_2-SiO_2 catalyst for synthesis of dimethyl carbonate." *Chin. J. Catal*., 22, 353–357.

Li, H., Eddaoudi, M., O'Keeffe, M., and Yaghi, O.M. (1999). Nature 1999, 402, 276.

Li, J.R., Kuppler, R.J., Zhou, H.C. (2009). "Selective gas adsorption and separation in metal–organic frameworks." *Chem. Soc. Rev.*, 38, 1477–1504.

López-Ortiz, A., Rivera, N.G.P., Rojas, A.R., and Gutierrez, D.L. (2004). "Novel carbon dioxide solid acceptors using sodium containing oxides." *Sep. Sci. Technol*., 39, 3559–3572.

Lu, C., Bai, H., Wu, B., Su, F., and Fen-Hwang, J. (2008). "Comparative study of CO_2 capture by carbon nanotubes, activated carbons, and zeolites." *Energy & Fuels*, 22, 3050–3056.

Maddox, R.N., Mains, G.J., and Rahman, M.A. (1987). "Reactions of Carbon Dioxide and Hydrogen Sulfide with Some Alkanolamines." *Ind. Eng. Chem. Res*., 26(1), 27–31.

Matranga, C., and Bockrath, B. (2005). "Hydrogen-bonded and physisorbed CO in single-walled carbon nanotubes bundles." *J. Phys. Chem. B*, 109, 4853–4864.

Maurin, G. Llewellyn, P.L., and Bell, R.G. (2005). "Adsorption mechanism of carbon dioxide in Faujasites: Grand canonical Monte Carlo simulations and microcalorimetry measurements" *J. Phys. Chem.*, 109, 16084–16091.

McDonald, T.M., D'Alessandro, D.M., Krishna, R., and Long, J.R. (2011). "Enhanced carbon dioxide capture upon incorporation of N,N′- dimethylethylenediamine in the metal–organic framework CuBTTri." *Chem. Sci.*, 2, 2022–2028.

MCCI (McKinsey Climate Change Initiative). (2008). *Carbon Capture & Storage: Assessing the Economics*, McKinsey & Company, 2008.

McLean, G.F., Niet, T., Prince-Richard, S., and Djilali, N. (2002). "An assessment of alkaline fuel cell technology." *International Journal of Hydrogen Energy*, 27(5), 507–526.

Melien, T. (2005). *CCP. Final cost estimation and economics common economic model team summary report*. In Thomas, D.C. (ed.). *Carbon Dioxice Capture for Storage in Deep Geologic Formations—Results from the CO_2 Capture Project,*

Capture and Separation of Carbon Dioxide from Combustion Sources. Volume 1, 47–87 (Elsevier).

Menéndez, A., Phillips, J., Xia, B., and Radovic, L.R. (1996). On the modification and characterization of chemical surface properties of activated carbon: In the search of carbons with stable basic properties." *Langmuir*, 12, 4404–4410.

MHIR (Mitsubishi Heavy Industries Research) (2001). Available at < http://www.mhi.co.jp/machine/recov_co2/index.htm> (accessed July 2001).

Millward, A.R., and Yaghi, O.M. (2005). "Metal-organic frameworks with exceptionally high capacity for storage of carbon dioxide at room temperature." *J. Am. Chem. Soc.*, 127, 17998–17999.

Mimura, T., Suda, T., Iwaki, I., Honda, A., and Kumazawa, H. (1998). "Kinetic of reaction between carbon dioxide and sterically hindered amines for carbon dioxide recovery from power plant flue gases." *Chem. Eng. Commun.*, 170, 245–260.

Mindrup, E.M., and Schneider, W.F. (2010). "Computational comparison of the reactions of substituted amines with CO_2." *ChemSusChem*, 3, 931–938.

Mosqueda, H.A., Vazquez, C., Bosch, P., and Pfeiffer, H. (2006). "Chemical sorption of carbon dioxide (CO_2) on lithium oxide (Li_2O)." *Chem. Mater.*, 18(9), 2307–2310.

Niswander, R.H., Edwards, D.J., DuPart, M.S., and Tse, J.P. (1993). "A more energy efficient product for carbon dioxide separation." *Sep. Sci. Technol.* 1993, 28 (1–3), 565–578.

Oliveira, E.L.G., Grande, C.A., and Rodrigues, A.E. (2008). "CO_2 sorption on hydrotalcite and alkali-modified (K and Cs) hydrotalcites at high temperatures." *Sep. Purif. Technol.*, 62, 137–147.

Pan, L., Adams, K.M., Hernandez, H.E., Wang, X.T., Zheng, C., Hattori, Y., Kaneko, K. (2003). "Porous lanthanide-organic frameworks: Synthesis, characterization, and unprecedented gas adsorption properties." *J. Am. Chem. Soc.*, 125, 3062–3067.

Pauley, C.R., Simiskey, P.L., Haigh. S. (1984). "N-ReN recovers CO_2 from flue gas economically." *Oil Gas J.*, 82(20), 87–92.

Pelkie, J.E., Concannon, P.J., Manley, D.B., and Poling, B.E. (1992). "Product distributions in the CO_2-NH_3-H_2O system from liquid conductivity measurements." *Ind. Eng. Chem. Res.*, 31, 2209–2215.

Perry, R.J., Grocela-Rocha, T.A., O'Brien, M.J., Genovese, S., Wood, B.R., Lewis, L.N., Lam, H., Soloveichik, G., Rubinsztajn, M., Kniajanski, S., Draper, S., Enick, R.M., Johnson, J.K., Xie, H-B., and Tapriyal, D. (2010). "Aminosilicone solvents for CO_2 capture." *ChemSusChem*, 3(8), 919–930.

Pevida, C., Plaza, M.G., Arias, B., Fermoso, J., Rubiera, F., and Pis, J.J. (2008). "Surface modification of activated carbon for CO_2 capture." *Appl. Sur. Sci.*, 254, 7165–7172.

Phan A., Doonan, C.J., Uribe-Romo, F.J., Knobler, C.B., O'Keeffe, M., Yaghi, O.M. (2010). "Synthesis, structure, and carbon dioxide capture properties of zeolitic imidazolate frameworks." *Acc. Chem. Res.*, 43(1), 58–67.

Philipp, R., and Fujimoto, K. (1992). "FTIR spectroscopic study of carbon dioxide adsorption/desorption on magnesia/calcium oxide catalysts" *J. Phys. Chem.*, 96, 9035–9038.

Plaza, M.G., Pevida, C., Arenillas, A., Rubiera, F., and Pis, J.J. (2007). "CO_2 capture by adsorption with nitrogen enriched carbons." *Fuel*, 86, 2204–2212.

Pohl, M., and Otto, A. (1998). "Adsorption and reaction of carbon dioxide on pure and alkali-metal promoted cold-deposited copper films." *Surface Science*, 406, 125–137.

Porcheron, F., Gibert, A. Jacquin, M., Mougin, P., Faraj, A., Goulon, A., Bouillon, P.-A., Delfort, B., Le Pennec, D., and Raynal, L. (2011). "High Throughput Screening of amine thermodynamic properties applied to post-combustion CO_2 capture process evaluation." *Energy Procedia,* 4, 15–22.

Puxty, G., Rowland, R., Allport, A., Yang, Q., Bown, M., and Burns, R. (2009). "Carbon dioxide post combustion capture: A novel screening study of the carbon dioxide absorption performance of 76 amines." *Environ. Sci. Technol.*, 43, 6427–6433.

Rabbani, R.G. (2012). "High CO_2 uptake and selectivity by triptycene-derived benzimidazole -linked polymers." *Chem. Commun.*, 48, 1141–1143.

Radosz, M., Hu, X., Krutkramelis, K., and Shen, Y. (2008). "Flue-gas carbon capture on carbonaceous sorbents: toward a low-cost multifunctional carbon filter for "green" energy producers." *Ind. Eng. Chem. Res.*, 47, 3783–3794.

Ram Reddy, M.K., Xu, Z.P., Lu, G.Q., and Diniz da Costa, J.C. (2006). "Layered double hydroxides for CO_2 capture: structure evolution and regeneration." *Ind. Eng. Chem. Res.*, 45(22), 7504–7509.

Rao, A.B. (2001). *Performance and Cost Models of an Amine-Based System for CO2 Capture and Sequestration.* Report to DOE/NETL, from Center for Energy and Environmental Studies, Carnegie Mellon University, Pittsburgh, PA, 2001.

Rao, A.B., and Rubin, E.S. (2002). "A technical, economic, and environmental assessment of amine-based CO_2 capture technology for power plant greenhouse gas control." *Environ. Sci. Tecnol.*, 36, 4467–4475.

Reddy, S., Scherffius, J., Freguia, S. and Roberts, C. (2003). "Fluor's econamine FG PlusSM technology–an enhanced amine-based CO_2 capture process." 2nd Annual Conference on Carbon Sequestration, Alexandria, VA, USA, 5–8 May,

Rochelle, G.T. (2009). "Amine scrubbing for CO_2 capture." *Science*, 325(5948), 1652–1654.

Romeo, L.M., Bolea, I., and Escosa, J.M. (2008). "Integration of power plant and amine scrubbing to reduce CO_2 capture costs." *Appl. Therm. Eng.*, 28, 1039–1046.

Rutherford, S.W., and Coons, J.E. (2003). "Adsorption dynamics of carbon dioxide in molecular sieving carbon." *Carbon*, 41, 405–411.

Samanta, A., Zhao, A., Shimizu, G.K.H., Sarkar, P., Gupta, R. (2012). "Post-combustion CO_2 capture using solid sorbents: a review." *Ind. Eng. Chem. Res.*, 51, 1438–1463.

Sanchez, L.M.G., Meindersma, G.W., and de Haan, A.B. (2007). "Solvent properties of functionalized ILs for CO_2 adsorption." *Chem. Eng. Res. Des.*, 85, 31–39.

Sartori, G., and Savage, D.W. (1983). "Sterically hindered amines for CO_2 removal from gases." *Ind. Eng. Chem. Fundam.*, 22, 239–249.

Satyapal, S., Filburn, T., Trela, J., and Strange, J. (2001). "Performance and Properties of a Solid Amine Sorbent for Carbon Dioxide Removal in Space Life Support Applications." *Energy Fuels*, 15, 250–255.

Schladt, M.J., Filburn, T.P., and Helble, J.J. (2007). "Supported Amine Sorbents under Temperature Swing Absorption for CO2 and Moisture Capture." *Ind. Eng. Chem. Res.*, 46, 1590–1597.

Serna-Guerrero, R., Da'na, E., and Sayari, A. (2008). "New insights into the interactions of CO_2 with amine-functionalized silica." *Ind. Eng. Chem. Res.*, 47, 9406–9412.

Shale, C.C., Simpson, D.G., and Lewis, P.S. (1971). "Removal of sulfur and nitrogen oxides from stack gases by ammonia." *Chem. Eng. Prog. Symp. Ser.*, 67, 52–58.

Shiflett, M.B., Drew, D.W., Cantini, R.A., and Yokozeki, A. (2010). "Carbon dioxide capture using ionic liquid 1-butyl-3-methylimidazolium acetate, *Energy Fuels*, 24, 5781–5789.

Simons, K., Nijmeijer, K., Mengers, H., Brilman, W., and Wessling, M. (2010). "Highly selective amino acid salt solutions as absorption liquid for CO_2 capture in gas-liquid membrane contactors." *ChemSusChem.*, 3(8), 939–947.

Sircar, S. (2006). "Basic research needs for design of adsorptive gas separation processes." *Ind. Eng. Chem. Res.*, 45, 5435–5448.

Siriwardane, R.V., Shen, M.S., Fisher, E.P., and Poston. J.A. (2001). "Adsorption of CO_2 on molecular sieves and activated carbon." *Energy & Fuels*, 15, 279–284.

Siriwardane, R.V., Shen, M.S., Fisher. E.P. (2005). "Adsorption of CO_2 on zeolites at moderate temperatures." *Energy Fuels*, 19(3), 1153–1159.

Smiglak, M., Metlen, A., and Rogers, R.D. (2007). "The second evolution of ionic liquids: from solvents and separations to advanced materials-energetic examples from the ionic liquid cookbook." *Acc. Chem. Res.*, 40, 1182–1192.

Søybe Grønvold M., Falk-Pedersen, O., Imai, N., and Ishida, K. (2005). "KPS membrane contactor modlule combined with Kansai/MHI advanced solvent, KS-1 for CO_2 separation from combustion flue gases, Carbon Dioxide Capture for Storage in Deep Geologic Formations-Results from the CO_2 Capture Project." *Capture and Separation of Carbon Dioxide from Combustion Sources*. Volume 1, 137–155 (Elsevier).

Steeneveldt, R., Berger, B., and Torp, T.A. (2006). CO_2 capture and storage: Closing the knowing–doing gap." *Chem. Eng. Res. Des.*, 84, 739–763.

Suda, T., Fujii, M., Yoshida, K., Iijima, M., Seto, T., and Mitsuoka, S. (1992) "Section 2. CO_2–recovery: Development of flue gas carbon dioxide recovery technology." *Energy Conversion and Management*, 33(5–8), 317–324.

Suh, M.P., Cheon, Y.E., and Lee, E.Y. (2008). "Syntheses and functions of porous metallosupramolecular networks." *Coord. Chem. Rev.*, 252, 1007–1026.

Tai, J.R., Ge, Q.F., Davis, R.J., and Neurock, M. (2004). "Adsorption of CO_2 on model surfaces of cesium oxides determined from first principles." *J. Phys. Chem. B*, 108, 16798–16804.

Tang, J., Tang, H., Sun, W., Plancher, H., Radosz, M., Shen, Y. (2005). "Poly(ionic liquid)s: a new material with enhanced and fast CO_2 absorption." *Chem. Commun.*, 26, 3325–3327.

Thallapally, P.K., Tian, J., Kishan, M.R., Fernandez, C.A., Dalgarno, S.J., McGrail, P.B., Warren, J.E., and Atwood, J.L. (2008). "Flexible (breathing) interpenetrated metal–organic frameworks for CO_2 separation applications." *J. Am. Chem. Soc.*, 130(50), 16842–16843.

Tolstoguzov, A.B., Bardi, U., and Chenakin, S.P. (2008). "Study of the corrosion of metal alloys interacting with an ionic liquid." *Bull. Russ. Acad. Sci. Phys.*, 72, 605–608.

Tsuda, T., Fujiwara, T., Taketani, Y. and Saegusa, T. (1992). "Amino silicagels acting as a carbon dioxide absorbent." *Chem. Lett.*, 2161–2164.

Tutuianu, M., Inderwildi, O.R., Bessler, W.G., and Warnatz, J. (2006). "Competitive adsorption of NO, NO_2, CO_2, and H_2O on BaO(100): A quantum chemical study." *J. Phys. Chem. B*, 110, 17484–17492.

US Patent (2012). "Aliphatic amine based nanocarbons for the absorption of carbon dioxide." Pub. No.: US 2012/0024153 A1 (Feb. 2, 2012).

Vaidya, P.D., and Kenig, E.Y. (2007). "CO_2-alkanolamine reaction kinetics: A review of recent studies." *Chem. Eng. Technol.*, 30, 1467–1474.

Vaidhyanathan, R. Iremonger, S.S., Shimizu, G.K.H., Boyd, P.G., Alavi, S., and Woo, T.K. (2010). "Direct observation and quantification of CO_2 binding within an amine-functionalized nanoporous solid." *Science*, 330, 650–653.

Wallace, J., and Krumdieck, S. (2005). "Carbon dioxide scrubbing from air for alkaline fuel cells using amine solution in a packed bubble column." *Proceedings of the Institution of Mechanical Engineers, Part C: Journal of Mechanical Engineering Science*, 219, 1225–1233.

Walton, K.S., Abney, M.B., and LeVan, M.D. (2006). "Adsorption of CO_2 in Y and X Zeolites modified by alkali metal cation exchange." *Microporous and Mesoporous Materials*, 91, 78–84.

Wang, Y., Lin, S., and Suzuki, Y. (2007). "Study of limestone calcination with CO_2 capture: Decomposition behavior in a CO_2 atmosphere." *Energy Fuels*, 21, 3317–3321.

Wang, B., Côté, A.P., Furukawa, H., O'Keeffe, M., Yaghi, O.M. (2008). "Colossal cages in zeolitic imidazolate frameworks as selective carbon dioxide reservoirs." *Nature*, 453, 207–211

Wei, J., Shi, J., Pan, H., Zhao, W., Ye, Q., and Shi, Y. (2008). "Adsorption of carbon dioxide on organically functionalized SBA-16." *Microporous Mesoporous Mater.*, 116, 394–399.

White, C.M., Strazisar, B.R., Granite, E.J., Hoffman, J.S., and Pennline, H.W. (2003). "Separation and capture of CO_2 from large stationary sources and sequestration in geological formations–coalbeds and deep saline aquifers." *Journal Air and Waste Management*, 53, 645–715.

Wilson, M.A., Wrubleski, R.M., and Yarborough, L. (1992). "Recovery of CO_2 from power plant flue gases using amines," *Energy Convers. Mgmt.* 33(5–8), 325–331.

Xu, X., Song, C., Miller, B.G., and Scaroni, A.W. (2005). "Adsorption separation of carbon dioxide from flue gas of natural gas-fired boiler by a novel nanoporous "molecular basket" adsorbent." *Fuel Processing Technology*, 86, 1457–1472.

Yang, R.T. (1997). *Gas Separation by Adsorption Processes*, Imperial College Press, London, 1997.

Yang, Z.Z., He, L.N., Gao, J., Liu, A.H., and Yu, B. (2012). "Carbon dioxide utilization with C–N bond formation: Carbon dioxide capture and subsequent conversion." *Energy Environ. Sci.*, 5, 6602–6639.

Ye, Q., Jiang, J., Wang, C., Liu, Y., Pan., H., and Shi, Y. (2012). "Adsorption of low-concentration carbon dioxide on amine-modified carbon nanotubes at ambient temperature." *Energy Fuels*, 26, 2497–2504.

Yeh, J.T., Pennline, H.W., and Resnik, K.P. (2001). "Study on CO_2 absorption and desorption in a packed column." *Energy Fuels*, 15, 274–278.

Yeh, J.T., Pennline, H.W., and Resnik, K.P. (2002). "Ammonia process for simultaneous reduction of CO_2, SO_2, and NO_x." Presentation given at the 19th Annual International Pittsburgh Coal Conference, Pittsburgh, PA, Sept. 23–27, 2002, paper 45–1.

Yeh, J.T., Resnik, K.P., Rygle, K.P., and Pennline, H.W. (2005). "Semi-batch absorption and regeneration studies for CO_2 capture by aqueous ammonia." *Fuel Process. Technol.*, 86, 1533–1546.

Yong, Z., Mata, M., and Rodrigues, A.E. (2000). "Adsorption of carbon dioxide on basic alumina at high temperatures." *J. Chem. Eng. Data*, 45, 1093–1095.

Yong, Z., and Rodrigues, A.E. (2002). "Hydrotalcite-like compounds as adsorbents for carbon dioxide." *Energy Convers. Manage.*, 43, 1865–1876.

Yong, Z., Mata, M., Rodrigues, A.E. (2002). "Adsorption of carbon dioxide at high temperature–a review." *Separation and Purification Technology*, 26(2–3), 195–205.

You, J.K., Park, H., and Yang, S.H. (2008). "Influence of additives including amine and hydroxyl groups on aqueous ammonia absorbent for CO_2 capture." *J. Phys. Chem. B*, 112, 4323–4328.

Yu, K.M.K., Curcic, I., Gabriel, J., and Tsang, S.C.E. (2008). "Recent advances in CO_2 capture and utilization." *ChemSusChem*, 1, 893–899.

Zhang, S.J., Zhang, X.P., Zhao, Y.S., Zhao, G.Y., Yao, X.Q., and Yao, H.W. (2010). "A novel ionic liquids-based scrubbing process for efficient CO_2 capture." *Sci. China Chem.*, 53(7), 1549–1553.

Zheng, X.Y., Diao, Y.F., He, B.S., Chen, C.H., Xu, X.C., and Feng, W. (2003). "Carbon dioxide recovery from flue gases by ammonia scrubbing." *Greenhouse Gas Control Technologies, Proc. of the 6th International Conference on Greenhouse Gas Control Technologies (GHGT-6)*, 1–4 Oct. 2002, Kyoto, Japan, Gale, J., and Kaya, Y. (eds.). Elsevier Science Ltd, Oxford, UK, 193–200.

Zhou, S., Chen, X., Nguyen, T., Voice, A.K., and Rochelle, G.T. (2010). "Aqueous ethylenediamine for CO_2 capture." *ChemSusChem*, 3(8), 913–918.

Zukal, A., Dominguez, I., Mayerová, J., and Cejka, J. (2009). "Functionalization of delaminated zeolite ITQ-6 for the adsorption of carbon dioxide." *Langmuir*, 25, 10314–10321.

Zulfiqar, S., Karadas, F., Park, J., Deniz, E., Stucky, G.D., Jung, Y., Atilhan, M., and Yavuz, C.T. (2011). "Amidoximes: promising candidates for CO_2 capture." *Energy Environ. Sci.*, 4, 4528–4531.

CHAPTER 10

Carbon Sequestration via Mineral Carbonation: Overview and Assessment

Mausam Verma, Satinder K. Brar, R. D. Tyagi and Rao Y. Surampalli

10.1 Introduction

Carbon dioxide (CO_2) has never been considered a pollutant as it is already present in the atmosphere for such a long time. However, continuous industrialization and population increase has put tremendous pressure on the environment, and the concentration of CO_2 in the earth's atmosphere has been increasing linked to global warming which has further caused frantic measures to be adopted for reduction of CO_2 emissions. Fossil-fuel-fired power plants account for approximately one third of the total CO_2 emissions. As burning carbon must yield CO_2, it is impossible to modify fossil fuel-fired power plants to stop producing CO_2. Therefore, if CO_2 emissions from fossil fuel-fired power plants are to be reduced, the CO_2 produced must be captured and stored.

Proposed methods to sequester CO_2 include terrestrial and ocean sequestration, geological sequestration, and mineral carbonation (Zhang and Surampalli 2013). Geological sequestration or underground storage is one of the proposed solutions for mitigating CO_2 emissions. The problem is that in many instances it is simply not feasible, geographically or economically, to bury CO_2 underground due to the subsequent long term environmental, human risks and their associated costs.

More than 8,100 power generation and industrial facilities in the world each emit more than 100,000 tons of CO_2 to the atmosphere each year. The sheer size of the potential market and its geographic scope is enough to bring out the importance of potential for carbon sequestration technologies to contribute to climate change mitigation. By speeding up natural processes, scientists propose that rocks from the Earth's deep interior, exposed on the surface by plate tectonics and erosion, may be able to capture and store billions of tons of CO_2 per year. When compared to the total human output of CO_2 to the atmosphere, currently about 30 billion tons of CO_2 per year is being produced, this could make a significant difference in the overall CO_2 budget of the planet until alternative energy sources replace global fossil fuel use. Due to the increasing concern about CO_2-driven global warming and ocean acidification, CO_2 capture and storage ideas are receiving increasing attention.

Since 1990, scientists and engineers considered using naturally occurring minerals that react with CO_2 to form carbonate minerals, as a means to capture CO_2 from the atmosphere and store it in solid form where it will remain stable for thousands or millions of years (Seifritz 1990). In fact, looking at the global spread of different industrial facilities that contribute to CO_2 emissions, there is certainly a large potential for carbon sequestration by mineralization as can be seen from Fig. 10.1. The mineral olivine (Mg_2SiO_4) has been the focus of the most research. Olivine forms the carbonate mineral magnesite via reactions such as:

$$Mg_2SiO_4 + 2CO_2 \Leftrightarrow 2MgCO_3 \text{ (magnesite)} + SiO_2 \text{ (quartz)} \quad (10.1)$$

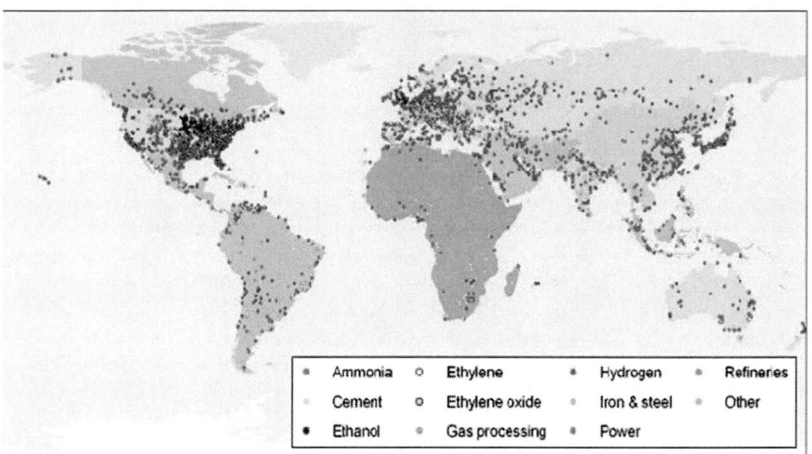

Figure 10.1. Different industrial facilities over the world responsible for carbon dioxide emissions representing the potential for carbon sequestration by mineralization (Dooley et al. 2006)

Olivine is abundant, has more magnesium than silicon, and reacts readily as it is far from equilibrium with the atmosphere and surface waters. Olivine forms 60 to 80% of the Earth's upper mantle, which in turn comprises about 1/4 of the Earth. Collisions of tectonic plates cause mantle rocks to be thrust onto the continents, where they are later exposed by erosion. Initially, engineers focused on using olivine to capture and store CO_2 via "ex-situ" methods. Olivine-rich rocks were to be quarried, transported to power plants, ground to a fine powder, and mixed with water plus purified CO_2 in reaction vessels at high pressure and temperature. These methods for olivine carbonation "at the smokestack" have proved to be relatively expensive, in both financial and energy terms, though engineers continue to seek methods to improve their efficiency (Mazzotti et al. 2005). Meanwhile, most proposed CO_2 capture and storage methods rely on injection of pure CO_2 in dense, supercritical liquid form into pore space in sub-surface rocks, or even into "puddles" of dense CO_2 on the seafloor.

There is an incentive to look for alternatives, since storage of CO_2 in inert, solid carbonate minerals could be safer, and easier to monitor and verify, than storage of CO_2 liquid in pore space. Mineral carbonation can be compared with other carbon sequestration technologies as given in Table 10.1. Some studies have focused on "in-situ" mineral carbonation, leaving rocks in the ground and using methods such as injection of CO_2 to increase reaction rates. Much of this work has been on carbonation of an abundant type of lava, basalt, which contains minor amounts of olivine together with large proportions of alumino-silicate minerals such as plagioclase feldspar. Basalt carbonation is slower than carbonation of olivine, but basalt is abundant near many large power plants (Goldberg et al. 2008). The idea is to pump CO_2 into pore space in basaltic lava, with the expectation that over decades the CO_2 will combine with olivine and feldspar to produce solid carbonate minerals.

Carbon dioxide may also be permanently stored in solid form via mineral carbonation. In this process, CO_2 is combined with the alkaline metal ions as silicate minerals to form thermodynamically stable carbonate minerals. Such carbonation reactions occur naturally on geological time scales, but the process can be greatly accelerated through the use of carbonation reactors. Several mineral types have been identified as candidates for use in an industrial-scale mineral carbonation process. The process has already been successfully demonstrated with wollastonite, a highly reactive calcium silicate mineral as presented in Fig. 10.2.

In a mineral carbonation strategy, alkaline earth metals react with CO_2 to form relatively stable and benign carbonate minerals as follows:

$$(Ca, Mg)SiO_3 + 2CO_2 + 3H_2O \rightarrow (Ca, Mg)CO_3 + H_4SiO_4 + H_2O + CO_2 \qquad (10.2)$$

Natural geologic processes of silicate weathering contribute to carbonation; however, the reaction rates are slow (on geologic time scales) and economic feasibility of wide-spread application is not yet fully known. The weathering process has been mimicked and catalyzed in the laboratory due to the stability of the end-products (i.e., carbonates). The majority of mineral carbonation research to date have examined sequestration in mined silicate minerals (e.g., serpentine, olivine) (Lackner et al. 1995; Fauth et al. 2000; 2002; Chen et al. 2006; Gerdemann et al. 2007). Mining operations and subsequent physical and chemical processing are required to produce a mineral form suitable for sequestration reactions. Both the mineral acquisition and pre-processing steps require energy inputs, reducing the overall efficiency of the process in terms of net carbon reduction, thus encompassing both advantages and disadvantages as given in Table 10.2.

Table 10.1. Comparison of mineral carbonation with other CO_2 sequestration options

Comparison parameter	**Geological storage**	**Methane hydrates**	**Mineral carbonation**
Potential quantities	High, 100 % of captured CO_2 emissions of a power plant	Not known, estimation can be high	High, 100 % of captured CO_2 emissions of a power plant
Additional energy requirement	Additional energy required for CO_2 transport and injection	Unknown	Currently, 30% of the power plant output
Storage time	Depending on storage; approx. 1000 a	Dependent on oceanographic conditions, approx. 1000 a	Mineral storage, geological time scale
State of technological development	Pilot trials; industrial scale 2020 at the earliest	Research and development stage	Research stage
Value addition	None	Methane	None
Time horizon	Medium-term	Long-term	Long-term
Public acceptance	Problems foreseen for underground storage in densely populated areas	Neutral if far away from the coasts	Neutral, harmless storage products
Research needs	Collection and independent validation of geological criteria	-Geological exploration of storage sites - Development of extraction process - Development of efficient gas-to-liquid technology	-Optimization of process parameters - Process evaluation , including comprehensive energy balance - Proof-of-concept on pilot scale using fly ash

However, more readily available oxide mineral sources may be available through the reuse of industrial solid wastes and residues as shall be discussed in subsequent sections in the chapter.

10.2 Choice of Minerals

Alkali and alkaline earth metals can be carbonated. Of the alkaline earth metals, calcium and magnesium are by far the most common in nature. Magnesium and calcium comprise ~2.0 and 2.1 mol% of the earth's crust. Thus, calcium and magnesium are generally selected for mineral CO_2 sequestration purposes. Although carbonation of calcium is easier, for mineral carbonation the use of magnesium-based minerals is favoured, as they are available worldwide in large amounts and in relatively high purity. Furthermore, the amount of oxide required to bind carbon dioxide from burning one ton of carbon also favours magnesium oxide at 3.3 ton compared to 4.7 ton calcium oxide. Therefore, most attention is paid to magnesium-containing minerals. Of the non-alkali and non-alkaline earth metals, only a few metals can be carbonated (e.g. Mn, Fe, Co, Ni, Cu and Zn). However, most of these elements are too rare or too valuable and would incur high costs during mineral carbonation.

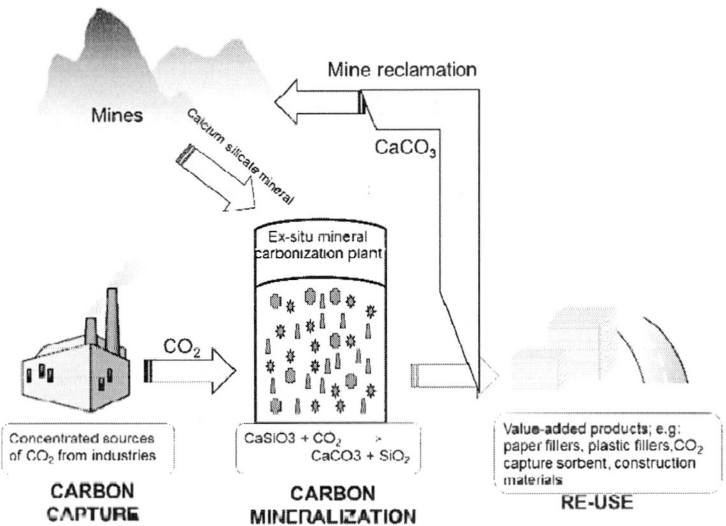

Figure 10.2. Ex-situ carbon mineralization using a calcium silicate mineral (wollastonite) (adapted from Kakizawa et al. 2001)

Table 10.2. Advantages and disadvantages of carbon mineralization

Advantages	Disadvantages
Well understood and can be applied at small scales	The technology is actually at an early phase of development for sequestering large amounts of captured CO_2
Simple process	A large-scale mineral carbonation process needs a large mining operation to provide the reactant minerals in sufficient quantity
Good ex-situ process	Estimating the actual amount of CO_2 that could be sequestered by this technique is difficult
	Often recycling of mineralization agents is not possible

The key parameter in mineral carbonation is the carbonation potential. Initially, Goff et al. (2000) used the molar concentration of Mg in a serpentine sample to calculate the theoretical number of moles of CO_2 that could be converted to magnesite ($MgCO_3$) by reaction with the serpentine. Later on, a modification to this method was carried out by Penner et al. (2003a), where the carbonation potential, RCO_2, was calculated from the total molar concentration of Ca, Fe^{2+}, and Mg in the feed, and was defined as the mass ratio of rock or mineral necessary to convert a unit mass of CO_2 to the solid carbonate (Eq. 10.3). By this definition, a low RCO_2 is preferable to a high RCO_2.

$$R_{CO_2} = \frac{100}{(\sum Ca^{2+} + Fe^{2+} + Mg^{2+})MW_{CO_2}} \tag{10.3}$$

where: $\sum Ca^{2+} + Fe^{2+} + Mg^{2+}$ = the sum of the molar concentrations for the specified cations; MW_{CO2} = molecular weight of CO_2. While the R_{CO2} is an inherent property of a rock or mineral, based strictly on its chemical composition, the carbonation reactivity of that rock or mineral is dependent on numerous factors, including the mineral composition, pre-treatment, and solubility at the specific carbonation conditions of time, temperature, and pressure.

In order to be able to react with acid CO_2, the mineral has to provide alkalinity. This alkalinity is derived from oxides or hydroxide. Although it is easier to convert carbonates into bicarbonates than to carbonate a silicate mineral (Lackner 2002), oxides and hydroxides are preferred. Controlled storage is only possible for carbonates, as carbonates are almost insoluble in water while bicarbonates are fairly soluble. Calcium and magnesium do not commonly occur as binary oxides in nature. They are typically found in silicate minerals. These minerals are capable of being carbonated as carbonic acid is a stronger acid than silicic acid (H_4SiO_4). Thus, silica present in the mineral is exchanged with carbonate and the mineral is carbonated. Mostly, igneous rocks are particularly suitable for CO2 fixation as they are essentially free of carbonates. The main candidate magnesium-rich ultramafic rocks are dunites, peridotites and serpentinites. The first two can be mined for olivine, a solid solution of forsterite (Mg_2SiO_4) and fayalite (Fe_2SiO_4). Ore grade olivine may contain alteration products, such as serpentine ($Mg_3Si_2O_5(OH)_4$) and talc ($Mg_3Si_4O_{10}(OH)_2$). Serpentine can take the form of antigorite, lizardite and chrysotile. The main calcium-containing candidate is wollastonite ($CaSiO_3$). Serpentine is found in large deposits as large reservoirs worldwide, for example, on both the East and West Coast of North America and in Scandinavia. The worldwide resources that can actually be mined are, however, unknown (Lackner et al. 2000).

An alternative source of alkalinity could be the use of solid alkaline waste materials, which are available in large amounts and are generally rich in calcium. Possible candidates, among others, could be asbestos waste, iron and steel slag and coal fly ash (NETL 2001). The carbonation of alkaline waste materials has two potential advantages: these materials constitute an inexpensive source of mineral matter for the sequestration of CO_2, and the environmental quality of the waste materials (i.e. the leaching of contaminants) can be improved by the resulting pH-neutralization and mineral nouveau formation.

10.3 Process Thermodynamics

Carbonate is the lowest energy-state of carbon. Different energy states of carbon are presented in Fig. 10.3. The carbonation reaction with gaseous CO_2 proceeds very slowly at room temperature and pressure. Increasing the temperature increases the reaction rate. However, due to the entropy effects, the chemical equilibrium favours gaseous CO_2 over solid-bound CO_2 at high temperatures (calcination reaction). The highest temperature at which the carbonation occurs spontaneously thus, depends on the CO_2 pressure and the type of mineral.

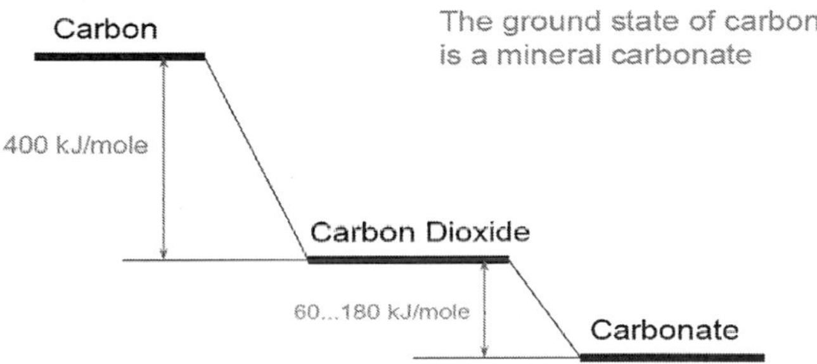

Figure 10.3. Different energy states of carbon

10.4 Pre-treatment

Activation of the mineral reactants has been achieved by both thermal and mechanical means, although the mechanism for this activation is not clearly understood. Most of the mineral dissolution reactions are surface controlled. Thus, the two pretreatment methods proved successful primarily due to increased surface area. Mechanical pretreatment reduces the mean particle size of the minerals, while thermal pretreatment removes chemically-bound water, which may increase the porosity and the resulting surface area. It is likely that both phenomena are responsible to some degree for the improvements in mineral reactivity achieved by pretreatment. However, the energy necessary to achieve such activation is most critical to the viability of any mineral carbonation process.

Mechanical activation was investigated by use of conventional rod and ball milling techniques, as well as ultra-fine grinding using a scalable stirred-media detritor (SMD) mill (Penner et al. 2003b). The carbonation test conditions varied by mineral, using the best demonstrated reaction parameters: 1 hour duration (all minerals); 185°C (olivine), 155°C (serpentines), 200°C (wollastonite); PCO_2 of 150 atm (olivine and serpentines), 40 atm (wollastonite); carrier solution of 1 M NaCl, 0.64 M $NaHCO_3$ (olivine and serpentines), distilled water (wollastonite). While olivine showed a nearly linear relationship between mechanical energy input and mineral reactivity, wollastonite activation peaked at a much lower energy input, with no gain at higher energies. In contrast, both serpentine minerals showed virtually no increase in mineral reactivity at energies up to nearly 400 kW•h/ton.

Thermal activation of the hydrated Mg-silicate species is mostly accomplished by the addition of a heat treatment stage in the mineral pretreatment process. Zhang et al. (1997) reported the enhancement of acid extraction of Mg and Si from serpentine by mechanochemical treatment, although the 6-hour grinding times utilized would likely make the methodology extremely energy intensive. McKelvy et al. (2001) described a meta-stable serpentine phase that forms in roughly the same temperature range, and suggested that heating above 800°C was undesirable, as this leads to a phase transformation to the non-hydrated Mg-silicate phases, forsterite and enstatite. This phase transformation is marked by an exotherm at just over 800°C. Mostly, pre-treatment options are avoided in order to save energy costs.

10.5 Carbonation Processes

Main carbonation processes via mineralization can be divided into several categories as illustrated in Fig. 10.4.

Direct Carbonation. It is the simplest approach where a suitable feedstock, e.g., serpentine or a Ca/Mg rich solid residue is carbonated in a single process step. For an aqueous process, both the extraction of metals from the feedstock and the subsequent reaction with the dissolved carbon dioxide to form carbonates takes place in the same reactor.

Direct Gas-Solid Carbonation. The particulate metal oxides are brought into contact with gaseous CO_2 at a particular temperature and pressure (for various temperature and pressure ranges applied). The dry process has the potential of producing high temperature steam or electricity while converting CO_2 into carbonates. Process integration with mining activities may be very advantageous from an economic point of view of the cost and energy, possibly allowing for, *e.g.*, higher valuable metal extraction rates (Kohlmann et al. 2002). The reaction rates of such a process have been too slow and the process suffers from thermodynamic limitations limiting further studies in this direction. Meanwhile, recent developments suggested that indirect/multi-step gas-solid carbonation routes hold a promising future (Zevenhoven et al. 2004; 2006a, 2006b, 2006c).

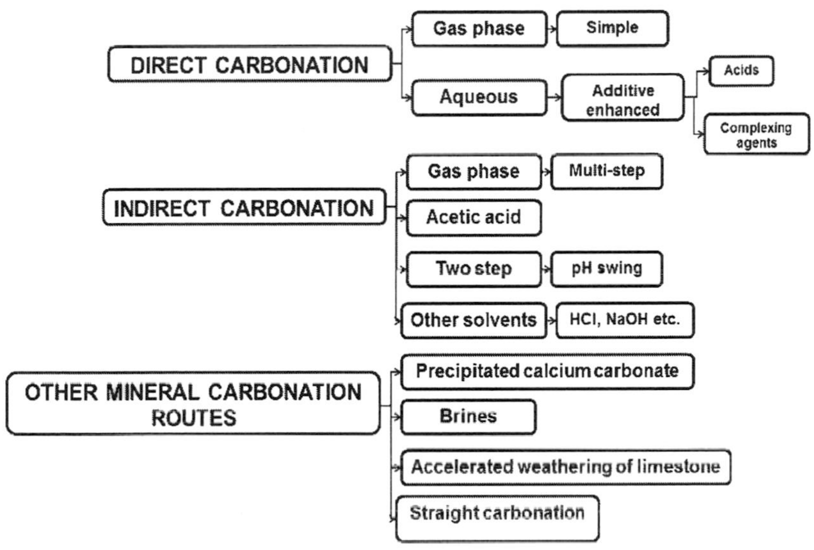

Figure 10.4. Main carbonation processes via mineralization

Industrial solid residues have been used as feedstock for carbonation instead of minerals as given for a typical cement plant in Fig. 10.5. The possibility of simultaneously binding CO_2 and lowering the hazardous nature of e.g. municipal solid waste incinerator ash makes this carbonation route interesting (Rendek et al. 2006). However, the potential CO_2 storage capacity for this option is limited as the quantity of material that may be carbonated is too small (Huijgen and Comans 2007; Rendek et al. 2006).

The direct gas-solid carbonation of minerals has remained unviable for industrial purposes and research has moved on to investigate indirect or multi-step gas-solid carbonation options.

Direct Aqueous Carbonation. Direct aqueous mineral carbonation can be further divided into two-step and three-step alternatives, depending on the type of solution used. Studies focusing on carbonation in pure aqueous solutions have made way for additive-enhanced carbonation experiments consisting of 0.64 M $NaHCO_3$ + 1.00 M NaCl (O'Connor et al. 2000). It is extremely important to recycle the additives added in carbonation processes due to large quantity of these additives being added during the process (Huijgen and Comans 2005). Some studies have been conducted on recycling of spent additive (e.g. NH_4Cl) but without any major breakthroughs.

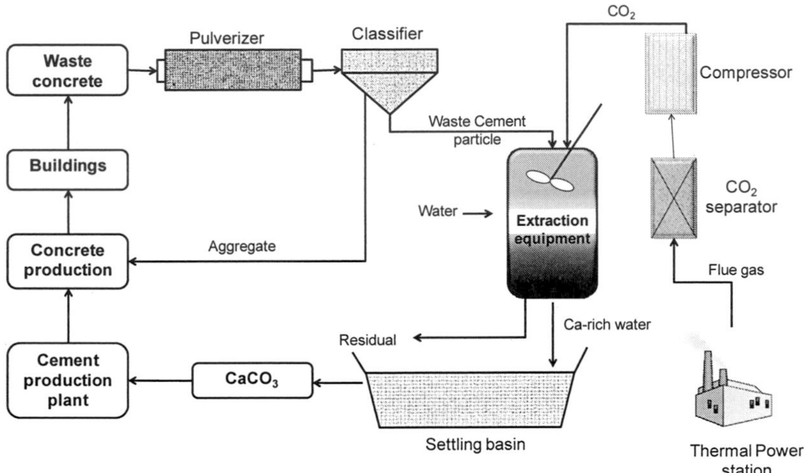

Figure 10.5. A typical carbon sequestration process in a cement plant by mineralization (based on Rendek et al. 2006)

Recently, in-situ carbonation has also been used as an add-in technology for the aqueous carbonation to overcome the problem of recycling of the added chemicals (Hansen et al. 2005; McKelvy et al. 2006; Soignard et al. 2006; Chizmeshya et al. 2007). It was found that minimizing porosity loss and maximizing permeability were beneficial for carbonation, which could be achieved using a CO_2-rich aqueous fluid, i.e. by controlling the input gas composition. Other studies providing tools for improving mineral carbonation have been provided by Hänchen et al. (2006; 2007a,b; 2008) who have developed models for both olivine dissolution and precipitation.

Indirect Carbonation. In indirect carbonation processes, mineral carbonation occurs in several steps. In this case, the reactive component (usually Mg or Ca) is first extracted from the feedstock (as oxide or hydroxide) in one step and then, in another step, it is reacted with carbon dioxide to form the desired carbonates. It has been found that the carbonation of MgO is significantly slower than the carbonation of $Mg(OH)_2$ (Butt et al. 1996). Zevenhoven et al. (2004; 2006b) suggested that the direct gas-solid carbonation process for $Mg(OH)_2$ production from serpentine (due to thermodynamic limitations) should be divided into three-steps: 1) MgO production (Equation 10.4) in an atmospheric reactor; followed by 2) MgO hydration (Equation 10.5); and 3) carbonation (Equation 10.6) at elevated pressures according to the following reactions:

$$Mg_3\,Si_2\,O_5\,(OH)_4\,(s) \rightarrow 3MgO(s) + 2SiO_2\,(s) + 2H_2O \tag{10.4}$$

$$MgO(s) + H_2O \leftrightarrow Mg(OH)_2\,(s) \tag{10.5}$$

$$Mg(OH)_2\,(s) + CO_2 \leftrightarrow MgCO_3\,(s) + H_2O \tag{10.6}$$

Experiments performed in fluidized bed reactors and preliminary tests have shown that product/carbonate removal is possible due to particle collisions (attrition, abrasion) and the following weight reduction, allowing the flue gas to carry the particles out of the reactor. However, the experimental results have been achieved using pure (97%) $Mg(OH)_2$. Thus, more detailed study needs to be carried out at the industrial scale to prove the feasibility of the process.

The use of acetic acid for the extraction of calcium from a calcium-rich feedstock has been also investigated in order to speed up the aqueous carbonation process (Kakizawa et al., 2001). The overall process consists of two-steps as given in Equations 10.7 and 10.8:

$$CaSiO_3 + 2CH_3COOH \rightarrow Ca^{2+} + 2CH_3\,COO^- + H_2O + SiO_2 \tag{10.7}$$

$$Ca^{2+} + 2CH_3COO^- + CO_2 + H_2O \rightarrow CaCO_3 + 2CH_3COOH \tag{10.8}$$

Equation 10.7 describes the extraction step and Equation 10.8 the precipitation step. The acetic acid used in the extraction step could be recovered in the following precipitation step.

A problem mostly encountered during the acetic acid route is that other elements, such as heavy metals, may also leach out during the Ca-extraction phase, leading to impure carbonate precipitate (Teir et al. 2007a, 2007b). Another problem with this route has been the need for an additive, NaOH, in order to precipitate carbonates. This additional chemical makes recycling of acetic acid impossible and, as a result, the chemical costs for this process route are too high for any large scale application.

The pH-swing process developed in Japan (and later also presented in a patent by Yogo et al. 2005) is another two-step aqueous carbonation process where at first the pH of the solution is lowered thereby enhancing the extraction of divalent metal ions. In the second step the pH is raised to enhance the precipitation of carbonates. The principal reactions taking place inside the extractor (Equation 10.9) and the precipitator (Equation 10.10) are:

$$4NH_4Cl + 2CaO{\cdot}SiO_2 \rightarrow 2CaCl_2 + NH_3 + 2H_2O \tag{10.9}$$

$$4NH_3 + 2CO_2 + 2H_2O + 2CaCl_2 \rightarrow 2CaCO_3 + 4NH_4Cl \tag{10.10}$$

Equation 10.10 consists of both CO_2 absorption and $CaCO_3$ precipitation. Various other acids and bases were tested (HCl, H_2SO_4, HNO_3, HCOOH, CH_3COOH, NaOH, KOH, NH_3, NH_4Cl, $(NH_4)_2SO_4$, and NH_4NO_3), but the basic solutions were not effective at extracting magnesium (Teir et al. 2007a). For the precipitation experiments conducted by Teir et al. (2007c), Mg extraction was achieved using solutions of HCl and HNO_3. In the extraction experiments, sulfuric acid was found to be the best extraction agent of all the chemicals tested, but none of the acids were able to extract Mg selectively. For this selectivity, the ammonium salts performed better and no Fe or Si could be measured in the solution after 1 h. However, the amount of Mg extracted remained low, only 0.3–0.5%. The effect of the particle size did not influence the Mg extraction rate significantly in the range of 125–500 μm and in 2 h and 70 °C all mineral acids (2 M HCl, H_2SO_4, HNO_3) were able to extract 100% Mg from the serpentine sample.

Although there have been several studies in literature which have demonstrated the possibility of use of different acids to extract calcium and magnesium by speeding up the reaction rate, yet the studies are still at its early stage. Moreover, the problem of effectively recycling the extraction agent remains unsolved and more research is warranted before this route can be considered feasible for long-term CO_2 storage.

Other Processes. The production of valuable products (e.g. precipitated calcium carbonate, PCC) by utilizing CO_2 has been the objective of many studies in recent years (Teir et al. 2005; Katsuyama et al. 2005; Domingo et al. 2006; Feng et al. 2007). Direct aqueous carbonation using additives (terpineol 0.1 and 1 vol-%, EDTA 0.25 and 1%-wt) in order to determine their effect on the precipitated calcium carbonate has been studied by Feng et al. (2007). Other variables included in the study were CO_2 bubble size (with frit pore size of 17–40 or 101–160 μm), CO_2 gas flow rate (3.5 and 4 l/min), CO_2 concentration (25 and 100 vol-%) and reaction temperature (25 and 80 °C). It appeared that the size of the carbonated particles was slightly smaller with smaller CO_2 bubble size and CO_2 concentration. The effect of the CO_2 gas flow rate and temperature was altered by the addition of additives, but in general the process was quicker at 80 °C than 25 °C.

Alternatively, use of brine solution is demonstrated in Fig. 10.6. Brine is a saline-based solution that is formed as a waste product during oil or natural gas extraction and as such it can be found stored in vast quantities in above-ground storage tanks. The large amount and relatively high concentration of metals capable of forming carbonates (mainly Ca and Mg) provides a carbonation process option for carbon dioxide storage. However, despite the fact that brine is capable of forming carbonates, an industrial scale operation is currently limited by slow reaction kinetics. Raising the pH of the brine speeds up the carbonation process, but uncertainties concerning the parameters (brine composition, temperature, pressure and pH) need to be further investigated (Druckenmiller et al. 2005). Soong et al. (2007) investigated the possibility of using fly ash in order to raise the pH of brine, thereby allowing for the precipitation of carbonates from the solution. The results of the experiments proved the feasibility of this concept, and 0.546 mol/l of CO_2 was sequestered in 2 h during a one stage approach via fly ash. However, the concept was tested only at the laboratory level; the actual application of the technology in the pilot plant level is still a point to be investigated.

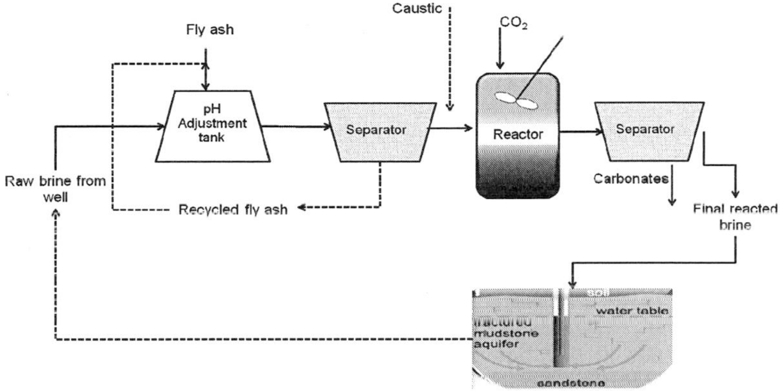

Figure 10.6. Carbon sequestration by carbon mineralization using brine and fly ash (based on Soong et al., 2007)

Another option that can be considered is carbon dioxide capture and storage by accelerated weathering of limestone (AWL). This option imitates the natural carbonate weathering according to the following reaction (Rau and Caldeira 1999):

$$CO_2\,(g) + H_2O\,(l) + CaCO_3\,(s) \rightarrow Ca^{2+}\,(aq) + 2HCO_3^-\,(aq) \qquad (10.11)$$

The product of an AWL plant would be a calcium bicarbonate solution that could readily be released and diluted into the ocean with a minimal or even a positive environmental impact (Rau et al. 2007). However, further research is needed before this alternative can be applied on any larger scale as there are still many issues to deal with, such as the energy demand of transporting large amounts of calcium containing (waste or mineral) material to the AWL plant that preferably should be located near a CO_2 point source as well as a possible disposal site (e.g. the ocean). In an ideal case (with access to free limestone, e.g. waste fines, and a "free" water source, e.g. power plant cooling water). The environmental effects of bicarbonate solution disposal into the ocean were also discussed by Rau et al. (2007). While direct CO_2 injection into the ocean lowers the pH, releasing a bicarbonate calcium ion containing solution could actually counteract the ongoing ocean pH reduction. In order to avoid negative impacts to the ocean, the CO_2 containing flue gas should be free of impurities, such as heavy metals. Despite the potential positive effect of bicarbonate disposal, further research is needed to fully understand the impacts of AWL effluent disposal in the ocean.

Also, straightforward carbonation has been suggested by simply spreading e.g., olivine on land where acidity is a problem and would simultaneously increase the pH of the soil (i.e. improve soil quality) and capture CO_2 from the surrounding air in a relatively short time frame (~30 years). This simple approach has been suggested by Schuiling and Krijgsman (2006) who emphasize that this method, even if simple, should initially be applied with caution in order to verify the impact of spreading large amounts of rock material on the ground. The amount of CO_2 that could be sequestered in this way is principally limited by available/suitable surface area and the theoretical binding capacity is given by the following reaction:

$$(Mg, Fe)_2SiO_4 + 4CO_2 + H_2O \rightarrow 2(Mg, Fe^{2+}) + 4HCO_3^- + H_4SiO_4 \qquad (10.12)$$

An important point to be noted is that the above reaction is highly dependent on natural environmental conditions, such as rainfall, soil type, (CO_2 pressure), temperature and type of rock, which limits its applicability. Another simple approach to CO_2 sequestration is the alternative of carbonation in underground cavities, such as caves. Schuiling (2007) has discussed the alternative of sequestering CO_2 by filling e.g. an opencast mine with olivine containing rock material and injecting CO_2. The benefits of such a solution are that no expensive reactor equipment would be required and that the reaction kinetics would not be of major importance. In addition, the heat of reaction generated by the reaction between CO_2 (and H_2O) and olivine could be recovered by placing heat exchangers in the olivine. Here is a caveat in this technology that unless the kinetics is fast enough, the system of olivine and CO_2 would reach thermal equilibrium with the surrounding rock material, and heat recovery would not be possible. Mine tailings are also prone to significant carbonation without any external intervention, but standard methods have not been suitable to measure the amount of CO_2 trapped within the tailings. In this context, Wilson et al. (2006) developed a method that enables the quantification of carbonates in serpentine-rich mine tailings.

10.6 Techno-Economic and Environmental Evaluation of Mineral Carbonation

The large amount of materials required for mineral carbonation causes practical concerns. Firstly, the scale of the mining needed is great and much greater than that of coal, as more serpentine is needed to sequester the carbon dioxide resulting from the combustion of an equal amount of coal. In addition, transportation is an important factor. In order to avoid the transport of large quantities of rock, transportation of the carbon dioxide is preferable. This implies that the sequestration facilities have to be placed near the mining locations, which imposes geographical constraints on the mineral sequestration option. Even the selection of a specific reactor type is arduous. The necessary long residence time complicates the selection of an appropriate reactor for an industrial process.

Mineral CO_2 sequestration is in general more expensive than other sequestration routes due to the additional equipment and the more complicated nature of the process needed. Potentially, costs can be limited by the exothermic nature of the reactions. Furthermore, mining for mineral sequestration can create valuable by-products, such as magnesium, silicon, chromium, nickel and manganese. Finally, re-use of the resulting products could improve the economic returns of the process. Costs can be divided in two ways. First, costs can be allocated to the various stages in the sequestration process: (1) CO_2 capture and separation from the flue gas; (2) compression and transport; and (3) sequestration. For mineral CO_2 sequestration the sequestration costs can be further divided into costs for the pre-treatment of minerals and for the sequestration process itself. Second, costs can be split up in fixed (investment costs) and variable costs (energy consumption, raw materials needed, etc.).

As mentioned earlier, the amounts of minerals needed as feedstock for mineral CO2 sequestration and the produced (by)-products are very large, which leads to environmental concerns. A single 500 MW power plant generates about 10,000 ton CO_2/day. To sequester this amount of CO_2 via the mineral route would require approximately 23,000 ton/day of magnesiumsilicate ore. Feedstock minerals are produced as by-products at several existing mines, but a massive increase in mining activities would be necessary to facilitate large-scale mineral CO_2 sequestration. The carbonate products can be returned to the mines to restore the landscape. Due to the increased volume of the carbonated minerals, part of the material cannot be used for mine reclamation, but has to be disposed of or re-used otherwise. The impact of such storage would be limited on a global scale, but could have serious consequences on a local scale in terms of environmental or land-use constraints. In addition to these considerations, mineral CO_2 sequestration technologies involve other environmental impacts, such as the consumption of extra energy. Assuming that the required energy is provided by fossil fuels, mineral sequestration would enhance the use of fossil fuels. When mineral sequestration is applied at a power plant, the extra energy consumption would consequently decrease its efficiency. Thus, research in future must be envisaged to further reduce the energy consumption to limit environmental and energy costs. Additionally, the use of hydrochloric acid as solvent in the HCl extraction route needs to be examined. In the process the acid is recycled. However, some hydrochloric acid probably escapes from the process and ends up in the (by-) products or the environment. This causes environmental concerns, and extra HCl has to be produced with additional energy costs. Therefore, process routes using large amounts of hydrochloric acid are less attractive. Besides reduction of the greenhouse effect, mineral sequestration also has positive environmental effects. Using alkaline solid waste as feed material can cause a decrease of the contamination of soil and groundwater and an enhanced re-use of materials. It would also contribute towards reducing the mining of primary minerals. Furthermore, some metal oxides are known to be able to bind acid gases, such as SO_2 in a process similar to mineral carbonation.

Compared to the other sequestration options, mineral carbonation is a longer-term option. It has some fundamental advantages, such as the permanent nature of the carbon dioxide storage and its theoretically vast capacity, but at this stage there is insufficient knowledge to conclude whether a cost-effective and energetically acceptable process will be feasible.

10.7 Benefits of CO_2 Sequestration by Mineral Carbonation

Carbon sequestration by reacting naturally occurring Mg and Ca containing minerals with CO_2 to form carbonates has many unique advantages as given below:

1) Long term stability: Mineral carbonation is a natural process that is known to produce environmentally safe and stable material over geological time frames. The production of mineral carbonates insures a permanent fixation rather than temporary storage of the CO_2; thereby guaranteeing no legacy issues for future generations.
2) Vast capacity: Raw materials for binding the CO_2 exist in vast quantities across the globe. Readily accessible deposits exist in quantities that far exceed even the most optimistic estimate of coal reserves (~10,000 x 10^9 tons).
3) Potential to be economically viable: The overall process is exothermic and, hence, has the potential to be economically viable. In addition, its potential to produce value-added products during the carbonation process may further compensate its costs.
4) Possibility of retrofit: At a single site and scale that is consistent with current industrial practice, the process can handle the output of one to several large power plants. It is directly applicable to advanced power plants or to existing

power plants; thereby providing an additional degree of flexibility for future implementation.

10.8 Future Research Directions

Nowadays, there is thrust on the use of CO_2 in organic chemical polymer and plastics production. However, the drivers are generally cost, elimination of hazardous chemical intermediates and the elimination of toxic wastes, rather than the storage of CO_2. Mineral carbonation has the advantage of sequestering carbon in solid, stable minerals that can be stored without risk of releasing carbon to the atmosphere over geologic time scales. Mineral carbonation involves three major activities: (1) preparing the reactant minerals—mining, crushing, and milling—and transporting them to a processing plant, (2) reacting the concentrated CO_2 stream with the prepared minerals, and (3) separating the carbonate products and storing them in a suitable repository. However, currently a broader reactivity of the minerals is sought, but for more efficient mineralization a narrow range of reactivity will be desirable as given in Fig. 10.7.

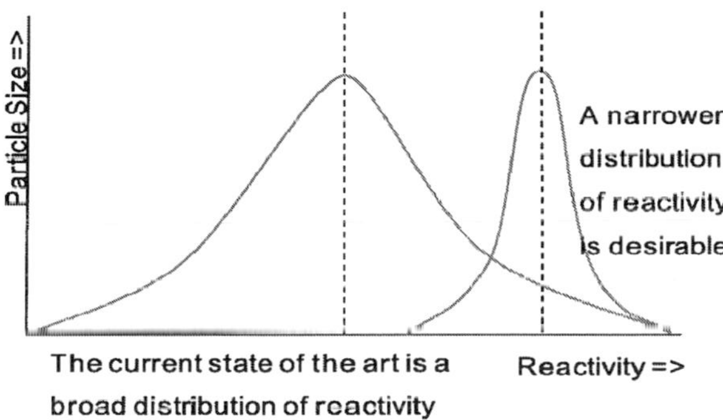

Figure 10.7. Range of reactivity of different mineralizing agents required for future mineralization reactions

Magnesium would be a better choice as a mineral carbonation agent due to following reasons:

1) Magnesium minerals are relatively abundant. At 2.09% of the crust, magnesium is the 8th most abundant element.
2) Due to the low molecular weight of magnesium, magnesium oxide, which hydrates to magnesium hydroxide and then carbonates, is ideal for scrubbing CO_2 out of the air and sequestering the gas into the built environment.

3) More CO_2 is captured than in calcium systems as the calculations below show:

$$\frac{CO_2}{MgCO_3} = \frac{44}{84} = 52\% ; \frac{CO_2}{CaCO_3} = \frac{44}{101} = 43\%$$

4) Magnesium minerals are potentially low cost. This will enable low cost simple non fossil fuel calcination with CO_2 capture of magnesium carbonate.
5) Large quantities of carbonates (the binder in cements) are not produced much from MgO. (The volumetric expansion from MgO to $MgCO_3 \cdot 5H_2O$ is 811%).
6) Magnesium oxide is easy to make using non fossil fuel energy.
7) Reactive, low lattice energy forms of magnesium oxide are most suitable as they are easier to get into solution, and efficiently absorb CO_2.
8) A high proportion of CO_2 and water means that a little MgO goes a long way.
9) In terms of sequestration or binder produced for starting material in cement, cements produced using magnesium carbonate will be nearly six times more efficient.
10) Use for sequestration directly and in the built environment would result in new and exciting markets for the magnesium compounds industry.

10.9 References

Butt, D.P., Lackner, K.S., Wendt, C.H., Conzone, S.D., Kung, H., Lu, Y.-C., and Bremser, J.K. (1996). "Kinetics of thermal dehydroxylation and carbonation of magnesium hydroxide." *J. Amer. Ceramic Soc.*, 7, 1892–1898.

Chen, Z.Y., O'Connor, W.K., and Gerdemann, S.J. (2006). "Chemistry of aqueous mineral carbonation for carbon sequestration and explanation of experimental results". *Environ. Prog.*, 25 (2), 161–166.

Chizmeshya, A.V.G., McKelvy, M.J., Marzke, R., Ito, N., Wolf, G., Béarat, H., and Doss, B. (2007). *Investigating geological sequestration reaction processes under in situ process conditions,* 32nd International Technical Conference on Coal Utilization & Fuel Systems, June 10–15, 2007, (paper 44), 441.

Domingo, C., Loste, E., Gómez-Morales, J., García-Carmona, J., and Fraile, J. (2006). "Calcite precipitation by a high-pressure CO_2 carbonation route." *The J. Supercritical Fluids*, 36, 202–215.

Dooley, J.J., Dahowski, R.T., Davidson, C.L., Wise, M.A., Gupta, N., Kim, S.H., and Malone, E.L. (2006). *Carbon Dioxide Capture and Geologic Storage: A Core Element of A Global Energy Technology Strategy to Address Climate Change,* a technology report from the second phase of the global energy technology strategy program. Global Energy Technology Strategy Program. Available at <http://www.epa.gov/air/caaac/coaltech/2007_02_battelle.pdf> (accessed July 2014).

Druckenmiller, M.L., and Maroto-Valer, M.M. (2005). "Carbon sequestration using brine of adjusted pH to form mineral carbonates." *Fuel Process. Technol.*, 86, 1599–1614.

Fauth, D.J., Goldberg, P.M., Knoer, J.P., and Soong, Y. (2000). "Carbon dioxide storage as mineral carbonates." *Abstracts of Papers of the American Chemical Society 220*, U395–U1395.

Fauth, D.J., Soong, Y., and White, C.M. (2002a). "Carbon sequestration utilizing industrial solid residues." *Abstracts of Papers of the American Chemical Society 223*, U565–U1565.

Fauth, D.J., Baltrus, J.P., Soong, Y., Knoer, J.P., Howard, B.H., Graham, W.J., Maroto-Valer M.M., and Anderson, J.M. (2002b). *Carbon storage and sequestration as mineral carbonates*. Environmental Challenges and Greenhouse Gas Control for Fossil Fuel Utilization in the 21st Century, Kluwer Academic/Plenum.

Feng, B., Yong, A.K., and An, H. (2007). "Effect of various factors on the particle size of calcium carbonate formed in a precipitation process." *Materials Sci. and Engg. A*. (445–446), pp. 170–179.

Gerdemann, S.J., O'Connor, W.K., Dahlin, D.C., Penner L.R., and Rush, H. (2007). "Ex-situ aqueous mineral carbonation." *Environ. Sci. Technol.*, 41 (7), 2587–2593.

Goff, F., Guthrie, G., Lipim, B., Chipera, S., Counce, D., Kluk, E., and Ziock, H. (2000). *Evaluation of the Ultramafic Deposits in the Eastern United States and Puerto Rico as sources of Magnesium for Carbon Dioxide Sequestration.* Los Alamos National Laboratory, LA-13694-MS, 36 pp.

Goldberg, D.S., Takahashi, T., and Slagle, A.L. (2008). "Carbon dioxide sequestration in deep-sea basalt." *PNAS*, 105(29), 9920–9925.

Hänchen, M., Prigiobbe, V., Storti, G., Seward, T.M., and Mazzotti, M. (2006). "Dissolution kinetics of fosteritic olivine at 90–150 °C including effects of the presence of CO_2." *Geochimica et Cosmochimica Acta*, 70, 4403–4416.

Hänchen, M., Krevor, S., Mazzotti, M., and Lackner, K. (2007a). "Validation of a population balance model for olivine." *Chem. Engg. Sci.*, 62, 6412–6422.

Hänchen, M., Prigiobbe, V., Storti, G., and Mazzotti, M. (2007b). "Mineral carbonation: Kinetic study of olivine dissolution and carbonate precipitation." 8th International Conference on Greenhouse Gas Control Technologies, 19–22 June, 2007, Trondheim, Norway, paper 01-_013_04.

Hänchen, M., Prigiobbe, V., Baciocchi, R., and Mazzotti, M. (2008). "Precipitation in the Mg carbonate system–effects of temperature and CO_2 pressure." *Chem. Engg. Sci.*, 63, 1012–1028.

Hansen, L.D., Dipple, G.M., Gordon, T.M., and Kellett, D.A. (2005). "Carbonated serpentinite (listwanite) at Atlin, British Columbia: A geological analogue to carbon dioxide sequestration." *Can. Mineral.*, 43, 225–239.

Huijgen, W.J.J., and Comans, R.N.J. (2005). *Carbon dioxide sequestration by mineral carbonation: Literature review update 2003-2004,* ECN-C--05-022, Energy Research Centre of The Netherlands, Petten, The Netherlands.

Kakizawa, M., Yamasaki, A., and Yanagisawa, Y. (2001). "A new CO_2 disposal process via artificial weathering of calcium silicate accelerated by acetic acid". *Energy*, 26, 341–354.

Katsuyama, Y., Yamasaki, A., Iizuka, A., Fujii, M., Kumagai, K., and Yanagisawa, Y. (2005). "Development of a process for producing high-purity calcium carbonate (CaCO3) from waste cement using pressurized CO_2." *Environ. Prog.*, 24, 162–170.

Kohlmann, J., Zevenhoven, R., Mukherjee, A.B., and Koljonen, T. (2002). *Mineral carbonation for long-term storage of CO_2 from flue gases,* TKK-ENY-9, Helsinki University of Technology, Energy Engineering and Environmental Protection, Finnish National Research Programme CLIMTECH (1999–2002).

Lackner, K.S., Wendt, C.H., Butt, D.P., Joyce, E.L., and Sharp, D.H. (1995). "Carbon-dioxide disposal in carbonate minerals." *Energy*, 20(11), 1153–1170.

Lackner, K.S., and Ziock, H.J. (2000). "From low to no emissions." *Modern Power Systems,* 20/3, 31–32.

Lackner, K.S. (2002). "Carbonate chemistry for sequestering fossil carbon." *Annual Rev. Energy and the Environ.,* 27, 193–232.

Mazzotti, M., Abanades, J.C., Allam, R., Lackner, K.S., Meunier, F., Rubin, E., Sanchez, J.C., Yogo K., Zevenhoven, R. (2005). "Mineral carbonation and industrial uses of CO_2." In: Metz, B., Davidson, O., de Coninck, H., Loos, M., and Meyer, L. (editors), *IPCC Special Report on Carbon Dioxide Capture and Storage. Cambridge*, UK: Cambridge Univ. Press, pp. 319–338.

McKelvy, M.J., Chizmeshya, A.V.G., Bearat, H., Sharma, R., and Carpenter, R.W. (2001). "Developing mechanistic understanding of CO_2 mineral sequestration reaction processes." *Proc. 26th International Tech. Conference on Coal Utilization & Fuel Systems*, Clearwater, FL, March 5–8, pp. 777–788.

McKelvy, M.J., Chizmeshya, A.V.G., Soignard, E., Marzke, R., Wolf, G.H., Béarat, H., and Doss, B. (2006). "Laboratory investigation of fluid/solid sequestration reaction processes under in situ sequestration process conditions." 31st International Technical Conference on Coal Utilization & Fuel Systems, May 21-26, 2006, Clearwater, Florida, USA, (paper 41), pp. 384–400.

NETL (2001). *Proceedings of Workshop NETL Mineral CO_2 Sequestration.* 2001 Conference Proceedings.

O'Connor, W.K., Dahlin, D.C., Nilsen, R.P., and Turner, P.C. (2000). "Carbon dioxide sequestration by direct mineral carbonation with carbonic acid." *Proceedings of the 25th International Technical Conf. On Coal Utilization & Fuel Systems*, Coal Technology Assoc. March 6–9, Clear Water, FL, Albany Research Center, Albany, Oregon.

Penner, L.R., O'Connor, W.K., Gerdemann, S.J., and Dahlin, D.C. (2003a). "Mineralization strategies for carbon dioxide sequestration," 20th Annual International Pittsburgh Coal Conference, Pittsburgh, PA, September 15–19, 19 pp.

Penner, L.R., Gerdemann, S.J., Dahlin, D.C., O'Connor, W.K., and Nilsen, D.N. (2003b). "Progress on continuous processing for mineral carbonation using a prototype flow loop reactor." *Proc. of the 28th Inter. Tech. Conf. on Coal Util. & Fuel Systems*, Clearwater, FL, March 9–13, 12 pp.

Rau, G.H., and Caldeira, K. (1999). "Enhanced carbonate dissolution: A means of sequestering waste CO_2 as ocean bicarbonate." *Energy Conversion and Management*, 40, 1803–1813.

Rau, G.H., Knauss, K.G., Langer, W.H., and Caldeira, K. (2007). "Reducing energy-related CO_2 emissions using accelerated weathering of limestone." *Energy*, 32, 1471–1477.

Rendek E., Ducom G., Germain P. (2006). Carbon dioxide sequestration in municipal solid waste incinerator (MSWI) bottom ash. *Journal of Hazardous Materials*, (B128), 73–79.

Schuiling, R.D., and Krijgsman, P. (2006). "Enhanced weathering: An effective and cheap tool to sequester CO_2." *Climatic Change,* 74, 349–354.

Schuiling, R.D. (2006). "Mineral sequestration of CO_2 and recovery of the heat of reaction." In: Badescu, V., Cathcart, R.B., and Schuiling, R.D, (eds.), *Macro-engineering –A challenge for the future*, Springer Dordrecht, the Netherlands, pp. 21–29 (chapter 2).

Seifritz, W. (1990). "CO_2 disposal by means of silicates." *Nature*, 345, 486.

Soignard, E., Robert Marzke, R., Piwowarczyk, J., Diefenbacher, J., and McKelvy, M.J. (2006). "Application of NMR to the direct measurement of diffusivities and activities of CO_2 and other dissolved species under geological and mineral sequestration conditions." 31st International Technical Conference on Coal Utilization & Fuel Systems, May 21–26, 2006, Clearwater, Florida, USA, (paper 42), p. 401.

Soong, Y., Fauth, D.L., Howard, B.H., Jones, J.R., Harrison, D.K., Goodman, A.L., Gray, M.L., and Frommell, E.A. (2006). "CO_2 sequestration with brine solution and fly ashes." *Energy Conversion and Management*, 47, 1676–1685.

Surampalli, R., Zhang, T. C., Ojha, C. S. P., Tyagi, R. D., and Kao, C. M. (eds.) (2013). *Climate Change Modeling, Mitigation and Adaptation,* ASCE, Reston, Virginia, 2013.

Teir, S., Eloneva, S., and Zevenhoven, R. (2005). "Production of precipitated calcium carbonate from calcium silicates and carbon dioxide." *Energy Conversion and Management*, 46, 2954–2979.

Teir, S., Revitzer, H., Eloneva, S., Fogelholm, C.-J., and Zevenhoven, R. (2007a). "Dissolution of natural serpentinite in mineral and organic acids." *Int. J. of Mineral Processing*, 83, 36–46.

Teir, S., Eloneva, S., Fogelholm, C.-J., and Zevenhoven, R. (2007b). "Dissolution of steelmaking slags in acetic acid for precipitated calcium carbonate production." *Energy*, 32, 528–539.

Teir, S., Kuusik, R., Fogelholm, C.-J., and Zevenhoven, R. (2007c). "Production of magnesium carbonates from serpentinite for long-term storage of CO_2." *Int. J. of Miner. Proc.*, 85, 1–15.

Wilson, S.A., Raudsepp, M., and Dipple, G.M. (2006). "Verifying and quantifying carbon fixation in minerals from serpentine-rich mineral tailings using the Rietveld method with X-ray powder diffraction data." *Amer. Mineral.*, 91, 1331–1341.

Yogo, K., Eikou, T., and Tateaki, Y. (2005). "Method for fixing carbon dioxide." Patent, JP2005097072, 14.4.2005.

Zevenhoven, R., Eloneva, S., and Teir, S. (2006a). "Chemical fixation of CO_2 in carbonates: Routes to valuable products and long-term storage." *Catalysis Today*, 115, 73–79.

Zevenhoven, R., Teir, S., and Eloneva, S. (2006b). "Heat optimisation of a staged gas-solid mineral carbonation process for long-term CO_2 storage." *Proceedings of ECOS 2006*, 12–14 July 2006, Crete, Greece, pp. 1661–1669, in revised form Energy (33), 2008, pp. 362–370.

Zevenhoven, R., and Tier, S. (2004). "Long term storage of CO_2 as magnesium carbonate in Finland." *Proceedings of the Third Annual Conference on Carbon Capture and Sequestration*, May 3–6, 2004, Alexandria (VA), USA, (paper 217).

Zevenhoven, R., Eloneva, S., and Teir, S. (2006c). "A study on MgO-based mineral carbonation kinetics using pressurised thermogravimetric analysis." 8th International Conference on Greenhouse Gas Control Technologies, 19–22 June, 2006, Trondheim, Norway, paper P02_01_09.

Zhang, Q., Sugiyama, K., and Saito, F. (1996). "Enhancement of acid extraction of magnesium and silicon from serpentine by mechanochemical treatment." *Hydrometallurgy*, 45, 23–331.

Zhang, T.C., and Surampalli, R.Y. (2013). "Carbon capture and storage for mitigating climate changes." Chapter 20 in *Climate Change Modeling, Mitigation and Adaptation,* Surampalli, R., Zhang, T.C., Ojha, C.S.P., Gurjar, B.R., Tyagi, R.D., and Kao, C.M. (eds.), ASCE, Reston, Virginia, February 2013.

CHAPTER 11

Carbon Burial and Enhanced Soil Carbon Trapping

Indrani Bhattacharya, T. T. More, J. S. S. Yadav, L. Kumar, Song Yan, Rojan P. John, R. D. Tyagi, Rao Y. Surampalli, and Tian C. Zhang

11.1 Introduction

The human activities, which produce greenhouses gases, are unequivocally held responsible for the global climate change or warming (Houghton et al. 2001; IPCC 2007; Sundquist et al. 2008; Friedlingstein et al. 2010). The human activities have led to an increase in the global surface temperature of 0.74 °C in the last 100 years (Solomon et al. 2007). This steep continuous increase in the global temperature, if not dramatically reduced, will be catastrophic to many species on the earth including human. At the global scale, the key greenhouse gases (GHGs) emitted by human activities are carbon dioxide (CO_2, 77%), methane (CH_4, 14%), Nitrous Oxide (N_2O, 8 %) and fluo gases (1%) (Marland et al. 2009). The CO_2 is the largest GHG and 57% of CO_2 emissions come from fossil fuel use, followed by 17% of CO_2 emissions from deforestation and decay of biomass, respectively. The CO_2 levels in the atmosphere were 280 ppm (parts per million) in pre-industrial era (Houghton et al. 2001). After industrial revolution to date, the CO_2 levels are now around 380 ppm (IPCC 2007). This is due to the burning of massive amounts of fossil fuels (Oelkers and Cole 2008).

To mitigate GHG emissions, the CO_2 emissions must be reduced immediately. The CO_2 sequestration can be achieved by either natural process and/or by humans. Natural processes such as vegetation and oceanic carbon cycles are the largest and most efficient methods of CO_2 sequestration. But the natural cycles cannot keep the check on human CO_2 emission rate, which is over 30 billion tons a year today (Houghton et al. 2001; Schrag 2007). To mitigate the global warming effects, global GHG emissions should be reduced substantially (by 25 to 40% below 1990 levels) by 2020, especially in industrialised countries (IPCC 2007). If this target is achieved, it would stabilize CO_2 concentrations in the atmosphere to 450 ppm. Meeting this target would mean reducing carbon emissions by about 10 billion tons a year. Today, most of the countries have a consensus on developing technologies or managing the human interference in ecosystem to reduce and/or stabilize excess CO_2 and other GHGs released into the atmosphere and hence mitigate the global warming (Houghton et al. 2001; IPCC 2007; Sundquist et al. 2008; Friedlingstein et al. 2010).

Table 11.1 shows carbon production and sequestration in different industries. The CO_2 emissions can be managed by reducing the fossil fuel usage, developing new fuels with low carbon sink and carbon sequestration by implementing various technologies (Lal 2008a; Sundquist et al. 2008; Marland et al. 2009; Rai et al. 2010). There are several processes and technological options that can be implemented for sequestration of atmospheric CO_2. These processes are including physical-chemical and biological processes (Table 11.1). Physical-chemical processes include natural carbon burial, carbonates formation and leaching, with different options such as geological sequestration, oceanic injection, chemical scrubbing and mineralization (Lal 2008a, b; Sundquist et al. 2008; Marland et al. 2009; Rai et al. 2010). Biological processes include organic carbon sequestration (woody plants, char and biomass soils, wetlands, seeding ocean with ferric ions, etc.), inorganic carbon sequestration (formation of secondary carbonates) and carbon sequestration via biofuels such as bioethanol, biodiesel and hydrogen fuel cells (Lal 2008a; Sundquist et al. 2008; Marland et al. 2009; Rai et al. 2010; Zhang and Surampalii 2013).

The objective of this chapter is to discuss the process and technological options (physical, chemical and biological) of carbon burial to mitigate/reduce the global warming by reducing the rate of CO_2 release from three major sources such as industrial combustion, biomass and organic wastes. Moreover, soil carbon trapping potential and various strategies to enhance the terrestrial carbon sequestration in soils are also discussed in the chapter.

11.2 Carbon Burial

There are several technological options for CO_2 capture and sequestration (CCS). The selection of single or a combination of two or more technologies is necessary to formulate energy policies by keeping in mind the future economic growth and development at regional, national and global scales. The major carbon burial options are discussed briefly in the following sections.

11.2.1 Carbon Capture and Storage: Industrial CO_2

Recently, there is rapid progress in developing/testing CCS technologies (Kerr 2001; Zhang and Surampalli 2013). CCS technology can be used with conventional fuels-coal or natural gas, which is available in many parts of the world especially in energy-intensive countries such as U.S., Russia, China and India, which hold about 67%

of the world's coal reserves. The CO_2 produced during different industrial activities are first captured by physical and chemical reactions and engineering techniques before it can escape to the atmosphere. The captured CO_2 is then turned into a fluid and injected deep underground in geological formations or in deep ocean. The CCS in oceanic and geological structures has received considerable attention (Freund and Ormerod 1997) because theoretically these methods are considered to possess a higher capacity of carbon sink than the other methods of carbon sequestration. Figure 11.1 shows the overview of the typical industrial CCS systems.

Table 11.1. Carbon production and sequestration in different industries

Industry	Country/ company/ project name	CO_2 generated per year	Current CO_2 or C storage	Future CO_2 storage	Where C is mainly stored	Remarks	References
Coal Mining	Allegheny Energy (In Association with U.S. Office of Surface Mining and Piedmont Energy Association (PEA)	204.3 pounds of CO_2 per million Btu (British thermal unit) when coal is completely burned in United States	3000 billion tonnes of Carbon	-	Most of the carbon is stored in oceans about 40 000 gigatons of carbon	PEA has reclaim an abandoned mine land with trees; as it is assumed that reforestation would have higher levels of CO_2 sequestration due to yearly tree growth	(Hong and Slatick 1994; EPA 2012)
Petroleum and natural Gas industries	Industries in USA	25,000 metric tons	48 million metric tons of CO_2	-	In the land	-	(EPA 2009)
	Sleipner Project in the North Sea	-	2,800 tonnes of CO_2 per day	-	In the seabed of North Sea	Potential problems while maintaining the reservoir was of chemical reactions following CO2 injection. Injection of CO2 can lower the pH of the water in the reservoir, dissolving minerals such as $CaCO_3$.	(Folger 2009)
Power plants	ALSTOM Power Komoti project in northern Greece	480 kg/MWh	Deep saline reservoirs (estimates range from 400 to 10,000 Gt of CO_2)	--	In the Ocean	Concerns are there that excess amount of CO_2 be poisoning the water in the form of $CaCO_3$ rock.	(Marion et al. 2001)
Biofuels	Algae.Tec in Australia	-	270,000 tonnes of that CO_2 per year	1.3 million tonnes	-	-	(ENS, 2013)
Bio-ethanol	Artenay and Toury refineries	45 000 tons from the fermentation unit and 60 500 tons from the cogeneration unit	100 000 tons of CO_2	-	Reservoirs	-	(Laude et al 2011)
	ADM's ethanol plant in Decatur	907 000 tonnes per year	-	-	-	-	(Gollakota and McDonald 2012)
Cement/ steel/ refineries	-	-	-	1.7 to 2.5 Gt (estimated) per year	Geological storage in saline aquafiers	-	(UNIDO 2010)

Geological Storage Options. Geological carbon sequestration includes CO_2 capture from industrial processes, liquefaction of the gaseous CO_2, and transportation of liquefied CO_2 via pipelines to predetermined underground storage sites and finally injection of CO_2 into deep geological formations. Various burial options for the industrial CO_2 are presented in Figure 11.2. The CO_2 may be injected in deep coal seams, old depleted oil and gas wells (to increase yield), stable rock strata (e.g., basalt) or saline aquifers (Tsang et al. 2002; Klara et al. 2003; Gale 2004). Saline aquifers are underground strata of very porous sediments filled with brackish (saline) water located below the freshwater reservoirs sandwiched between impermeable layers. The industrial CO_2 is first pumped into the aquifer. The CO_2 sequestration occurs at the aquifer hydrodynamically as well as by carbonation. At the aquifer, CO_2 reacts with dissolved salts to form carbonates. The CO_2 in a supercritical state is injected at deep underground aquifers (Figure 11.3). The CO_2 in supercritical state has significantly lower density and viscosity than the liquid brine present originally in the aquifer (that injected CO_2 displaces). At the aquifer, it forms a gas-like phase as well as dissolves in the aqueous phase. This creates a multiphase multicomponent environment. Injecting CO_2 into reservoirs in which it displaces oil or gas could be an economic strategy of enhanced oil recovery (Buyanovsky and Wagner 1998). Production from oil and gas fields, which has been in decline, is raised by CO_2-enhanced recovery (Klusman 2003). This strategy of CO_2 sequestration is used in Texas, USA, to inject 20 million Mg of CO_2/yr at a price of \$10–15/Mg (Lackner 2003). Similar technique has also been used in Norway in offshore oil wells. If the CO_2 injected is extracted from the underground, it cannot be called as a carbon sequestration. The CO_2 can also be injected into unmineable coal seams where CH_4 is absorbed. Injected CO_2 is absorbed onto coal twice as much as CH_4, and the process enhances the gas recovery of coal bed CH_4.

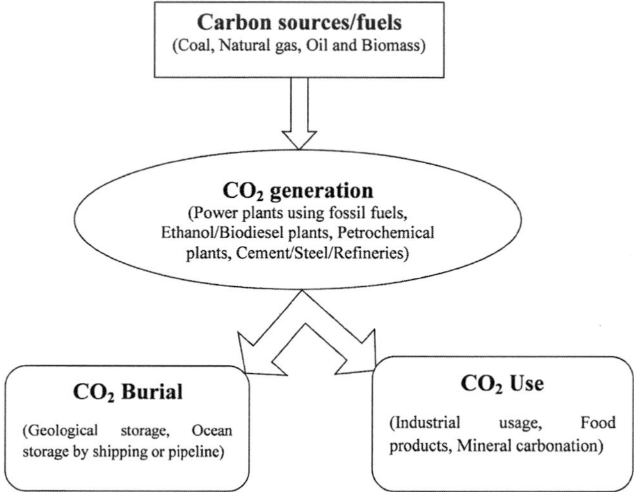

Figure 11.1. Overview of the industrial CCS systems

The high cost is the most critical issue of geological sequestrations. Moreover, there are serious environmental safety concerns of geological sequestration as finding a reliable geological formation is very crucial (Kintisch 2007; Schrag 2007). There are controversial opinions on the risks involved in geological carbon storage. Some studies have argued that risks of leakage are low. However, a very limited field experience is available on the issue. Because of the low density and viscosity of the CO_2 under supercritical conditions, there is high risk of CO_2 leakage through confining strata as compared to the liquid wastes (Tsang et al. 2002).There are some attempts of direct injection of CO_2 at a commercial scale such as Regional Carbon Sequestration Partnership project of US DOE is planning for several demonstrations during 2008-2009 (Lal et al. 2008b). To make the geological carbon storage a reality, the costs involved in the projects must be reduced. Moreover, there must be surety about the leakage proof geological storage sites based upon extensive field studies. The guidelines for appropriate regulatory and monitoring controls of the geological sequestration are needed based upon the increased knowledge of the current and future issues.

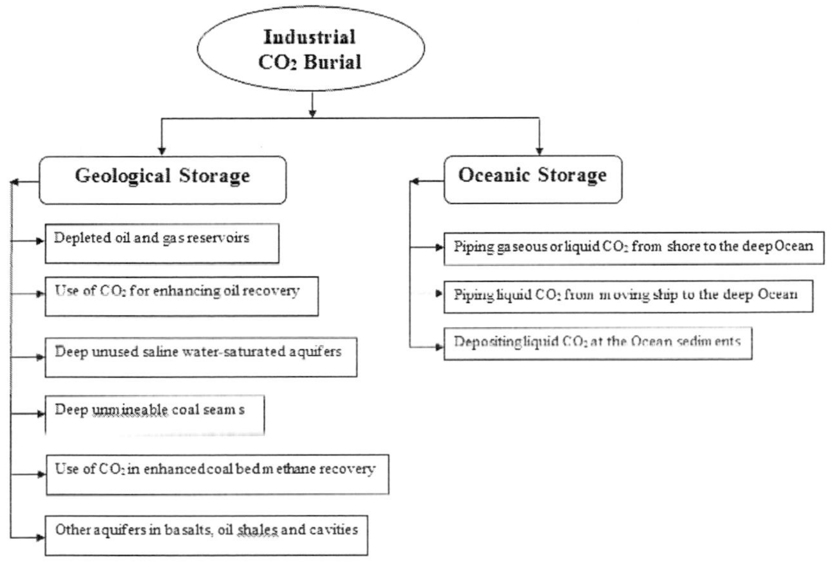

Figure 11.2. Two major industrial CO_2 burial options

Oceanic Storage/Disposal. Injection of a pure CO_2 stream deep in the ocean has been considered as one of the potential carbon sequestration technique. Since last few decades, this technique has significantly progressed (Lal et al. 2008a, b). There are several strategies of oceanic injection of CO_2 (Fig. 11.1). The liquefied CO_2 separated from industrial sources is generally injected below 1 km from a manifold lying at the ocean floor. The injected CO_2 rises in the form of droplet plume because of its lighter weight than the water (Lal et al. 2008a, b). The CO_2 can also be injected as a denser CO_2-seawater mixture at 0.5–1 km depth. After injection, this mixture settles down at deeper ocean. In another way, the captured CO_2 can be discharged from a large pipe towed behind a ship. This injected CO_2 can be deposited into oceanic sediment floor, which forms a lake like CO_2 storage place. At an ocean depth of 3 km or more, the injected liquefied CO_2 is considered to remain stable. The estimated oceanic sink capacity is 5000–10000 Pg of Carbon, which is much higher than the estimated fossil fuel reserves (Herzog et al. 2000). The CO_2 injection can also be detrimental to the biology of the deep ocean (Seibel and Walsh 2001). Similar to the geological storage methods, economy and stability of the carbon storage are the main concerns of oceanic method of carbon sequestration. In addition to the economics, the issue of stability of such an injection must be addressed owing to the increased stratification of the ocean water column and its turn over through natural processes.

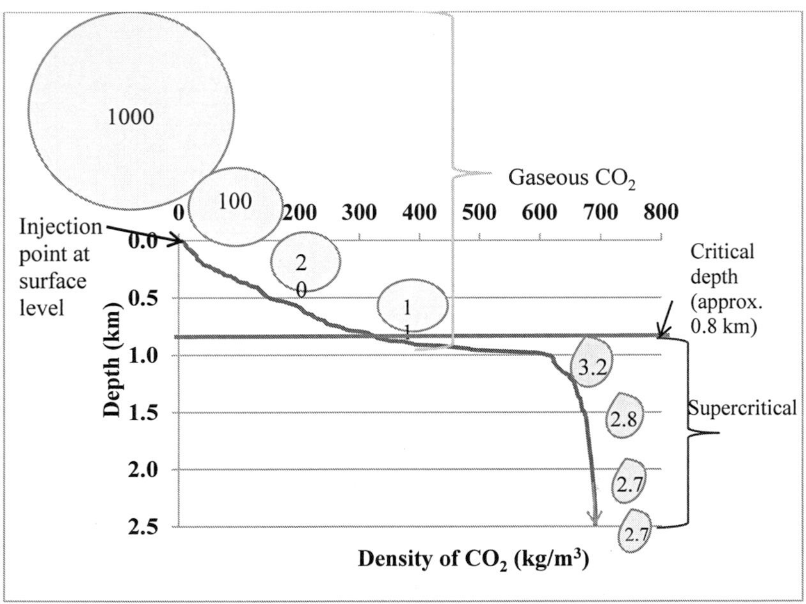

Figure 11.3. CO_2 as a gas at the surface and as a supercritical fluid at depth

Concerns and Challenges. Geological and oceanic storage technologies have capacity to significantly reduce industrial CO_2 emission release into the atmosphere and may help in mitigating global warming effects. However, there are several concerns and challenges for these two options (Zhang and Surampalli 2013):

1) For geological C-sequestration technologies:
   - Little is known about saline aquifers compared to oil fields; as the salinity of the water increases, less CO_2 can be dissolved into the solution. Larger uncertainty about the saline aquifers exists if the appraisal study is limited.
   - Liquid CO_2 is nearly incompressible with a density of ~1000 kg/m^3; over pressuring and acidification of the reservoir may cause i) changes in the pore/mineral volume, ii) saline brines (or water) moving into freshwater aquifers or uplift; old oil wells may provide leak opportunities.
   - In CO_2-ECBM, the key reservoir screening criteria include laterally continuous and permeable coal seams, concentrated seam geometry, minimal faulting and reservoir compartmentalization, of which not much is known.
   - New technologies are needed to ensure CO_2 stays in place forever. Need thorough site appraisal studies to reduce harmful effects.
2) For oceanic C-sequestration technologies:
   - Unknown impact on ecosystems (ocean acidification, wildlife, O_2 supply).
   - Difficult to certify the dissolution, leakage and location of CO_2.
   - Unknown impact on microbial carbon pump and biological carbon pump.
   - Need to i) make reliable predications of the technical feasibility and storage times, ii) understand how to predict and minimize any environmental impact; and iii) making reliable cost estimates and assess the net benefit.

It should be noted that the costs involved in construction, maintenance and operation of CCS facility is quite expensive with current technology. However, with the increase in knowledge and technological developments these costs would be lowered significantly. Therefore, it is possible for the coal-fired power plants with CCS technology to be more economical than the other renewable energy options such as solar energy. Currently, there is scarcity of funds globally to support CO_2 capture and storage projects. In future, with development in cost effective CCS technologies, the industrial involvement in such projects is expected to increase.

11.2.2 Biomass

Biomass is biological material derived from living, or recently living organisms (plant/vegetable derived or animal derived material). Biomass is composed of a mixture of organic molecules containing mainly carbon, hydrogen, nitrogen, oxygen and others such as alkali, alkaline earth and heavy metals (Rahman et al. 2013). These molecules are often found in functional molecules such as the porphyries, including magnesium-containing chlorophyll. Most of the biomass consists of plant-based material, and therefore, generally it is termed as lignocelluloses biomass. Biomass can be classified as virgin wood, energy crops, agriculture residues, food waste and industrial waste and co-products (Table 11.2).

Wood. Table 11.3 shows an estimation of an average carbon stock in different woody biomass. Forest accounts for nearly 90% of exchange rate of carbon between the land and the atmosphere. Simultaneously, this is one of the ways of removing CO_2 from the atmosphere. It has even been estimated that soil carbon emissions are higher in moist and lower altitude forests. Considering this fact, in the northern part of the central United States, carbon sinks are static over a period of time now (Woodall 2010). When various carbon pools are considered in the world, in the northern part of America, temporal forests of the U.S.A. and Canada have the largest carbon pool (Petrokofsky et al. 2012). Apparently, carbon losses are observed in Canada's boreal forests. Among the European nations, notable average pools are observed in countries such as Austria, Germany and France; whereas Russia has a much larger carbon pool size (Myneni et al. 2001). According to Barredo et al. (2012), higher amounts of carbon are sequestered in live biomass in European nations, and nearly 10% of entire GHG emissions are removed from the atmosphere (= 430 million tonnes of CO_2).

Table 11.2. Categories of biomass materials

Category	Origin
Virgin wood	Forestry, arboricultural activities, wood processing
Energy crops	High yield crops grown specifically for energy applications
Agricultural residues	Residues from agriculture harvesting or processing
Food waste	Food and drink manufacture, preparation and processing, and post-consumer waste
Industrial waste and co-products	Manufacturing and industrial processes

Table 11.3. Mean carbon stock available in various forest biomasses

Biomass Stock	Mean C Stock (Mg/ha)	Reference
Live biomass	21.3–36.48	Bradford et al. 2012; Domke et al. 2013
Standing dead	2.7–3.8	Woodall 2010
Woody debris (coarse)	3.82	Woodall 2010
Woody debris (fine)	2.91	Woodall 2010

It has been observed that carbon stocks are lower in very young plants and then the live biomass. However carbon stock gradually decreases as the tree ages, and if the ecosystem has been disturbed (Bradford et al. 2012). Carbon accumulated in leaves return back to the atmosphere, whenever the fallen leaves start decomposing whereas 50% of carbon is from dry wood.

Table 11.3 indicates that standing dead and woody debris are the major biomass sources for wood burial. In a variety of forms, wood debris (wood processing residues) is generated, which includes bark, round-offs, end cuts, trimmings, sawdust, shavings and rejected lumber (Morris 2002; Morris 2008). In the United States, nationwide wood processing residues remain the prominent solid biomass fuel source used. At a primary sawmill, 50% of the entire biomass content of a typical saw log converts into residue (Shukla 1998; Morris 2002; Morris 2008). To use a portion of this material, a variety of secondary forest products industries have been developed simultaneously. Sawmills residues are taken to the markets of the highest value. However, still a considerable quantity of the sawmill residues, which is an overall of 15–20% of the total biomass from a saw log available, has no useful product application and must be disposed of (Morris 2002; Morris 2008).

Down and dead woody materials or DDW are often considered as the relevant material for carbon sinks in forests. Dead woods include non-living biomass that cannot be included into the litter and the dead plants that are standing upright, or lying on the forest ground (including tree roots and stumps). It has been estimated that in the United States, total forest carbon sink of vegetation and dead organic material are ranging from 14–35%. Apart from the wood processing residues, coarse woody debris (CWD) and fine woody debris (FWD) are the essential co-factors for carbon sinks (Woodall et al. 2008; Merganičová and Merganič 2010). Carbon density of CWD is not forest's entire standing tree carbon stocks (Domke et al. 2013). In New Zealand, the entire forestry sector is dedicated to wood, where burial of wood processing residues has been proved beneficial (Garrett and Walker 2009).

Energy Crops. Any plant material, which can be used to produce energy in the form of heat, fuels, electricity, etc. is termed as energy crop. These crops have large capacity to produce biomass and possess high energy potential. Currently energy crops are the fourth largest energy source (more than 55 EJ/year), especially in developing countries where biomass is the main energy source (Hall and House, 2004). Energy crops are considered as future main renewable energy sources. Contribution of certain energy crops, which can transfer atmospheric CO_2 into the biomass and soil carbon pools, is considered as very attractive options to reduce the dependency on fossil fuel.

Energy crops capture an amount of carbon in the harvested biomass that is usually equivalent to the carbon released during combustion. This makes energy crops a carbon-neutral energy source (Hansen, 1993). Energy crops consist of herbaceous bunch-type grasses and short-rotation woody perennials (Lemus and Lal 2005). Table 11.4 shows the classification of energy crops and their important examples. These crops can be grown in marginal soils. The energy crops have capability to improve soil quality, enhance nutrient cycling and carbon sequestration. Since last two decades, the need for agricultural involvement in GHG mitigation has been widely advocated globally. Recently there is growing interest in herbaceous and woody species as energy crops and source of biofuel stock. Cultivation of bioenergy crops in degraded soils is very attractive and promising agricultural option, which has carbon sequestration potential of 0.6–3.0 Mg Carbon/Ha/year (Lemus and Lal 2005). The energy crops usage has relatively lower emissions as compared to the fossil fuels. For example, the CO_2 emissions are 1.9 kg carbon G/j from using switch grass as energy crop, whereas the CO_2 emissions are 13.8, 22.3 and 24.6 kg C G/j from using gas, petroleum and coal, respectively (Turhollow and Perlack, 1991). This clearly indicates that the bioenergy crops can play very important role in GHG mitigation.

Table 11.4. Different categories of energy crops

Category		**Important examples**
Herbaceous Bunch type grasses	Switchgrass	*Panicum* virgatum L
	Elephant grass	*Pennissetum* purpureum Schum
	Tall fescue	*Fetusca* arundinacea L.
Short-rotation woody perennials	Polar	*Populus* spp.
	Willow	*Salix* spp.
	Mesquite	*Prosopis* spp.

According to one estimate, bioenergy crops can possibly sequester 318 Tg carbon/year in the US and 1631 Tg carbon/year worldwide (Lemus and Lal 2005). This estimation is based on biomass yields, the land area dedicated to crop production, the estimated carbon sequestration potential, and the conversion efficiency. In the case of energy crops cultivated to decrease the carbon emissions from fossil fuel, the estimated carbon sequestration per unit land is very large. Therefore, increasing agricultural land into energy crops has potential to increase CCS in terms of soil organic matter, and this brings net reductions in GHG emissions. The potential of energy crops to offset CO_2 emissions through soil carbon sequestration depends on i) the rate of soil carbon additions, ii) the long-term capacity of soil carbon storage, and iii) the stability of carbon sequestered over time (McLaughlin et al. 2002). Energy crops must replace annual row crops to achieve net gains in carbon sequestration over the fossil fuels (Lemus and Lal 2005).

Understanding of soil carbon dynamics in energy crops is very important because biomass production, ecosystem sustainability, soil fertility and structure are depended on soil carbon dynamics. There is lack of information about the capability of perennial energy crops for soil organic carbon sequestration; although short-term studies have advocated that soil organic carbon can be significantly improved with perennial biomass production. A broad base of knowledge of combined physical properties, management practices and biological interactions and plant species is necessary to obtain more reliable quantities estimates of CCS in energy crops and the strategies to improve CCS capability. Mitigation of GHG emissions requires increases usage of energy fuels along with reduction in fossil fuel combustion.

Agricultural and Cropland Residues. Agricultural residues can be categorized in a wide variety of forms, not all of which are suitable for being used as the power plant fuel (Schubert 2009). Agricultural residues, such as pits, shells, orchard and vineyard removals, field straws and stalks, are suitable for fuel (Morris 2008; Schubert 2009). In state of California, agriculture being a multibillion-dollar enterprise generates a huge volume of biomass residues. Nearly 1/3 of California's biomass energy plants were constructed in the state's agricultural regions in order to enhance these residues as fuel (Morris 2008). Agricultural fuels currently substitute nearly 20% of the state's biomass fuel. Most of the state's biomass facilities receive emissions offsets for pollutants. It can be avoided when biomass residues that would have been burned open otherwise are used for the energy production (Witzke et al. 2008).

Food Waste. According to the estimates of The Food and Agriculture Organization (FAO), one third of all food produced for human consumption in the world is lost or wasted. The global volume of food wastage is estimated to be 1.6 Gtonnes of "primary product equivalents," while the total wastage for the edible part of food is 1.3 Gtonnes. This amount can be weighed against total agricultural production (for food and non-food uses), which is about 6 Gtonnes. The food wastage causes the economic loss of $750 billion annually, a figure that excludes wasted fish and seafood.

11.2.3 Organic Wastes

Landfills and composting are the two major carbon burial options for organic wastes. In a study by Couth and Trois (2010), in Africa certain waste management practices were followed to assess methods to reduce carbon emissions. It is seen that an average organic content for urban Municipal Solid Waste in Africa is around 56% and its degradation is a major contributor to GHG emissions. In order to reduce carbon emissions, segregating wastes at collection points, removing compost and the remaining biogenic carbon waste by using the maturated compost as a substitute fertilizer, and disposing the remaining fossil carbon waste in controlled landfills are viable options. In addition, municipal solid waste control was introduced to improve landfill soil cover properties and for enhancing CH_4 oxidation. For example, the landfill cover soil was leached (washed) for maintaining appropriate moisture content and for CH_4 oxidation reactions. It exhibited landfill gas emission control (Tanthachoon et al. 2007). The status of the current practice and related issues in using these two techniques are described as follows.

Landfills. Landfill disposal of wood-processing residues are not so desirable. Unlike other forms of biomass in the landfill environment, waste wood has a snail-pace decay rate, and thus is slow to stabilize (Morris 1999). Therefore, it is suggested that landfill dispose of organic wastes should be avoided whenever it is possible. For example, the state of California strongly encourages reduction of the amount of material being buried in the state's landfills, and the introduction of a sizable new waste stream which would make compliance with the California state's recycling regulations which is otherwise impossible for the other countries (Morris 2008). A minor traceable quantity of the sawmill residues would be composted, while the remaining residues destined for landfill disposal. Biomass fuels such as wood-processing residues are the cheapest which generate and deliver biomass to power plants.

Sources of organic wastes in landfills. Biomass residues in landfills at consecutive years have been seen gradual increase from forest to agriculture next comes urban and mill biomass residues in landfills. The data usually is represented in biomass residues as bdt (Bone Dry Ton), and bdt is referred to as thousands of bone dry ton equivalents, which is the measure of the dry weights of biomass fuels. In the year 2005, maximum bdt has been seen in the case of urban and mill residues which is 55 and 63 (Mann and Brown 2008; Morris 2008). Approximately half of the biomass fuels used in the state of California in 2005 otherwise would have been buried in landfills, if other options wouldn't have been available, in the absence of energy production. Maximum of the agricultural residues used for energy production otherwise would have been being openly burned. Irrespective of the fact that California state policy is basically oriented towards reducing the total deposit of material disposed of in sanitary landfills, it is evident to assume that the probable alternative fate for most of the state's urban waste wood that is currently used for energy production would be landfill disposal. A petite amount of these residues would be composted (Morris 2008). Over the years, biomass residues in agriculture has shown significant increase especially in the year 2007 and during the year 2008, forest biomass residues have shown significant increase, irrespective of the biomasses from urban and mill biomass residues. Consecutively, at the year 2008, maximum bdt has been reported for forest biomass residues in landfills as 83.8, whereas urban and mill biomass residues are been nearly nil or zero (Placer Country CEO and TSS Consultants 2008).

In-forest biomass residues includes two main categories of materials: 1) in the forest, where residues are generated where timber is harvested for wood products, and it is called slash; and 2) material which are naturally found in forests in the form of overgrowth material, whose physical removal will provide benefits of health and environmental to the forest (Morris 2008). Harvesting residues are combination of limbs and leaves of the harvested trees, bark that are cut and removed during harvesting operations. To leave it in the forest as it is produces is the cheapest form of management for this material, but it is considered as the management practice from a forestry perspective not to be considered, as leaving harvesting residues in the field hampers the forest re-growth (Ferrari 2007).

Burning and mechanical thinning are the two prescribed techniques which are used to mitigate the biomass overloading especially in forests which are standing. Tonnage of most of the forest overgrowth biomass material on and near the floor of the forest is usually called as ground fuel. The condition of pre-extensive exploitation can be understood as periodic fires. California forests tended to have primarily ground-fuel fires. These naturally occurring ground fires control the build-up of excess forest fuels (Morris 2008). Ground biomass is left for accumulation for a prolonged period of time. Certain biomass is present underground, which otherwise grows into grass/shrubs/weeds and called as ladder fuels. A mechanism is provided by ladder fuels to transfer ground fires to the top of mature trees in the forest, and thus, severely increasing the damage caused by the forest fire, converting benign ground fire into out-of-control and destructive wildfires of the forest. The fuel-overloading problem in the forest can be exacerbated by the traditional commercial harvesting operations, because neither ground nor ladder fuels are removed systematically (Morris 2008).

Apparently, biomass residues which were generated in the year 2005 in California, measured in th.bdt was 780 thousands of bone dry ton (th, bdt) in case of open burning, 408 th.bdt for forest accumulation, 25 th.bdt for spreading, 337 th.bdt for composting and 429 th.bdt from the firewoods (Morris 2008). In California's municipal landfills, nearly 15–20% (by weight) of the wood residues that traditionally is disposed of is often clean. Such materials as waste wood specifically come from a variety of sources, such as construction contractors, old and damaged pallets, land clearing, public and private tree trimmers and landscapers, industrial manufacturers (including packing materials and trimmings, furniture, crates), etc. (Morris 1999; Morris 2008). Residues are placed in landfills *e.g.* woods of different forms, including loads of chipped wood and brush from public and private tree trimmers and land clearers, debris boxes, other wood products like demolition wood waste, painted or treated wood (Morris 2008). Landfill operators segregate out loads containing fuel-useable materials as they process the material to produce a high-quality fuel product (Morris 2008). Currently, a loophole in current state of California solid-waste regulations permits landfill operators to use wood chips as better alternative to daily cover for the operations, on the privilege of counting the biomass as a material that was disposed of and decomposed in the landfills (Morris 2008).

Landfill gases. The partial fraction of the biomass that enters landfills can be bifurcated at the gate, processed, and shipped as the solid biomass fuel. However, a large quantity of the biomass that has been buried in landfills already enhances diversion initiatives. Rather certain fraction of society's biomass wastes will continue to be disposed of in landfills. Landfills produce landfill gas, which is nearly 50–60% of CH_4 and 40–50% of CO_2 (Morris 2008). While the landfill gas is biogenic carbon, the fact that when left uncared and uncontrolled, nearly half of the biogenic carbon in landfill gas is generated in the form of the more potent carbonaceous GHG, CH_4, which is of substantial concern from the perspective of climate and climate change. Apart from that, in order to contain large volumes of CH_4, landfill gas also contains noxious VOCs, which have similar per-carbon potency as CH_4 and contribute to the formation of the ground-level ozone layer. As GHGs become subject to regulation over the coming several years, the threshold for landfills that are subjected to control is expected to be further extended to smaller landfills. The other fates for the landfill gases are: a) no control, b) control to meet regulations, and c) control beyond regulation to stimulate and maximize the energy production (Morris 2008; Leggett et al. 2009).

Landfill gas control integrates installing a gas-collection system in landfill cells to collect the landfill gas that is generated as biomass wastes degrade. The collected gas can be burned in engines for energy production. Combustion does the conversion from the reduced carbon in the landfill gas to CO_2, which has much lower GHG potency (Wanichpongpan et al. 2004). By installing a more extensive gas-collection system, GHG emissions are reduced, which is being motivated apart from the regulations by motivating a landfill that is not subjected for the regulation to install a gas-collection system. Designing of the modern landfills relies on the collection of an increasing percentage of the landfill gas, but some of the gas cannot be captured. The reason is landfills are porous in nature, allowing fugitive gases to escape inevitably. Nevertheless, gas collection and conversion to energy greatly would significantly reduce the GHGs emissions (Leggett et al. 2009). Federal and state regulations for the state of California require huge landfills for the collection and combustion of the gases regardless of the gas that is used for energy production (Morris 2008).

Composting. Composting followed by land applications is one of the major economical ways for the final disposal of many organic wastes as it combines material recycling and disposal of organic wastes at the same time.

As a biological process decomposing organic wastes under aerobic or oxygen rich conditions, composting can occur naturally under a wide range of conditions. However, to achieve rapid decomposition specific conditions are required (Table 11.5). When these conditions are met, the microbial populations will increase rapidly, resulting in elevated temperatures in the composting mix. During composting aerobically decomposition of microbial colonies transforms organic substrates into a stable, humus-like material.

Table 11.5. Rapid composting conditions

Condition	Reasonable range	Preferred range
Carbon to nitrogen (C:N) ratio	20:1 to 40:1	25:1 to 30:1
Moisture content (%)	40 to 65	50 to 60
O_2 concentration	> 5%	Much > 5%
Particle size (d in inches)	1/8 to 1/2	Varies
pH	5.5 to 9.0	6.8 to 8.0
Temperature (°C)	43 to 65	54 to 60

(adapted from Rynk, 1992; VanDevender and Pennington 2004)

There are several ways to use composite for organic wastes treatment and management. Small scale composting which is backyard composting happens only in the

USA can be used to handle food scraps, yard trimmings and mixed organics. An alternative approach can be expressed as organic burial composting (OBC). For example, cattle mortality can be buried with a sufficient amount of organic carbon source (e.g., sawdust, hay) to ensure that decomposing takes place in an environmental-friendly manner. Out of the various carbon sources, green sawdust performs the best when it comes to rapid decomposition. However, waste hay and waste silage even works in a significantly reduced rate (VanDevender and Pennington 2004).

Co-composting means composting of two or more materials together. Co-composting processes are reported for composting of agro-industrial along with wastewaters. Along with solid waste, wastewater treatment plant sludge can be co-composted with agricultural, forestry and some agro industry residues. These residues would behave as bulking agent, and they might improve the pile structure by allowing air circulation. Addition of sawdust reduces pH levels of masses of the final compost. Sawdust as it has low content of heavy metals decreases heavy metals in the final compost mass. Mixed compost after the second stage and addition of sawdust, concentrations of heavy metals such as chromium, cadmium, copper and zinc has been become minimal (Hasanimehr et al. 2011).

Composting animal manure has the potential to reduce emissions of N_2O and CH_4 from agriculture. Agriculture has been recognized as a major contributor of GHG, releasing an estimated of 81% and 70% of the anthropogenic emissions of N_2O and CH_4 respectively. A significant amount of CH_4 is emitted along the storage of liquid manure, whereas N_2O is emitted from the storage of manure, from soil following manure or fertilizer application. Composting animal manure can reduce GHG emissions in two ways: i) reducing N_2O and CH_4 emissions during manure storage and application; and ii) reducing the amount of manufactured fertilizers and the GHG associated with their production and use (Paul et al. 2001). Luo et al. (2014) estimated that it is possible to transfer 29% of total nitrogen, 87% of phosphorus, 34% of potassium and 75% of magnesium to the compost, and to reduce the total acidification potential, eutrophication potential and global warming potential of manure management on the farm by 64.1%, 96.7% and 22%, respectively, compared with the current system.

Concerns and challenges. During composting, microbes degrade the original waste materials into organic compounds through a variety of pathways, and ~80% of the initial organic matter is emitted as CO_2. While the remainder of the organic compounds eventually stabilizes and become resistant to further rapid microbial decomposition. The mature form of compost is characterized as containing a high percentage of these stable, humic substances. When the compost is mature, nearly all of the water-soluble compounds (such as dissolved organic carbon) will have leached out.

While the USEPA is currently researching the mechanisms and magnitude of carbon storage, the USEPA's Waste Reduction Model has assumed that carbon from compost remains stored in the soil through two main mechanisms: i) direct storage of carbon in depleted soils, and ii) carbon stored in non-reactive humus compounds. The USEPA's Waste Reduction Model calculates the carbon storage impact of each carbon storage path separately and then adds them together to estimate the carbon storage factor associated with each short ton of organics composted (EPA 2008). As being modeled in the USEPA's Waste Reduction Model, composting results in carbon storage as well as minimal carbon dioxide (CO_2) emissions from transportation and mechanical turning of the compost piles.

In a study by Maeda et al. (2011), it is seen that 74.3% of the CH_4 emission from manure compost could be mitigated by mixing dried grass into the manure. This is because mixing dried grass within the piles enhances the availability of O_2, and thus, inhibits CH_4 production by these anaerobic methanogens. It was estimated that after this strategy was introduced, the Japanese dairy sector could be expected to mitigate its GHG emissions by 1907 Gg of CO_2 equivalent per year. Favoino and Hogg (2008) suggest that compost can only store carbon in soils temporarily. As the carbon will be released, in the long run, into the atmosphere; this will help to reduce emissions of CO_2 in the short to medium term. Nevertheless, composting can contribute in a positive way to restore soil quality and sequester carbon in soils as applying compost as fertilizer can build-up soil organic carbon over time and reduce the depleting rate of organic matter in soils. Detailed studies on life cycle assessment are needed.

Compost vs. Landfills. A study by Luske (2010) has shown that composting of organic waste and compost usage result in a significant reduction of GHG emissions when compared with landfills. The reduction is mainly reached due to avoiding CH_4 emissions from landfills or dumping the organic waste in the landfill. The same accounts for the emission of N_2O during composting where composting results in 90% less emission than the landfills which is 149 kg CO_2/ton. At times, major CH_4 emissions may occur depending upon the type of compost. The amount of CH_4 emitted is highly dependent on the management of the composting process.

Another aspect is estimating the amount of GHG emissions that can be reduced by diverting household organic waste into a centralized community composting system rather than sending it to a landfill. When organic waste is placed in a landfill it decomposes gradually over decades and creates a significant amount of methane due to the absence of oxygen. In contrast, when organic waste is composted it decomposes within a year and predominantly creates CO_2 if O_2 is made available. As CH_4 is 21 times more potent than CO_2 on a 100 year global warming potential basis, composting can reduce GHG emissions by over 90% in contrast to sending the same waste to a landfill.

These reductions are dependent on the type of organics diversion projects and the efficiency of a landfill gas collection system (BCME 2013).

11.3 Enhanced Soil Carbon Trapping

Biological sequestration of CO_2 is one of the natural and cost effective technologies for carbon capture and sequestrations (CCS). Biological technologies for CCS include: 1) photosynthetic system of microorganism or higher plants (e.g., algae, microbes, ocean flora); 2) sustainable practices (e.g., soil conservation, development of grasslands and peat bogs, reforestation/afforestation); and 3) use of biomass/residues (e.g., biomass-fuelled power plant, production of biofuels, and biochar). These processes normally affect the cycle of carbon in the planet. Since plants and soils naturally absorb CO_2, preventing outright deforestation and managing forests and agricultural lands as carbon sinks can remove significant amounts of greenhouse gases (GHGs). Algal biomass can sequestrate carbon and be part of soil sink for long duration. Higher plants including trees can fix CO_2 into their biomass and prevent it from easily releasing to the environment till the death and degradation. Many of the microbes and some of the higher plants can fix CO_2 in non-photosynthetic pathway. So, biotic community, water and soil in all ecosystems take part in CCS and other parts of the carbon cycle.

Currently, considerable studies have been conducted on carbon sequestration in terrestrial ecosystems, i.e., sequestering carbon with photosynthesis and soil sequestration. Many reviews have been published about biotic carbon sequestration, wetlands, soil carbon sequestration (Bruce et al. 1999; Lal 2008a, b). This section discusses biological processes and technologies related to enhanced soil carbon burial, with specific attention being paid to the metabolisms involved, and carbon sequestration at the molecular, organism and ecosystems levels.

11.3.1 Carbon and Biological Activities in Environments

Carbon is one of the major macro-elements of living organisms. The flow of carbon from one ecological system to another is termed as carbon cycle. Biotic communities from microorganisms to very large plants and animals take part in this biogeochemical cycle. Biota is mainly involved in two main processes, i.e., photosynthesis (capturing of CO_2) and respiration (release of CO_2), in carbon cycle (Fig. 11.4). Different biotic communities make different but unique contributions: 1) algal bloom and other photosynthetic organisms in the ocean, other water bodies and land play a great role in fixation of CO_2. The photosynthate is used as the energy source of producers to synthesize biomass and storage materials, and to maintain routine activities; 2) consumers of different trophic levels depend on the producers for their energy and body development; 3) dead matter of both producers and consumers can be converted into simpler forms or into body components by decomposers; and 4) all these organisms generate energy in anaerobic or aerobic processes and release of CO_2 and less degradable compounds such as lignin in order to enhance sequestration of carbon as a part of terrestrial or oceanic pool.

Phytoplankton photosynthesis is one of the greatest oceanic sequestration mechanisms. According to Lal (2008a), the movement of CO_2 into biotic and pedologic carbon pools is called terrestrial carbon sequestration. The living and dead organic matter can act as a carbon sink in biosphere. The soil organic carbon pool has doubled the amount of carbon than the atmospheric carbon pool, and thus, it is considered a potential major source to drive global climate change (Alewell et al. 2009). According to Alewell et al. (2009), soil erosion is the major form of soil degradation as it relocates a large quantity of carbon in lower altitude areas such as valleys, water bodies, sediments and generates CO_2 during and after the relocation processes. Approximately one-thirds of the total increase in atmospheric CO_2 is a result of loss of soil organic carbon due to land use change such as the clearing of forests and the cultivation of land for food production (Lal 2004b).

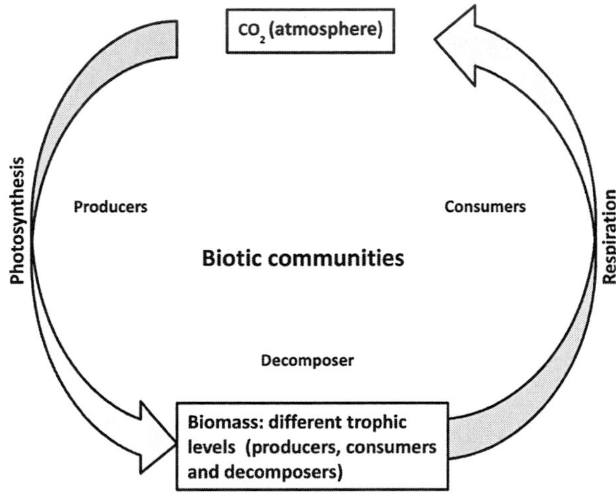

Figure 11.4. Diagrammatic of biological processes involved in carbon cycle

11.3.2 Soil Environment, Soil Carbon and Climate Change

Soil is a thin layer of material on the Earth's surface and consists of layers (horizons) of mineral and other components (Fig. 11.5). Soils are dynamic and have been forming continuously since the ecological periods. There are different types of soil, mainly depending on the parent materials and the influence of other environmental factors. Besides parent material, the climate, organisms living under or on the soil and the duration of soil formation processes have influence on the structure of soil. Soil components consist of mineral particles, organic matter, soil water (capillary, hygroscopic, etc.), soil air and soil biota (microorganisms, such as bacteria and fungi and animals including nematodes, arthropods, etc.).

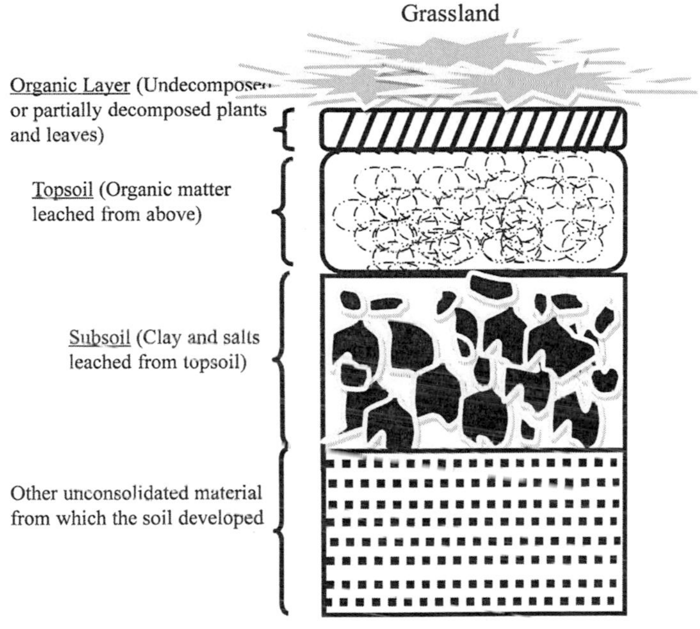

Figure 11.5. Soil profile showing typical soil horizons

Soil Formation and Horizons. Soil is formed by the weathering of rocks and it happens for a long duration. There are three types of weathering: chemical, biological and physical weathering (Durgin 1977). The physical weathering occurs by the mechanical forces such as wind, pressure, heat and cold. The physical weathering breaks the rock into small portions, but there is no change in the rock constituents. However, chemical weathering can change the chemical property of the rocks by chemical reaction through external factors. Acid rain, influences of microbial metabolites, etc. can act as chemical weathering agents. The biological weathering is the release of chelating compounds (i.e., organic acids, siderophores) and of acidifying molecules (i.e., H^+, organic acids) by living or dead plants or microbial activities (e.g., the symbiotic mycorrhizal fungi associated with root systems) so as to break down aluminium and iron containing compounds in the soils beneath them or to cause chemical weathering.

Soil horizon is a specific layer in the soil; each one differs from the others on their physical, chemical and biological characters. Horizon formation is termed as horizonation, and a horizon is formed by many geological, chemical and biological processes. The parent material degrades and favors the growth of living organisms. Plants such as lichen may be pioneer living organisms, and their growth results in the accumulation of organic residues and serves as substrate for higher plants such as mosses, pteridophytes, angiosperms, etc. and later animal communities (ecological succession). The accumulated organic matter makes this layer more fertile and this layer is called the O-horizon. O-horizon consists of the dead and decaying plant and animal matters. A-horizon is formed by the humification of the organic layer. The top most part of A-horizon is highly organic, and this part helps in the plant growth by providing nutrients. Many of the soil organisms including earth worms, arthropods, nematodes, bacteria and fungi are dominated in this range. This layer together with O-horizon is called top soil. Below this is subsoil layer (B-horizon). Between A- and B-horizons there may be an eluviated layer from which mineral and organic content has been leached out and deposited in B-horizon. This layer is often called illuviated layer or accumulation layer. Roots of large trees can reach this layer. There is a less weathered parent material below B-horizon called C-horizon. Some of these horizons are again subdivided according to different geographical area. The roots of plants and leaves can release organic compounds such as terpenes and was found in mineral zones (Morand 2013). Below that is a bed rock and is partially weathered called R-horizon.

Soil Properties. Soils have different properties in different locality, and these properties severely affect the biota of the soil. The ability of water to move through the soil (permeability of soil) and water holding capacity of the soil are affected by the structure, size and number of pores and are relative to the soil texture, compaction and organic matter. Water contained in the capillary pores is the available water to plants and other soil microorganisms. The pores, especially macropores, influence the exchange of air in soil, which is essential for the effective growth of organisms in soil. The soil texture is relative to the percentage of sand, silt or clay. The fine-textured soil has more water holding capacity, but is not ideal for aeration, and coarse-textured is reverse. Improvement of sandy soils and clay soils can be done by the addition of organic matter such as, compost. Biochar as a fertilizer can increase the soil carbon (Fowles 2007). The humification of the supplementing carbon can be sequestered in the upper layer of soil (Lal 2008a).

Soil Carbon and Mitigation of Climate Change. Soil is a major natural body of the Earth's ecosystem, which is composed of solid phase of the organic and inorganic matter as well as a porous phase that holds gases and moisture. Soil consists of several layers of minerals that differ in texture, structure, consistency, and color; with various chemical and biological characteristics. Soil is the end product of climate, organisms, original minerals accumulated over time. Soil acts as an engineering medium, a habitat for soil organisms, a recycling system for nutrients and organic wastes, a regulator of water quality, a modifier of atmospheric composition, and a medium for plant growth. Soil intercepts between the lithosphere, hydrosphere, atmosphere and biosphere. Soil is the part of the climate change problem; however it is also an integral part of the solution.

The soil carbon is a major part of the Earth's carbon cycle. Soil contains approximately 2500 gigatons of the carbon (Gt-C), which is nearly 80% of the total carbon present in terrestrial ecosystems of the Earth (Lal 2008a; Ontl and Schulte 2012). Soil carbon can be either organic (1550 Gt-C) or inorganic carbon (950 Gt-C). The latter consists of elemental carbon and carbonate materials such as calcite, dolomite, and gypsum (Lal 2004b). The organic soil carbon is four times the amount of carbon in terrestrial biota and three times the atmosphere (800 Gt-C) (Oelkers and Cole 2008). The amount of carbon found in living plants and animals (560 Gt-C) is comparatively small relative to that found in soil. Only the ocean has a larger carbon pool (38,400 Gt-C), mostly in inorganic forms (Houghton 2007).

According to Lal (2004a, b, c), the rate of soil organic carbon sequestration by biomass or agriculture depends on texture and structure of soil, rainfall, temperature, farming system, and management of soil. Lal (2004a, b, c) suggested that several approaches could increase the soil carbon pool, such as soil restoration and woodland regeneration, no-tillage, cover crops, nutrient management, green manuring and sludge application, improved grazing, water conservation and harvesting, efficient irrigation, agroforestry practices, and cultivation of fuel crops on unused lands. Management of forest can maintain the CO_2 to a harmless level. The conversion of natural vegetation to cropland leads to a loss of soil carbon up to 50 %, which is mainly due to annual tilling, and annual tilling increases the rate of decomposition by aerating undecomposed organic matter (Pacala and Socolow 2004). Afforestation (in grasslands, shrublands and croplands) is one of the speediest processes for sequestration of carbon but the large scale planting leading to the reduction in soil water, losses in stream flow, soil acidification and salination (Jackson et al. 2005; Lal 2008a). There is a chance of minimization of flood protection, nutrient retention, pollination, biological control, etc. due to the monoculture plantations (Bunker et al. 2005). Wetlands and wetland soil constitute ~450 Pg of carbon, and wetland soil contains about 200 times more carbon than the vegetation of that area. The human activities such as agriculture and forestry in wetland ecosystems increased the CO_2 concentration (Lal 2008a).

Enhancing soil carbon sequestration with proven management practices includes i) converting marginal land to productive grassland or forest, ii) increasing productivity on crop and forest land with residue management, iii) reduced C loss with conservation practices (e.g., no-till, residue mulching, cover cropping, crop rotation, efficient use of fertilizer, pesticide, and water, and other technologies). As shown in Table 11.6 and Fig. 11.6, these conservation practices (sometimes called regenerative agriculture) have great potential for carbon-sequestration.

Soil carbon sequestration can be efficiently done by two ways, namely storing carbon after fixation and prevention of stored carbon erosion. Mulching, cover cropping, addition of carbon containing stable material such as charcoal to fields, utilization of biological material as alternative for petroleum fuels, utilization of biological fertilizers for agriculture, etc. are some of the methods to improve carbon sequestration. The soil porosity can be affected by the cropping system. Large quantities of sediments and soil organic carbon are moving laterally over earth surface due to agricultural erosion. The change in tillage-required crop to non-required one can change the arrangement of soil particle and pore types. The biotic community including earthworm, residue cover, etc. can influence these factors. Earth worm can make different sized pores and increases the fertility by adding the digested organic matter to the soil. The vertical pores in bare land varied from the land covered with organic matter. Burrowing animals can make holes and push out some internal soil matter. Ants and other small creatures living in burrows or holes can influence soil structure. The main reason for soil degradation is the land use for agricultural purposes, and many studies are underway on the loss of soil quality generated by agricultural operations.

Table 11.6. C-sequestration potential of major ecosystems (adapted from Metting et al. 2002)

Ecosystem	**Method to increase C-sequestration**	**Potential CS (Gt-C/yr)**
Agricultural lands	Management (H)	0.85–0.90
Biomass crop lands	Management (H)	0.5–0.8
Grasslands	Management (M)	0.5
Rangelands	Management (M)	1.2
Forests	Management (M)	1–2
Wetlands	Restoration and management (M)	0.1–0.2
Urban forest and grass lands	Creation and management (H)	< 0.1
Deserts & Degraded lands	Management (H)	0.8–1.3
Sediments and aquatic systems	Protection (L)	0.6–1.5
Tundra and taiga	Protection (L)	0.1–0.3
Total		~5.5–8.7

Note: i) the sustained, long-term annual sequestration of 5.5–8.7 Gt-C is speculative. Today's terrestrial systems sequester only about half this amount and the other half is uncertain; and ii) the primary C sequestration method is rated with High (H), Medium (M), and Low (L) levels of sustained management intensity required over the long term. Global potential C sequestration (CS) rates were estimated that might be sustained over a period of up to 50 years.

Currently, worldwide overgrazing is substantially reducing grassland's performance as carbon sinks. It was suggested that, if practiced on the planet's 3.5 billion tillable acres, regenerative agriculture could sequester up to 40% of current CO_2 emissions. The U.S. CO_2 emissions from fossil fuel combustion were ~6170 million metric tons in 2006 (EWR 2009). If a 2,000 (lb/ac)/year sequestration rate was achieved on all 434,000,000 acres (1,760,000 km^2) of cropland in the US, about one quarter of the country's total fossil fuel emissions would be sequestered per year (Zhang and Surampalli 2013).

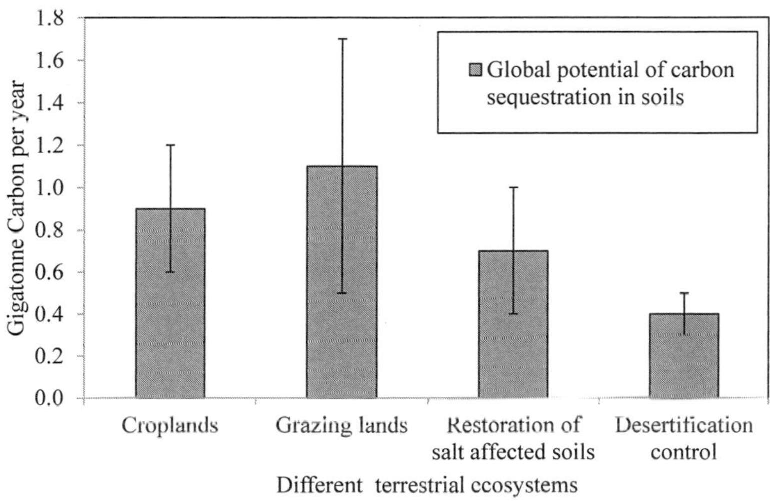

Figure 11.6. C-sequestration potential in different terrestrial ecosystems (IPCC 1995)

11.3.3 Enhanced Soil Carbon Sequestration

Soil carbon sequestration includes improving the soil carbon (organic and/or inorganic) by adopting eco-friendly and sustainable soil management. This includes: i) proper land-use conversion and improved activities/processes in agricultural, pastoral and forestry ecosystems; and ii) restoration of degraded and drastically disturbed soils (Lal 2008b). Soil carbon sequestration also includes production and use of biochar as a fertilizer (Fowles 2007). Carbon is stored or buried in the soil at 0.5–1.0 m depth via natural processes of humification. Natural ecosystems contain more soil organic carbon than the managed ecosystems such as cultivated soils because of depletion of soil organic carbon due to oxidation/mineralization, leaching and erosion. Newly created agricultural ecosystem contains 50–75% of the original soil organic carbon stock. The reduction in soil organic carbon occurs in these newly created agriculture ecosystem mainly during their first 5–10 years (in tropical region) or 20–50 years (in temperate regions) of conversion (Lal 2001).

The soils, which are degraded, degradable, and vulnerable to degrade in future, are very important for the soil carbon trapping. The soil degradation occurs by erosion, nutrient depletion, and acidification and leaching, structural decline and pollution or contamination. Prevention and reduction in soil degradation as well as restoring degraded soils and ecosystems are sustainable approaches. These approaches yield several benefits to the water quality, biomass productivity and, more importantly, to reduction of net CO_2 emission. Strategies for enhancing soil carbon trapping and enhancing soil productivity are listed in Table 11.7.

According to Grainger (1996), tropical region has about 750 Mha of degraded lands, which has a terrestrial carbon sequestration potential of 1.1 Pg C/year (including that of soil organic carbon and biomass). This could be achieved by several strategies including afforestation and soil quality enhancement. According to Lal (2001), controlling desertification of soils in arid and semi-arid regions could sequester 0.4–0.7 Pg C/year from the soil.

Table 11.7. Management of soil properties and relevant activities to enhance the productivity and carbon sequestration

Managing soil properties	Activities/process management	References
Drought-tolerant crops	Managing hydrological cycle	Janzen et al. 1998; Sauerbeck 2001; Post et al. 2004; Lal 2010; Ontl and Schulte 2012
Stress emission signals by crops	Water harvesting/recycling	Studdert and Echeverria 2000; Lal 2010
Remote sensing of drought stress	Soil-water conservation	Witzke et al. 2008; Lal 2010
Drip sub-irrigation	Ground water recharge	Witzke et al. 2008; Lal 2010
Condensation irrigation	Integrated nutrient management	Uhlen and Tveitnes 1995; Lal 2010
Carbon sequestration in soils	Precision farming	Sauerbeck 2001; Post et al. 2004; Lal 2006, 2010
Stable micro-aggregates	Biofertilizers	Uhlen 1991; Rasmussen et al. 1998; Lal 2010
Translocation of carbon into sub-soil	Strengthening recycling	Buyanovsky and Wagner 1998; Lal 2010
Coupled cycling of carbon and water	Increasing use efficiency	Uhlen and Tveitnes 1995; Lal 2010
Protecting C by physical/chemical/biological mechanisms	Nano-enhanced fertilizers	Post et al. 2004; Lal 2010
Creating positive carbon budget	Slo release formulations	McCarl and Schneider 2001; Post et al. 2004; Lal 2010
Conservation agriculture	Creating positive nutrient budgets	Liebig et al. 2002; Lal 2010
Cover cropping	Balanced nutrients applications	Uhlen and Tveitnes 1995; Studdert and Echeverria 2000; Lal 2010
Agroforestry	Coupled cycling of C, N, P, S.	Rasmussen et al. 1998; Lal 2010
Biochar and other amendments	Carbon burial in soils	Lal 2010; Ontl and Schulte 2012

According to West and Post (2002) adopting no-till farming soil organic carbon sequestration rates could be improved. As per their assessment, by adopting sustainable till methods, soil organic carbon sequestration of 570 ± 140 Kg C/ha yr could be maintained for next 40–60 years. The global soil organic carbon sequestration potential is 0.4–1.2 Pg C/year, which is 5–15% of the global fossil fuel emissions (Lal 2004a, b). As per Pacala and Socolow (2004), by adopting no-till farming on 1600 Mha of cropland along as well as adopting conservation-effective measures on the same cropland, soil organic carbon sequestration of 0.5–1 Pg C/year can be achieved by 2050.

Soil nutrients play an important role in carbon sequestration. Lack of important soil nutrients (N, P, S etc.) can significantly reduce the soil humification process (Himes 1998). Unbalanced C and N in the soils reduce the carbon sequestration efficiency (Paustian et al. 1997). By increasing the rate of biomass carbon application to the soil, the soil carbon and N sequestration rates can be enhanced (Campbell et al. 1991; Janzen et al. 1998; Halvorson et al. 1999; Halvorson et al. 2002). According to Liebig et al. (2002), application of N at a high rate, the soil carbon sequestration rates were increased by 1.0–1.4 Mg C ha/yr as compared with the unfertilized controls. The studies conducted by Bowman and Halvorson (1998), Studdert and Echeverria (2000) and Jacinthe et al. (2002) have also supported the positive impacts of nutrients management on soil carbon sequestration. According to Malhi et al. (1997), the source of N and the rate at which it is applied to the soil is very critical to the soil carbon sequestration. The effect of nutrients on the carbon present in the upper soil layer of 0–10 cm were studied by Ridley et al. (1990) in Victoria, Australia. Their study revealed that the application of P and lime had increased the soil carbon sequestration by 11.8 Mg/ha over a 68-year period at an average rate of 0.17 Mg C/ha/year.

Soil carbon sequestration can also be enhanced by the application of manures and other organic amendments (e.g., composting). Soil organic carbon sequestrations rates were higher for the application of organic manures as compared to the chemical fertilizers (Jenkinson et al. 1990; Witter et al. 1993; Christensen 1996; Körschens and Müller 1996). Soil carbon sequestration in the 0–30 cm soil depth was increased by 10% for the long-term use of manure as compared to the chemical fertilizers over 100 years in Denmark (Christensen 1996), 22% over 90 years in Germany (Körschens and Müller 1996), 100% over 144 years at Rothamsted, UK (Jenkinson et al. 1990) and 44% over 31 years in Sweden (Witter et al. 1993). The soil applied with manure had 44.6 Mg/ha more soil carbon than the soils without manure application (Anderson et al. 1990). In a study conducted by Arends and Casth (1994) on manured soils in Hungary, it was observed that the application of manure had increased the soil organic carbon by 1.0–1.7%. Smith et al. (1997) assessed that the application of manure at the rate of 10 Mg/ha to cropland in Europe would increase the SOC pool by 5.5% over 100 years. According to the estimates of Uhlen (1991) and Uhlen & Tveitnes (1995), soil organic carbon sequestration can be increased at the rate of 70–227 Kg/ha-yr by the manure application in Norway over a period of 37–74 years.

Agricultural lands with single culture cropping systems contain significantly lesser soil organic carbon as compared to those with diverse cropping systems (Buyanovsky and Wagner 1998; Drinkwater et al. 1998). Soil carbon losses can be minimised by avoiding fallow cropland, especially in semi-arid regions (Rasmussen et al. 1998). Soil carbon sequestration can be enhanced by growing a winter cover crop. The grass-ley set-aside has significant potential contribution to the soil conservation. According to Fullen & Auerswald (1998), the grass leys set aside had increased soil carbon by 0.02% per year for 12 years in the UK. In another study by Grace & Oades (1994) in Australian crops, it was observed that the soil organic carbon was proportional to the frequency of pasture in the crop rotation cycle, in the 0–10-cm layer.

In arid regions, the rate of soil carbon sequestration is always either zero or in negative side whereas, in arid and hot climates it varied up to approximately 1000 Kg C/ha-yr under humid and temperate climates (Lal 2004c, 2005a,b). Normal soil carbon sequestration in agricultural soils occurs at the rate of 300–500 Kg C/ha-yr. Soil carbon sequestration occurs at a higher rate for no-till farming, crop residue retention as mulch, planned crop rotation and agroforestry, soil nutrient management by manuring and through afforestation.

Adoption of advanced soil management on agricultural and forest soils is an eco-efficient and sustainable solution for improving soil productivity and soil carbon sequestration (Follett 2001; Lal et al. 2003; Lal 2004a, b, 2006). Advanced soil management increases water quality, decreases water pollution by reducing dissolved and sediment loads from soils to water resources and decreases the net CO_2 emission along with enhancing soil quality and productivity. This makes the advanced soil management a natural process (Marris 2006). To make the soil management strategy successful, the knowledge of several critical aspects is must. These critical aspects include: i) the soil stabilizing mechanisms (Six et al. 2002), ii) the biophysical limitations for carbon sequestration (Schlesinger 1999; Sauerbeck 2001), and iii) the regulatory, economy and policy considerations (McCarl and Schneider 2001). Moreover, soil carbon trading requires several permits (at federal, state and local levels) and marketing procedures.

11.4 Conclusions

Natural carbon sequestration is not capable of assimilating enough CO_2 emissions. To mitigate the global warming effects caused mainly by CO_2 emissions, the immediate development, adaptation and systematic implementation of various (physical, chemical and biological) carbon sequestration technologies is the need of time. Direct injection of industrial CO_2 in geological formations and oceanic strata is considered as emerging promising carbon sequestration technology. Presently, these engineering technologies are at their infancy, and their wider acceptance and implementation mainly depends upon their cost effectiveness, environmental friendly nature and viability in long run. While implementing any new carbon sequestration strategy social, legal and regulatory aspects are required to bring consensus considering present and future ecological and economic considerations. Therefore, extensive research is warranted for the development of such technologies in the near future. Most importantly proper policy and regulatory measures need to be developed with regards to measurement, monitoring, residence time and trading of carbon credits. The carbon sequestration capability of forests, soils and wetlands can be enhanced through land use, forest and agricultural crops management. Increasing scientific understanding and adaptation of appropriate management practices for water use in agriculture and maintaining soil nutrients balance are the absolute requirements of enhancing soil productivity and soil carbon trapping capability. This can be achieved by adaptation of desired regulatory measure and identification and implementation of relevant policy incentives.

11.5 Acknowledgements

Sincere thanks are due to the Natural Sciences and Engineering Research Council of Canada (Grant A 4984, Canada Research Chair) for their financial support. Views and opinions expressed in this article are those of the authors.

11.6 Abbreviations

BDT	Bone Dry Ton
CCS	Carbon Capture and Storage
CWD	Coarse Woody Debris
DDW	Down and Dead Woody materials
ECMB	Enhanced Coalbed Methane Recovery
GHG	Green house gases
N_2O	Nitrous Oxide
ppm	Parts per million
Th.bdt	Thousands of bone dry ton
UNIDO	United Nations Industrial Development Organization
VOCs	Volatile Organic Compounds

11.7 References

Alewell, C., Schaub, M., and Conen, F. (2009). "A method to detect soil carbon degradation during soil erosion." *Biogeosciences*, 6(11), 2541–2547.

Anderson, S., Gantzer, C., and Brown, J. (1990). "Soil physical properties after 100 years of continuous cultivation." *Journal of Soil and Water Conservation*, 45 (1), 117-121.

Arends, T., and Casth, P. (1994). "The comparative effect of equivalent amounts of NPK applied in farmyard manure or in fertilizers, as a function of soil properties." *Agrok mas Talajtan*, 43, 398–407.

Barredo, J.I., San-Miguel-Ayanz, J., Caudullo, G., and Busetto, L. (2012). *A European Map of Living Forest Biomass and Carbon Stock.* EUR 25730EN, Joint Research Centre of the European Commission,

Bowman, R.A., and Halvorson, A.D. (1998). "Soil chemical changes after nine years of differential N fertilization in a no-till dryland wheat-corn-fallow rotation." *Soil Science*, 163(3), 241–247.

Bradford, J.B., Fraver, S., Milo, A.M., D'Amato, A.W., Palik, B., Shinneman, D.J. (2012). "Effects of multiple interacting disturbances and salvage logging on forest carbon stocks." *Forest Ecology and Management*, 267, 209–214.

BCME (British Columbia, Ministry of Environment) (2013). "Community energy and emission plans." Retrieved from http://www.surrey.ca/files/Ceep-nov2013-final.pdf (Jan19, 2014).

Bruce, J.P., Frome, M., Haites, E., Janzen, H., Lal, R., and Paustian, K. (1999). "Carbon sequestration in soils." *Journal of Soil and Water Conservation*, 54, 382–389.

Bunker, D.E., DeClerck, F, Bradford J.C., Colwell, R.K., Perfecto, I., Phillips, O.L., Sankaran, M., and Naeem. S. (2005). "Species loss and aboveground carbon storage in a tropical forest." *Science*, 310, 1029–1031.

Buyanovsky, G.A., and Wagner, G.H. (1998). "Carbon cycling in cultivated land and its global significance." *Global Change Biology*, 4(2), 131–141.

Campbell, C., Biederbeck, V., Zentner, R., and Lafond, G. (1991). "Effect of crop rotations and cultural practices on soil organic matter, microbial biomass and respiration in a thin Black Chernozem." *Canadian Journal of Soil Science*, 71(3), 363–376.

Christensen, B.T. (1996). "The Askov long-term experiments on animal manure and mineral fertilizers." *Springer Berlin Heidelberg*, 301–312.

Couth, R., and Trois, C. (2010). "Carbon emissions reduction strategies in Africa from improved waste management: A review." *Waste management*, 30(11), 2336–2346.

Domke, G.M., Woodall, C.W., Walters, B.F., and Smith, J.E. (2013). "From models to measurements: Comparing downed dead wood carbon stock estimates in the US forest inventory." *PloS one*, 8(3), e59949.

Drinkwater, L.E., Wagoner, P., and Sarrantonio, M. (1998). "Legume-based cropping systems have reduced carbon and nitrogen losses." *Nature*, 396(6708), 262–265.

Durgin P.B. (1977). "Landslides and the weathering of granitic rocks." *Reviews in Engineering Geology*, 3, 127–131.

ENS (2013). "Australia to build first CO_2 capture for algae biofuel." *Environment News Service*, http://ens-newswire.com/2013/07/05/australia-to-build-first-co2-capture-for-algae-biofuel/ (Jan19, 2014).

EPA (2008). *Municipal Solid Waste Generation, Recycling, and Disposal in the United States: Facts and Figures*. Retrieved from http://www.epa.gov/osw/nonhaz/municipal/pubs/msw2008rpt.pdf.

EPA (2009). Greenhouse gas emissions reporting from the petroleum and natural gas industry. Retrieved from http://www.epa.gov/ghgreporting/documents/pdf/2010/Subpart-W_TSD.pdf. (accessed on Dec. 2013).

EPA (2012). *Carbon sequestration through reforestation: A local solution with global implications*. OSRTI, Abandoned Minelands Team, United States Environmental Protection Agency Office of Superfund Remediation and Technology Innovation (OSRTI), 1–21.

Favoino, E., and Hogg, D. (2008). "The potential role of compost in reducing greenhouse gases." *Waste Management and Research*, 26(1), 61–69.

Ferrari, O.D. (2007). *Coal and Biomass-to-Liquids a Comparative Analysis*. Thesis, presented to the Nicolas School of the Environment and Earth Sciences of Duke University for the degree of Masters in Environmental Management.

Folger, P. (2009). *Carbon Capture and Sequestration (CCS)*. Report ADA502774, Library of Congress Washington DC congressional research service.

Follett, R.F. (2001). "Soil management concepts and carbon sequestration in cropland soils." *Soil and Tillage Research*, 61(1–2), 77–92.

Fowles, M. (2007). "Black carbon sequestration as an alternative to bioenergy." *Biomass and Bioenergy*, 31(6), 426–432.

Freund, P., and Ormerod, W.G. (1997). "Progress toward storage of carbon dioxide." *Energy Conversion and Management*, 38, 199–204.

Friedlingstein, P., Houghton, R., Marland, G., Hackler, J., Boden, T.A., Conway, T., Canadell, J., Raupach, M., Ciais, P., and Le Quere, C. (2010). "Update on CO_2 emissions." *Nature Geoscience*, 3(12), 811–812.

Fullen, M.A. (1998). "Effects of grass ley set-aside on runoff, erosion and organic matter levels in sandy soils in east Shropshire, UK." *Soil and Tillage Research*, 46(1), 41–49.

Gale, J. (2004). "Why do we need to consider geological storage of CO_2." *Geological Society*, 233(1), 7–15.

Garrett, T., and Walker, W. (2009). *An Investigation into the Carbon Storage Potential of Pinus Radiata Wood Processing Residue Burial for Carbon Credits*. Report presented to University of Canterbury, in partial fulfilment of the requirements for the BE (Hons) degree in Natural Resources Engineering.

Gollakota, S., and McDonald, S. (2012). "CO_2 capture from ethanol production and storage into the Mt Simon Sandstone." *Greenhouse Gases: Science and Technology*, 2(5), 346–351.

Grace, P., and Oades, J. (1994). "Long-term field trials in Australia." *Long-term Experiments in Agricultural and Ecological Sciences*. Wallingford, CAB International, 53–81.

Grainger, A. (1996). "Modelling anthropogenic degradation of drylands and the potential to mitigate global climate change." *Combating Global Warming by Combating Land Degradation*, edited by V. Squires and V.A. Glenn, Combating Global Warming by Combating Land Degradation, United Nations Environment Programme, Nairobi, 193–214.

Hall, D. O., and House, J. I. (2004). "Biomass energy development and carbon dioxide mitigation options." *Biomass Energy Development*. Retrieved from http://www.uccee.org/CopenhagenConf/hall.htm (Jan. 19, 2014).

Halvorson, A.D., Reule, C.A., and Follett, R.F. (1999). "Nitrogen fertilization effects on soil carbon and nitrogen in a dryland cropping system." *Soil Science Society of America Journal*, 63(4), 912–917.

Halvorson, A.D., Wienhold, B.J., and Black, A.L. (2002). "Tillage, nitrogen, and cropping system effects on soil carbon sequestration." *Soil Science Society of America Journal*, 66(3), 906–912.

Hansen, E.A. (1993). "Soil carbon sequestration beneath hybrid poplar plantations in the north Central United States." *Biomass and Bioenergy*, 5, 431–436.

Hasanimehr, M.H., Rad, H.A., Babaee, V., and Baei, M.S. (2011). "Use of municipal solid waste compost and waste water biosolids with co-composting process." *World Applied Sciences Journal*, 14, 60–66.

Herzog, H., Eliasson, B., and Kaarstad, O. (2000). "Capturing greenhouse gases." *Scientific American*, 282(2), 72–79.

Himes, F. (1998). "Nitrogen, sulfur, and phosphorus and the sequestering of carbon." *Soil Processes and the Carbon Cycle*, edited by R. Lal, CRC Press, Boca Raton, FL, 315–315.

Hong, B.D., and Slatick, E.R. (1994). *Carbon Dioxide Emission Factors for Coal.* Quarterly Coal Report *DOE/EIA-0121(94/Q1)*, Energy Information Administration, Washington, D.C., 1–8.

Houghton, J.T., Ding, Y., Griggs, D.J., Noguer, M., van der Linden, P.J., Dai, X., Maskell, K., and Johnson, C. (2001). *Climate Change 2001: The Scientific Basis*, *IPCC*, Cambridge University Press, Cambridge.

Houghton, R.A. (2007). "Balancing the global carbon budget." *Annual Review of Earth and Planetary Sciences*, 35(1), 313–347.

Panel on Climate Change (IPCC) (1995). *Climate Change 1995, Impacts, Adaptations and Mitigation of Climate Change: Scientific-Technical Analyses*, Cambridge University Press: Cambridge, 1996.

IPCC (2007). *Climate Changes in Climate and Their Effects. Contribution of Working Groups I, II, III to the Fourth Assessment Report of the Intergovernmental Panel on Climate Change.* Report of the Intergovernmental Panel on Climate Change, Geneva, Switzerland.

Jacinthe, P.A., Lal, R., and Kimble, J.M. (2002). "Effects of wheat residue fertilization on accumulation and biochemical attributes of organic carbon in a central Ohio Luvisol." *Soil Science*, 167(11), 750–758.

Jackson, R.B., Jobbágy, E.G., Avissar, R., Roy, S.B., Barrett, D.J., Cook, C.W., Farley, K.A., Maitre, D.C.L., McCarl, B.A., and Murray, B.C. (2005). "Trading water for carbon with biological carbon sequestration." *Science*, 310(5756), 1944–1947.

Janzen, H.H., Campbell, C.A., Izaurralde, R.C., Ellert, B.H., Juma, N., McGill, W.B., and Zentner, R.P. (1998). "Management effects on soil C storage on the Canadian prairies." *Soil and Tillage Research*, 47(3–4), 181–195.

Jenkinson, D.S., Andrew, S.P.S., Lynch, J.M., Goss, M.J., Tinker, P.B. (1990). *The Turnover of Organic Carbon and Nitrogen in Soil [and Discussion].* Philosophical Transactions of the Royal Society of London.

Kerr, R.A. (2001). "Bush backs spending for a 'Global Problem'." *Science*, 292(5524), 1978.

Kintisch, E. (2007). "Report backs more projects to sequester CO_2 from coal." *Science*, 315(5818), 1481.

Klara, S.M., Srivastava, R.D., and McIlvried, H.G. (2003). "Integrated collaborative technology development program for CO_2 sequestration in geologic formations-United States Department of Energy R and D." *Energy Conversion and Management*, 44(17), 2699–2712.

Klusman, R.W. (2003). "Evaluation of leakage potential from a carbon dioxide EOR/sequestration project." *Energy Conversion and Management*, 44(12), 1921–1940.

Körschens, M., and Müller, A. (1996). "The static experiment bad Lauchstädt, Germany." *Evaluation of soil organic matter models*, 38, 369–376.

Lackner, K.S. (2003). "A guide to CO_2 sequestration." *Science*, 300(5626), 1677–1678.

Lal, R. (2001). "World cropland soils as a source or sink for atmospheric carbon." *Advances in Agronomy*, 71, 145–191.

Lal, R., Follett, R.F., and Kimble, J.M. (2003). "Achieving soil carbon sequestration in the United States: A Challenge to the Policy Makers." *Soil Science*, 168(12), 827–845.

Lal, R. (2004a). "Agricultural activities and the global carbon cycle." *Nutrient Cycling in Agroecosystems*, 70(2), 103–116.

Lal, R. (2004b). "Soil carbon sequestration impacts on global climate change and food security." *Science*, 304(5677), 1623–1627.

Lal, R. (2004c). "Soil carbon sequestration in natural and managed tropical forest ecosystems." *Journal of Sustainable Forestry*, 21(1), 1–30.

Lal, R. (2005a). "Forest soils and carbon sequestration." *Forest Ecology and Management,* 220(1–3), 242–258.

Lal, R. (2005b). "World crop residues production and implications of its use as a biofuel." *Environment International*, 31(4), 575–584.

Lal, R. (2006). "Enhancing crop yields in the developing countries through restoration of the soil organic carbon pool in agricultural lands." *Land Degradation and Development*, 17(2), 197–209.

Lal, R. (2008a). "Carbon sequestration." *Philosophical Transactions of Royal Society* B, 363, 815–830.

Lal, R. (2008b). "Crop residues as soil amendments and feedstock for bioethanol production." *Waste Manage*, 28, 747–758.

Lal, R. (2010). "Enhancing eco-efficiency in agro-ecosystems through soil carbon sequestration." *Crop Science*, 50, 120–131.

Laude, A., Ricci, O., Bureau, G., Royer-Adnot, J., and Fabbri, A. (2011). "CO_2 capture and storage from a bioethanol plant: Carbon and energy footprint and economic assessment." *International Journal of Greenhouse Gas Control*, 5(5), 1220–1231.

Leggett, J.A., Lallanzio, R.K., Ek, C., and Parker, L. (2009). *An Overview of Greenhouse Gas (GHG) Control Policies in Various Countries.* CRC Report for Congress R40936, Congressional Research Service.

Lemus, R., and Lal, R. (2005). "Bioenergy crops and carbon sequestration." *Critical Reviews in Plant Sciences*, 24(1), 1–21.

Liebig, M.A., Varvel, G.E., Doran, J.W., and Wienhold, B.J. (2002). "Crop sequence and nitrogen fertilization effects on soil properties in the Western Corn Belt." *Soil Science Society of America Journal*, 66(2), 596–601.

Liski, J., Zhou, L., Alexeyev, V., and Hughes, M.K. (2001). "A large carbon sink in the woody biomass of Northern forests." *Proceedings of the National Academy of Sciences*, 98(26), 14784–14789.

Luo, Y., Stichnothe, H., Schuchart, F., Li, G., Huaitalla, R.M., and Xu, W. (2014). "Life cycle assessment of manure management and nutrient recycling from a Chinese pig farm." *Waste Management & Research*, 32(1), 13–23.

Luske, B. (2010). *Reduced GHG Emissions due to Compost Production and Compost Use in Egypt*. Louis Bolk Institutit.

Maeda, K., D. Hanajima, S. Toyoda, N. Yoshida, R. Morioka, and T. Osada. (2011). "Microbiology of nitrogen cycle in animal manure compost." *Microbial Biotechnology*, 4, 700–709.

Malhi, S.S., Nyborg, M., Harapiak, J.T., Heier, K., and Flore, N.A. (1997). "Increasing organic C and N in soil under bromegrass with long-term N fertilization." *Nutrient Cycling in Agroecosystems*, 49(1–3), 255–260.

Mann, M., and Brown, E. (2008). *Initial Market Assessment for Small-scale Biomass-based CNP.* Report NREL/TP-640-42046, NREL National renewable Energy Laboratory, U. S. Department of Energy, U.S.A.

Marion, J., Nsakala, N., Griffin, T., and Bill, A. (2001). "Controlling power plant CO_2 emissions: a long range view." *1st National Conference on Carbon Sequestration*, National Energy and Technology Laboratory, 14–17, USA.

Marland, G., Boden, T.A., and Andres, R.J. (2009). "Global, regional, and national fossil fuel CO_2 emissions." In *Trends: A Compendium of Data on Global Change.* Carbon Dioxide Information Analysis Center, Oak Ridge National Laboratory, U.S. Department of Energy, Oak Ridge, Tenn., U.S.A.

Marris, E. (2006). "Putting the carbon back: Black is the new green." *Nature*, 442(7103), 624–626.

McCarl, B.A., and Schneider, U.A. (2001). "Greenhouse gas mitigation in U.S. agriculture and forestry." *Science,* 294(5551), 2481–2482.

McLaughlin, S. B., De La Torre-Ugarte, D. G., Garten, C. T., Lynd, L. R., Sanderson, M. A., Tolbert, V. R., and Wolf, D. D. (2002). "High-value renewable energy from prairie grasses." *Environ. Sci. Technol.*, 36, 2122–2129.

Merganičová, K., and Merganič, J. (2010). "Coarse woody debris carbon stocks in natural spruce forests of Babia hora." *Journal of Forest Science*, 56(11), 397–405.

Metting, F. B., Jacobs, G. K., Amthor, J. S., and Dahlman, R. (2002). "Terrestrial carbon sequestration potential." *Management*, 1, 1–2.

Morand, D.T. (2013). "The world reference base for soils (WRB) and soil taxonomy: an appraisal of their application to the soils of the Northern Rivers of New South Wales." *Soil Research*, 51(3), 167–181.

Morris, G. (1999). *The Value of the Benefits of U.S. Biomass Power*. Report NREL/SR-570-27541, National Renewable Energy Laboratory, U.S. Department of Energy Laboratory, Colorado.

Morris, G. (2008). *Bioenergy and Greenhouse Gases*. Report 05-JV-11272164-003, Green Power Institute, the Renewable Energy Program of the Pacific Institute, Berkeley, California.

Morris, G. (2002). *Biomass Energy Production in California 2002: Update of the California Biomass Database*. Report NREL/SR-510-33111, National Renewable Energy Laboratory (NREL), Green Power Institute, Berkeley, California.

Myneni, R.B., Dong, J., Tucker, C.J., Kaufmann, R.K., Kauppi, P.E., Liski, J., Zhou, L., Alexeyev, V., and Hughes, M.K. (2010). "A large carbon sink in the woody biomass of Northern forests." *Proceedings of the National Academy of Sciences*, 98(26), 14784–14789.

Oelkers, E.H., and Cole, D.R. (2008). "Carbon dioxide sequestration a solution to a global problem." *Elements*, 4(5), 305–310.

Ontl, T.A., and Schulte, L.A. (2012). "Soil carbon storage." *Nature Education Knowledge,* 3(10), 35.

Pacala, S., and Socolow, R. (2004). "Stabilization wedges: Solving the climate problem for the next 50 years with current technologies." *Science*, 305(5686), 968–972.

Paul, J.W., Wagner-Riddle, C., Thompson, A., Fleming, R., and MacAlpine, M. (2001). "Composting as a strategy to reduce greenhouse gas emissions." *Climate Change*, 2, 3–5.

Paustian, K., Collins, H.P., and Paul, E.A. (1997). "Management controls on soil carbon." In *Soil Organic Matter in Temperate Ecosystems: Long-term Experiments in North America,* Paul, E.A., Paustian, K, Elliott, E.T., Cole, C.V., (eds.), CRC/Lewis Pblishers, Boca Raton, FL, 15–49.

Petrokofsky, G., Kanamaru, H., Achard, F., Goetz, S.J., Joosten, H., Holmgren, P., Lehtonen, A., Menton, M.C.S., Pullin, A.S., and Wattenbach, M. (2012). "Comparison of methods for measuring and assessing carbon stocks and carbon stock changes in terrestrial carbon pools. How do the accuracy and precision of current methods compare? A systematic review protocol." *Environ. Evidence*, 1(6), 1–21.

Placer County Chief Executive Office and TSS Consultants. (2008). *Forest Biomass Removal on National Forest Lands*. Sierra Nevada Conservancy, Placer County, and the Placer County Air Pollution Control District, 1–27.

Post, W.M., Izaurralde, R.C., Jastrow, J.D., McCarl, B.A., Amonette, J.E., Bailey, V.L., Jardine, P.M., West, T.O., and Zhou, J. (2004). "Enhancement of carbon sequestration in US soils." *Bioscience*, 54(10), 895–908.

Rahman, S.R., Mahmud, N.A., Rahman, M., Hussain, M.Y., and Ali, M.S. (2013). "Overview of biomass energy." *International Journal of Engineering Research and Technology*, 2(11).

Rai, V., Victor, D.G., and Thurber, M.C. (2010). "Carbon capture and storage at scale: Lessons from the growth of analogous energy technologies." *Energy Policy*, 38(8), 4089–4098.

Rasmussen, P.E., Albrecht, S.L., and Smiley, R.W. (1998). "Soil C and N changes under tillage and cropping systems in semi-arid Pacific Northwest agriculture." *Soil and Tillage Research*, 47(3–4), 197–205.

Ridley, A., Slattery, W., Helyar, K., and Cowling, A. (1990). "The importance of the carbon cycle to acidification of a grazed annual pasture." *Australian Journal of Experimental Agriculture*, 30(4), 529–537.

Rynk, Robert (ed.). (1992). *On-Farm Composting Handbook.* Publication NRAES-54. Northeast Regional Agricultural Engineering Service, Cornell Cooperative Extension, Ithaca, NY.

Sauerbeck, D.R. (2001). "CO_2 emissions and C sequestration by agriculture–perspectives and limitations." *Nutrient Cycling in Agroecosystems*, 60(1–3), 253–266.

Schlesinger, W.H. (1999). "Carbon sequestration in soils." *Science*, 284(5423), 2095.

Schrag, D.P. (2007). "Preparing to capture carbon." *Science*, 315(5813), 812–813.

Schubert, P.J. (2009). "Removing crop residues without hurting soil." *Biomass Magazine*, 11, 1–2.

Seibel, B.A., and Walsh, P.J. (2001). "Potential impacts of CO_2 injection on deep-sea biota." *Science*, 294(5541), 319–320.

Shukla, P.R. (1998). "Biomass energy in India: Transition from traditional to modern." *The Social Engineer*, 6(2), 1–20.

Six, J., Conant, R.T., Paul, E.A., and Paustian, K. (2002). "Stabilization mechanisms of soil organic matter: Implications for C-saturation of soils." *Plant and Soil*, 241(2), 155–176.

Smith, P., Powlson, D., Glendining, M., and Smith, J.O. (1997). "Potential for carbon sequestration in European soils: preliminary estimates for five scenarios using results from long-term experiments." *Global Change Biology*, 3(1), 67–79.

Solomon, S., Qin, D., Manning, M., Chen, Z., Marquis, M., Averyt, K., Tignor, M., and Miller, H. (2007). *The Physical Science Basis. Contribution of Working Group I to The Fourth Assessment Report of The Intergovernmental Panel on Climate Change,* Climate Change 2007: The Physical Science Basis, New York, USA.

Studdert, G.A., and Echeverria, H.E. (2000). "Crop rotations and nitrogen fertilization to manage soil organic carbon dynamics." *Soil Science Society of America Journal*, 64(4), 1496–1503.

Sundquist, E.T., Burruss, R., Faulkner, S., Gleason, R., Harden, J., Kharaka, Y., Tieszen, L., and Waldrop, M. (2008). "Carbon sequestration to mitigate climate change." US Department of the Interior, US Geological Survey. Retrieved from: http://pubs.usgs.gov/fs/2008/3097/pdf/CarbonFS.pdf.

Tanthachoon, N., Chiemchaisri, C., and Chiemchaisri, W. (2007). "Utilisation of municipal solid waste compost as landfill cover soil for reducing greenhouse gas emission." *International Journal of Environmental Technology and Management*, 7(3), 286–297.

Tsang, C.F., Benson, S., Kobelski, B., and Smith, R. (2002). "Scientific considerations related to regulation development for CO_2 sequestration in brine formations." *Environmental Geology*, 42(2–3), 275–281.

Turhollow A.F., and R.D. Perlack. (1991). "Emissions of CO_2 from energy crop production." *Biomass and Bioenergy*, 1(3), 129–135.

Uhlen, G. (1991). "Long-term effects of fertilizers, manure, straw and crop rotation on total-N and total-C in soil." *Acta Agriculturae Scandinavica*, 41(2), 119–127.

Uhlen, G., and Tveitnes, S. (1995). "Effects of long-term crop rotation, fertilizers, farm manure and straw on soil productivity." *Norwegian Journal of Agricultural Sciences*, 9, 143–161.

UNIDO (2010). *Global Technology Roadmap for CCS in Industry.* Policy Workshop, http://www.unido.org/fileadmin/user_media/Services/Energy_and_Climate_Change/Energy_Efficiency/CCS/annexes.II.pdf (Jan19, 2014).

VanDevender, K., and Pennington, J.A. (2004). "Organic burial composting of cattle mortality." *FSA1044,* Cooperative Extension Service, University of Arkansas Division of Agriculture, US Department of Agriculture.

Wanichpongpan, W., Gheewala, S.H., and Towpraynoon, S. (2004). "Environmental evaluation of energy production from landfill gas in a life cycle perspective. A case study Nakhon Ratcharima." *The Joint International Conference on Sustainable Energy and Environment (SSE)*, Hua Hin, Thialand, 595–599.

West, T.O., and Post, W.M. (2002). "Soil organic carbon sequestration rates by tillage and crop rotation: A global data analysis." *Soil Science Society of America Journal*, 66(6), 1930–1946.

Witter, E., Mårtensson, A.M., and Garcia, F.V. (1993). "Size of the soil microbial biomass in a long-term field experiment as affected by different n-fertilizers and organic manures." *Soil Biology and Biochemistry*, 25(6), 659–669.

Witzke, P., Banse, M., Gömann, H., Heckelei, T., Breuer, T., Mann, S., Kempen, M., Adenäuer, M., and Zintl, A. (2008). "Modelling of energy-crops in agricultural sector models–a review of existing methodologies." *JRC42597*, European Commission, Joint Research Centre (JRC), Institute for Prospective Technological Studies (IPTS), Spain.

Woodall, C.W., Heath, L.S., and Smith, J.E. (2008). "National inventories of down and dead woody material forest carbon stocks in the United States: Challenges and opportunities." *Forest Ecology and Management*, 256(3), 221–228.

Woodall, C.W. (2010). "Carbon flux of down woody materials in forests of the North Central United States." *International Journal of Forestry Research*, 2010, 1–10.

Zhang, T.C., and Surampalli, R.Y. (2013). "Carbon capture and storage for mitigating climate changes." Chapter 20 in *Climate Change Modeling, Mitigation and Adaptation,* Surampalli, R., Zhang, T.C., Ojha, C.S.P., Gurjar, B.R., Tyagi, R.D., and Kao, C.M. (eds.), ASCE, Reston, Virginia, February 2013, 538–569.

CHAPTER 12

Algae-based Carbon Capture and Sequestrations

Xiaolei Zhang, Song Yan, R. D. Tyagi, Rao Y. Surampalli and Tian C. Zhang

12.1 Introduction

Increasing emission of greenhouse gases caused by the growing need in the amount of fossil fuel has received a considerably attention because of the predicted relationship with global warming (Surampalli et al. 2013). Carbon capture and sequestration (CCS), mainly refers to carbon dioxide capture and sequestration, is an efficient technology for mitigating climate change (Zhang and Surampalli 2013). Various sequestration methods such as ocean sequestration, soil sequestration, terrestrial and marine vegetation sequestration, and geologic formation sequestration, have been reported (Pan et al. 2003; Adhikari et al. 2009; Nicot et al. 2009; Strand and Benford 2009; Zhang and Surampalli 2013). Ocean carbon sequestration and geologic carbon sequestration are considered to be the most acceptable carbon sequestration methods in comparison with other methods because of the abundantly available sources (oceans and underground geologic formations). However, the possibility of the carbon leakage in ocean and/or geologic carbon sequestration has inhibited their applications. On the other hand, vegetation carbon sequestration, also called carbon dioxide biofixation, is receiving a great deal of interest. Vegetation carbon sequestration is a process in which vegetation (e.g., trees, crops, grasses, and algae) uses carbon dioxide as a carbon source to form energy-rich organic compounds (for growth) through photosynthesis (Eq. 12.1). It is a method that simultaneously decreases the amount of carbon dioxide in the atmosphere and creates value-added products for human beings.

$$6CO_2 + 12H_2O \xrightarrow{Light} C_6H_{12}O_6 + 6O_6 \qquad (12.1)$$

Among all the vegetation, algae are superior to others in carbon sequestration due to its fast growth rate (10 to 50 times faster than other plants such as trees and crops) and the possibility of using them for producing green energy such as biodiesel, protein, etc. (Usui and Ikenouchi 1997; Borowitzka 1999; Bush and Hall 2006; Chisti 2007; Li et al.

2008; Meng et al. 2009). In addition, most of algae grow in the aqua environment, which leads the algae-based carbon sequestration more favorable than other vegetation carbon sequestration because of the land saving. Moreover, it is reported that algae could also absorb the gases such as SO_X and NO_X, which are the major cause of acid rain (Rushing 2008). Various macro- and micro-algae, such as *Chlorella sorokiniana*, *Chlorella vulgaris*, *Chlorella pyrenoidosa*, *Chaetomorpha linum*, *Haematococcus Pluvialis*, *Pterocladiella capillacea*, *Scenedesmus obliquus*, *Spirulina platensis*, cultivated in either open ponds or closed photoreactors, have been reported in carbon sequestration (Jeong et al. 2003; Aresta et al. 2005; Moheimani 2005; Huntley and Redalje 2007; de Morais and Costa 2007; Liebert 2008; Oilgae 2011). Although these algae are mainly applied for environmental control such as the purification of flue gases (Douskova et al. 2009; Suryata et al. 2010) and wastewater treatment (Benemann 2002; Wang et al. 2010), some algae (e.g., *Chlamydomonas reinhardtti*) have hundreds of genes that are uniquely associated with carbon dioxide capture and generation of biomass as indicated by studies on genes in the algal genomes (Oilgae 2011).

Recently, considerable studies have concluded that algae, the third generation feedstock researched for bio-energy production, is the best agent for the biologically capturing carbon dioxide from large-scale emitters such as power plants and industries (e.g., Oilgae 2011; PPCCS 2013). Algae are also a sensible choice with regard to their fast proliferation rates, extensive tolerance to wild, extreme environments, and their potential for comprehensive cultures (Oilgae 2011). In this chapter, the principle and carbon cycle of algae-based carbon dioxide sequestration are described first, and then influence factors are considered, followed by the applications of algae-based carbon sequestration and cost estimation. The chapter ends with some discussions about future perspective and conclusions.

12.2 Principle and Carbon Cycle

As any other terrestrial plant-based carbon sequestration, algae accomplish carbon dioxide fixation through photosynthesis. As shown in Eq. 12.2, algae can convert carbon dioxide to biomass and oxygen under sufficient supply of nutrients, photons, and water through photosynthesis. Some studies used $CH_{1.8}N_{0.17}O_{0.56}$ to represent the composition of typical algae (Bayless et al. 2003, 2009); thereby, according to Eq. 12.2, every 44 g of carbon dioxide can produce 25 g of algae biomass through algae photosynthesis, which is a rather efficient way to fix carbon dioxide.

$$CO_2 + H_2O + Nutrients + Photons \rightarrow Algae\ biomass + O_2 \qquad (12.2)$$

As shown in Eq. 12.2, water, nutrients, and light also take significant parts in the process. Algae grow in high moisture conditions (usually in aqua conditions), and therefore, water is considered to be always sufficient during the process. Nutrients (nitrogen and phosphorus) can be added by adding chemicals or wastewater which contains abundant nutrients (Benemann 2002). Photons are from the sun or illumination should a cultivation reactor be used (Hon-nami and Kunito 1998; Aresta et al. 2005).

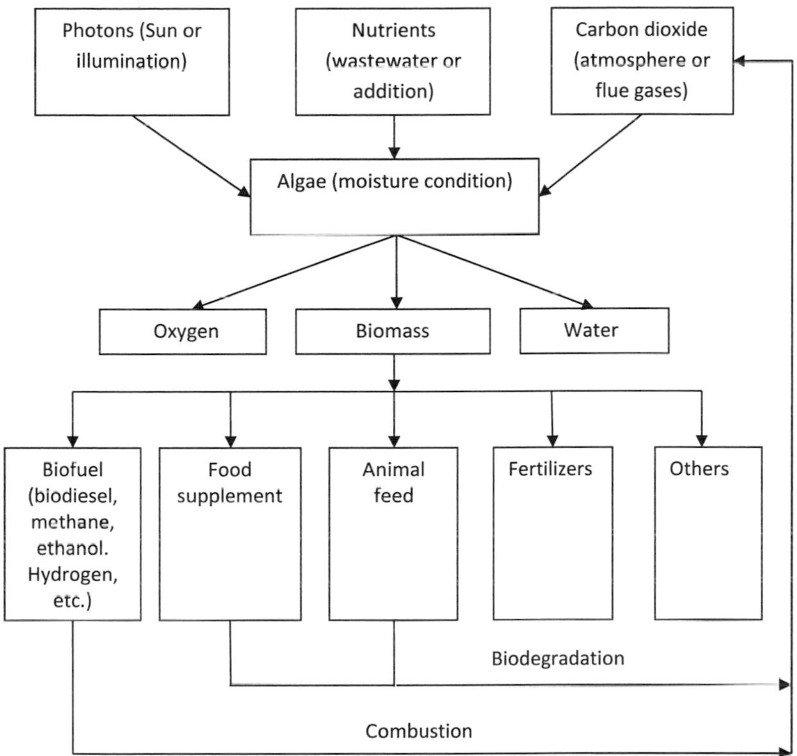

Figure 12.1. Carbon dioixde fixation by algae and possible usage of algae. CO_2 is first 'eaten' by algae for biomass growth, which will be collected/used to produce valuable substances. After the algal substances being consumed, sequestered CO_2 will finally be released to the atmosphere as CO_2, and thereafter, be captured by algae as food again

The products of Eq. 12.2 are algae biomass and oxygen. The increased biomass can be used for human food supplements, animal feed, or raw materials for making ethanol, methane, and biodiesel according to the algae properties, and finally will be converted to carbon dioxide again through combustion (biofuels) or biodegradation (food supplements and animal feed) (Becker 2004). Oxygen will go back to the atmosphere and be used for respiration of living beings. The whole process of algae carbon sequestration is simply exhibited in Figure 12.1. As shown in Figure 12.1, the carbon cycle in algae-based carbon sequestration is an endless chain of carbon dioxide fixation and release without net carbon dioxide impact.

12.3 Effects of Major Factors

The efficiency of algae-based carbon sequestration is affected by many factors such as algae species, cultivation condition, carbon dioxide concentration, and other components of the feeding gas. In this section, the effect of some major factors on algae-based carbon dioxide sequestration is discussed.

12.3.1 Algae Strains Effect

Algae are known as a rich source of protein (50–60% dry weight), lipid (2%–50% dry weight), and vitamins. Various algae, including high vitamin content algae, high CO_2 tolerance algae, high lipid content algae, etc., have been reported to be capable of fixing carbon dioxide (Hanagata et al. 1992; Becker 1994; Graham and Wilcox 2000; Pedroni et al. 2004; Aresta et al. 2005; Rengel 2008). Usually, algae selection is based on the project purpose. For example, when algae are prepared for vitamin supplement, algae with high productivity of target vitamins will be used in carbon dioxide sequestration. In addition, algae growth rate is significantly important in determining the use of algae for carbon dioxide sequestration as well. Algae with high growth rates also refer to having high carbon dioxide sequestration rate as well as high production rate on protein, lipid, or vitamins. Additionally, it is also considered to be beneficial in saving the cultivation area. Stromgren (1984) observed that, among three algae, *Pelvetia fastigiata*, *Pterocladia capillacea* and *Z. farlowii*, *Pterocladia capillacea* had a greater growth rate than the other two algae under the similar cultivation condition. The report suggests that algae selection be necessary in order to achieve high growth rate for carbon dioxide sequestration.

In the early 1930s, researchers have observed that algae (*Chlorococcurn* sp.) were capable of producing vitamins (Gunderson and Skinner 1934). Berg-Nilsen (2006) pointed out that *Spirulina* and *Chlorella* had high value in protein production. *Chlamydomonas reinhardtii* was reported being a good source of therapeutic proteins (protein content > 50%) (Rasala et al. 2010). Oleaginous algae have been extensively studied for biodiesel production. Several algae species such as *Botryococcus* sp., *Chlorella* sp., *Oedogonium* sp., and *Spirogyra* sp. show great potential in biodiesel production (Banerjee et al. 2002; Hossain and Salleh 2008; Francisco et al. 2010). Velea et al. (2009) investigated 35 strains of microalgae to sequester carbon dioxide and to produce oil for biodiesel production, and they found that the lipid content of some microalgae such as *Chlorobotrys* species was up to 70% dry weight and the protein content was around 40–43 %. The studies reveal that selection of algae in carbon dioxide sequestration should comply with the expecting utilization of algae biomass.

12.3.2 Reactor Effect

Bioreactors used for algae carbon dioxide cultivation include lakes, raceway ponds, oceans, plate photobioreactor, carboy photobioreactor, and tubular photobioreactor (Jeong et al. 2003; Ono and Cuello 2004; Moheimani 2005). Compared to open algae cultivation systems (lakes, raceway ponds, oceans), closed photobioreactors provide higher algae productivities and better control on the cultivation condition such as temperature, medium pH, salinity, light intensity, and concentrations of nutrients (Jeong et al. 2003). However, in photobioreactors artificial light is usually used, which increases the cost of carbon sequestration, while open algae cultivation systems mostly rely on natural light (sunlight), which can avoid the addition of the energy that is usually needed in closed photobioreactors (Douskova et al. 2009). Because the cultivation condition has a significant impact on the composition and productivity of algae (Dauta et al. 1990; Suryata et al. 2010), photobioreactors are preferred. In order to reduce the cost of algae-based carbon dioxide sequestration using closed photobioreactors, researchers employed solar electrical energy generation systems to convert sunlight into electricity, which then was used to supply light for the photobioreactors. To some extend the algae cultivation cost could be reduced through the application of solar electrical energy generation. Additionally, a natural light and artificial light combination system is also an efficient method for reducing the cultivation cost. For instance, the greenhouse photobioreactor system placed in a greenhouse can take the advantage of the sunlight in day and generate artificial light at night. Therefore, compared to the cultivation system which depends on either net sunlight (which works only in day) or net artificial light (which consumes a large amount of energy), the greenhouse photobioreactor system prolongs the cultivation time from around 12 hours (for the outdoor system) to 24 hours and save some extra energy (Berg-Nilsen 2006).

12.3.3 Effects of Cultivation Conditions

Algae cultivation conditions, such as pH, salinity, nutrient concentrations of the medium, temperature and light intensity of the cultivation system, are essential factors in algae-based carbon sequestration because they determine the algae productivity and the process efficiency. Furthermore, they also affect the accumulation of protein, vitamins, and lipid, which are associated with algae-based carbon sequestration cost when the final use of the algae is considered as a part of the cost of algae cultivation. Table 12.1 shows some major growth parameters for algae. For flue-gas CO_2 capture, the species that survive best in acidic conditions and high temperature are more desirable.

Table 12.1. Major growth parameters for algae (Oilgae 2011)

Species	Temp (°C)	pH	CO_2%	DT (h)[a]	Features
Chlorococcum sp.	15–27	4–9	≤ 70	8	High CO_2 fixation rate, densely culturable
Chlorella sp.	15–45	3–7	≤ 60	2.5–8	High growth ability, high temperature tolerance
Euglena gracilis	23–27	3.5	≤ 100	24	High amino acid content; good digestibility (effective fodder); grow well under acidic conditions; not easily contaminated
Galdieria sp.	≤ 50	1–4	≤ 100	13	High CO_2 tolerance, like acidic cultures
Viridiella sp.	15–42	2–6	≤ 5	2.9	Accumulates lipid granules inside the cell; high temperature and CO_2 tolerance; like acidic cultures
Synechococcus lividus	44–55	≤ 8.2	≤ 70	8	High pH tolerance

[a]DT = doubling time.

pH. Studies reported that pH requirement varied according to the algae species (Chen and Durbin 1994; Yang and Gao 2003). Chen and Durbin (1994) studied the pH effect on the growth of marine algae, *Thalassiosira pseudonana*, and observed that the algae grew best at pH of 8.8 to 9.4, which was due to the pH effect on algae metabolism (EPOCA 2009). In algae-based carbon dioxide sequestration, the pH of the algae cultivation system becomes very important as it also affects the solubility and availability of carbon dioxide in the system besides the effect on algae metabolism (Moheimani 2005). The form of carbon dioxide in aqua solution could be free carbon dioxide, bicarbonate, and carbonate, depending on the pH. Most of the researchers agreed that free carbon dioxide was the major available carbon source in algae photosynthesis, which was predicted that only free carbon dioxide could bind with enzyme ribulose bisphosphate carboxylase to accomplish the photosynthesis process (Blackman and Smith 1911; Rabinowitch 1945; Cooper et al. 1969; Riebesell et al. 1993). Moreover, some studies reported that bicarbonate also impacted the photosynthesis of algae and predicted that certain algae were capable of dehydrating bicarbonate into carbon dioxide in cells (Badger et al. 1980, Kaplan et al. 1980, Beardall and Raven 1981; Falkowski, 1991). However, it is still debatable that bicarbonate is one form of the substrates in the algae photosynthesis process because the studies on bicarbonate limitation in algae cultivation displayed that photosynthesis was not limited by bicarbonate limitation (Riebesell et al. 1993; Chen and Durbin 1994). Therefore, as the pH of the algae cultivation system increases, free carbon dioxide concentration will be decreased, and thus, the photosynthesis process will be inhibited due to the carbon source (carbon dioxide) limitation. Low pH also could inhibit algae growth due to the enhanced toxicity to the algae from heavy metals, which usually exists in the cultivation system (Luderitz and Nicklisch 1989) and the decrease in photosynthetic activity of algae (Baker et al. 1983).

In short, pH effect on algae carbon dioxide sequestration is mainly due to the effect of pH on algae metabolism, the type of carbon species in the cultivation systems, and algae photosynthetic activity. The optimal pH of the cultivation system should be determined before cultivation according to the algae species and the supplied carbon dioxide concentration.

Temperature. As any living beings, algae have a range of suitable growth temperature. Usually, algae could functionalize normally between 10 °C and 30 °C, and the growth rate of algae increases as temperature increases within this temperature range (Laws et al. 1988; Suryata et al. 2010). Temperature effect on the activity of photosynthesis enzymes of algae is the main reason of temperature effect on algae growth because low or high temperature could inactive enzyme activities (Daniel et al., 2008). However, studies revealed that some algae can tolerate low temperature (< 10 °C) or high temperature (> 30 °C). Hence, the algae which can bear high temperature such as *Cyanidium caldarium* and *Synechococcus elongatus* could be used for sequestering the flue-gas CO_2 which are usually with high temperatures (around 120 °C); the algae which can bear low temperature could be used for sequestering carbon dioxide in cold regions (Seckbach et al. 1971; Miyairi 1995). Temperature is controllable in indoor cultivation systems while the temperature varies in the day and year in outdoor algae carbon dioxide sequestration systems. This is one of the reasons that closed photobioreactor has a higher carbon dioxide sequestration efficiency than outdoor open ponds.

Salinity. Salinity effect on algae growth is mainly due to the effect of osmotic stress and ion stress on algae cell growth. Generally, salinity should be controlled according to the algae species. Suryata et al. (2010) reported that the optimal salinity for blue-green microalgae was around 1.5%. Up to date, salinity effect on algae carbon sequestration has not been paid sufficient attention. However, it has been stimulated recently to study the salinity effect on algae-based carbon sequestration because it has been accepted by the public that using algae growing in oceans could be the most profitable and practical way to sequester carbon dioxide due to the large available ocean area.

Light. Light is one of the key factors in algae carbon sequestration as it is the base of energy conversion (luminous energy to chemical energy) (Bouterfas et al. 2002, 2006). Light intensity is usually used to evaluate light effect on algae-based carbon dioxide sequestration because it, to some extent, determines the penetration and distribution of light (Meseck et al. 2005). Bouterfas et al. (2002) reported that the appropriate light intensity for algae growth was around 400 μmol photons/($m^2 \cdot s$). For the cultivation system with a light source from artificial light, the light effect can be avoided by light adjustment and reactor design. However, for the cultivation system with the light source from natural light (the Sun), the light effect is significant in algae-based carbon dioxide sequestration. Generally, the appropriate reactor design can improve the light penetration and distribution, and hence decrease the light impact on algae cultivation (Bayless et al.2003). Furthermore, stirring is also a method to improve the light penetration and distribution to the algae cultivation system with the light source from natural light since the stirring would offer an equal light exposure opportunity to the algae in deep position of the reactor (pond or lake). However, the stirring strategy is with the risk of reduce the productivity of algae when the shear stress is higher than algae tolerance (Ogbonna and Tanaka 1997; Stepan et al. 2002). Some studies employed lenses, glass or plastic cones, optical fibers, and quartz or acrylic rods to enhance light receive of algae (Tredici and Zittelli 1998).

Apart from light intensity, light/dark cycle, also called photoperiod, is another essential element in algae growth (Bouterfas et al. 2006). The optimal light/dark cycle for algae growth varies according to the algae species (Brand and Guillard 1981). Bouterfas et al. (2002) reported that *Selenastrum minutum*, *Coelastrum micropurum f. astroidea*, and *Cosmarium subprotumidum* showed the best growth under the light/dark cycle of 15h/9h.

Nutrients. The basic nutrients in algae growth are nitrogen and phosphorus. For large scale cultivation systems, it is not practical to add nutrients in the system due to the cost consideration. Therefore, large scale algae-based carbon dioxide sequestration can cooperate with treatment of wastewater which is rich in nitrogen and phosphorus (Lau et al. 1996). For small scale cultivation systems, compared to cooperating with wastewater treatment, the addition of nutrients into the reactor is a more efficient way to supply nutrients to algae (Stepan et al. 2002). In fact, the nutrient source and addition amount should be determined according to the reality. For instance, in carbon dioxide sequestration from flue gases of power plants, some of the nutrients, mainly referring to nitrogen, are available in the flue gases, and thus, nutrient addition should be adjusted according to the composition of the flue gases (Brown 1996; Matsumoto et al. 1997).

12.3.4 Mixing Effect

Mixing is significantly important for open pond algae cultivation systems because the algae in deep location may not be able to obtain enough photons to complete the photosynthesis without mixing due to the shade from the top layer algae in the pond. In addition, mixing also influences the transfer of substances, including carbon dioxide, nitrogen, phosphate, and minerals. However, an appropriate mixing intensity is required because too much mixing intensity could harm algae, while too less mixing intensity could not reach the demand to assist substance transfer (Ogbonna and Tanaka 1997; Jungo et al. 2001; Stepan et al. 2002). In addition, mixing could influence pH stability of the cultivation system (Persoone et al. 1980). Usually, direct injection of carbon dioxide into the algae cultivation system is the most common way for carbon dioxide addition. However, the addition of a large amount of carbon dioxide into the cultivation system would cause a decrease in local pH. Therefore, appropriate mixing is required in algae-based carbon dioxide sequestration. Moreover, mixing would also enhance oxygen diffusion into the atmosphere and thus promote the photosynthesis process (Eq. 12.2). Drapcho and Brune (2000) observed that the carbon fixation rate increased 50% when a proper mixing was performed. Moheimani (2005) reported that the growth rate and productivity of algae, *Pleurochrysis carterae*, showed an increasing trend when the mixing speed was increased from 0 to 200 rpm, while when the mixing speed was beyond 200 rpm, the growth rate and productivity of algae started to decrease. It was also observed that algae started dying when the mixing speed was up to 600 rpm. Marshall and Huang (2010) used a numerical model to stimulate the mixing effect on algae productivity and reported that a proper mixing would significantly enhance the algae production rate. Moreover, mixing would also improve the transfer rate of carbon dioxide to liquid because turbulence can decrease the thickness of the liquid boundary layer (Yang and Cussler 1986).

12.3.5 Carbon Dioxide Concentration and Transportation Effect

As discussed in Section 12.3.3, free carbon dioxide is the major substrate in an algae photosynthesis process. Researchers reported that carbon dioxide concentration had a great impact on carbon metabolism and photochemical properties of algae cells (Badger et al. 1980; Kaplan et al. 1980; Spalding et al. 1984). Hence, carbon dioxide concentration would significantly impact algae productivity. Low carbon dioxide concentration will inhibit algae productivity due to the limited supply of the substrate (carbon dioxide). For instance, using algae to fix carbon dioxide in the air would not be effective because the concentration of carbon dioxide in the air is only around 0.03–0.06%. It was also reported that high carbon concentration would inhibit algae growth because of the possibility of pH decrease when a high concentration of free carbon dioxide presents in the cultivation system.

Although CO_2 concentrations vary, depending on the flue gas source, 15–20% (v/v) is a typically-assumed amount of concentration (PPCCS 2013). Kumar et al. (2010) used algae, *Spirulina platensis*, to fix carbon dioxide from gas mixtures with carbon dioxide contents of 15%, 30%, and 100% (v/v), respectively, and observed that algae productivity was reduced, and even the death of algae occurred during the long term cultivation with these high carbon dioxide concentrations. On the contrary, some studies reported that some algae such as *Chlorella vulgaris*, *Chlorococcum littorale*, and *Scenedesmus* sp., could tolerance to high carbon dioxide concentration (13%, 60%, and 80% (v/v)) (Hanagata et al. 1992; Kodama et al. 1993; Douskova et al. 2009). These studies suggest that the capability for algae to tolerance to high carbon dioxide concentration varies with the algae species. Therefore, in algae carbon dioxide sequestration, algae can be selected according to the carbon dioxide concentration in the target gases. For instance, the algae with best productivity under carbon dioxide concentration from 5% to 15% (v/v) can be used to fix the carbon dioxide of flue gases from power plants because the carbon dioxide contents in power plant flue gases are usually between 5% and 15%.

The rate of carbon dioxide transfer from target gases to algae cultivation systems also shows great effect on algae productivity. Gas bubbling is the most common way for feeding carbon dioxide to the algae cultivation system. The transfer rate of carbon dioxide from the target gases to the cultivation system is associated with the solubility of carbon dioxide in the cultivation system. Carbon dioxide solubility is affected by several factors, including the pH, temperature, and composition of the dissolving solution, the composition of and the carbon dioxide fraction in the gases to be treated, and the reaction between the components of the target gases and the absorption solution. Based on Eq. 12.3, at low pH, carbon dioxide transfer will be inhibited. It is known that high temperature will decrease the solubility of gases in the solution. Therefore, the temperature of flue gases should be lowered when algae are used for sequestration of carbon dioxide in order to obtain a high efficiency. The compositions of dissolving solution and target gases and the composition of the solution after target gases being treated would also affect the solubility of carbon dioxide (Stepan et al. 2002).

$$CO_2 + H_2O \rightarrow HCO_3^- + H^+ \tag{12.3}$$

On the other hand, indirect carbon dioxide feeding, which first fixes carbon dioxide from the target gases through chemical reaction and then uses the latter as substrate for algae growth, has also been studied (Merrett et al. 1996, Emma et al. 2000). For instance, carbonates such as Na_2CO_3 converted from carbon dioxide can be used as substrate by certain algae such as *Nannochloropsis oculata* which has the capacity to decompose the carbonates into free carbon dioxide through enzyme carboanhydrase, and thus, completes carbon dioxide sequestering (Merrett et al. 1996; Emma et al. 2000).

12.3.6 Others Factors

Generally, algae-based carbon dioxide sequestration aims to treat the gases of a high content of carbon dioxide, such as industrial exhausted gases, which usually also contains NO_X, SO_X, and oxygen. The presence of sulfur oxides (SO_2 is the most common sulfur oxides in exhausted gases) as well as nitrogen oxides (NO is the most common nitrogen oxide in exhausted gases) in the target gases can result in a decrease in the pH of the algae cultivation system, and then impacts algae growth. Stepan et al. (2002) pointed out that the pH of the cultivation system could decrease to less than 4 within a day when the SO_2 concentration in the exhausted gases was up to 400 ppm. In addition to the influence of SO_2 on pH, the presence of SO_2 in the target gases would harm algae health as well. In order to minimize the impact of SO_2 on algae growth, NaOH can be used to maintain an optimal pH for algae growth (Matsumoto et al. 1997).

Moreover, selecting the algae with great tolerance to SO_2 is also an alternative. Zeiler et al. (1995) and Brown (1996) revealed that *Nannochloris* sp. could perform normal even under the SO_2 concentration of 50 ppm.

On the other hand, nitrogen oxides could cause only a slight change in pH. However, the contribution of nitrogen oxides on nutrient supply is significant when oxygen is sufficient in the cultivation system or in the target gases as NO can be oxidized into NO_2 whose concentration in the algae cultivation system is proportional to algae productivity (Brown 1996; Matsumoto et al. 1997). However, the presence of the high concentration of oxygen in the system will hinder algae growth because of the photorespiration (Richmond et al. 1993; Moheimani 2005).

Contamination is a severe problem in algae open pond cultivation systems. The growth of other organisms such as unwanted algae, fungi, bacteria, and yeast in the algae cultivation system will compete to the nutrients with wanted algae, and thus, result in reduction of the quality and quantity of the yield. Furthermore, it may cause the loss of the culture (Richmond et al. 1990; Borowitzka 1999). In order to reduce the contamination risk, close cultivation systems should be used as needed, and additionally, periodic cleaning should be carried out (Richmond 2004).

Harvesting period is also considered as an important factor for algae growth. With a long harvesting time in open pond cultivation systems, the algae growth rate will be negatively affected because of the shade from the top layer algae, which will cause the light limitation to the algae in deep location of the systems. Moreover, nutrient limitation is another problem. The algae growth will be hampered when a long harvesting period is carried out due to the large amount of algae present in the cultivation system in which nutrients could be limited.

The design on the reactor could also influence algae productivity since the reactor design (e.g., the surface to volume ratio of the cultivation system) would significantly impact light penetration and distribution (Bayless et al. 2003). In addition, it also affects the carbon dioxide transfer rate in the system, removal of toxic components (sulfur oxides) and accumulation of oxygen in the system. As described in Section 12.3.2, open pond cultivation systems and closed photoreactor cultivation systems are the mainly used algae cultivation systems. Open pond cultivation systems suffer from the difficulty in controlling cultivation conditions and carbon dioxide loss to the atmosphere. Closed photoreactor cultivation systems can be better controlled and thus usually lead greater algae productivity than open pond cultivation systems. However, there are also shortcomings, such as (1) the accumulation of toxic gases such as SO_2 and oxygen, and (2) limited cultivation space, which limits the yield of algae within a certain period time in compared to the open pond system.

Various types of photobioreactors have been used in algae cultivation. Tubular photobioreactors showed better performance in light penetration and distribution than other reactors such as spiral, helical, inclined plate, and flat panel reactors (Watanabe and Hall 1995; Watanabe and Hall 1996; Tredici and Zittelli 1998; Moheimani 2005). Kumar et al. (2010) enhanced algae productivity through enhancing the carbon dioxide transfer rate due to the increase of interfacial area by using a hollow fiber membrane. However, the use of membrane to enhance the carbon dioxide transfer rate may cause an increase in cost because of the need in membrane replacement. In order to obtain a cost-effective cultivation system, the design on the cultivation system should be improved.

In short, many factors have been shown affecting algae-based carbon dioxide sequestration. These factors usually affect one another. For instance, the change in pH generally correlate with the change in temperature and dissolved oxygen of the system (Chen and Durbin 1994). Therefore, selection of cultivation systems should take into account of the relationships among different factors in order to obtain a high efficiency sequestration system.

12.4 Applications

Applications of algae-based carbon dioxide sequestration mainly include treating flue gases from power industry for greenhouse gas emission control, treating wastewater (for nitrogen removal), producing raw materials for biofuels production (e.g., ethanol, biodiesel), recovering fertilizer (effluent from agriculture irrigation), and controlling eutrophication. Some of these applications are described below. Detailed information on live projects of algae-based CO_2 capture is summarized in Table 12.2.

12.4.1 Application in Greenhouse Gas Emission Control

Researchers predicted that carbon dioxide concentration in the atmosphere would continually increase from the current value of 385 parts per million by volume (ppmv) to 600 ppmv at the end of the century (Solomon et al. 2009; Wang et al. 2010; Surampalli et al. 2013). Carbon dioxide from burning fuels in power plants is the major contributor of carbon dioxide to the atmosphere (Lisbona et al. 2010; Wang et al. 2010). In order to mitigate carbon dioxide emission from fuel utilization, capturing carbon dioxide from the flue gases becomes extremely necessary. Using algae to fix the carbon dioxide from flue gases have been widely studied (Kadam 1997; Kumar et al. 2010; Lisbona et al. 2010). Figure 12.2 shows a typical process used for capturing carbon dioxide from flue gases. After pretreatment to remove harmful gases (NO_x and SO_2), flue dust, and so on, flue gases can be directly injected into the algae cultivation system in which the algae have high resistance to temperature (Section 12.3.3). Otherwise, cooling can be performed to bring the high temperature flue gases to an acceptable value for algae, and then the flue gases are injected into the algae cultivation system. As discussed above, solubility of carbon dioxide in the cultivation system has great effects on algae productivity. Therefore, a process for enhancing carbon dioxide solubility in the solution can be set up before flue gases enter the algae cultivation system.

Table 12.2. Live projects of algae-based carbon dioxide capture (PPCCS 2013)

Location/Co. name	CO_2 capturer	Cultivation	Algae application
Portland/Columbia Energy Partners and Portland General Electric	Algae	Photo-bioreactor	Biodiesel and ethanol production
Australian/MBD Energy	Algae	-	high grade plastics, transport fuel, and livestock feed
Israel/Seambiotic & Israel Electric Company	Algae	Open pond	Animal, fish, and human food source
Italy/Enitechnologie	Algae	Raceway pond	Methanol
Canada/Trident Exploration and Menova	Algae	Photo-bioreactor	Biofuel
Arizona/ Arizona Public Service Company	Algae	Photo-bioreactor	Biofuel
Germany/ RW Energy	Algae	Photo-bioreactor	-
Hamburg/ E.ON Hanse	Algae	Photo-bioreactor	Animal feed and biofuel
California/ Carbon Capture Corporation	Algae	Photo-bioreactor	Biodiesel, butanol, biomethane, and jet fuel propellant

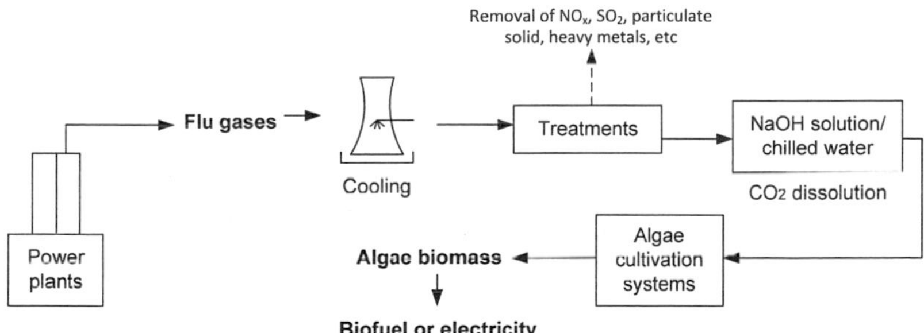

Figure 12.2. Algae-basd CO_2 cultivation near Power plants (PPCCS 2013)

Negoro et al. (1993) used *Nannochloropsis* sp and *Phaeodactylum* sp. cultivated in a raceway pond inside a glass greenhouse to fix carbon dioxide emitted from Tohoku Electric Power Company's Shin-Sendal power station by directly injecting the flue gas, SO_x-free flue gas, and flue gas with compressed (12%) carbon dioxide into the pond, respectively. It was observed that the algae productivity was similar (~10 g/(m^2·d)) for the three situations, indicating that the two algae were not affected by direct injection of the flue gas into the cultivation system. It suggests that the two algae could be used to fix carbon dioxide of flue gas without complex pretreatment. A study on the fixation of the flue-gas carbon dioxide from a natural gas combined cycle power plant with *Tetraselmis suecica* cultivated in both the open pond and the closed photobioreactor, showed that algae productivity of the open pond system was higher than that of the closed photoreactor system (Pedroni et al. 2004). It reveals that large-scale algae carbon dioxide fixation from flue gases is feasible; open pond algae cultivation systems are of the most potential for large scale algae cultivation. Feng (2008) studied fixation of carbon dioxide from the flue gas of a coal-fired power plant with algae *Botryococcus braunii*, which had a high hydrocarbon content. Feng (2008) then extracted hydrocarbon from algae harvested from the cultivation system fed with the flue gas and found that the algae contained 14% of carbon dioxide. These studies show that algae application for removing carbon dioxide from flue gases could be a cost-effective method because the further use of algae as a raw material for value-added products could offset part of the treatment cost.

The algae fixation of carbon dioxide from flue gases generated by the power plant provides an effective way for carbon dioxide removal. However, complex pretreatment on flue gases for minimizing the harm to algae from heavy metals, toxic gases, and organic compounds often results in an increase in capital cost. In addition, flue gases with high concentrations of carbon dioxide and high temperature would exclude many algae, which may prevent us from using algae as raw materials for the production of biofuels, protein, vitamins, etc.

12.4.2 Application in Wastewater Treatment

Using algae to treat wastewater in ponds is well established (Oswald et al. 1953; Oswald 2003). Algae-based treatment of dairy and piggery waste also has been investigated (e.g., An et al. 2003; Craggs et al. 2004; Kebede-Westhead et al. 2006; Mulbry et al. 2008). As discussed in Section 12.3.3, nutrients (nitrogen and phosphate) are essential substances for algae growth. Usually, the ratios of carbon:nitrogen (C:N 3.5:1) and carbon:phosphorus (C:P 20:1) in domestic sewage and dairy lagoon water (C:N 3:1; C:P 10:1) are low compared to typical ratios in rapidly growing algae biomass (C:N 6:1; C:P 48:1) (Woertz et al. 2009). Adding CO_2 in the flue gas could enhance algae production and complete assimilation of wastewater nutrients by algae, making the combination of alga-based carbon dioxide sequestration and wastewater treatment a perfect process to reduce the cost of both carbon dioxide sequestration and wastewater treatment. According to the composition of wastewater (nitrogen and phosphorus concentrations), carbon dioxide sequestration can be accomplished through directly cultivating algae in the wastewater or the mixture of the medium and nutrients-rich wastewater.

CO_2 supplementation to promote algae productivity has been studied for many years (Burlew 1953; Benemann et al. 1980). However, using the flue gas as a CO_2 source for algae culture started in 1990s (e.g., Yun et al. 1997; Straka et al. 2000). Yun et al. (1997) investigated simultaneous fixation of the flue-gas carbon dioxide and ammonia removal from the wastewater of a steel-making plant by cultivating algae, *Chlorella vulgaris*, in the wastewater. It was found that the carbon dioxide fixation rate was around 26 g/(m^3-h), and the ammonia removal rate was around 1 g/(m^3-h). Even through the ammonia removal rate is still not comparable with traditional activated sludge wastewater treatment (around 5–50 g/m^3-h) (Sotirakou et al. 1999; Abd El-Hady et al. 2001; Joo et al. 2007), it is possible to enhance ammonia removal rate by appropriate selection of algae (Joo et al. 2007). Deviller et al. (2004) employed a high-rate algae pond to reduce nutrient of the wastewater discharged from a fish rearing system; they observed that the treated water was qualified for recirculating to the fish rearing system. Kumar et al. (2010) reported that the removal efficiency of carbon dioxide from flue gas with an initial carbon dioxide concentration of 15% v/v and nitrogen from wastewater with an initial nitrogen concentration of 412 mg NO_3-N/L reached 85% and 68%, respectively, in a membrane photoreactor algae cultivation system.

So far, many studies have been reported on the potential of simultaneous carbon dioxide fixation and wastewater treatment using algae (Oswald 1973; An et al. 2003; Gomez-Villa et al. 2005). The strategy of using the contaminants (ammonia and phosphorus) in wastewater as nutrients to feed algae for fixing carbon dioxide should save energy and mitigate carbon dioxide emission simultaneously.

12.4.3 Application in Biofuel Production

Currently, biofuel production has granted an extensive interest due to the predication of energy crisis. Diesel, one of the most widely used fuels, can be produced from oil of seeds and plants, and animal fat, which is called biodiesel. However, the competition with food industry results in an increase in the cost of biodiesel production from conventional feed stocks (oil of seeds and plants or fat of animal). Algae are a promising raw material for biofuel production, such as biodiesel, ethanol, methane, hydrogen, biomass (where algae biomass used directly for combustion), and other hydrocarbon fuel variants (e.g., JP-8 fuel, gasoline, biobutanol) (PPCCS 2013). Algae are capable of producing oil suitable for conversion to biodiesel with an areal productivity 20–40 times that of oilseed crops, such as soy and canola (Sheehan et al. 1998). The U.S. Department of Energy's Aquatic Species Programme (ASP) undertook over a decade of research (between 1978 and 1996) and found that algae were only economically viable as a biofuel at oil prices of more than $60 a barrel (PPCCS 2013).

As shown in Fig. 12.3, production of biodiesel from algae includes algae cultivation (which can be combined with CO_2 sequestration), algae harvesting, lipid/oil extraction, biodiesel conversion, and transferring algae biomass to biogas for power generation. Whatever be the final energy product(s), the following represent the stages involved in the algae-to-fuel process: 1) strain selection; 2) cultivation/growth; 3) harvesting/drying; 4) extraction; and 5) conversion to an energy product (PPCCS 2013).

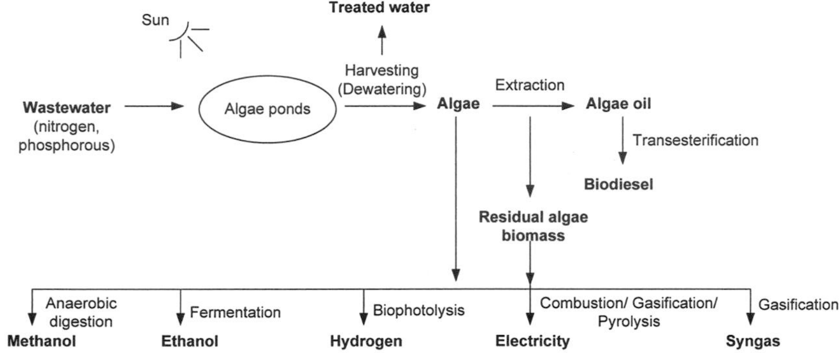

Figure 12.3. Process train for algae wastewater treatment and biofuel production (Adapted from Woertz et al. 2009)

Huntley and Redalje (2007) reported that *Haematococcus pluvialis*, cultivated in a photoreactor coupled with an open pond, was a great candidate in producing microbial oil. It was observed that the oil production rate of the algae was greater than 420 GJ/ha-year (50–400 GJ/ha-year for terrestrial plants) with the maximum production rate of 1014 GJ/ha-year. Francisco et al. (2010) investigated the quality of biodiesel derived from microalgae obtained from carbon dioxide sequestration and observed that the fuel properties of the microalgae biodiesel complied with the US standard (ASTM 6751) and the European Standard (EN 14214). However, Aresta et al. (2005) pointed out that using algae for biodiesel production wouldn't be a cost-effective method unless carbon dioxide emission could be reduced during the process after they evaluate the energetic balance for producing biodiesel from macro-algae using computing software. It indicates that the simultaneous biodiesel production and carbon dioxide sequestration would be an alternative for energy generation and reduction in carbon dioxide emission.

Other fuels such as ethanol can also be produced through fermentation of algae biomass. Hirano et al. (1997) used the algae with high starch content (37%), *Chlorella vulgaris*, to produce ethanol and obtained a 65% ethanol production rate. It was found that algae starch was a good source for ethanol production through fermentation.

Methane has also been reported to be produced from algae used for carbon dioxide sequestration (Hansson 1983). Woertz et al. (2009) investigated lipid productivity and nutrient removal by green algae grown during treatment of dairy farm and municipal wastewaters supplemented with CO_2. For dairy wastewater, maximum lipid productivity peaked at day 6 of batch growth, with a volumetric productivity of 17 mg/day-L of reactor and an aerial productivity of 11,000 L/ha-yr (1,200 gallons/acre-year) if sustained year-round. After 12 days, ammonium and orthophosphate removals were 96% and > 99%, respectively. Municipal wastewater was treated in semi-continuous indoor cultures with 2–4 day hydraulic residence times (HRTs). Maximum lipid productivity for the municipal wastewater was 24 mg/day-L, observed in the 3-day HRT cultures. Over 99% removal of ammonium and orthophosphate was achieved. The results from both types of wastewater suggest that CO_2-supplemented algae cultures can simultaneously remove dissolved nitrogen and phosphorus to low levels while generating a feedstock potentially useful for liquid biofuels production.

Biofuel production from algae used for carbon dioxide sequestration offers a way to compensate the cost of carbon dioxide sequestration. On the other hand, producing other products such as protein, vitamins, and food supplements could also be an alternative to balance the cost of carbon dioxide sequestration using algae.

12.4.4 Application in Fertilizer Recovery and Eutrophication Control

Large amounts of nitrogen and phosphorus from agriculture irrigation are discharged into rivers or lakes every year, which causes eutrophication and severe problems on the living beings in the receiving water bodies. It is known that algae are efficient in nitrogen and phosphorus storing because the N and P content could reach 10% and 1% of the dry weight of the algae, respectively, which is several times higher than in other plants, and thus, makes algae be a competitive candidate for nutrient removal. The harvested algae biomass can be used for biofuel production, and the residue can be reapplied to agriculture as fertilizers. Therefore, using algae cultivation systems to remove nutrients from agriculture irrigation effluent, in concert with carbon dioxide sequestration from the flue gas or atmosphere would be an alternative for fertilizer recovery and eutrophication control. Benemann et al. (2006) evaluated the use of algae for carbon dioxide sequestration as well as for fertilizer recovery from agriculture drains in Southern California where 10,000 tons of nitrogen and phosphorus are discharged annually from irrigated agriculture into the sea. They found that the cost of carbon dioxide sequestration could be only \$10/ton of CO_2-C equivalent if biodiesel production and fertilizer recovery could also be achieved. Furthermore, it was also revealed that carbon dioxide emission from fossil fuel burning could be mitigated about several hundred thousand tons if such a process was used. Certainly, it would also avoid the eutrophication risk.

Apart from the applications of algae carbon dioxide sequestration in the above aspects, there are many other applications such as in animal waste treatment (Fallowfield et al. 1999) and electricity generation (Wang et al. 2010). It can be found that the applications of algae-based carbon dioxide sequestration could reduce the cost of algae-based carbon dioxide sequestration, and would also benefit the environment, energy and agriculture.

12.5 Economic Analysis

Currently, algae-based carbon dioxide sequestration is mainly applied in carbon dioxide removal from flue gases, and is considered being an expensive carbon dioxide sequestration method because of the nutrient addition, costly algae harvesting, and low carbon dioxide fixation efficiency (Benemann et al. 2006; Francisco et al. 2010; Kumar et al. 2010). However, its cost would be reduced if algae biomass obtained from carbon dioxide sequestration could be used for production of biofuel, fertilizer, or other substances (Hirano et al. 1997; Benemann et al. 2006; Huntley and Redalje 2007). Compared to open pond algae cultivation, closed photoreactors generally provide higher algae productivity (Bayless et al. 2001; Huntley and Redalje 2007). However, Huntley and Redalje (2007) reported that the average annual capital cost of closed photobioreactors for algae-based carbon dioxide sequestration is $1,000,000/ha, which is nine times higher than the cost of open pond systems ($94,000/ha). It is apparent that the open pond cultivation system is more cost-effective, and thus, more feasible and sustainable for carbon dioxide sequestration than the closed photobioreactor system. Based on the literature, the average algae productivity is 30 g/m^2-day (110 ton dry weight/ha-year) in open pond cultivation systems (Stepan et al. 2002; Ono and Cuello 2004; Huntley and Redalje 2007). Therefore, the cost of algae-based carbon dioxide sequestration can be roughly assessed based on the valuable material content of algae. For instance, it is known that the average lipid content of algae is 30 % (dry weight of algae), and thus, it can be calculated that approximate 33 ton oil will be annually generated per hectare. The current price of biodiesel is around $160/bbl; hence, the annual offset from biodiesel production for the removal of the flue-gas carbon dioxide will be $38,544/ha. Based on this, the cost of algae-based carbon dioxide sequestration from flue gases can be reduced from $94,000/ha-year to $55,454/ha-year, saving over 40% of the sequestration cost. In fact, the residue of the algae after biodiesel production can be used for producing methane or ethanol, which would also benefit the reduction on the cost of carbon dioxide sequestration from flue gases. In addition, if the process is coupled with wastewater treatment, the cost would be further reduced due to the cost saving in nutrient addition in algae cultivation and wastewater treatment.

In short, conventional CCS methods cost about $30–50 to capture, transport and store 1 T of CO_2. Currently, the algae-based carbon capture (and partial sequestration) method costs ~\$175 per T of CO_2. However, under optimal conditions, not only are all the costs covered through the revenues generated from biodiesel and other by-products, the CO_2 capture process might even provide an attractive business opportunity by actually being profitable (Oilgae 2011; PPCCS 2013).

12.6 Limitation and Future Perspectives

All human activities together may emit ~30 billion ton of CO_2 every year, of which power plants alone emit about 10 billion ton of CO_2. Theoretically, one ton of algae biomass requires about 1.8 T of CO_2, which implies that out of 10 billion ton of CO_2 emitted by the power plants, we can get ~5.5 billion ton of algae biomass (Oilgae 2011). The right strains of algae may have ~30% of oil by weight. Thus, 5.5 billion ton of algae will result in about 1.65 billion tons of oil, which is > 40% of the total yearly world consumption of oil (~4.2 billion ton/year). While the above analyses indicate the potential to provide sustainable solutions for alternative biofuels and CO_2 mitigation, important challenges faced in algae-based carbon capture include: 1) how to increase carbon dioxide uptake; 2) how to cope with the availability of water source and land near power plants; 3) obtaining suitable algae strains that tolerate raw flu gases for CO_2 capture; 4) engineering challenges of CO_2 capture in large algae cultivation system; 5) developing economic and efficient algae harvesting and drying technologies; 6) using power plants wastewater for algae growth; 7) evaluating regional climate impact on algae-based CO_2 capture; 8) how to minimize energy required for algae-based CO_2 capture; and 9) lack of information on energy and economy of algae carbon dioxide capture (PPCCS 2013).

High cost or low efficiency has been the main obstacle for using algae to sequester carbon dioxide. As mentioned previously, algae-based carbon dioxide sequestration usually occurs either in open ponds or in closed photoreactor. However, each of them is suffering from problems. The closed photobioreactor is more efficient due to the better control on algae cultivation. However, it is also more costly than the open pond system (Bayless et al. 2003; Huntley and Redalje 2007). Compared to the closed photobioreactor, an open pond is more feasible for large scale algae carbon dioxide sequestration. However, the low efficiency has hindered open pond's application. Moreover, the requirement of large land area makes the application of open pond algae-based carbon dioxide sequestration difficult in the regions with a high population density and limited land. For example, a 50-MW 50% base-load natural gas-fired electrical generation plant operating 18 h/day over a 240-day season would produce 216 million kWh/season, releasing 30.3 million kg-C/season of fossil-fuel CO_2. An algal process designed to capture 70% of the flue-gas CO_2 would require an area of 880 ha of high-rate algal ponds operating at a productivity of 20 g VS/m^2-day, which would produce 42.4 million kg algal dry wt/season (Brune et al. 2009). If 100% of the algal biomass were harvested and used for replacing biogas methane usage, soybean feed replacement, and biodiesel production, the gross greenhouse gas reduction would be about 36%; the net parasitic energy cost to harvest and process the algal biomass would be about 10% of plant total energy output, resulting in a new greenhouse gas reduction of 26% (Brune et al. 2009).

Algae have shown great potential for removing carbon dioxide from flue gases of power plants. However, the presence of toxic substances and high temperature of the gas steam strongly require appropriate pretreatments, which would not be necessary if using other methods to remove carbon dioxide from flue gases, such as injecting the flue gas to $Ca(OH)_2$ solution. On the other hand, using algae to treat the flue gas with high carbon dioxide concentrations demands a careful selection on algae tolerant to high concentration carbon dioxide, high temperature, or even high-salinity. Isolating fast strains and strains that grow in hyper saline environments are alternatives to monocultures or are well suited to an environment with vast and rapid temperature changes would either improve carbon capture and biomass accumulation abilities without the need for genetic modification, mitigate costs required to deal with contamination, or minimize the amount of heating and cooling necessary to keep a culture alive (PPCCS 2013).

Coupling algae-based removal of flue-gas carbon dioxide with wastewater treatment, biofuel production, and/or fertilizer recovery could be a cost-efficient way for removal of flue-gas carbon dioxide. However, there is still a difficulty in selecting algae which is tolerant to high carbon dioxide concentrations and have a high oil content. Furthermore, the transport of flue gases/wastewater/effluent of agriculture irrigation would increase the treatment cost because the algae cultivation system may not be located near both the power plants and wastewater treatment plants/agriculture land.

For enhancing the application of algae-based carbon dioxide sequestration, the following aspects need to be studied:

(1) Reduction of the capital cost of closed photobioreactor cultivation systems through using wastewater as nutrient source, and artificial light derived from the electricity generated from solar generators should be performed.

(2) Application of membranes for efficient separation of carbon dioxide from toxic gases of flue gases so as to reduce the cost of pretreatment.

(3) Selection of algae that have high values in producing biofuels, protein, vitamins, and other substances in order to offset the process cost.

(4) The efficient use of algae biomass for compensating the cost of algae carbon dioxide sequestration, such as evaluation on the use of algae biomass obtained from wastewater for protein production or fertilizer recovery to find out which process is more cost-efficient.

(5) Design of open pond algae cultivation systems to enhance the process efficiency.

(6) Selection of algae capable of using other forms of carbon as substrate such as bicarbonate and carbonate so as to enhance carbon dioxide capture to overcome the problem related to the low solubility of carbon dioxide in the solution.

12.7 Summary

Algae-based carbon sequestration is one of the latest methods of biological sequestration vastly exploited in CO_2 emitting industries. Using algae for carbon dioxide sequestration is an endless chain with no net carbon dioxide emission, that is, 1) algae use carbon dioxide as substrate for biomass growing through photosynthesis; 2) the carbon dioxide fixed on the biomass returns to the environment through decomposition (if used as animal feed) or combustion (if converted to biodiesel, ethanol, and methane); and 3) the released carbon dioxide will be again captured and used by algae. The efficiency of algae-based carbon dioxide sequestration is affected by various factors which include the algae species, cultivation conditions, design of the cultivation system (the open pond or closed photoreactor), carbon dioxide transfer, etc. It is important to study the factors in order to optimize the process and extend their applications. The algae-based process has been widely applied to remove carbon dioxide from flue gases of power plants. However, the high capital cost has hampered the application. In order to reduce the cost, the use of algae biomass to produce high value substances such as biodiesel, protein, food supplement and vitamins has been studied; to some extend these products can offset the process cost. However, the current cost ($35–150/ton carbon dioxide removed by algae from flue gases from power plants) is still much higher than the expected one ($10/ton). Algae-based CCS is still on its infant. Much work is needed to achieve high carbon dioxide sequestration efficiency with low cost.

12.8 Acknowledgements

Sincere thanks are due to the Natural Sciences and Engineering Research Council of Canada (Grant A 4984, Canada Research Chair) for their financial support. The views and opinions expressed in this chapter are those of the authors.

12.9 References

Abd El-Hady, H.M., Grunwald, A., Vlckova, K., and Zeithammerova, J. (2001). "Clinoptilolite in drinking water treatment for ammonia removal." *Acta Polytechnica*, 41(1), 41–45.

Adhikari, S., Bajracharaya, R.M., and Sitaula, B.K. (2009). "A review of carbon dynamics and sequestration in wetlands." *J. Wetlands Ecology*, 2(1–2), 42–46.

An, J.Y., Sim, S.J., Lee, J.S., and Kim, B.W. (2003). "Hydrocarbon production from secondarily treated piggery wastewater by the green alga *Botryococcus braunii*." *Journal of Applied Phycology*, 15(2–3),185–191.

Aresta, M., Dibenedetto, A., and Barberio, G. (2005). "Utilization of macro-algae for enhanced CO_2 fixation and biofuels production: Development of a computing software for an LCA study." *Fuel Processing Technology*, 86(14–15), 1679–1693.

Badger, M.R., Kaplan, A., and Berry, J.A. (1980). "Internal inorganic carbon pool of *Chlamydomonas reinhardtii*: Evidence for a carbon dioxide concentrating mechanism." *Plant Physiology*, 66(3), 407–413.

Baker, M.D., Mayfield, C.I., Inniss, W.E., and Wong, P.T.S. (1983). "Toxicity of pH, heavy metals and bisulfite to a freshwater green alga." *Chemosphere*, 12(1), 35–44.

Banerjee, A., Harma, R.S., Chisti, Y., and Banerjee, U.C. (2002). "*Botryococcus braunii*: A renewable source of hydrocarbons and other chemicals." *Critical Reviews in Biotechnololgy*, 22(3), 245–279.

Bayless, D.J., Kremer, G.G., Prudich, M.E., Stuart, B.J., Vis-Chiasson, M.L., Cooksey, K., and Muhs, J. (2003). "Enhanced practical photosynthetic CO_2 mitigation." In *Proceedings of the 1st National Conference on Carbon Sequestration*, 2001, 1–14.

Bayless, D.J., Kremer, G., Vis, M., Stuart, B., and Shi, L. (2009). "Photosynthetic CO_2 mitigation using a novel membrane-based photobioreactor." Available at <http://www.ohio.edu/ohiocoal/research/upload/Enhanced%20Practical%20Photosynthetic%20CO2%20Mitigation.pdf> (accessed Oct. 2012).

Beardall, J., and Raven, J.A. (1981). "Transport of inorganic carbon and the 'CO_2 concentrating mechanism' in *Chlorella emersonii* (*Chlorophyceae*)." *Journal of Phycology*, 17(2), 134–141.

Becker, E.W. (1994). *Microalgae: Biotechnology and microbiology. Cambridge University Press, Cambridge.*

Becker, E.W. (2004). "Micro-algae for human and animal consumption." In Richmond, A. (ed.), *Handbook of microalgal culture*. Blackwell, Oxford (2004), 312–351.

Benemann, J.R., Koopman, B.L., Weissman, J.C., Eisenberg, D.M., and Goebel, R. (1980). "Development of microalgae harvesting and high rate pond technologies in California." *Algae biomass: Production and use*, Shelef, G., and Soeder, C.J. (eds.), Elsvier North, Amsterdam, The Netherlands, 457–496.

Benemann, J.R. (2002) *A Technology Roadmap for Greenhouse Gas Abatement with Microalgae.* Report to the U.S. Department of Energy, National Energy Technology Laboratory, and the International Energy Agency Greenhouse Gas Abatement Programme.

Benemann, J.R., van Olst, J.C., Massingill, M.J., Weissman, J.C., and Brune D.E. (2006). "The controlled eutrophication process: Using microalgae for CO_2 utilization and agricultural fertilizer recycling." <http://www.oilgae.com/blog/2006/10/microalgae-for-co2-utilization.html> (accessed Oct. 2012).

Berg-Nilsen, J. (2006). *Production of Micro Algae-based Products*. Project Report, Aug. 2006. Available at <http://www.climatebabes.com/documents/Algae%20 market.pdf> (accessed Oct. 2011).

Blackman, F.F., and Smith, A.M. (1911). "Experimental researches on vegetable assimilation and respiration. IX. on assirnilation in submerged water-plants and its relation to the concentration of carbon dioxide and other factors." *Proceedings of the Royal Society*, 83, 389–412.

Borowitzka, M.A. (1999). "Commercial production of microalgae: ponds, tanks, tubes and fermenters." *Journal of Biotechnology*, 70(1–3), 313–321.

Bouterfas, R., Belkoura, M., and Dauta, A. (2002). "Light and temperature effects on the growth rate of three freshwater algae isolated from a eutrophic lake." *Hydrobiologia*, 489, 207–217.

Bouterfas, R., Belkoura, M., and Dauta, A.L. (2006). "The effects of irradiance and photoperiod on the growth rate of three freshwater green algae isolated from a eutrophic lake." *Limnetica*, 25(3), 647–656.

Brand, L.E., and Guillard R.R.L. (1981). "The effects of continuous light and light intensity on the reproduction rates of twenty-two species of marine phytoplankton." *Journal of Experimental Marine Biology and Ecology*, 50(2–3), 119–132.

Brown, L.M. (1996). "Uptake of carbon dioxide from flue gas by microalgae." *Energy Conversion Manage*, 37(6–8), 1363–1367.

Brune, D.E., Lundquist, T.J., and Benemann, J.R. (2009). "Microalgal biomass for greenhouse gas reductions: Potential for replacement of fossil fuels and animal feeds." *J. Environ. Eng.*, 135(11), 1136–1144.

Burlew, J.S. (1953). *Algal Culture: From Laboratory to Pilot Plant*. Carnegie Institution of Washington Publication 600, Washington, D.C.

Bush, R.A., and Hall, K.M. (2006). "Process for the production of ethanol from algae." *United States Patent* 7135308.

Chen, C.Y., and Durbin, E.G. (1994). "Effects of pH on the growth and carbon uptake of marine phytoplankton." *Marine Ecology Progress Series*, 109, 83–94.

Chisti, Y. (2007). "Biodiesel from microalgae." *Biotechnol. Adv.*, 25(3), 294–306.

Cooper, T. G., Filmer, D., Wishnick, M., and Lane, M.D. (1969). "The active species of "CO_2" utilized by ribulose bisphosphaten carboxylase." *Journal of Biology and Chemistry*, 244, 1081–1083.

Craggs, R.J., Sukias, J.P., Tanner, C.T., and Davies-Colley, R.J. (2004). "Advanced pond system for dairy-farm effluent treatment." *N. Z. J. Agric. Res.*, 47, 449–460.

Daniel, R., Danson, M.J., Eisenthal, R., Lee, C.K., and Peterson, M. (2008). "The effect of temperature on enzyme activity: new insights and their implications." *Extremophiles*, 12(1), 51–59.

Dauta, A., Devaux, J., Piquemali, F., and Boumnich, L. (1990). "Growth rate of four freshwater algae in relation to light and temperature." *Hydrobiologia*, 207(1), 221–226.

de Morais, M.G., and Costa, J.A.V. (2007). "Biofixation of carbon dioxide by *Spirulina sp.* and *Scenedesmus obliquus* cultivated in a three stage serial tubular photobioreactor." *Jounral of Biotechnology*, 129(3), 439–445.

Deviller, G., Aliaume C., Nava, M.A.F., Casellas, C., and Blancheton J.P. (2004). "High-rate algal pond treatment for water reuse in an integrated marine fish recirculating system: Effect on water quality and sea bass growth." *Aquaculture*, 235(1–4), 331–344.

Douskova, I., Doucha, J., Livansky, K., Machat, J., Novak, P., Umysova, D., Zachleder, V., and Vitova, M. (2009). "Simultaneous flue gas bioremediation and reduction of microalgal biomass production costs." *Appl. Microbiol. Biotechnol.*, 82(1), 179–185.

Drapcho, C.M., and Brune, D.E. (2000). "The partitioned aquaculture system: Impact of design and environmental parameters on algal productivity and photosynthetic oxygen production." *Aquacultural Engineering*, 21(3), 151–168.

Emma, H.I., Colman, B., Espie, G.S., and Lubian, L.M. (2000). "Active transport of CO_2 by three species of marine microalgae." *J. Phycology*, 36(2), 314–320.

EPOCA (European Project on OCean Acidification). (2009) "Testing the effects of ocean acidification on algal metabolism: considerations for experimental designs." Available at < http://oceanacidification.wordpress.com>, published 12/29/2009 (accessed Jan. 2013).

Falkowski, P.G. (1991). "Species variability in the fractionation of ^{13}C and ^{12}C by marine phytoplankton." *Journal of Plankton Research*, 13, S21–S28.

Fallowfield, H.J., Martin, N.J., and Cromar, N.J. (1999). "Performance of a batch-fed high rate algal pond for animal waste treatment." *European Journal of Phycology*, 34(3), 231–237.

Feng, M. (2008). "Microalgae cultivation in bioreactors for CO_2 mitigation from power plant flue gas and fuel production by supercritical CO_2 extraction." *The 2008 Spring National Meeting*, New Orleans, LA, April 6, 2008.

Francisco, E., Neves, D., Jacob-Lopesb, E., and Francoa, T.T. (2010). "Microalgae as feedstock for biodieselproduction: Carbon dioxide sequestration, lipid production and biofuel quality." *Chemical Technology and Biotechnology*, 85(3), 395–403.

Gomez-Villa, H., Voltolina, D., Nieves, M., and Pina, P. (2005). "Biomass production and nutrient budget in outdoor cultures of *Scenedesmus obliquus* (*Chlorophyceae*) in artificial wastewater, under the winter and summer conditions of Mazatlan, Sinaloa, Mexico." *Vie et Milieu*, 55,121–126.

Graham, L.E., and Wilcox, L.W. (2000). *Algae.* Prentice-Hall, Inc., Upper Saddle River, NJ.640.

Gunderson, M.F. and Skinner, C.E. (1934). "Production of vitamins by a pure culture of *Chlorococum* grown in darkness on a synthetic medium. " *Plant Pysiology*, 9(4), 807–815.

Hanagata, N., Takeuchi, T., Fukuju, Y., Barnes, D.J., and Karube, I. (1992). "Tolerance of microalgae to high CO_2 and high temperature." *Phytochemistry*, 31(10), 3345–3348.

Hansson, G. (1983). "Methane production from marine, green macro-algae." *Resources and Conservation*, 8(3), 185–194.

Hirano, A., Ueda, R., Hirayama, S., and Ogushi, Y. (1997). "CO_2 fixation and ethanol production with microalgal photosynthesis and intracellular anaerobic fermentation." *Energy*, 22(2–3), 137–142.

Hon-nami, K., and Kunito, S. (1998). "Microalgae cultivation in a tubular bioreactor and utilization of their cells." *Chinese Journal of Oceanology and Limnology*, 16(1), 75–83.

Hossain, S.A.B.M., and Salleh, A. (2008). "Biodiesel fuel production from algae as renewable energy." *American Journal of Biochemistry and Biotechnology*, 4(3), 250–254.

Huntley, M.E., and Redalje, D.G. (2007). "CO_2 mitigation and renewable oil from photosynthetic microbes: A new appraisal." *Mitigation and Adaptation Strategies for Global Change*, 12(4), 573–608.

Jeong, M.L., Gillis, J.M., and Hwang, J.-Y. (2003). "Carbon dioxide mitigation by microalgal photosynthesis." *Bulletin of Korean Chemical Society*, 24(12), 1763–1766.

Joo, H.-S., Hirai, M., and Makoto, S. (2007). "Improvement in ammonium removal efficiency in wastewater treatment by mixed culture of *Alcaligenes faecalis* No. 4 and L1." *Journal of bioscience and bioengineering*, 103(1), 66–73.

Jungo, E., Visser, P. M., Stroom, J., and Mur, L.R. (2001). "Artificial mixing to reduce growth of the blue-green alga *Microcystis* in lake Nieuwe Meer, Amsterdam: An evaluation of 7 years of experience." *Water Science and Technology: Water Supply*, 1(1), 17–23.

Kadam, K.L. (1997). "Power plant flue gas as a source of CO_2 for microalgae cultivation: Economic impact of different process options." *Energy Conversion and Management*, 38, S505–S510.

Kaplan, A., Badger, M.R., and Berry, J.A. (1980). "Photosynthesis and the intracellular inorganic carbon pool in the blue green alga *Anabaena variabilis*: response to external CO_2 concentration." *Planta*, 149(3), 219–226.

Kebede-Westhead, E., Pizarro, C., and Mulbry, W. (2006). "Treatment of swine manure effluent using freshwater algae: Production, nutrient recovery, and elemental composition of algal biomass at four effluent loading rates." *J. Appl. Phycol.*, 18(1), 41–46.

Kodama, M., Ikemoto, H., and Miyachi, S. (1993). "A new species of highly CO_2-tolreant fast-growing marine microalga suitable for high-density culture." *Journal of Marine Biotechnology*, 1, 21–25.

Kumar, A., Yuan, X., Sahu Ashish, K., Dewulf, J., Ergas, S.J., and Van Langenhoveb, H. (2010). "A hollow fiber membrane photo-bioreactor for CO_2 sequestration from combustion gas coupled with wastewater." *Journal of Chemical Technology and Biotechnology*, 85, 387–394.

Lau, P.S., Tam, N.F.Y., and Wong, W.S. (1996). "Wastewater nutrients removal by *Chlorella vulgaris*: Optimization through acclimation." *Environmental Technology*, 17(2), 183–189.

Laws, E.A., Taguchi, S., Hirata, J., and Pang, L. (1988). "Optimization of microalgal production in a shallow outdoor flume." *Biotechnology and Bioengineering*, 32(2), 140–147.

Li, Y., Horsman, M., Wu, N., Lan, C.Q., and Dubois-Calero, N. (2008). "Biofuels from microalgae." *Biotechnology Progress*, 25(4), 294–306.

Liebert, T.C. (2008). "CO_2 sequestration by algae reactors calling it unproved technology understates its problems." Available at <http://kansas.sierraclub.org/Wind/ AlgaeReactors.htm.> (accessed Oct. 2012).

Lisbona, P., Martinez, A., Lara, Y., and Romeo, L.M. (2010) "Integration of carbonate CO_2 capture cycle and coal-fired power plants. A comparative study for different sorbents." *Energy Fuels*, 24(1), 728–736.

Luderitz, V., and Nicklisch, A. (1989). "The effect of pH on copper toxicity to blue-green algae." *International Revue der Gesamten Hydrobiologie*, 74(3), 283–291.

Matsumoto, H., Hamasaki, A., Sioji, N., and Ikuta, Y. (1997). "Influence of CO_2, SO_2, and NO in flue gas on microalgae productivity." *Journal of Chemical Engineering of Japan*, 30(4), 620–324.

Marshall, J.S., and Huang, Y. (2010). "Simulation of light limited algae growthing in homogeneous turbulence." *Chemical Engineering Science*, 65(12), 3865–3875.

Meng, X., Yang, J., Xu, X., Zhang, L., Nie, Q., and Xian, M. (2009). "Biodiesel production from oleaginous microorganisms." *Renewable Energy*, 34(1), 1–5.

Merrett, M.J., Nimer, N.A., and Dong, L.F. (1996). "The utilization of bicarbonate ions by the marine microalga *Nannochloropsis oculata* (Droop) Hibberd." *Plant, Cell, and Environment*, 19(4), 478–484.

Meseck, S.L., Alix, J.H., and Wikfors, G.H. (2005). "Photoperiod and light intensity effects on growth and utilization of nutrients by the aquaculture feed microalga. *Tetraselmis chui* (*PLY429*)." *Aquaculture*, 246(1–4), 393–404.

Miyairi, S. (1995). "CO_2 assimilation in a thermophilic cyanobacterium." *Energy Conversion and Management*, 36(6–9), 763–766.

Moheimani, N.R. (2005). "The culture of coccolithophorid algae for carbon dioxide bioremediation." *Doctoral thesis of Philosophy of Murdoch University*.

Mulbry, W., Kondrad, S., and Buyer, J. (2008). "Treatment of dairy and swine manure effluents using freshwater algae: Fatty acid content and composition of algal biomass at different manure loading rates." *J. Appl. Phycol.*, 20(6), 1079–1085.

Negoro, M., Hamasaki, A., Ikuta, Y., Makita, T., Hirayama, K., and Suzuki, S. (1993). "Carbon dioxide fixation by microalgae photosynthesis using actual flue gas discharged from a boiler." *Applied Biochemistry and Biotechnology. Part A: Enzyme Engineering and Biotechnology*, 51–52, 681–692.

Nicot, J.-P., Oldenburg, C.M., Bryant, S.L., and Hovorka, S.D. (2009). "Pressure perturbations from geologic carbon sequestration: Area-of-review boundaries and borehole leakage driving forces." *Energy Procedia*, 1(1), 47–54.

Ono, E., and Cuello, J.L. (2004). "A selection of optimal microalgae species for CO_2 sequestration." Available at <http://www.netl.doe.gov/publications/proceedings /03/carbon-seq/PDFs/158.pdf> (accessed Jan. 2013).

Ogbonna, J.C., and Tanaka, H. (1997). "Industrial-sized photobioreactors." *Chemtech*, 27(7), 43–49.

Oilgae (2011). *The Comprehensive Guide for Algae-based Carbon Capture*. Free sample report available upon request at admin@oilgae.com (requested March 2013).

Oswald, W.J., Gotaas, H.B., Ludwig, H.F., and Lynch, V. (1953). "Algae symbiosis in oxidation ponds: Photosynthetic oxygenation." *Sewage Ind. Waste.*, 25(6), 692–705.

Oswald, W.J. (1973). "Productivity of algae in sewage disposal." *Solar Energy*, 151(1), 107–117.

Oswald, W.J. (2003). "My sixty years in applied algology." *J. Appl. Phycol.*, 15, 99–106.

Pan, G., Li L., Wu, L., and Zhang, X. (2003). "Storage and sequestration potential of topsoil organic carbon in China's paddy soils." *Global Change Biology*, 10(1), 79–92.

Pedroni, P.M., Lamenti, G., Prosperi, G., Ritorto, L., Scolla, G., Capuano, F., and Valdiserri, M. (2004). "Enitecnologie R&D project on microalgae biofixation of CO_2: Outdoor comparative tests of biomass productivity using flue gas CO_2 from a NGCC power plant." Available at <http://uregina.ca/ghgt7/PDF/papers/ nonpeer/075.pdf> (accessed Oct. 2012).

Persoone, G., Morales, J., Verlet, H., and de Pauw, N. (1980). "Air-lift pumps and the effect of mixing on algal growth." *In: Shelef, G. et al. (Eds.) (1980) Algae Biomass, Elsevier/North Holland Biomed, Amsterdam*, 506–522.

PPCCS (PowerPlantCCS) (2013). *Comprehensive Guide for Algae-based Carbon Capture*. Available at <http://www.powerplantccs.com/ccs/cap/fut/alg/alg.html> (accessed March, 2013).

Rabinowitch, E. (1945). *Photosynthesis and Related Processes. Vol. I. Chemistry of Photosynthesis, Chemosynthesis and Related Processes in vitro and in vivo.* Interscience Publishers, Inc., New York.

Rasala, B.A., Muto, M., Lee, P.A., Jager, M., Cardoso Rosa, M.F., Behnke, C.A., Kirk, P., Hokanson, C.A., Crea, R., Mendez, M., and Mayfield, S.P. (2010). "Production of therapeutic proteins in algae, analysis of expression of seven human proteins in the chloroplast of *Chlamydomonas reinhardtii.*" *Plant Biotechnology of Journal*, 8(6), 719–733.

Rengel, A. (2008). "Promissing technologies for biodiesel production from algae growth systems." 8th European IFSA Symposium, Clermont-Ferrand, 6–10 Jul. 2008.

Richmond, A., Boussiba, S., Vonshak, A., and Kopel, R. (1993). "A new tubular reactor for mass production of microalgae outdoors." *Journal of Applied Phycology*, 5(3), 327–332.

Richmond, A., Lichtenberg, E., Stahl, B., and Vonshak, A. (1990). "Quantitative assessment of the major limitation on the productivity of *Spirulina platensis* in open raceways." *Journal of Applied Phycology*, 2(3), 195–206.

Richmond, A. (2004). "Principles for attaining maximal microalgal productivity in photobioreactors: An review." *Hydrobiologia*, 512, 33–37.

Riebesell, U., Wolf-Gladrow, D.A., and Smetacek, V. (1993). "Carbon dioxide limitation of marine phytoplankton growth rates." *Nature*, 361, 249–251.

Rushing, S.A. (2008). "Carbon dioxide sequestration via algae biofuels: an overview." Available at <http://www.biofuelsdigest.com/blog2/2008/08/21/carbon-dioxide-sequestration-via-algae-biofuels-an-overview> (accessed Oct. 2012).

Seckbach, J., Gross, H., and Nathan, M.B. (1971). "Growth and photosynthesis of *Cyanidium caldarium* cultured under pure CO_2." *Israel J.Botany*, 20, 84–90.

Sheehan, J., Dunahay, T., Benemann, J., and Roessler, P. (1998). *A Look Back at the U.S. Department of Energy's Aquatic Species Program Biodiesel from Algae*, National Renewable Energy Laboratory, Golden, Colo.

Solomon, S., Plattner, G.K., Knutti, R., and Friedlingstein, P. (2009). "Irreversible climate change due to carbon dioxide emissions." *Proceedings of National Academy of Science U.S.A.*, 106(6), 1704–1709.

Sotirakou, E., Kladitis, G., Diamantis, N., and Grigoropoulou, H. (1999). "Ammonia and phosphorus removal in municipal wastewater treatment plant with extended aeration." *Global Nest: the International Journal*, 1(1), 47–53.

Spalding, M.H., Critchley, C., Govindjee, and Orgren, L.W. (1984). "Influence of carbon dioxide concentration during growth on fluorescence induction characteristics of the green alga *Chlamydomonas feinhardii*." *Photosynthesis Research*, 5(2), 169–176.

Stepan, D.J., Shockey, R.E., Moe, T.A., and Dorn, R. (2002). *Production of Micro Algea-Bbased Products.* Project report, Nordic Innovation Centre project #: 03109.

Straka, F., Doucha, J., and Livansky, K. (2000). "Flue-gas CO_2 as a source of carbon in closed cycle with solar culture of microalgae." *Proc., 4th European Workshop on Biotechnology of Microalgae*, 29–30.

Strand, S.E., and Benford, G. (2009). "Ocean sequestration of crop residue carbon: recycling fossil fuel carbon back to deep sediments." *Environment Science and Technology*, 43(4), 1000–1007.

Stromgren, T. (1984). "Diurnal variation in the length growth-rate of three intertidal algae from the pacific west coast." *Aquatic Botany*, 20(1–2), 1–10.

Surampalli, R., Zhang, T.C., Ojha, C.S.P., Gurjar, B.R., Tyagi, R.D., and Kao, C.M. (eds.) (2013). *Climate Change Modeling, Mitigation and Adaptation.* ASCE, Reston, Virginia, February 2013.

Suryata, I., Svavarsson, H.G., Einarsson, S., Brynjolfsdottir A., and Maliga G. (2010). "Geothermal CO_2 bio-mitigation techniques by utilizing microalgae at the blue lagoon, Iceland." *Proceedings*, Thirty-Fourth Workshop on Geothermal Reservoir Engineering Stanford University, California, February 1–3, 2010.

Tredici, M.R., and Zittelli, G.C. (1998). "Efficiency of sunlight utilization: Tubular versus flat photobioreactors." *Biotechnology and Bioengineering*, 57(2), 187–197.

Usui, N., and Ikenouchi, M. (1997). "The biological CO_2 fixation and utilization project by RITE(1)-highly-effective photobioreactor system." *Energy Conversion and Management*, 38(1), S487–S492.

Velea, S., Dragos, N., Serban, S., Ilie, L., Stalpeanu, D., Nicoara, A., and Stepan, E. (2009). "Biological sequestration of carbon dioxide from thermal power plant emissions, by absorbtion in microalgal culture media." *Romanian Biotechnological Letters*, 14(4), 4485–4490.

Wang, X., Feng, Y., Liu, J., Lee, H., Li, C., Li, N., and Ren, N. (2010). "Sequestration of CO_2 discharged from anode by algal cathode in microbial carbon capture cells (MCCs)." *Biosensors and Bioelectronics*, 25(12), 2639–2643.

Watanabe, Y., and Hall, D.O. (1995). "Photosynthetic CO_2 fixation technologies using a helical tubular bioreactor incorporating the filamentous *Cyanobacterium spirulina platensis*." *Energy Conversion and Management*, 36(6–9), 721–724.

Watanabe, Y., and Hall, D.O. (1996). "Photosynthetic CO_2 conversion technologies using a photobioreactor incorporating microalgae-energy and material balances." *Energy Conversion and Management*, 37(6–8), 1321–1326.

Woertz, I., Feffer, A., Lundquist, T., and Nelson, Y. (2009). "Algae grown on dairy and municipal wastewater for simultaneous nutrient removal and lipid production for biofuel feedstock." *J. Environ. Eng.*, 135(11), 1115–1122.

Yang, M.C., and Cussler, E.L. (1986). "Designing hollow fiber contactors." *American Institute of Chemical Engineering Journal*, 32(11), 1910–1916.

Yang, Y., and Gao, K. (2003). "Effects of CO_2 concentrations on the freshwater microalgae *Chlamydomonas reinhardtii*, *Chlorella pyrenoidosa* and *Scenedesmus obliquus (Chlorophyta)*." *Journal of Applied Phycology*, 15(5), 379–389.

Yun, Y.-S., Lee, S.B., Park, J.M., Lee, C.-I., and Yang, J.-W. (1997). "Carbon dioxide fixation by algal cultivation using wastewater nutrients." *Chemical Technology and Biotechnology*, 69(4), 451–455.

Zeiler, K.G., Heacox, D.A., Toon, S.T., Kadam, K.L., and Brown, L.M. (1995). "The use of microalgae for assimilation and utilization of carbon dioxide from fossil fuel-fired power plant flue gas." *Energy Conversion and Management*, 36(6–9), 707–712.

Zhang, T.C., and Surampalli, R.Y. (2012). "Carbon capture and storage for mitigating climate changes." Chapter 20 in *Climate Change Modeling, Mitigation and Adaptation,* Surampalli, R., Zhang, T.C., Ojha, C.S.P., Gurjar, B.R., Tyagi, R.D., and Kao, C.M. (eds.), ASCE, Reston, Virginia, February 2013.

CHAPTER 13

Carbon Immobilization by Enhanced Photosynthesis of Plants

Klai Nouha, Archana Kumari, Song Yan, R. D. Tyagi, Rao Y. Surampalli and Tian C. Zhang

13.1 Introduction

It is well known that carbon capture, storage and sequestration (CCS) can play a central role in the mitigation of greenhouse gas (GHG) emissions (Surampalli et al. 2013). Currently, CO_2 capture from non-point sources can be realized via four major biologically-related methods, i.e., via trees/organisms, ocean flora, biomass-fueled power plant/biofuel/biochar, and sustainable practices (e.g., soils/grassland, peat bogs). In many cases, these methods are called biosequestration (i.e., the capture and storage of the atmospheric CO_2 by biological processes). Of all these options, the method of using tree and photosynthesizing organisms for CO_2 capture seems to have big potential as trees and other photosynthesizing organisms perform CO_2 capture routinely, and they widely exist in our planet (Zhang and Surampalli 2013). Forests account for ~ 2 times the amount of carbon in the atmosphere, and remove ~30% of all carbon dioxide emissions from fossil fuels every year. Therefore, an increase in the overall forest cover around the world would tend to mitigate global warming (Wikipedia 2013).

Canadell et al. (2008) proposed four major strategies to mitigate carbon emissions through forestry activities: 1) increase the amount of forested land through a reforestation process; 2) increase the carbon density of existing forests at a stand and landscape scale; 3) expand the use of forest products that will sustainably replace fossil fuel emissions; and 4) reduce carbon emissions that are caused from deforestation and degradation. Currently, considerable studies and activities have been focused on implementation of these strategies. One activity is to fulfill the Kyoto Protocol requirement of mandatory land use, land use change and forestry (LULUCF) accounting for afforestation (no forest for last 50 years), reforestation (no forest on Dec. 31, 1989), and deforestation. Another activity is to enhance reforestation while prohibit deforestation via Best Management Practices (BMP) in forestry operations. Another activity centers on selection and genetic improvement of plants/crops for CO_2 capture and biofuel production, that is, on increasing the Earth's proportion of C4 carbon fixation photosynthetic plants. This is because these plants account for ~30% of terrestrial carbon fixation even though they only represent about 5% of Earth's plant biomass (Osborne and Beerling 2006). Wheat, barley, soybeans, potatoes and rice (all C3 staple food crops) can be genetically engineered with the photosynthetic apparatus of C4 plants. For example, ribulose-1,5-bisphosphate carboxylase/oxygenase (RubisCO) is the enzyme responsible for CO_2 fixation during photosynthesis. Modifying or replacing C3 plants' RubisCO genes with more efficient forms, such as those found in some algae, could increase the specificity of RubisCO for CO_2 relative to O_2 and thus, may increase the efficiency of CO_2 capture of the crop plants (Beerling 2008).

The objective of this chapter is to overview the current status of implementation of the aforementioned four major strategies to mitigate carbon emissions through forestry activities. After the introduction section, the chapter briefly describes the concepts and activities related to deforestation, reforestation and afforestation, and then moves into its major section about genetic engineering to increase C4 plants for carbon dioxide fixation. The chapter also discusses future trend and possible challenges associated with the four strategies. Understanding these strategies and related issues would help us to overcome the associated challenges in the future.

13.2 Deforestation and Reforestation

13.2.1 Concept of Deforestation

Deforestation is defined as the destruction of forested land. It has proved to be a major problem all over the world. The rate of deforestation is particularly high in the tropics. The term deforestation has different meanings to different people. For some it's the conversion of forest land to other land use practices, whereas for others it includes all activities that destroy forest land. Some definitions include: loss of any kind of closed forest (FAO/UNEP 1981); conversions of forest to another land use or the long-term reduction of the tree canopy cover below a minimum 10 percent threshold (FAO 1990); the loss of original forest for temporary or permanent clearance of forest for other use purposes (Grainger 1993). Deforestation, for some, describes a situation of complete long-term removal of tree cover (Kaimowitz and Angelsen 1998), while for others it entails permanent destruction of indigenous forests/woodlands (Collin 2001).

Other definitions of deforestation include: 1) any activity that disrupts the natural ecology of the virgin forest; 2) an inevitable result of the current social and economic policies being carried out in the name of development (Revington 2008); and 3) complete destruction of forest cover along with removal of, or unsurvivable injury to, the great majority of trees (Myers 1994).

13.2.2 Global Deforestation

The topic of deforestation has become one of the major global issues since the 1980s. Between 1923 and 1985, at least 26 different calculations of closed forestland were made, and they ranged from 2400 million hectares (ha.) to 6500 million hectares (Mathews et al. 2000). According to the FAO (1993), since 1979 the total loss of forest land increased from 75,000 to 126,000 square kilometers. According to this estimate, Africa, South-East Asia and the 14 developing countries in South America, have already lost more than 250,000 hectares of tropical forests.

In the period between 1980 and 1985, total deforestation in developing countries was about 200 million hectares. In Asia and Africa, these figures stand at 60 and 55 million hectares, respectively. Worldwide, however, Latin American had the highest deforestation rate and over that time 85 million hectares of forestland was lost, much of it in Brazil (FAO 1997). Indeed, between May 2000 and August 2006, Brazil lost nearly 150,000 square kilometers of forest (Butler 2008). Both the FAO and Global Forest Resources Assessment (GFRA 2005) estimated that the world's net loss of forests in the 1990s was 94 million hectares; they projected that if this rate continues, in just over 400 years all of the world's 3,869 million hectares of forest will be gone.

Goodland and Pimentel (2000) reported that 20 to 30 percent of the world's forest land has already been converted to agricultural land use, and this conversion accounts for about 60 percent of worldwide deforestation. The World Resources Institute (WRI) statistics indicate that once tropical forest occupied 16 million square kilometers where now only about 8-9 million square kilometers exist. Current FAO estimates indicate that approximately 10.16 million hectares of tropical forest were lost between 1990 and 2000. This increased to 10.4 million hectares from 2000 to 2005. From these estimates, it is clear that, every year the rate of deforestation has been increasing, and thus far, we have already lost almost one-half of the earth's forest cover.

13.2.3 Causes of Deforestation

The main cause of deforestation differs between countries, depending on the socio-economic, political and physical structure of the country. The most common causes, however, are logging, agricultural expansion, wars, and mining. Based on the nature of the study, micro, regional or macro, different causes of deforestation have been identified and examined (Mahapatra 2001). High levels of wood production are an important cause of deforestation in developing countries (Allen and Barnes 1985). According to Myers (1989), cattle ranching and commercial logging are the chief causes of deforestation whereas logging is a secondary cause of deforestation in Southeast Asia (Kummer and Sham 1994; Panayotou and Sungsuwan 1994). The timber trade has been cited as a contributing factor to deforestation worldwide, and the commercial exploitation of forests has been identified as an important factor in the destruction of forests (Repetto and Holmes 1983; Capistrano 1990). Commercial logging operations seriously deplete forest stocks (Duraiappah 1996), accounting for about 20 percent of forest loss, while the rest is chiefly due to fuelwood collection and other uses, such as urban development and infrastructure (Shafik 1994; Southgate 1994). In a separate study, Shafik (1994) considered the harvesting of timber, the removal of other forest products and the clearing of forest for livestock and agricultural production to be the chief factors leading to deforestation.

Increased fuelwood demand, burning and grazing, and weak forest protection institution (Brown and Pearce 1994) contribute to deforestation. The World Resources Institute (WRI 1994) cited increasing human and livestock, populations, poverty, the demand for fuel wood and high levels of consumption by industrialized nations to be important causes of deforestation. The United Nation's Food and Agriculture Organization (FAO 1997) estimates show that 1.5 billion of the 2 billion people worldwide who rely on fuelwood for cooking and heating are overcutting forests. Indeed, the demands for fuelwood by subsistence agricultural households may be the leading cause of world deforestation (Amacher et al. 1996).

13.2.4 Deforestation and its Effects

Because of deforestation, many problems are growing, and today's world is facing many serious problems, such as erosions, loss of biodiversity through extinction of plants and animal species, and increased atmospheric carbon dioxide. There have been many discussions on the effects of deforestation in our environment. These effects range from alarming to catastrophic. There are increasing concerns about the potential ecological impacts of the large-scale clearing of forests (Whitmore 1997; Groombridge and Jenkins 2000; Wright 2005). Two sets of effects have been identified that can have potentially deleterious consequences on biota: 1) the effects from habitat loss, and 2) the effects from the fragmentation of remaining habitats (Fahrig 2003). Habitat loss includes the agricultural conversion of forest to crops or pasture for grazing and has the strongest effect (Whitemore 1997; Groombridge and Jenkins 2000; Fahrig 2003). The fragmentation of remaining forest patches is also thought to be a serious threat to the long term viability of biota, although these effects are more subtle (Harrison and Bruna 1999; Fahrig 2003).

There are three main factors contributing to the effect of habitat fragmentation but these are difficult to separate as they are often confounded and can interact. These are: 1) a reduction in habitat area, 2) interference in dispersal of individuals and propagules among habitat fragments, and 3) increase in edge effects (Saunders et al. 1991; Murcia 1995; Tscharntke et al. 2002; Fahrig 2003; Henle et al. 2004). An additional effect is habitat degradation within patches, although some aspects of this overlap with edge effects (Harrison and Bruna 1999; Haila 2002; Laurance et al. 2002). Habitat degradation can include within patch alterations to nutrient levels, hydrological flows and fire regimes (Saunders et al. 1991; Murcia 1995). A reduction in habitat area is likely to result in reduced population sizes, which increases the risk of extinction (Henle et al. 2004). If distance provides a barrier to the dispersal of individuals or propagules between habitat fragments, local populations within fragments have an increased risk of going extinct (Henle et al. 2004). Edge effects are likely to have very different impacts on different organisms because they encompass a range of both abiotic and biotic effects (Saunders et al. 1991; Murcia 1995). Increases in the amount of edge will alter abiotic microclimatic properties such as the amount of sunlight, humidity, wind velocity, and temperature extremes (Chen et al. 1999). Some organisms may respond directly to such abiotic changes in a positive or negative way. Other organisms may show a direct response to edge effects by responding to the changes in abiotic conditions (Harrison and Bruna 1999; Laurance et al. 2002), and to differing life expectation among organisms (Henle et al. 2004). Other forms of environmental degradation, such as logging and climate change, may also interact with habitat fragmentation to impact on biota deleteriously (Grove 2002; Opdam and Wascher 2004).

13.2.4.1 Technological/Best Management Practices

BMPs include water pollution control measures in widespread use in forestry operations throughout the world. These management practices involve a variety of locally appropriate erosion control measures which help prevent pollution in surface waters resulting from forestry activities and deforestation (Mutchler et al. 1994). BMPs are important because they prevent or minimize environmental problems associated with forestry activity, such as turbidity, nutrient transport, and runoff of herbicides (Beckie 2006; Green 2007), insecticides and fungicides into surface waters affecting drinking water, fisheries and aquatic habitats, flooding, siltation of dams and irrigation systems, and crop damage from siltation on leaves from irrigation water (Frisvold et al. 2009; Monsanto 2009a, 2009b).

Many different specific control technologies, or BMPs, are available, including preharvest planning to minimize runoff and erosion from roads and harvest areas into streams, use of streamside buffer or management areas (areas along surface waters where the vegetative cover is left) to reduce runoff from upslope activities and trap sediments, use of road construction, maintenance, and post-harvest revegetation techniques that minimize erosion, and use of effective erosion control devices, as locally appropriate, such as sediment control devices like silt fences, riprap, and sediment traps or check dams (Caltrans 2003; Wright Water Engineers, Inc. and Denver Regional

Council of Governments 1999). Other control technologies include timber harvesting techniques that minimize erosion like cable yarding and aerial harvesting, particularly for dispersed high value timber. Fire management is important in preventing erosion, particularly on steep slopes near streams. Careful management of cosmetics used in forestry is important in reducing environmental damage. Aerial applications of pesticides may pose the greatest risk to water quality, but streamside buffer zones have been found to minimize the effects of pesticide application. Studies have shown prompt revegetation of disturbed areas effectively reduces erosion.

An indication of the extent to which different BMPs are used in the U.S. is provided in a U.S. Environmental Protection Agency study. The study reported that over 80% of the states had state BMP regulations or manuals. Most state BMPs addressed preharvest planning (over a third), road construction and maintenance (all), timber harvesting, streamside buffer or management zones (almost 60%), site preparation, chemical management (over 40%), revegetation (almost 70%), prescribed burning, and drainage structures (USEPA and Tetra Tech, Inc. 1993). A recent summary by USEPA of the effectiveness of various forestry management measures indicates revegetation, roads, and streamside buffer or management zones offer the greatest opportunities for pollution reduction.

The establishment of various categories of protected forest areas has been used effectively by many countries to retain important benefits of forests and prevent environmental damage from deforestation. In some countries, the only remaining forests are those with protected status. Protected forest areas have existed since the 4th century BC in India, and hunting reserves existed in Europe for hundreds of years. Most protected areas were established in the late 19th century. The International Union for the Conservation of Nature has developed a standard classification system of ten types of protected areas. Using these criteria, 169 countries have protected sites covering over 5% of the world's land area. Of this amount, about 9% is in subtropical/temperate rainforests/woodlands, about 5% in tropical humid forests, 4.7% in tropical dry forests/woodlands, 4.7% in evergreen sclerophyllous forests, about 3% in temperate broad-leaf forests, and about 2.9% in temperate needle-leaf forests/woodlands (World Conservation Monitoring Centre 1990). Because of the range of different forest types protected, preservation of biodiversity is a major benefit.

A different type of ban/protected area was established by the government of Thailand. A full commercial logging ban on government forests was imposed after uncontrolled runoff from rains caused landslides, and destroyed the homes of 40,000 people. However, between 1985 and 1988, forest cover fell from 29% to 19% (Robert 1990). A logging ban was also imposed in Ecuador to reduce deforestation. A tree planting program was initiated to increase the forest area, and the government has started giving villages their own forest plots to manage (Lean 1990). Other types of logging bans which have been used include bans on steep slopes (e.g. over 30% grade), bans on logging near surface waters (streamside buffer areas), and logging bans in government reserves (extractive reserves) (Manuel et al. 1993). A critical component of such program is balancing the extraction of non-wood products with the maintenance of biodiversity, and to avoid over-harvesting.

13.2.4.2 Economic Options

Economic options use market forces to encourage activities reducing deforestation and/or forestry activities. Such options include tax policies that reduce assessments for conservation land. Government assistance for reforestation, tax incentives and government subsidies for turbidity control and other BMPs, and extending the life of timber concessions to provide an incentive for protection and maintenance of the reforested area until the new growth are well established. Other economic options include changing laws inadvertently causing deforestation, provision of secure land tenure for forest residents protecting the forest, the development of community forestry programs, and programs for timber theft prevention.

An example of an economic option is the establishment of conservation land areas in Lincoln, Massachusetts, USA. Tax rates are set at a lower level for forest lands of conservation interest to the town. Another example is to change the forest pricing methods. Options include raising forest sale fees to market levels, simplifying overly complex procedures, adjusting for inflation, increasing collection rates, using market mechanisms (e.g. competitive bids) for concession allocation, and reducing wasteful logging through payment per tree or volume of trees felled (rather than removed) (Robert 1990).

The report by the World Bank on Forest Pricing contains detailed recommendations in each of these areas (Grut et al. 1991). Community forestry programs and land titling programs work with local populations and their economic interests to increase forest protection. For example, the Awa reserve was created in 1982 in Ecuador to protect 1700 hectares of forest from deforestation by developing a multi-faceted program, including land titling for local residents, inventorying forest resources and developing a program for effective forest use (Judith and Russell 1988).

13.2.4.3 Voluntary Options

Voluntary approaches are widely used in the forestry area to encourage compliance with environmental goals. Voluntary options include education and technical assistance, timber certification programs, and awards programs. Examples include many of the BMP programs in the U.S., which are voluntary and depend heavily on education and technical assistance efforts by forestry staff. For instance, a study of the effectiveness of the U.S. State programs directed at private landowners indicated that technical assistance programs were judged most effective (Antony and Paul 1993).

As an individual or as a part of the global community, we can do something to help prevent deforestation as follows:

- Reforestation is the positive event that should take place to counter deforestation effects. We should plant trees and we can begin doing this in our own backyard. Trees capture the carbon dioxide that humans and animals exhale and give off oxygen that is essential for human existence.
- Do not cut small or baby trees. Cut down only mature trees and for every tree plant one as a replacement.
- Farmer should rotate crops. This practice also helps in maintaining the soil fertility. By the crop rotation, not only are there harvest different varieties each year, but also there is an increase in the possibility of land age.
- Do not waste any books, papers, toilet paper, shopping bags, so that new raw material would be required to replace them. Always use recycled items.
- Use coal instead of firewood to heat up during the winter season. Think twice before use firewood as a tree take years to grow but it takes only few hours to consume the firewood.
- Follow the laws and programs which were made for forest protection and to stop any deforestation. Programs such as the Tropical Forestry Action Plan has done a major difference on the way deforestation is looked at today.
- We should give awareness to children to save tree and explain why trees are important for our environment.

No matter what you deed may be, the important thing is that "every act can make a difference." Deforestation can be prevented and you can be an active force in achieving that. Finally, prevention of deforestation is not only strongly linked to climate change but also to other global problems, most importantly the loss of global biodiversity. Tropical forests are home to more than half of the estimated global biodiversity, and forest conversion and degradation are major contributions to its rapid decline (UNEP 2001).

13.2.5 Benefits of Reforestation

The quickest solution to deforestation would be to simply stop cutting down trees or to manage forest resources. Reforestation efforts are a vital way to undo some of the damage that has already been done. In this case, reforestation will provide many benefits on a chemical, social, and biological level as well as social dimensions of the global ecosphere (Peter et al. 2004). During the planting process, special care is taken to preserve the soil, and improve quality of life in the area. The hope is that their efforts will improve overall health and welfare of the human and forest population (Peter et al.

2004). Reforestation is a great way to offset CO_2 emissions (Raj et al. 2010). As forests grow they absorb CO_2 and store that CO_2 for the long term. Trees thrive on carbon gases, using carbon molecules to produce everything from sugar during photosynthesis to cellulose and wood as they grow. Planting trees, particularly young ones, effectively sequesters substantial amounts of atmospheric carbon, because, as trees grow, they transform carbon in the air into biomass (wood, foliage, glucose, etc.). Because of this, reforestation is a very effective strategy for decreasing the amount of carbon gas in the atmosphere, an important step in the fight against global climate change.

It has been suggested that reforestation can increase habitat area, reduce the influence of edge effects by creating buffers, improve connectivity among habitat patches, and provide functional benefits such as improvements to water quality (Hobbs 1993, Saunders et al. 1993; Lamb et al. 1997; Tucker 2000). Reforestation may be deliberate, such as for conservation (ecological restoration) or timber production (mono-multi-species plantations), or it may be unintended such as in abandoned old fields (Parrotta and Knowles 1997). Such a diversity of management influences can result in reforested stands with strongly differing structural and floristic characteristics (Kanowski et al. 2003). Consequently the ability of reforested areas to provide habitat for biota is likely to vary enormously. Specific factors that may influence faunal recruitment to reforested areas include: developmental (e.g., stand age or stage), spatial (e.g., patch size, surrounding land cover, and degree of isolation), and physical context (e.g., soil type, topography, bioregion, and climate) (Majer 1989; Catterall et al. 2004; Lindenmayer and Hobbs 2004). The relative importance of these factors is likely to vary among organisms although habitat quality and suitability are likely to be particularly important. For example, the most common type of reforestation practised globally is forestry plantation using monoculture; it can only provide habitat for a subset of forest biota due to the absence of structural complexity (amongst other reasons) (Lindenmayer and Hobbs 2004; Kanowski et al. 2005). Currently, the degree to which different reforestation styles can support biota and the relative importance of developmental and spatial factors in influencing faunal recruitment remain poorly understand.

Reforestration can provide both ecosystem and resource benefits and has the potential to become a major carbon sink. The concept of forests as carbon sinks has drawn attention around reforestation as a possible tool in the fight against global climate change. Because trees draw carbon dioxide from the atmosphere in the process of photosynthesis, they can potentially remove GHGs from the atmosphere and help fight global warming. Reforestation also impacts climate change. Trees play an important role in absorbing the GHGs that fuel global warming and also anchor soil with their roots. When they are removed, the soil is transported and deposited in other areas releasing carbon into the air. Adding to forests means smaller amounts of GHGs entering the atmosphere. It should be kept in mind that planting a single tree won't offset the damage of removing another as trees digest CO_2 quite slowly. It could possibly take a century for an expanding tree to extract all the CO_2 released every time a mature wood is lessening!

13.3 Genetic Engineering to Increase C4 Plants

Ribulose-1,5-bisphosphate carboxylase/oxygenase, most commonly known by the shorter name RuBisCO, is an enzyme that is used in the Calvin cycle to catalyze the first major step of carbon fixation, a process by which the atoms of atmospheric carbon dioxide are made available to organisms in the form of energy-rich molecules such as sucrose. RuBisCO catalyzes either the carboxylation or oxygenation of ribulose-1,5-bisphosphate (also known as RuBP) with carbon dioxide or oxygen. RuBisCO is the key enzyme responsible for photosynthetic carbon assimilation in catalysing the reaction of CO_2 with ribulose 1,5-bisphosphate (RuBP) to form two molecules of D-phosphoglyceric acid (PGA). It also initiates photorespiration by catalysing the reaction of oxygen, also with RuBP, to form one molecule of phosphoglycolate and of PGA. RuBisCO is a complex enzyme and catalyses these reactions at rather slow rates. It constitutes about 30% of the total protein in many leaves. Therefore, it is of considerable interest in relation to the nitrogen nutrition of plants. Biochemists have shown much interest in RuBisCO because of the catalytic mechanism, lack of specificity, regulation, and turnover aspects. Physiologists have concerned because of the properties of RubisCO for the gas exchange characteristics of photosynthetic tissues and because of the consequences of the amount of nitrogen tied up in the enzyme and its recycling upon senescence of leaves.

RuBisCO is apparently the most abundant protein in leaves, and it may be the most abundant protein on Earth (Cooper and Geoffrey 2000; Dhingra et al. 2004; Feller et al. 2008). Given its important role in the biosphere, there are currently efforts to "improve on nature" and genetically engineer crop plants so as to contain more efficient RuBisCO. RuBisCO is rate limiting for photosynthesis in plants. It may be possible to improve photosynthetic efficiency by modifying RuBisCO genes in plants to increase its catalytic activity and/or decrease the rate of the oxygenation activity (Spreitzer et al. 2002). Approaches that have been investigated include expressing RuBisCO genes from one organism in another organism, increasing the level of expression of RuBisCO subunits, expressing RuBisCO small chains from the chloroplast DNA, and altering RuBisCO genes so as to increase specificity for carbon dioxide or otherwise increase the rate of carbon fixation (Parry et al. 2003).

13.3.1 Evolution of Manipulation of RuBisCO in Plants

The advent of technology for genetically transforming the small circular genome of the plastids of tobacco (Svab et al. 1990) opened the door to molecular manipulation of higher-plant RuBisCO. Previously, such manipulation was limited to varying the content of RuBisCO in leaf cells by transforming the nuclear genome with antisense genes directed at the RbcS message for the nuclear-encoded small subunits (Rodermel et al. 1988; Hudson 1992). Even the less ambitious manipulation of higher-plant RuBisCO in heterologous hosts, such as *Escherichia coli*, was (and still is) blocked by unknown requirements for folding and assembly that are not satisfied in the bacterial host and resulted in the production of aggregated, nonfunctional protein (Gatenby et al. 1984; Cloney 1993). Plastid transformation has circumvented this impediment, allowing deletion (Kanevski and Maliga 1994), mutation (Whitney et al. 1999) and replacement (Kanevski et al. 1999; Whitney and Andrews 2001b) of the rbcL gene for the large catalytic subunit.

Currently, C3 cycles is genetically improved at different times mainly through the following six approaches (Table 13.1): (i) overexpression of cyanobacterial ictB gene in C3 plants like *Nicotiana tabacum* and *Arabidopsis thaliana* to reduce the oxygenation of RubisCO; (ii) introduction of bacterial glycolate pathway (glcD, glcE, glcF) into C3 plants; (iii) introduction of thermostable RubisCO activase (rca) to reduce the inhibition of photosynthesis when plants are subjected to mild heat stress; (iv) introduction of sedoheptulose-1,7-bisphosphatase (SBP), Fru bisphosphatase (FBP), and Tranketolase (TK) to regenerate the CO_2 acceptor molecule of RuBP; (v) synthesizing an active form of RubisCO *in vitro* using unfolded large subunits with the chaperone proteins GroEL/ES, RbcX, and ATP and then expressing RubisCO with large and small subunits as a single active holoenzyme (fusion protein) in chloroplast; and (vi) transefering C4 photosynthesis in C3 plants by introducing prokaryotic carboxysome and eukaryaotic algal pyranoid which can act as CO_2-concentrating mechanisms (CCMs) to reduce photorespiration.

Table 13.1 Gene targets to accelerate C3 pathway

Organism	Host Organism	Target Gene	Regulation[1]
Genes targeted in the past to accelerate C3 pathway:			
Cyanobacteria	*Nicotiana tabacum*	*ictB*	reduce oxygenation reaction of Rubisco [1]
Anabaena PCC7120	*Arabidopsis thaliana*	*ictB*	reduce oxygenation reaction of Rubisco [1]
Escherichia coli	*Arabidopsis thaliana*	*glcD, glcF, glcF*	photorespiratory bypass in the chloroplast [2]
Arabidopsis thaliana	*E. coli*	*Rca*	Rubisco activation [3]
Synechococcus	*Nicotiana tabacum*	*FBP/SBP*	Gegeneration of C3 cycle [4]
Potato, Yeast	*Nicotiana tabacum*	*TK* (Tranketolase)	Gegeneration of C3 cycle [5]
Potential gene targets can be selected to accelerate C3 pathway:			
Cyanobacteria	NA	*RbcX* (Large subunit of Rubisco)	Reengineering Rubisco [6]
Synechococcus	*Escherichia coli*	*rbcL-S* (operon)	Fusing subunits of Rubisco [7]
C4 plants (Eukaryotic algae)	C3 plants	Carboxysome, pyrenoid	CO_2-concentrating mechanisms [8-10]

[1]References: [1] = Lieman-Hurwitz et al. (2003); [2] = Kebeish et al. (2007); [3] = Kurek et al. (2007); [4] = Miyagawa et al. (2001); [5] = Henkes et al. (2001); [6] = Liu et al. (2010); [7] = Whitney and Sharwood (2007); and [8] = Badger et al. (2006); [9] = Moroney andYnalvez (2007); and [10] = Price et al. (2008).

13.3.2 Manipulation Methods of RuBisCO in Plants

Nuclear Transformation. In plants, RuBisCO's small subunits are encoded by a multi-gene family of RbcS genes in the nuclear genome. This restricts genetic manipulation to suppression of the expression of the small subunit by antisense RNA or supplementation with additional sense RbcS genes. RbcS expression has been suppressed with antisense genes in tobacco (Rodermel et al. 1988; Hudson 1992), rice (Makino et al. 1997) and *Flaveria bidentis* (Furbank et al. 1996). The amounts of sense RbcS transcript, small subunits, and holoenzyme were reduced, sometimes to very low levels. For plants grown in dim illumination, whole-leaf photosynthesis measured in the same illumination was not affected until more than half of the control's RuBisCO content had been removed (Quick et al. 1991a). However, photosynthesis measured at higher illumination was more linearly correlated with the RuBisCO content (Stitt et al. 1991). A similar near-linear correlation was observed when the plants were grown in strong natural illumination and measured at high light levels (Hudson 1992; Krapp et al. 1994; Makino et al. 1997). The reduced RuBisCO content was compensated only partly by an increase in the degree of activation of the remaining enzyme (Quick et al. 1991a). The reduced RuBisCO activity resulted in lower steady state pools of its product, 3-phosphoglycerate (Quick et al. 1991b; Mate et al. 1996). Reduction of the RuBisCO content to 35% of that of the wild type reduced photosynthesis sufficiently such that, at 1000 μmol quanta $m^{-2}s^{-1}$ illumination, it was never limited by RuBP regeneration, even at high CO_2. This allowed measurement of RuBisCO's kinetic parameters *in vivo* by leaf gas exchange. The resulting data agreed quite well with *in vitro* measurements with isolated RuBisCO (Caemmerer et al. 1994).

Introduction of a pea RbcS gene into the nuclear genome of *Arabidopsis thaliana* resulted in production of a heterologous small subunit, which was transported to the plastid and assembled into the RuBisCO holoenzyme. However, the catalytic rate of the hybrid enzyme and its ability to bind ligand appeared to be impaired equally, and the extent of the impairment was approximately proportional to the amount of foreign small subunits present (Getzoff et al. 1998). This illustrates the incompatibility of mismatched RuBisCO subunits. These data serve to highlight the impediments to manipulating RuBisCO in plants by nuclear transformation only. Not only is the manipulation restricted to the small subunit, but also the presence of multiple RbcS genes makes precise mutagenesis or targeted replacement cumbersome and arduous.

Plastid Transformation. The surgical precision of the homologous recombination mechanism that enables plastome manipulation (Maliga 2002) stands in marked contrast to the limitations of nuclear transformation. Both site-specific mutagenesis (Whitney et al. 1999) and total replacement (Kanevski et al. 1999; Whitney and Andrews 2001a) of the plastid-encoded large subunits are now routinely in process. Although tobacco is the only higher plant in which plastid-transformation technology is applicable routinely, efforts to develop the procedure for other species are showing promise (Sikdar et al. 1998; Sidorov et al. 1999). In the meantime, tobacco plastid transformation provides an excellent vehicle for studying a wide range of aspects of RuBisCO biology.

Express rbcL and RbcS Genes. Higher-plant RuBisCO provides an excellent, and comparatively simple, example of the migration of genes from the original prokaryotic ancestor of the plastid to the nucleus. Migration of RbcS, but not rbcL, to the nucleus must have required solutions to problems of coordinated expression of the two genes, targeting of cystoplasmically synthesized small subunits to the plastid, and coordinated folding and assembly of the subunits synthesized in different cellular compartments. Separate transplantation of rbcL and RbcS to the opposite genome has provided unique insight into some aspects of these processes. Kanevski and Maliga (1994) excised the rbcL gene from the tobacco plastome, producing plants that contained neither RuBisCO subunit nor required supplementation with sucrose to grow (Fig 13.1B). They then transformed the nucleus of these plants with the same rbcL gene fused to the transit presequence of pea RbcS under the transcriptional control of the *Cauliflower mosaic* virus 35S promoter. The introduced nuclear gene complemented the plastomic rbcL deletion, allowing RuBisCO to accumulate to approximately 3% of the wild types content, which was enough to permit slow autotrophic growth (Fig. 13.1C). This shows that there is no impediment to nuclear expression of rbcL and that large-subunit precursors can be imported to the plastid and processed, folded, and assembled by the existing machinery. Expression of a bacterial rbcL similarly incorporated into the nuclear genome was unsuccessful, however (Madgwick et al. 2002). Whitney and Andrews (2001b) performed the reciprocal transplantation, transferring one member of the tobacco nuclear RbcS family, both with and without its transit presequence and equipped with tobacco plastid psbA promoter and terminator elements, back to its endosymbiotic origin in the plastome (Fig. 13.2D). In this case, the limitations to engineering the nuclear genome mentioned earlier prevented removal of the nuclear copies but attachment of a C-terminal polyhistidine tag allowed the small subunits synthesized in the plastid to be distinguished. Once again, the transplanted gene was active in its new compartment, producing abundant transcript, and its product was correctly processed and assembled into hexadecameric RuBisCO. However, again the amount of tagged small subunits found assembled into the RuBisCO complex was not large (approximately 1% of the total small subunits).

Competition from nucleus-encoded small subunits does not appear to be the cause of the scarcity of plastid-synthesized small subunits in this system because suppression of the abundance of nucleus encoded small subunits with a nuclear antisense small subunit gene did not improve the result (Zhang et al. 2002a). Both sets of transplantation experiments agree that there is no qualitative deficiency preventing expression of either subunit gene from either genome. However, the quantitative shortcomings are glaring in both cases. For the transplanted small subunits, at least, the limitation is not transcriptional (Whitney and Andrews 2001b; Zhang et al. 2002a). More studies of both systems will be required to pinpoint the causes of the quantitative restrictions. Translational difficulties, including codon-usage bias, plastid targeting and import, processing, folding, and assembly are all potential hurdles that could impede the expression of the transplanted gene. One possible post-translational bottleneck might apply, in reciprocal senses, to both transplanted subunits: the newly translated or newly transported and processed subunits might need to be presented to the plastid chaperone machinery in particular contexts. Thus the chaperone complexes that draw the small subunit precursor through the translocon complex in the plastid envelope may be different from those that receive the nascent large subunits from the plastid ribosomes despite, perhaps, having some subunits, such as Hsp70, in common. Although both types of chaperone complexes may pass their unfolded clients onto the plastid chaperonin 60/21 system, they may not carry out their roles with equal efficiency when their clients are exchanged. Any unfolded subunits lost from, or denied access to, the folding pathway would be degraded.

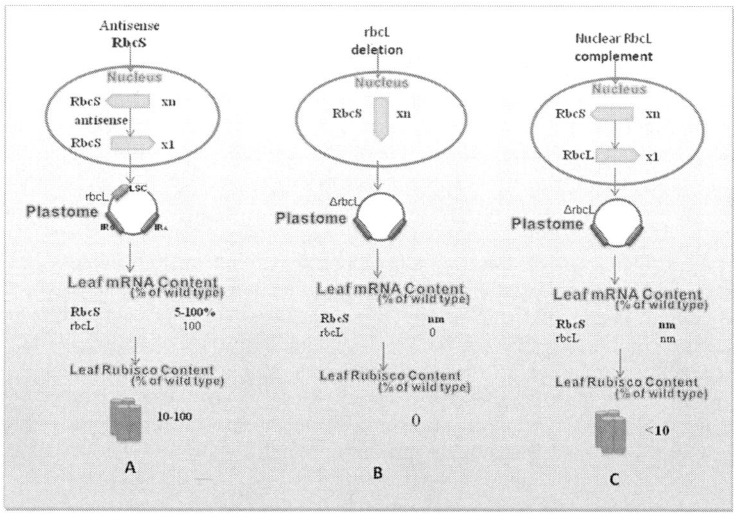

Figure 13.1. Types of manipulation of RuBisCO in tobacco conducted so far. (A) Suppression of RbcS by nuclear antisense. (B) Deletion of plastomic rbcL. (C) Complementation of "c" with nuclear RbcL (adapted from Whitney and Andrews 2001a)

13.3.3 Increase C4 Carbon Fixation Photosynthetic Plants

In the plant kingdom, there are three pathways of photosynthetic CO_2 fixation. However, the vast majority of plant species fix atmospheric CO_2 using the enzyme RubisCO in the Calvin-Benson cycle (Fig. 13.3). The first stable product of this cycle is a three-carbon compound, phosphoglycerate (3-PGA). For this reason this process is referred to as the C3 cycle. Plants utilizing this pathway are often referred to as C3 species. A major problem with the C3 cycle is the enzyme RubisCO because RubisCO is not only an inefficient enzyme with a low turnover number, but it also catalyzes two competing reactions: carboxylation and oxygenation (Portis and Parry 2007). The oxygenation reaction directs the flow of carbon through the photorespiratory pathway, and this can result in losses of between 25% and 30% of the carbon fixed.

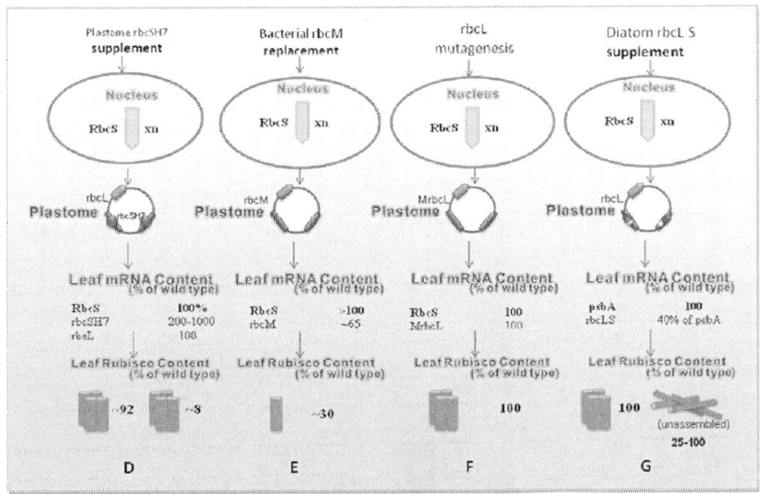

Figure 13.2 (D) Supplementation of RbcS with a plastomic hepta-His-tagged rbcSH7 copy. (E) Replacement of tobacco rbcL with rbcM of R. rubrum. (F) Directed mutation of rbcL. (G) Supplementation with plastomic copies of red-type rbcLS operons. The example of the red-type RuBisCO from a diatom is shown. Amounts of RuBisCO and mRNA are indicated as percentages of the amounts in the wild type. *n* denotes the multiple members of the *Rbc*S nuclear gene family. The plastomic *psb*A gene, which encodes the D1 protein of photosystem II, produces one of the most abundant mRNA transcripts in chloroplasts

The C3 cycle is the primary pathway of carbon assimilation in the majority of photosynthetic organisms. It is the single largest flux of organic carbon in the biosphere and assimilates about 100 billion tons of carbon a year (15% of the carbon in the atmosphere). Understanding the responses of the Calvin cycle to altered demands for photosynthate within the plant and to external environmental conditions is essential for attempts to increase yield and to redirect carbon into important products. The C3 cycle utilizes the products of the light reactions of photosynthesis, ATP and NADPH, to fix atmospheric CO_2 into carbon skeletons that are used to fuel the rest of plant metabolism (Stitt et al. 2010). The C3 cycle is initiated by the enzyme Rubisco that catalyzes the carboxylation of the CO_2 acceptor molecule ribulose-1,5-bisP (RuBP). 3-PGA formed by this reaction is used to form the triose phosphates glyceraldehyde phosphate (G-3-P) and dihydroxyacetone phosphate via two reactions that consume ATP and NADPH (Fig. 13.3b). The reductive phase of the cycle follows with two reactions catalyzed by 3-PGA kinase (PGK) and GAPDH, producing G-3-P. The G-3-P enters the regenerative phase catalyzed by aldolase (Ald) and either FBPase or SBPase, producing Fru-6-P (F-6-P) and sedoheptulose-7-P (S-7-P). Fru-6-P and sedoheptulose-7-P are then utilized in reactions catalyzed by TK, R-5-P isomerase (RPI), and ribulose-5-P (Ru-5-P) epimerase (RPE), producing Ru-5-P (Fig. 13.3b). The final step converts Ru-5-P to RuBP, catalyzed by PRK. The oxygenation reaction of Rubisco fixes O_2 into RuBP, forming 3-PGA and 2-phosphoglycolate (2PG), and the process of photorespiration releases CO_2 and 3-PGA (Fig. 13.3b). Therefore, the regenerative phase of the cycle involves a series of reactions that convert triose phosphates into the CO_2 acceptor molecule RuBP cycle and are essential for growth and development of the plant (Raines and Paul 2006). While the majority (five-sixths) of the triose phosphate produced in the Calvin cycle remain within the cycle to regenerate RuBP, one-sixth of the carbon exits the cycle for biosynthesis of a range of compounds. Triose and hexose phosphates are used to synthesize sucrose and starch, and erythrose-4-P (E-4-P) goes directly to the shikimate pathway for the biosynthesis of amino acids and lignin, G-3-P to the isoprenoid pathway, and Rib-5-P (R-5-P) for nucleotide, thiamine metabolism, and cell wall biosynthesis (Fig. 13.3b).

The significance of the inhibition of photosynthesis in many organisms by oxygen (Warburg 1920; Ogren 1984) became evident with the discovery of the oxygenation of RuBP and consequent stimulation of photorespiration (Bowes et al. 1971; Ogren and Bowes et al. 1971; Lorimer et al. 1981). Increased CO_2 concentration diminished the inhibitory effect of oxygen on photosynthesis. This also finds explanation in the properties of RuBisCO as a catalyst: the carboxylation and oxygenation reactions are catalysed at the same active site on the enzyme with CO_2 and O_2 being competitive substrates (Andrews and Lorimer 1978). Evolution in various environments, usually hot or deficient in available inorganic carbon (e.g., CO_2, HCO_3^-, CO_3^{2-}), has resulted in photosynthetic organisms that can concentrate CO_2 in cells or organelles containing their RuBisCO.

C4 plants have much higher rates of photosynthesis in warm conditions at high light intensities than C3 plants that have no CO_2 concentrating mechanism (Hatch 1976; Edwards and Walker 1983). Also, C3 plants in atmospheres with low O_2 or elevated CO_2 assimilate CO_2 and grow more quickly than in ambient conditions, provided that nutrients and temperature are not limiting. Mechanistic models of photosynthetic gas exchange based upon RuBisCO kinetics have proved very successful in representing the effects of light, temperature and atmospheric composition on assimilation of carbon by plants (Farquhar et al. 1980; Collatz et al. 1990).

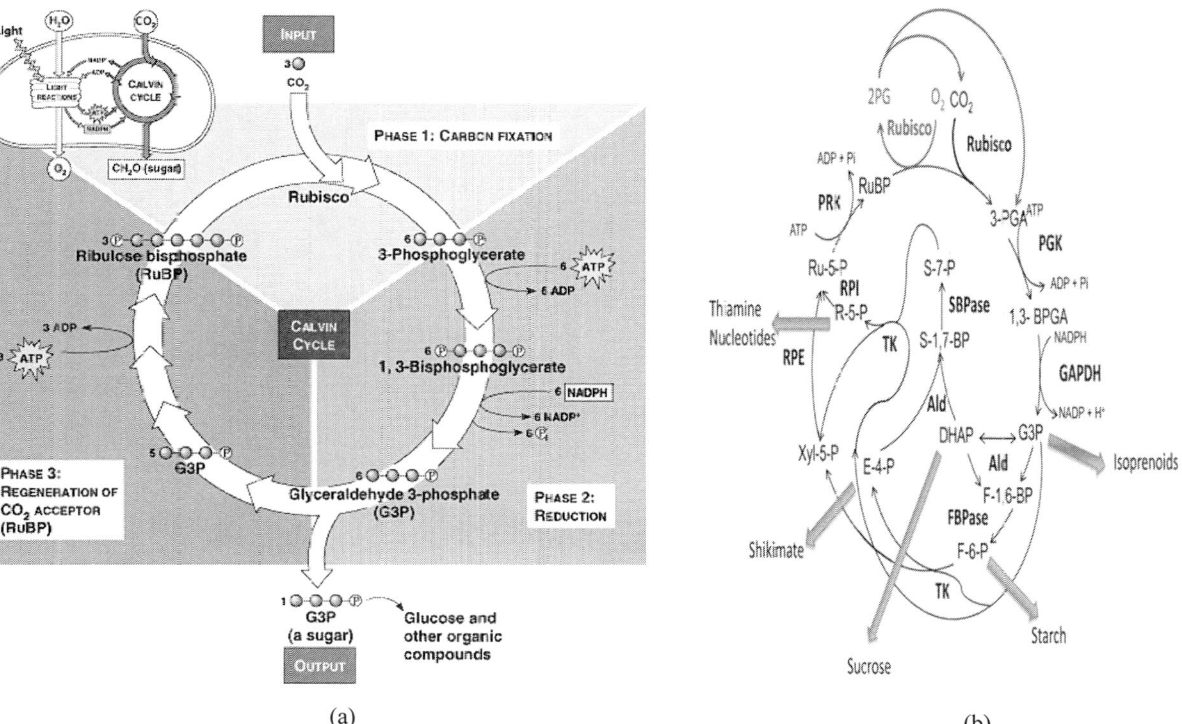

Figure 13.3 Photosynthesis: (a) pathway (adapted from Bassham et al. 1950) and (b) role of RubisCO as a catalyst in the Calvin cycle (adapted from Stitt and Usadel 2010)

Genetic manipulation of RuBisCO to double its specificity for CO_2 would theoretically increase CO_2 assimilation by perhaps 20%, and photosynthesis at sub-saturating light intensities would also be improved (Reynolds et al. 2000). Consequently, it has been accepted by many researchers that manipulating RuBisCO to decrease the inhibitory effect of oxygen and its competitive involvement in reaction with RuBP, as opposed to reaction with CO_2, is a worthwhile target to increase plant productivity.

Essential to the activity of RuBisCO is the carbamylation of an active site lysine residue (Lorimer and Miziorko 1980). The extent of this carbamylation depends on: i) the concentrations of CO_2 and Mg^{2+}; ii) the absence from the non-carbamylated sites of certain phosphorylated compounds and particularly RuBP; and iii) the activity of an enzyme called RuBisCO activase (Portis 1992). The activity of this latter enzyme is controlled by the ratio of ATP/ADP (Streusand and Portis 1987) and redox potential, in effect by light intensity (Zhang et al. 2002b). RuBisCO activase also facilitates the removal of 2-carboxyarabinitol 1-phosphate (CA1P) from carbamylated sites of RuBisCO (Robinson and Portis 1988). CA1P is a tight binding naturally occurring inhibitor of RuBisCO which is bound to the enzyme in many species at night. The significance of the presence of CA1P is subject to some debate. It could be a regulator of activity at low light intensities, but may be more important in protecting RuBisCO from degradation by proteases (Khan et al. 1999) when the natural substrate, RuBP, is present at low concentrations. Manipulation of the activity of RuBisCO activase or of the synthesis and breakdown of CA1P may be of value. These aspects are explored.

The large amount of RuBisCO in leaves has had consequences for the development of research on this enzyme. It has been estimated to be normally present at a concentration of 240 ± 1 mg/ml in the stroma of chloroplasts and constitutes about 30–50% of the soluble protein in the leaves of C3 plants (Kung 1976; Ellis 1979). There is so much present that not only can it sometimes be seen as crystals in the chloroplast stroma (Steer et al. 1966) but it also crystallizes very readily from relatively crude extracts (Chan et al. 1972).

Nevertheless, particularly in bright light RuBisCO t may exert considerable limitation over the rate of CO_2 fixation (Hudson et al. 1992). The fate of RuBisCO during leaf senescence has been intensively studied, and the nitrogen from this source has been shown to be extensively reutilized in the synthesis of proteins in seeds and perennating organs (Dalling et al. 1976; Peoples et al. 1983; Millard and Catt 1988). Thus the function of RuBisCO as a store of nitrogen has resulted in much speculation and research.

The genes for the RuBisCO polypeptide subunits from many species have been cloned and sequenced, as have genes for RuBisCO activase polypeptides. Furthermore, the crystal structure of RuBisCO from several species and the extensive homology of amino acid sequences have allowed the advance of genetic manipulation, protein engineering and transformation experiments (Spreitzer and Salvucci 2002).

One problem with the manipulation of RuBisCO in higher plants is that it is composed of eight large and eight small polypeptide subunits. The genes for the small subunits are in the nuclear genome (Kawashima and Wildman 1972), but those for the large subunits are encoded in the chloroplast genome (Chan and Wildman 1972; Ellis 1981). Problems have also been encountered in assembling large and small subunits into the hexadecameri choloenzyme following manipulation (Gutteridge and Gatenby 1995). Many protein engineering projects have, therefore, been conducted using cyanobacterial, algal and bacterial RuBisCOs for which assembly into the holoenzyme is less problematic. Mutagenesis *in vitro* has been used to make changes to DNA encoding both large and small subunits. The effects of such changes on the expressed protein have been used to increase understanding of the catalytic properties of RuBisCO and the extent to which the specificity and activity can be altered. The use of antisense constructs to alter the amount of expression of RuBisCO has been used both to determine whether the amount of RuBisCO in plants can be decreased to save nutrient nitrogen and to determine the extent to which RuBisCO controls the rate of photosynthesis. Transgenic plants expressing altered amounts of RuBisCO activase or RuBisCO activase polypeptides with mutations or from different species have also increased the understanding of the details of RuBisCO activation.

Many photosynthetic organisms, including cyanobacteria, algae, and land plants, have developed active CO_2-concentrating mechanisms to overcome Rubisco's inefficiencies (Badger et al. 1998), which led to the development of C4 photosynthesis, a biochemical CO_2-concentrating mechanism, among land plants. C4 photosynthesis arose multiple times in the past 60 million years in warm semi-arid regions, with early occurrences coinciding with low atmospheric CO_2 in the late Oligocene (Sage et al. 2011). During C4 photosynthesis, CO_2 is fixed within specialized leaf tissues known as mesophyll cells to produce C4 acids, which diffuse to and are decarboxylated in another type of specialized tissue, the bundle sheath cells. This process elevates the CO_2 concentration in the bundle sheath and inhibits Rubisco oxygenase activity, allowing Rubisco to operate close to its maximal rate. In comparison with C3 crops such as rice, C4 crops (such as maize and sorghum) have higher yields with increased water- and nitrogen-use efficiency (Langdale 2011). In an evolutionary context, the transition from C3 to C4 photosynthesis has occurred independently in more than 60 different plant taxa as shown in Table 13.2 (Sage et al. 2011). Genomic and transcriptional sequence comparisons of cell specific and leaf-developmental gradient transcription profiles between closely related C3 and C4 species are being used to identify C4-specific regulatory genes (Langdale 2011).

13.4 Future Trend and Perspectives

Plant and ecosystem responses to varying concentration of CO_2 are currently best understood at the physiological and ecological levels and on timescales of one generation or less. Investigating plant responses to varying atmospheric CO_2 concentration in an evolutionary context is important because variations of the same over geological timescales are believed to have played important roles in the evolution of ecologically and economically important traits in extant species. From a practical perspective, there is the possibility that despite major breeding successes, present elite crop varieties may not be adapted for optimal performance at the present and future CO_2 concentrations. Accordingly, improving plant productivity in high CO_2 concentration environments may be an open opportunity for biotechnology or breeding to improve crop performance now and in the future. The realization that crop yields are reaching a plateau, while the pace of population increases continues, has placed manipulation of photosynthesis in a central position to achieve the yield increases.

Table 13.2 Examples of projects converting C3 to C4 plants

Project	Focus
C4 Rice	Screens of mutagenized C4; *Sorghum bicolor* and *Setaria viridis* along with activation-tagged rice populations reveal genes in the C3-to-C4 switch that tested in transgenic rice and *S. viridi* (Caemmerer et al. 2012)
20:20 project announced for wheat	Converting C3 to C4 by Rothamsted Research in the UK in 2011: Appearance and accumulation of C4 carbon pathway enzymes (RuBPC, Pyruvate orthophosphate dikinase (PPDK), malic dehydrogenase, NADP specific and phosphoenolpyruvate carboxylase) in developing wheat leaves (Reynolds et al. 2012)
Potato, tobacco, Arabidopsis	Transfer of C4-like features to C3 plants: Overexpression of phosphoenolpyruvate carboxylase (PEPC) and chloroplastic malic enzyme or alternative decarboxylating enzymes for CO_2 fixation in potato and/or in *Arabidopsis. T* to complete C4-like cycles (Häusler et al. 2002)
Cotton	Introduction of rbcS promoter for the overexpression of Cry1Ac gene in cotton plants (Bakhsh 2010)

Manipulation of the C3 cycle offers an opportunity to increase photosynthesis and yield. A number of clear targets have been shown to have the potential to impact yield in 3- to 5-year period but to date have been tested only in model species. It is important to fully exploit this knowledge in crops. Clearly, the reduction of the RuBisCO oxygenase reaction remains a target for future improvement of photosynthesis. However, although the current strategies that might be exploited to achieve this goal are conceptually straightforward, all of these approaches will be technically demanding, requiring fundamental research to identify the genes involved. Improvement of the C3 cycle is not just about increasing CO_2 fixation but should also aim to increase both nitrogen use efficiency and water use efficiency while maintaining high productivity. Therefore manipulation of the C3 cycle to improve these parameters is also an important goal. The range of genetic and molecular techniques that are now available, together with the development and application of rapid *in vivo* techniques to allow in-field analysis of a wider range of species in their natural environments, will facilitate the wider analysis of natural variation in photosynthetic carbon assimilation.

Similarly, greater photosynthetic rates in C4 plants lead to more biomass being produced for a given amount of sunlight relative to C3 crops. Such findings are key drivers behind strategies to "supercharge" photosynthesis in C3 plants and improve crop yield potential. The focal point of these strategies is to overcome the catalytic inefficiencies of RuBisCO by emulating the carbon concentrating process found in C4 plants, which elevates CO_2 around RuBisCO to minimize photorespiration and its associated energy costs and carbon loss. Over the last 60 million years, C4 plants have evolved a variety of CO_2-concentrating strategies that enabled their RuBisCOs to persevere with lower CO_2 affinities while retaining enhanced CO_2 fixation rates. As a result, C4 plants maintain high photosynthetic rates with less RuBisCO (increasing nitrogen use efficiency) and can operate efficiently under low CO_2 levels, alleviating the need for wide stomata apertures, thereby reducing leaf water loss (Ghannoum et al. 2005). A variety of strategies for introducing CO_2 concentration approaches into C3 plants to minimize photorespiration are under way (Maurino and Peterhansel, 2010). These aim to introduce C4-like features into rice, improve productivity by introducing CO_2/HCO_3^- transporter proteins from cyanobacteria into chloroplast membranes, or engineer new pathways into plastids that bypass photorespiration and release CO_2 in the stroma. Each strategy will face challenges in their fine-tuning and integration into crops. Improvement in yields and water/nitrogen use efficiencies will likely follow the lead of C4 plants by increasing the vCO_2 of the inherent C3 RuBisCO.

13.5 Summary

The physiological response of plants to atmospheric CO_2 has received considerable attention because CO_2 is the substrate for photosynthesis. Today, rising CO_2 concentration is a key component of anthropogenic global environmental change that will impact plants and the ecosystem goods and services they deliver. Currently, there is limited evidence that natural plant populations have evolved in response to contemporary increases in CO_2 concentration in ways that increase plant productivity or fitness, and no evidence for incidental breeding of crop varieties to achieve greater yield enhancement from rising CO_2 concentration. RuBisCo, a bi-functional enzyme found in the chloroplasts of plants, catalyzes the initial carbon dioxide fixation step in the Calvin cycle and functions as an oxygenase in photorespiration. Evolutionary responses to the elevated CO_2 concentration have been studied by applying selection in controlled environments, quantitative genetics and trait-based approaches. Plant genetic transformation is a powerful application used to study gene expression in plants. With the functional replacement of RuBisCO in plants by plastid-transformation technology, a milestone in engineering photosynthesis to demonstrate the potential of RuBisCO manipulation has been passed. In this context, C4 photosynthesis was investigated by many biologists who have envisaged the introduction of the C4 photosynthetic pathway into C3 crops such as rice and soybeans. C4 photosynthesis is important for understanding the origin and function of the modern biosphere. The imperative to meet humanity's growing food, fuel, and fibre needs has promoted a resurgence in photosynthesis research in general, and C4-related research in particular. Recent advances in genomics capabilities and new evolutionary and developmental studies proved the discovery of the key genes controlling the expression of C4 photosynthesis.

13.6 Acknowledgment

Sincere thanks are due to the Natural Sciences and Engineering Research Council of Canada (Grant A 4984, Canada Research Chair) for their financial support. Views and opinions expressed in this article are those of the authors.

13.7 References

Aids, T.M. (2000). “Clues for tropical forest restoration.” *Restoration Ecology*. 8, 327.

Allen, J.C., and Barnes, D.F. (1985). “The causes of deforestation in developing countries.” *Departments of Geography and Environmental Studies*, University of California, Santa Barbara, CA.

Antony, C., and Paul, E. (1993). “State programs directed at the forestry practices of private forest landowners: Program administrators.” Assessment of Effectiveness, Minnesota Agricultural Experiment Station, University of Minnesota, S. Paul, MN. 1–57.

Amacher, G.S., Hyde, W., and Kanel, K.R. (1996). “Household fuelwood demand and supply in Nepal's Tarai and Mid-Hills: Choice between cash outlays and labor opportunity.” *World Development*, 24(11), 1725–1736.

Andrews,T.J., and Lorimer, G.H. (1978). “Photorespiration still unavoidable?” *FEBS Letters*. 90, 1–9.

Angelsen, A., and Wunder, S. (2003). “Exploring the forest-poverty link: Key concepts, issues and research implications.” *CIFOR Occasional* Paper No. 40, CIFOR.

Ayyad, M.A., and Ghabbour, S.I. (1985). “Hot deserts of Egypt and Sudan.” In: E.M, Noy-Meir I, Goodall, D.W. (eds.), *Hot Desert and Arid Shrublands,* B. Elsevier. 149–202.

Badger, M.R., Andrews, T.J., Whitney, S.M., Ludwig, M., Yellowlees, D.C., Leggat, W., and Price, G.D. (1998). “The diversity and co-evolution of Rubisco, plastids, pyrenoids and chloroplast-based CCMs in the algae.” *Canadian Journal of Botany,* 76, 1052–1071.

Bassham J., Benson A., and Calvin M. (1950). "The path of carbon in photosynthesis" (http:/ / www. jbc. org/ cgi/ reprint/ 185/ 2/ 781. pdf). *J. Biol. Chem.* 185(2). 781–787. PMID 14774424.

Beckie, H.J. (2006). “Herbicide-resistant weeds: Management tactics and practices.” *Weed Technology*, *20*(3), 793–814.

Brown, K., and Pearce, D. (1994): *The Cause of Tropical Deforestation: The Economicand Statistical Analysis of Factors Giving Rise to the Loss of the Tropical Forests.* Vancouver: UBC Press.

Bowes, G., Ogren, W.L., and Hageman, R. (1971). "Phosphoglycolate production catalysed by ribulose diphsophate carboxylase." *Biochemical and Biophysical Research Communications.* 45, 716–722.

Butler, R.A. (2008). "Deforestation in the Amazon." http//www/mongabay.com/brazil. Html. (Accessed Oct. 2013).

Byron, R.N., and Arnold, J.E.M. (1999). "What futures for the people of the tropical forests?" *World Development* 27(5), 789–805.

Caemmerer Susanne von, W. Paul Quick, and Robert T. Furbank (2012). "The development of C4 Rice: Current progress and future challenges." *Science*, 336(6089), 1671–1672.

Caltrans (2003). *Erosion Control New Technology.* Report. CTSW-RT-03-049, *California Department of Transportation*, Sacramento, CA.

Canadell, J.G., and Raupach, M.R. (2008). "Managing forests for climate change." *Science*, 320(5882), 1456–1457.

Capistrano, A.D. (1990). *Macroeconomic Influences on Tropical Forest Depletion: A Cross-Country Analysis.* Ph.D. Dissertation, Department of Food and Resource Economics, University of Florida, Gainsville.

Catterall, C., Kanowski, J., Wardell-Johnson, G., Proctor, H., Reis, T., Harrison, D., and Tucker, N. (2004). "Quantifying the biodiversity values of reforestation: Perpectives, design issues and outcomes in Australian rainforest landscapes." In *Conservation of Australia's Forest Fauna.*, Luuney, D. (ed.), 359–393. Royal Zoological Society of New South Wales.

Chan, P.H., Sakano, K., Singh, S., and Wildman, S.G. (1972). "Crystalline fraction 1 protein: Preparation in large yield." *Science*, 176, 1145–1146.

Chan, P.H., and Wildman, S.G. (1972). "Chloroplast DNA codes for the primary structure of the large subunit of Fraction 1 protein." *Biochimica et Biophysica Acta*, 277, 677–680.

Chen, J., Saunders, S.C., Crow, T.R., Naiman, R.J., Brosofske, K.D., Mroz, G.D., Brookshire, B.L., and Franklin, J.F. (1999). "Microclimate in forest ecosystem and landscape ecology." *Bioscience*, 49, 288–297.

Cloney, L.P., Bekkaoui, D.R., and Hemmingsen, S.M. (1993). "Co-expression of plastid chaperonin genes and a synthetic Rubisco operon in *Escherichia coli.*" *Plant Mol. Biol.*, 23, 1285–1290.

Collatz, G.J., Berry, J.A., Farquhar, G.D., and Pierce, J. (1990). "The relationship between the Rubisco reaction mechanism and models of photosynthesis." *Plant, Cell and Environment*, 13, 219–225.

Collins, J. (2001). "*Deforestation.*" Retrieved from WWC Enviro Facts index page http://www.botany.uwc.ac.za/envFacts/facts/deforestation.htm (Oct. 7, 2010).

Cooper, G.M. (2000). "The Chloroplast Genome." Chapter 10 in *The Cell: A Molecular Approach* (2nd ed.). Washington, D.C: ASM Press. ISBN 0-87893-106-6.

Dalling, M.J., Boland, G., and Wilson, J.H. (1976). "Relation between acid proteinase activity and redistribution of nitrogen during grain development in wheat." *Australian Journal of Plant Physiology*, 3, 721–730.

Dhingra, A., Portis, A.R., and Daniell, H. (2004). "Enhanced translation of a chloroplast-expressed RbcS gene restores small subunit levels and photosynthesis in nuclear RbcS antisense plants." *Proc. Natl. Acad. Sci.,* 101 (16), 6315–6320.

Dobson, A.P., Bradshaw, A.D., and Baker, A.J.M. (1997). "Hopes for the future: Restoration ecology and conservation biology." *Science,* 277, 515–522.

Duraiappah, A. (1996) "Poverty and environmental degradation: A literature review and analysis." *Creed Working Paper Series,* 8, 1–35

Edwards, G.E., and Walker, D.A. (1983). *C3, C4: Mechanisms, and Cellular and Environmental Regulation of Photosynthesis.* Blackwell Sci., Oxford, UK.

Ellis, R.J. (1979). "The most abundant protein in the world." *Trends in Biochemical Science*, 4, 241–244.

Ellis, R.J. (1981). "Chloroplast proteins: Synthesis, transport and assembly." *Annual Review of Plant Physiology*, 32, 111–137.

Fahrig, L. (2003). "Effects of habitat fragmentation on biodiversity." *Annual Review of Ecology and Systematics*. 34, 487–515.

FAO/UNEP. (1981). *Tropical Forest Resources Assessment Project (in the Framework of GEMS). Forest Resources of Tropical Africa, Part 1: Regional Synthesis.* Rome, http://www.fao.org/corp/statistics/en/ (Oct.7, 2010).

FAO. (1993). "Forest Resources Assessment 1990 Tropical countries." FAO Forestry Paper No. 112. Rome. www.fao.org/docrep/007/t0830e/t0830e00.htm.

FAO. (1997). *State of the World's Forests 1997.* Rome. www.fao.org/docrep/w4345e/w4345e00.htm. (Accessed Oct. 2011).

Farquhar, G.D., von Caemmerer, S., and Berry, J.A. (1980). "A biochemical model of photosynthetic CO_2 assimilation in leaves of C-3 species." *Planta,* 149, 78–90.

Feller, U., Anders, I., and Mae, T. (2008). "Rubiscolytics: Fate of Rubisco after its enzymatic function in a cell is terminated." *J. Exp. Bot.,* 59 (7), 1615–24.

Frisvold, G.B., Hurley, T.M., and Mitchell, P.D. (2009). "Adoption of best management practices to control weed resistance by corn, cotton, and soybean growers." *AgBioForum*, *12*(3 & 4), 370–381.

Furbank, R.T., Chitty, J.A., von Caemmerer, S., and Jenkins, C.L.D. (1996). "Antisense RNA inhibition of RbcS gene expression reduces Rubisco level and photosynthesis in the C4 plant *Flaveria bidentis*." *Plant Physiol.* 111, 725–734.

Gatenby, A.A. (1984). "The properties of the large subunit of maize ribulose bisphosphate carboxylase/oxygenase synthesized in *Escherichia* coli." *Eur. J. Biochem.,* 144, 361–366.

Getzoff, T.P., Zhu, G.H., Bohnert, H.J., and Jensen, R.G. (1998). "Chimeric *Arabidopsis thaliana* Rubisco containing pea small subunit protein is compromised in carbamylation." *Plant Physiol.,* 116, 695–702.

Ghannoum, O., Evans, J.R., Chow, W.S., Andrews, T.J., Conroy, J.P., and von Caemmerer, S. (2005) "Faster Rubisco is the key to superior nitrogen-use efficiency in NADP-malic enzyme relative to NAD-malic enzyme C4 grasses." *Plant Physiol.,* 137, 638–650.

GFRA (Global Forest Resources Assessment) (2005): "Progress towards sustainable forest Management." *Food and Agriculture Organization of the United Nations Rome*, FAO Forestry Paper. 147–148.

Goodland, R., and Pimentel, D. (2000). "Environmental sustainability and integrity in the agriculture sector." In: *Ecological Integrity: Integrating Environment, Conservation, and Health*, Pimentel, D.; Westra, L.; and Noss, R.F. (eds.), Washington: Island Press.

Grainger, A. (1993). "Rates of deforestation in the humid tropics: Estimates and measurements." *The Geographical Journal*, 159 (1), 33–44.

Green, J.M. (2007). "Review of glyphosate and ALS-inhibiting herbicide crop resistance and resistant weed management." *Weed Technology*, *21*(2), 547–558.

Groombridge, B., and Jenkins, M.D. (2000). *Global Biodiversity: Earth's Living Resources in the 21st Century*. World Conservation Press, Cambridge.

Grove, S.J. (2002). "Saproxylic insect ecology and the sustainable management of forests." *Annual Review of Ecology and Systematics*. 33, 1–23.

Grut, M., John, G., and Nicolas, E. (1991). *Forest Pricing and Concession Policies.* World Bank, Washington, D.C.

Gutteridge, S., and Gatenby, A.A. (1995). "Rubisco synthesis, assembly, mechanism and regulation." *The Plant Cell.* 7, 809–819.

Haila, Y. (2002). "A conceptual geneology of fragmentation research: from island biogeography to landscape ecology." *Ecological Applications*, 12, 321–334.

Harrison, S., and Bruna, E. (1999). "Habitat fragmentation and large-scale conservation: What do we know for sure?" *Ecography*. 22, 225–232.

Hatch, M.D. (1976). "Photosynthesis: the path of carbon." In: *Plant biochemistry*, 3rd ed., Bonner, J, and Varner, J. (eds.), New York, San Francisco, London: Academic Press, 797–844.

Häusler R.E., Hirsch H.J., Kreuzaler F., and Peterhansel C. (2002) "Overexpression of C4-cycle enzymes in transgenic C3 plants: A biotechnological approach to improve C3-photosynthesis." *Journal of Experimental Botany,* 53, 591−607.

Henle, K., Davies, K.F., Kleyer, M., Margules, C.R., and Settele, J. (2004). "Predictors of species sensitivity to fragmentation." *Biodiversity and Conservation,* 13, 207−251.

Hobbs, R (1993). "Can revegetation assist in the conservation of biodirversity in agricultural areas?" *Pacific Conservation Biology*, 1, 29−38.

Holl, K.D., and Howarth, R.B. (2000). "Paying for restoration." *Restoration Ecology*, 8, 260–267.

Hudson, G.S., Evans, J.R., von Caemmerer, S., Arvidsson, Y.B.C., and Andrews, T.J. (1992). "Reduction of ribulose-1,5-bisphosphate carboxylase oxygenase content by antisense RNA reduces photosynthesis in transgenic tobacco plants." *Plant Physiology*. 98, 294–302.

Janzen, D.H. (1994). "Priorities in tropical biology." *Tree*. 9, 365–368.

Judith, G. and Russell, G. (1988). "*Saving the Tropical Forests*." Earthscan Publications Ltd., London, England. 83–85.

Kaimowitz, D., and Angelsen A. (1998): *Economic Models of Tropical Deforestation A Review*. Center for International Forestry Research, Bogor, Indonesia

Kanevski, I., and Maliga, P. (1994). "Relocation of the plastid *rbcL* gene to the nucleus yields functional ribulose-1,5-bisphosphate carboxylase in tobacco chloroplasts." *Proc. Natl Acad. Sci. USA,* 91,1969–1973.

Kanevski, I., Maliga, P., Rhoades, D.F., and Gutteridge, S. (1999). "Plastome engineering of ribulose-1,s-bisphosphate carboxylaseloxygenase in tobacco to form a sunflower large subunitand tobacco small subunit hybrid." *Plant Physiol.,* 119,133–141.

Kanowski, J., Catterall, C., Wardell-Johnson, G., Proctor, H., and Reis, T. (2003). "Development of forest structure on cleared rainforest land in eastern Australia under different style of reforestation." *Forest Ecology and Management*. 183, 265–280.

Kanowski, J., Catterall, C., and Wardell-Johnson, G. (2005). "Consequences of broadscale timber plantation for biodiversity in cleared rainforest landscapes of tropical and subtropical Australia." *Forest Ecology and Management*, 208, 359–372.

Kawashima, N., and Wildman, S.G. (1972). "Studies of fraction protein. IV. Mode of inheritance of primary structure in relation to whether chloroplast or nuclear DNA contains the code for a chloroplastic protein." *Biochimica et Biophysica Acta*, 262, 42–49.

Khan, S., Andralojc, P.J., Lea, P.J., and Parry, M.A.J. (1999). "2-carboxy-Darabinitol 1-phosphate (CA1P) protects ribulose-1,5-bisphosphate carboxylase/oxygenase against proteolytic breakdown." *European Journal of Biochemsitry*, 266, 840–847.

Krapp, A., Chaves, M.M., David, M.M., Rodriques, M.L., Pereira, J.S., and Stitt, M. (1994). "Decreased ribulose-1,5-bisphosphate carboxylase/oxygenase in transgenic tobacco transformed with 'antisense' rdcS.VIIL Impact on photosynthesis and growth in tobacco growing under extreme high irradiance and high temperature." *Plant Cell Environ.,* 17, 945–953.

Krzywinski, K., and Pierce, R.H. (eds.) (2001). *Deserting the Desert a Threatened Cultural Landscape between the Nile and the Sea*. Bergen: *Alvheim og Eide Akademisk* Forlag.

Kummer, D.M., and Sham, C.H. (1994). "The causes of tropical deforestation: a quantitative analysis and case study from the Philippines." In: *The Causes of Tropical Deforestation*, Brown, K., and Pearce, D.W. (eds.), UCL Press, London.

Kung, S.D. (1976). "Tobacco fraction 1 protein: a unique genetic marker." *Science*. 191, 429–434.

Lamb, D., Parotta, J., Keenan, R., and Tucker, N. (1997). "Rejoining habitat remnants: Restoring degraded rainforest lands." *Tropical Forest Remnants: Ecology, Management, and Consevation of Fragmented Communities,* Laurance, W., and Bierregard, R. (eds.), pp. 366–385. The University of Chicago Press, Chicago.

Lamb, D. (1998). "Large-scale ecological restoration of degraded tropical forest land: the role of timber plantations." *Restoration ecology*, 6, 271–279.

Langdale J.A. (2011). "C4 cycles: Past, present, and future research on C4 photosynthesis." *Plant Cell*, 23, 3879–3892.

Laurance, W.F., Laovejoy, T.E., Vasconcelos, H.L., Bruna, E., Didham, R.K., Stouffer, P.C., Gascon, C., Bierregaard, R.O., Laurance, S.G., and Sampaio, E. (2002). "Ecosystem decay of Amazonian forest fragments: A 22 year investigation." *Conservation Biology*, 16, 605–618.

Lean, G. (1990). *World Wildlife Fund Atlas of the Environment*. Report. Prentice Hall Press, N.Y. 80–82.

Lindemayer, D.B. and Hobbs, R. (2004). "Fauna conservation in Australian plantation forests-a review." *Forest Ecology and Management*. 119, 151–168.

Lorimer, G.H., and Miziorko, H.M. (1980). "Carbamate formation on the ε-amino group of a lysyl residue as the basis for the activation of ribulosebisphosphate carboxylase by CO_2 and Mg^{2+}." *Biochemistry*, 19, 5321–5324.

Lorimer, G.H. (1981). "The carboxylation and oxygenation of ribulose 1,5-bisphosphate: The primary events in photosynthesis and photorespiration." *Annual Review of Plant Physiology*, 32, 349 383.

Madgwick, P.J., Colliver, S.P., Banks, F.M., Habash, D.Z., Dulieu, H., Parry, M.A.J., and Paul, M.J. (2002). "Genetic Manipulation of Rubisco: *Chromatium vinosum rbcL* is expressed in *Nicotiana tabacum* but does not form a functional protein." *Ann. Appl. Biol.,* 140, 13–19.

Mahapatra, R. (2001). "Betrayed: Nepal's forest bureaucracy prepares for the funeral of the much-hailed community forest management programme." *Down To Earth,* 9(22), 20.

Majer, J.D. (1989). "Long term colonization of fauna in rehabilitated land." In *Animals in Primary Succession: The rple of Fauna in Reclaimed Land,* Majer, J.D. (ed.), pp. 143–174. Cambridge University Press, Cambridge.

Makino, A., Shimada, T., Takumi, S., Kaneko, K., Matsuoka, M., Shimamoto, K., Nakano, H., Miyao-Tokutomi, M., Mae, T., and Yamamoto, N. (1997). "Does decrease in ribulose-1,5-bisphosphate carboxylase by antisense rbcS lead to a higher N-use efficiency of photosynthesis under conditions of saturating CO_2 and light in rice plants?" *Plant Physiol.* 114, 483–491.

Maliga, P. (2002). "Engineering the plastid genome of higher plants." *Curr. Opin. Plant Biol.,* 5, 164–172.

Manuel, P., Jeffrey, S., and Susanna, J. (1993). "El Extractivismo en America Latina, IUCN." Gland, Switzerland, 63−64.

Maryani, Z. (2010). "Protecting rainforest in time of development." The Brunei Times News online.

Mate, C.J., Caemmerer, S. von, Evans, J.R., Hudson, G.S., and Andrews, T.J. (1996). "The relationship between CO_2 assimilation rate, Rubisco carbamylation and Rubisco activase content in activase-deficient transgenic tobacco suggests a simple model of activase action." *Planta.* 198, 604−613.

Matthews, S.A., Shivakoti, G.P., and Chhetri, N. (2000). "Population forces and environmental change: Observations from Western Chitwan, Nepal." *Society and Natural Resources,* 13,763–775.

Maurino V.G., and Peterhansel C. (2010) "Photorespiration: current status and approaches for metabolic engineering." *Curr Opin Plant Biol.*, 13, 249–256.

Millard, P., and Catt, J.W. (1988). "The influence of nitrogen supply on the use of nitrate and ribulose1,5-bisphosphate carboxylase/oxygenase as leaf nitrogen stores for growth of potato tubers (Solanum tuberosum L.)." *Journal of Experimental Botany,* 39, 1−11.

Monsanto. (2009a). "Weed resistance management, practical approaches to managing weeds: Stewardship." St. Louis, MO: Author. Available on the World Wide Web: http://www.weedresistancemanagement.com/stewardship.html. (Oct.7, 2010).

Monsanto. (2009b). "Weed resistance management, practical approaches to managing weeds: Professional recommendations." St. Louis, MO: Author. Available on the World Wide Web: http://www.weedresistancemanagement.com/recommendations. html. (Oct.7, 2010).

Murcia, C. (1995). "Edge effects in fragmented forests: implications for conservation." *Trends in Ecology and Evolution*, 10, 58−62.

Mutchler, C.K., Murphree, C.E., and McGregor, K.C. (1994). "Laboratory and field plots for erosion research." In Lal (ed.), *Soil Erosion Research Methods,* 2nd ed. Soil and Water Conservation Society, 7515 Northeast Ankeny Road, Ankeny, IA 50021.

Myers, N. (1989). *Deforestation Rates in Tropical Forests and Their Climatic Implications.* London, Friends of the Earth, U.K.

Ogren, W.L., and Bowes, G. (1971). "Ribulose diphosphate carboxylase regulates soybean photorespiration." *Nature New Biology*, 230, 159−160.

Ogren, W.L. (1984). "Photorespiration: Pathways, regulation and modification." *Annual Review of Plant Physiology*, 35, 415−442.

Opdam, P., and Wascher, D. (2004). "Climate change meets habitat fragmentation: Linking landscape and biogeographical scale levels in research and conservation." *Biological Conservation*, 117, 285−297.

Panayotou, T., and Sungsuwan, S. (1989): "An econometric study of the causes of tropical deforestation: the case of northeast Thailand." Cambridge, Massachusetts, Harvard Institute for International Development. 32(2), (Development Discussion Paper No. 284)

Parrotta, J.A., Turnbull, J.W., and Jones, N. (1997). "Catalyzing native forest regeneration on degraded tropical lands." *Forest Ecology and Management*. 99, 1–7.

Parry, M.A., Andralojc, P.J., Mitchell, R.A., Madgwick, P.J., and Keys, A.J. "Manipulation of Rubisco: The amount, activity, function and regulation." *J. Exp. Bot.*, 54 (386), 1321–33.

Peoples, M.B., Pate, J.S., and Atkins, C.A. (1983). "Mobilization of nitrogen in fruiting plants of a cultivar of cowpea." *Journal of Experimental Botany*, 34, 563–578.

Peter, H.M., Emily, B., Fernando, V., and Manyu, C. (2004). "Local sustainable development effects of forest carbon projects in Brazil and Bolivia." *Environmental Economics Programme*, 1–132

Portis, A.R. (1992) "Regulation of ribulose-1,5-bisphosphate car- boxylase/oxygenase activity. "*Annu Rev Plant Physiol Plant Mo1. Biol.*, 43, 415–437.

Portis, A.R., and Parry, M.A.J. (2007) "Discoveries in Rubisco (ribulose 1,5-bisphosphate carboxylase/oxygenase): A historical perspective." *Photosynthesis Res.*, 94, 121–143.

Quick, W.P., Schurr, U., Fichtner, K., Schulze, E.-D., Rodermel, S.R., Bogorad, L., and Stitt, M. (1991a). "The impact of decreased Rubisco on photosynthesis, growth, allocation and storage in tobacco plants which have been transformed with antisense *rbcS*." *Plant Journal*, 1, 51–58.

Raj, S., Rattan, L., and Arjun, H. (2010). "Offsetting carbon dioxide emissions through minesoil reclamation." In: *Encyclopedia of Earth.* Cleveland, C.J. (eds.), Washington, D.C.: Environmental Information Coalition, National Council for Science and the Environment). http://www.eoearth.org/article/Offsetting_carbon_dioxide_emissions_through_minesoil_reclamation> (Oct.7, 2010).

Repetto, R., and Holmes, T. (1983), "The role of population in resource depletion in developing countries." *Population and Development Review*, 9(4), 609–632.

Revington, J. (1991). *The Cause of Tropical Deforestation*. Renaissance Universal.

Reynolds, M.P., Van Ginkel, M., and Ribaut, J.M. (2000). "Avenues for genetic modification of radiation use efficiency in wheat." *Journal of Experimental Botany*, 51, 459–473.

Robert, R. (1990). "Deforestation in the Tropics." Online article Scientific American, 39 42.

Robinson, S.P., and Portis, A.R. (1988). "Release of the nocturnal inhibitor, carboxyarabinitol-1-phosphate, from ribulose bisphosphate carboxylase/oxygenase by Rubisco activase." *FEBS Letters*, 233, 413–416.

Rodermel, S.R., Abbott, M.S., and Bogorad, L. (1988). "Nuclear-organelle interactions: nuclear antisense gene inhibits ribulose bisphosphate carboxylase enzyme levels in transformed tobacco plants." *Cell*. 55,673–681.

Sage, R.F., Christin, P.A., and Edwards, E.J. (2011). "C4 plant lineages of planet Earth." *Special Issue of Journal of Experimental Botany,* 62, 3155–3169.

Saunders, D.A., Hobbs, R.J., and Margules, C.R. (1991). "Biological consequences of ecosystem fragmentation: A review." *Conservation Biology,* 5, 18–32.

Saunders, D.A., Hobbs, R.J., and Ehrlich, P.R. (1993). "Reconstruction of fragmented ecosystems: Problems and possibilities." *Reconstruction of fragmented ecosystems,* Saunders, D.A., Hobbs, R.J., and Ehrlich, P.R. (eds.), Surrey Beaty and Sons, 305–310.

Southgate, D. (1994). "Tropical deforestation and agricultural development in Latin America." In: *The Causes of Tropical Deforestation,* Brown, K., and Pearce, D.W. (eds.), UCL Press, London.

Shafik, N. (1994). "Macroeconomic causes of deforestation: Barking up the wrong tree?" In: *The Causes of Tropical Deforestation,* Brown, K., and Pearce, D.W. (eds.), UCL Press, London.

Sidorov, V.A., Kasten, D., Pang, S.Z., Hajdukiewicz, P.T.J., Staub, J.M., and Nehra, N.S. (1999). "Stable chloroplast transformation in potato: Use of green fluorescent protein as a plastid marker." *Plant J.,* 19, 209–216.

Sikdar, S.R., Serino, G., Chaudhuri, S., and Maliga, P. (1998). "Plastid transformation in *Arabidopsis Thalian.*" *Plant Cell Rep.,* 18, 20–24.

Spreitzer, R.J., and Salvucci, M.E. (2002). "Rubisco: structure, regulatory interactions, and possibilities for a better enzyme." *Annual Review of Plant Biology*, 53, 449–475.

Steer, M.W., Gunning, B.E.S., Graham, T.A., and Carr, D.J. (1966). "Isolation and properties of fraction 1 protein from Avena sativa L." *Planta*, 79, 256–267.

Stitt, M., Quick, W.P., Schurr, U., Schulze, E.-D., Rodermel, S.R., and Bogorad, L. (1991). "Decreased ribulose-1,5-bisphosphate carboxylase-oxy genase in tobacco transformed with 'antisense' rbcS. 11 Fluxcontrol coefficients for photosynthesis in varying light, CO, and air humidity." *Planta,* 183, 555–566.

Stitt, M., Lunn, J., and Usadel, B. (2010) "Arabidopsis and primary photosynthetic metabolism–more than the icing on the cake." *Plant J.*, 61, 1067–1091.

Streusand, V.J., and Portis, A.R. (1987). "Rubisco activase mediates ATP dependent activation of ribulose bisphosphate carboxylase." *Plant Physiology*, 85, 152–154.

Sunderlin, W.D., Angelsen, A., Belcher, B., Burgers, P., Nasi, R., Santoso, L., and Wunder, S. (2005). "Livelihoods, forests, and conservation in developing countries: an overview." *World Development,* 33(9), 1383–1402.

Surampalli, R., Zhang, T.C., Ojha, C.S.P., Tyagi, R.D., and Kao, C.M. (eds.) (2013). *Climate Change Modeling, Mitigation and Adaptation,* ASCE, Reston, Virginia, 2013.

Svab, Z., Hajdukiewicz, P., and Maliga, P. (1990). "High-frequency plastid transformation in tobacco by selection for a chimeric aadA gene." *Proc. Natl Acad. Sci.,* 87, 8526–8530.

Tscharntke, T., Steffan dewenter, I., Kruess, A., and Thies, C. (2002). "Characteristics of insect populations on habitat fragments: A mini review." *Ecological Research,* 17, 229–239.

Tucker, N.I.J. (2000). "Linkage restoration: Interpreting fragmentation theory for the design of a rainforest linkage in the humid Wet Tropics of north-eastern Queensland." *Ecological Management and Restoration*, 1, 35–41.

UNEP. (2010). http://www.rrcap.unep.org/reports/soe/nepalsoe.cfm (Oct.7, 2010).

USEPA (U.S. Environmental Protection Agency) and Tetra Tech, Inc. (1993). *Summary of Current State Nonpoint Source Control Practices for Forestry.* USEPA Office of Wetlands, Oceans and Watersheds. 2–3.

Warburg, O. (1920). "UÈ ber die Geschwindigkeit der photochemischen KohlensaÃurezersetung in lebenden Zellen II." *Biochemische Zeitschrift*, 100, 188–217.

Whitemore, T.C. (1997). "Tropical forest disturbance, disappearance, and species loss." *Tropical Forest Remnants: Ecology, Management, and Consevation of Fragmented Communities,* Laurance, W., and Bierregard, R. (eds.), The University of Chicago Press, Chicago, pp. 3–12.

Whitney, S.M., and Andrews, T.J. (2001a). "Plastome-encoded bacterial ribulose-1,5-bisphosphate carboxylase/oxygenase (RubisCO) supports photosynthesis and growth in tobacco." *Proc. Natl. Acad. Sci.,* 98, 14738–14743.

Whitney, S.M., and Andrews, T.J. (2001b). "The gene for the ribulose-1,5-bisphosphate carboxylase/oxygenase (Rubisco) small subunit relocated to the plastid genome of tobacco directs the synthesis of small subunits that assemble into Rubisco." *Plant Cell,* 13, 193–205.

Whitney, S.M., Caemmerer, S. von, Hudson, G.S., and Andrews, T.J. (1999). "Directed mutation of the Rubisco large subunit of tobacco influences photorespiration and growth". *Plant Physiol.,* 121,579–588.

Whitney, S.M., Baldet, P., Hudson, G.S., and Andrews, T.J. (2001). "Form I Rubiscos from non-green algae are expressed abundantly but not assembled in tobacco chloroplasts." *Plant J.*, 26(5), 535–547.

Wikipedia (2013). "Reforestation." Available at <http: en.wikipedia.org/wiki/> (accessed Mar 2013).

Wildner, G.F. (1981). "Ribulose-1,5-bisphosphate carboxylaseoxygenase: Aspects and products." *Physiologia Plantarum*, 52, 385–389.

World Conservation Monitoring Centre with IUCN, UNEP, WWF and WRI (1990). *Global Biodiversity.* Report. Chapman & Hall, London, England. 447–452.

Write, S.J. (2005). "Tropical forests in a changing environment." *Trends in Ecology and Evolution*, 20, 553–560.

Wright Water Engineers, Inc. and Denver Regional Council of Governments (1999). *Mountain Driveway Best Management Practices Manual.* Document Prepared Under Clean Water Act Section 319 Nonpoint Source Grant Funding (Contract #WQC9808703).

Zhang, X.H., Ewy, R.G., Widholm, J.M., and Portis, A.R. (2002a). "Complementation of the nuclear antisense rbcS-induced photosynthesis deficiency by introducing an rbcS gene into the tobacco plastid genome." *Plant Cell Physiol.,* 43, 1302−1313.

Zhang, N., Kallis, R.P., Ewy, R.G., and Portis, A.R. (2002b). "Light modulation of Rubisco in Arabidopsis requires a capacity for redox regulation of the larger Rubisco activase isoform." *Proceedings of the National Academy of Sciences*, 99, 3330−3334.

Zhang, T.C., and Surampalli, R.Y. (2013). "Carbon capture and storage for mitigating climate changes." Chapter 20 in *Climate Change Modeling, Mitigation and Adaptation,* Surampalli, R., Zhang, T.C., Ojha, C.S.P., Gurjar, B.R., Tyagi, R.D., and Kao, C.M. (eds.), ASCE, Reston, Virginia, February 2013.

CHAPTER 14

Enzymatic Sequestration of Carbon Dioxide

Xiaolei Zhang, Song Yan, R. D. Tyagi, Rao Y. Surampalli and Tian C. Zhang

14.1 Introduction

Since the 1990s, the idea of using enzyme for carbon dioxide sequestration has emerged and has rapidly grabbed a great deal of interest. Enzymatic sequestration of carbon dioxide is a way to sequester carbon dioxide through transforming carbon dioxide into different compounds or valuable chemicals, such as 1) bicarbonate/ carbonate, 2) formate, 3) methanol, and 4) methane (CH_4). There is a great potential for using enzymes to sequestrate carbon dioxide from flue gases of industry (Salmon et al. 2009) because the methods may be rapid, environmentally friendly, offering permanent carbon dioxide disposal, and cost-effective (Bond et al. 2001a, 2003; Hamilton 2007; Figueroa et al. 2008; Dilmore et al. 2009). Although a few reviews have been made about the technology (Favre 2011; Pierre 2012; Shekh et al. 2012), an overview of using the technology to form the aforementioned four chemicals is not available. In addition, some new publications and new insights have been evolved due to the accelerated research pace in this area. Therefore, this chapter aims to provide a brief state-of-the-art overview on the technology. The type of enzymes used and the associated mechanisms for carbon dioxide sequestration are described. The current limitations and challenges to scale up the application of the technology are discussed, and the solutions are suggested.

14.2 Carbonic Anhydrase Catalytic Carbon Dioxide Sequestration

14.2.1 Mineralization of CO_2 via $CaCO_3$ Formation

Five basic equations are involved in mineralization of CO_2 via $CaCO_3$ formation. First, Eq. 14.1 represents the phase change of CO_2. Second, aqueous CO_2 is associated with water to form H_2CO_3 (Eq. 14.2). Third, H_2CO_3 is dissociated to HCO_3^- and further to CO_3^{2-} (Eq. 14.3–4). Finally, in the presence of Ca^{2+}, CO_3^{2-} is converted to $CaCO_3$ (Eq.

14.5). In the whole process, the H_2CO_3 formation step is the governing step as it has the lowest reaction rate (Mirijafari et al. 2007). To enhance the reaction rate, an enzyme catalyst has been reported (Ramanan et al. 2009).

$$CO_2 \text{ (gaseous)} \rightarrow CO_2 \text{ (aqueous)} \quad (14.1)$$

$$CO_2 \text{ (aqueous)} + H_2O \rightarrow H_2CO_3 \quad (14.2)$$

$$H_2CO_3 \rightarrow HCO_3^- + H^+ \quad (14.3)$$

$$HCO_3^- \rightarrow CO_3^{2-} + H^+ \quad (14.4)$$

$$Ca^{2+} + CO_3^{2-} \rightarrow CaCO_3 \quad (14.5)$$

The enzyme used in the process mainly refers to carbonic anhydrase (CA), which is a zinc metalloenyme and can be found in all living organisms such as humans, animals, plants, and microorganisms (Karlsson et al. 1998; Liu et al. 2005). In the bodies of living organisms, CA functions as a catalyst to accelerate the conversion of carbon dioxide to bicarbonate ion, which is easier to be transported between the inside and outside of the cells. Because of CA, the hydration rate increases from 14.1 reactions/second in the absence of CA (Smith 1988) to 10^4–10^6 reactions/second in the presence of CA at the same temperature of 25 °C (Heck et al. 1994). According to the CA's capacity to transform carbon dioxide into its hydration, researchers and engineers have studied to enhance carbon dioxide hydration with free CA, CA immobilized onto biodegradable materials (e.g., alginate and chitosan-based material), or surface modified CA (Simsek et al. 2001; Simsek-Ege et al. 2002; Yadav et al. 2010). Apart from CA, another type of enzyme, a tungsten-containing formate dehydrogenase enzyme, has been reported to be able to reduce carbon dioxide into formate which is stable and can be further used to produce methane and methanol (Reda et al. 2008). It provides an alternative for carbon dioxide sequestration.

14.2.2 Carbonic Anhydrase

Carbonic anhydrase, carbonate dehydratase and carbonate hydrolyase, being known as one of the most rapid enzymes are widely found in nature. Most of the CAs are identified to contain a zinc ion in the active site. Hence, they are also classified as metalloenzymes, in which zinc is coordinated by three histidines and a hydroxide (Domsic and McKenna 2010). The active site is believed consisting of a hydrophobic patch and a hydrophilic patch. So far, five types of CAs have been identified, namely, α-, β-, γ-, δ-, and ε-type CAs. These five types are evolved independently because no similarity on amino acid sequences is observed in the five CAs (Smith 1988). Table 14.1 shows the historical evolution of CA discovery (together with their use for CO_2 capture).

Among all CAs, α- and β-type CAs are the most known types. α-type CA mainly occurs in humans, animals, and eubacteria, while β-type CA mostly occurs in plants, eubacteria, and prokaryotic organisms (HewettEmmett and Tashian 1996; Karlsson et al. 1998). The other three types, γ-, δ-, and ε-type CAs, are present in methane-producing bacteria, diatoms, and chemolithotrophs and marine cyanobacteria, respectively (So et al. 2004). For humans and animals, CA plays a vital role in respiration as a catalyst to enhance carbon dioxide hydration to bicarbonate ion which is easier to be transported from red blood cells to lung. For plants and microorganisms, CA is very important in the photosynthesis process in which CA captures carbon dioxide from the environment and converts it into bicarbonate ion, which is a way that plants or microorganisms to concentrate carbon source for converting bicarbonate ion back to carbon dioxide as substrate when carbon dioxide is insufficient.

Table 14.1 Historical evolution of CA discovery and their use for CO_2 capture

Time[Ref]	Discovery/activities
1933[1]	The first CA when they were studying the reasons for a rapid transition of HCO_3^- from erythrocytes towards the lung capillary
1939[2]	CAs of plant origin were shown to be different from the previously known CAs
1940[3-5]	CA extracts were purified from bovine erythrocytes; CA was found to contain a Zn atom in their active site
1933–1948[10]	The electrochemical method was developed to measure the activity of CA for CO_2 capture.
1963[6]	CA enzymes were found in prokaryotes
1972[7]	CA enzymes extracted from prokaryotes (*Neisseria Sicca*) were purified
1988[11]	The immobilization of CA on silica beads and graphite rods was developed
1992–1993[8, 9]	The first genetic sequence of a purified CA of prokaryote origin (bacteria *Escherichia coli*) was established. This metalloenzyme was the first *β*-type CA, while the previous ones were classified in the *α* type
1990s & 2000s[10]	New CA varieties were discovered; the catalytic mechanism of these enzymes, particularly regarding the CA of human origin were understood
1999-2003[12-16]	CA was immobilized on chitosane and alginate beads, and used for CO_2 sequestration. The proof of principle was demonstrated
2003[17]	CA from the alga *Dunaliella salina* was expressed in *Escherichia coli*
2006 [18]	Carboxysomal Carbonic Anhydrase CsoSCA from *Halothiobacillus neapolitanus* was characterized
2008 [19]	CA with high activity was screened from *Enterobacter* and *Aeromonas* isolates
2009[20]	CA was purified to homogeneity from *C. freundii* SW3. Effects of the host metal ions, cations and anions were investigated on CA activity
2010[21]	Carbonic anhydrase (CA) purified from *Pseudomonas fragi*, *Micrococcus lylae* and *Micrococcus luteus* was used for CO_2 sequestration and compared with commercial bovine carbonic anhydrase (BCA)
2012[22]	CA was covalently immobilized onto OAPS (octa(aminophenyl)silsesquioxane)-functionalized Fe_3O_4/SiO_2 nanoparticles. The immobilized CA was used for CO_2 sequestration and the reusability was investigated
2013[23]	CA from the cyanobacterium *Synechocystis* sp was immobilized onto tetramethoxysilane and its maximum stability was studied. The immobilized CA was used for CO_2 sequestration

[1] = Meldrum and Roughton (1933). [2] = Neish (1939). [3–5] = Keilin and Mann (1939, 1940); Rowlett and Siverman (1982). [6] = Veitch and Blankenshi (1963). [7] = Adler et al. (1972). [8, 9] = Guilloton et al. (1992, 1993). [10] = Pierre (2012). [11] = Crumbliss et al. (1988). [12–16] = Bond et al. (1999a, b, 2001a, b, c, 2003). [17] = Premkumar et al. (2003). [18] = Heinhorst et al. (2006). [19] = Sharma et al. (2008). [20] = Ramanan et al. (2009). [21] = Sharma and Bhattacharya (2010). [22] = Vinoba et al. (2012). [23] = Chien et al. (2013).

14.2.3 Carbon Dioxide Sequestration

Mechanism. According to the similar theory as CA functions in living organisms, researchers pointed out that CAs could be used to grab and transfer carbon dioxide into bicarbonate or carbonate ions which thereafter can be stabilized by the addition of cations (Ca, Mg) (Bond et al. 2001a, b; Liu et al. 2005; Kanbar 2008). The mechanism of carbon dioxide sequestration with CAs is shown in Figure 14.1. As aforementioned, the active site of CA is composed of a half hydrophilic patch and a half hydrophobic patch. In the presence of CA, carbon dioxide first interacts with the hydrophobic side of the active site, and is then placed in the specific pocket in the active site. The hydroxide coordinating with zinc ion in the active site will then nuclophilically attack the carbon of the bound carbon dioxide, which leads to the formation of bicarbonate ion. Thereafter, the bicarbonate ion will be replaced by a water molecule due to the stronger coordination between water and the zinc ion than between bicarbonate ion and the zinc ion. A proton in the water molecule will be finally transferred into the bulk solvent by a prominent proton shuttle residue histidine, and a zinc-bound hydroxide will be regenerated to interact with carbon dioxide. The dissociated bicarbonate ion will be formed into carbonate by providing cation source (Fisher et al. 2007; Silverman and Mckenna 2007; Domsic and McKenna 2010).

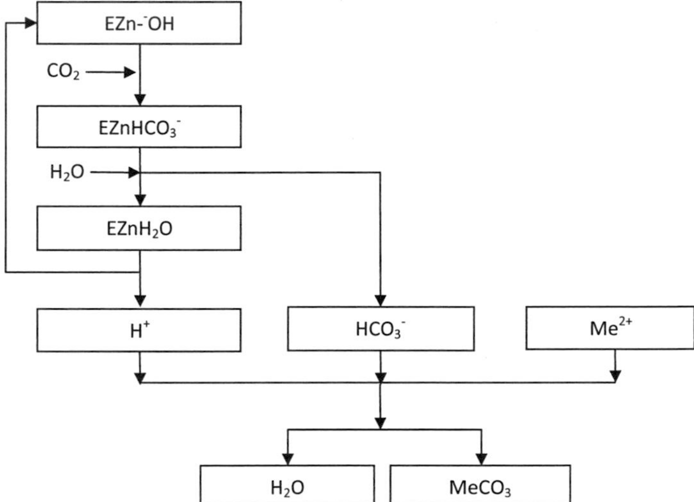

Figure 14.1 Mechanisms of CA catalytic carbon dioxide sequestration. E represents the three histidines; Me^{2+} represents metal ions (particularly Ca^{2+} and Mg^{2+}); and the CA can be free CA or immobilized onto supports

Factors Affecting CA Enzymatic Carbon Dioxide Sequestration. According to the mechanism, it can be understood that CA is the essential element in carbon dioxide sequestration. Therefore, any factor that appears in the sequestration process and affects CA activity will impact sequestration efficiency of carbon dioxide. Several factors have shown apparent effects on the CA catalytic carbon dioxide sequestration, including the type and the concentration of CA, temperature, pH, carbon dioxide concentration, cation species and concentration, and the buffer presence in the reaction system. In addition, CA immobilized onto some materials which are cost-effective and environmentally friendly is used in carbon dioxide sequestration; therefore, the materials used for supporting CA also affect the sequestration since the material may affect the activity and stability of CA. Descriptions on the effects of some major factors follow.

Temperature. It is well known that irreversible denaturation of protein can be caused when temperature reaches a certain value which varies according to the type of protein. Therefore, temperature is a significant parameter in the process of CA enzymatic carbon dioxide sequestration. Several studies have reported that erythrocyte CA could well maintain its activity as long as the temperature is not greater than 60 °C. However, the activity of CA starts to reduce at the temperature > 60 °C, and finally will be lost when temperature reaches 70 °C (Lavecchia and Zugaro 1991; Cioci 1995). The activity of bovine CA was stable at temperature 40 °C, but dramatically reduced as temperature increased from 40 °C to 60 °C, and was fully lost within 40 min once temperature reached 63 °C (Kanbar 2008). The studies suggest that appropriate temperature should be studied in order to prevent the activity loss of CA.

Mostly, CA used in carbon dioxide sequestration are not free CA but CA immobilized on solid supports such as polymer foams, chitosan alganite beads, and silica monoliths (Jovica and Kostic 1999; Drevon et al. 2003; Liu et al. 2005; Cheng et al. 2008; Kanbar 2008). Therefore, studies on temperature effects on the activity of immobilized CA seem to be more important in practice than that on the activity of free CA. Kanbar (2008) reported that the activity of immobilized bovine CA increased as temperature increased between 20 °C to 45 °C, and rapidly decreased after 45 °C, and finally was lost when temperature increased to 60 °C. Kanbar (2008) also estimated the stability of the activity of immobilized CA and found that the activity was rather stable under 40 °C and 50 °C; however, the activity began to loss at temperature above 50 °C, and at 60 °C the activity was completely lost in 10 to 20 min. Moreover, by comparing the stability of the activity of free CA and immobilized CA, it was found that immobilization enhanced the stability of CA activity (Kanbar 2008).

The fact that the solubility of gases decreases as temperature increases implies that the temperature effect on CA catalytic carbon dioxide sequestration is also contributed by the solubility change of carbon dioxide under different temperatures. Mirjafari et al. (2007) conducted carbon dioxide sequestration under temperature of 0,

30, and 50 °C through CA catalysis to finally form $CaCO_3$. It was observed that the quantity of the product, $CaCO_3$, evidently decreased as temperature increased under similar reaction conditions (0.2098 g at 0 °C, 0.1283 g at 30 °C, and 0.096 g at 50 °C).

pH and Carbon Dioxide Concentration. CA has showed the excellent performance at neutral pH (Supuran et al. 1999; Özensoy et al. 2005; Innocenti et al. 2009). Reports indicated that the carbon dioxide conversion to bicarbonate ions using CA as catalyst is a strong pH depending reaction process due to the pH effect on the activity of CA (Trachtenberg and Bao 2005). Acidic pH or alkaline pH can inhibit the CA activity, resulting in the reduction of the carbon dioxide sequestration efficiency (Innocenti et al. 2009). Apart from the pH effect on the activity of CA, the low pH could also lead the dissolution rather than precipitation of solid carbonate. Mirjafari et al. (2007) found that no $CaCO_3$ was detected in the absence of buffer in CA catalytic carbon dioxide sequestration. Therefore, using CA for enhancing carbon dioxide sequestration requires a good maintenance of pH at around 7. Carbon dioxide sequestration is usually carried out through injecting target gases (air or the gases wanted to be treated), which most likely induces a decrease in the pH of the system. In order to overcome the influence of carbon dioxide addition on pH, buffer or alkaline bicarbonate has been generally employed in using CA for carbon dioxide sequestration (Ge et al. 2002; Trachtenberg and Bao 2005; Mirjafari et al. 2007). Ge et al. (2002) designed an enzyme contained liquid membrane which is composed of two polypropylene membranes separated by CA containing phosphate buffer solution, to evaluate the carbon dioxide capture capacity of the system. They observed that the addition of buffer solution could well keep the pH constant and thus prevented the influence of pH to carbon dioxide capture. Trachtenberg and Bao (2005) used $NaHCO_3$ to minimize the pH change during carbon dioxide feed and obtained a stable carbon dixoide capture.

As aforementioned, carbon dixoide concentration has an extensive impact on the pH of the sequestration system without buffer or alkaline bicarbonate/carbonate adjustment. In addition, another consideration on carbon dixoide concentration in the CA catalystic carbon dixoide sequestration is the inhibition of carbon dixoide concentration on activity of CA. Some researchers reported that the carbon dioxide concentration above the atmospheric carbon dixoide level (0.03–0.05% (v/v)) would inhibit the activity of CA, whereas CA could perform normal under a low carbon dixoide concentration (Bahn et al. 2005). However, some other researchers found a contradictory result that the CA activity could be improved as the carbon dixoide concentration increased from the atmospheric level to 5% (v/v) (Ramanan et al. 2009). Ramanan et al. (2009) also observed that the CA activity reduced by half after the carbon dixoide concentration reached 10% (v/v), predicted that it could be due to the feedback inhibition from bicarbonate ion, and suggested that timely precipitation of bicarbonate ion would aviod the inhibition. These studies indicate that carbon dixoide concentration of the target gases plays an essential role in CA-based carbon dixoide sequestration.

Cations and Anions. Normally, flue gases contain heavy metals such as Hg^{2+}, Pb^{2+}, and Zn^{2+}, and anions such as Cl^-, NO_3^-, and SO_4^{2-} (Bond et al. 2001a). Understanding the effect of cations and anions on CA catalytic carbon dioxide sequestration is very important. The influence of cations on CA activity has been evaluated, such as Al^{3+}, Ca^{2+}, Cd^{2+}, Co^{2+}, Cu^{2+}, Fe^{3+}, Hg^{2+}, Mg^{2+}, Mn^{2+}, Ni^{2+}, Pb^{2+}, and Zn^{2+} (Supuran et al. 1999; Ramanan et al. 2009). The studies pointed out that Zn^{2+} could positively stimulate the enzyme activity of CA, which was most likely because CA was a zinc metalloenzyme; moreover, Cd^{2+}, Co^{2+}, Cu^{2+}, and Fe^{3+} were also observed to enhance the activity of CA. The strongest inhibition was from mercury ion which was one of the essential parts of flue gases, followed by lead ion. Other cations such as Ca^{2+} and Mg^{2+} expressed only slight inhibition on enzyme activity due to the possibility of precipitation with bicarbonate/carbonate ions. For anions, sulfonamides and sulfamates were the most known inhibitors, and Cl^-, HCO_3^-, and CO_3^{2-} were also potent inhibitors, while NO_3^- seemed to have rather weak inhibition on enzyme activity of CA, and SO_4^{2-} showed nearly no effect on the enzyme activity (Franchi et al. 2003; Vullo et al. 2006; Ramanan et al. 2009). The inhibition of anions on enzyme activity of CA was suggested due to the interaction between zinc ion and the anions or the transporter of the anions which would result in reducing or even losing the enzyme function of CA.

Enzyme. As discussed in Section 14.2.1, there are five types of CAs in nature, and each family of the CAs has several isoforms. For example, α-type of CA has the isoform of CA I, CA II, CA III, etc. So far, various types or isoforms of CA have been utilized in carbon dioxide sequestration, with Bovine CA being studied most (Liu et al. 2005; Mirjafari et al. 2007; Kanbar 2008). In addition, human CA II (HCA II) as the fastest isozyme has also been produced via bacterial overexpression and used in carbon dioxide sequestration (Bond et al. 2001c, 2003; Domsic and McKenna 2010). Ramanan et al. (2009) reported carbon dioxide sequestration by using β-type CA extracted from *Citrobacter freundii*. CA purified from bateria, *Pseudomonas fragi*, *Micrococcus lylae*, and *Micrococcus luteus*, have also showed great potential on carbon dixoide sequestration (Sharma and Bhattacharya 2010).

Up to date, the aspect on comparison of the influence of the type and isoform of enzyme CA on carbon dioxide sequestration has not been reported. However, the behaviors of different types and isoforms of CA under different reaction conditions such as reaction temperature, reaction pH, and the presence of cation and/or anion, could be used as basic criteria to select an appropriate CA for carbon dioxide sequestration. For instance, α-type CA, CA II and CA IV are more inhibited by the presence of cations such as Cu, Mn, and Mg than α-type CA, CA I (Supuran et al. 1999). β-type CA purified from bacteria was not affected by the SO_4^{2-} presence while that purified from animal showed an apparent effect by the SO_4^{2-} presence (Bond et al. 2001b; Ramanan et al. 2009). In addition, modified CA could improve the CA stability by several folds (Yadav et al. 2010). Yadav et al. (2010) modified CA by enclosing the CA with biopolymers to form particles in nanosize, 70–80 nm. The author observed that the half-life of the modified CA was up to 100 days while it was only 15 days for free CA at temperature of minus 20 °C.

The effect of CA concentration on sequestration has also been studied, and the results seemed debatable. Bond et al. (2001b) reported that CA performed better at low concentration than at high concentration. Unfortunately, no specific CA concentration was available in the article. Mirjafari et al. (2007) observed that similar $CaCO_3$ amount (around 0.2 g) was produced at the CA concentration of 3 μM and 6 μM in carbon dioxide sequestration. This is possibly due to three reasons. One could be that the CA concentration of 3 μM in the sequestration is excess, hence when the concentration increased to 6 μM, no effect was observed. One could be that the calcium ion was limited in the sequestration system and hence no more $CaCO_3$ than 0.2 g could be detected due to the reactant limit. Another reason could be the $CaCO_3$ formation rate had reached the maximum under the two concentration CA conditions. In order to understand the CA concentration effects, more studies are needed.

Materials for CA Immobilization. Immobilized CA is more favorable than free CA for carbon dioxide sequestration because the immobilization can minimize the loss of CA from the sequestration system and enhance the activity stability of CA (Bond et al. 2001c; Kanbar 2008). Polyurethane foam (PU), acrylamide, alginate, chitosan-alginate, and silica monoliths have been used as support materials for CA immobilization (Jovica and Kostic 1999; Bond et al. 2001c; Drevon et al. 2003; Liu et al. 2005; Cheng et al. 2008; Kanbar 2008). The selection of immobilizing materials is based on the consideration of cost, stabilization, effect on enzyme activity, and impact on the environment. Among all the supports reported, chitosan-alginate was considered more suitable in usage for supporting CA due to their characteristics (e.g., cheap, stable, safe, environmentally friendly, and biodegradable) (Bond et al. 2001c; Liu et al. 2005). Kanbar (2008) pointed out that PU, an environmentally friendly material, was a promising material for CA immobilization. It was found that all the isocyanate groups of PU would react with the amine and/or hydroxyl group of CA to form covalent bonds after immobilization, which would prevent the inhibition on CA from the unreacted isocyanate groups of PU.

Cation Source. After carbon dioxide is converted into bicarbonate/carbonate ions, an essential step is to rapidly transform bicarbonate/carbonate ions into precipitates ($CaCO_3$ or $MgCO_3$) because the accumulated bicarbonate ions in reaction solution would inhibit CA activity, and hence, affect the sequestration efficiency. Therefore, the cation supply becomes important in the sequestration. The cations can be directly supplied to the sequestration by adding chemicals (e.g., $CaCl_2$ or $MgCl_2$), but this is a rather expensive way. The sustainable cation sources are seawater, waste brines from desalination operations, brines from saline aquifers, and produced water (Bond et al. 2001c; Liu et al. 2005). However, cation source selection should also consider the source transportation cost. Thus, when no cost-effective cation source is available in the sequestration site, which implies that cation source transport is demanded, the evaluation should be performed to balance the cost of cation supply. For instance, when the cost (including transportation, labor cost, etc.) for supplying cations by an aforementioned alternative is lower than the cost for supplying cations by directly adding chemicals, then the alternative can be used as a cation source; otherwise, adding chemicals should be taken into consideration.

14.2.4 Applications

As carbon dioxide concentration in the atmosphere is rather low, capture carbon dioxide directly from the atmosphere is not practical unless the atmospheric carbon dioxide is concentrated first. Therefore, carbon dioxide sequestration for mitigate carbon dioxide effect on climate normally means to reduce or cut off the large carbon dioxide emission from human activities. In order to reduce the carbon dioxide emission, researchers and engineers set up reactors containing CA or producing CA to capture the carbon dioxide produced from the plants. National Aeronautics and Space Administration (NASA) has developed a thin liquid membrane system to capture CO_2 even from atmosphere (Cowan et al. 2003). The system consists of 2 microporous hydrophobic polypropylene membranes which are separated by a thin liquid layer containing CA. CO_2 flow can enter the liquid layer from any side of the microporous membrane. It then diffuses across the liquid layer. While CO_2 contacts with water in the presence of CA, it is captured to form H_2CO_3. Due to the concern of liquid layer drying in the long term operation, a similar process was developed by Carbozyme Company (Trachtenberg et al. 2009). Instead of liquid layer, microfibers immobilized with CA were inserted between microporous hydrophobic polypropylene membranes to capture CO_2. A company called CO_2 Solution in Quebec city of Canada tested the ability of a bioreactor, in which *Escherichia coli* capable of producing CA was grown onto the surface of a modified solid support, to capture carbon dioxide from a small municipal incinerator and an Alcoa aluminum smelter (Hamilton 2007), and after that, they also investigated the bioreactor potential on carbon dioxide capture from a power plant. Bond et al. (2001a, b, 2003) and Liu et al. (2005) designed carbon dioxide scrubbers which contained CA immobilized onto biodegradable chitosan-alginate beads, to capture carbon dioxide from coal-fired power plants by forming them into solid carbonate in which cations (Ca and Mg) were provided by adding produced water or brines depending on the sites of the power plants and cation sources. The problem of using CA to capture carbon dioxide from flue gases of industry is the temperature effect on CA activity because normally the flue gas stream has high temperature (> 100 °C). Usually, cooling systems should be coordinated with the sequestration system.

14.3 Other Enzyme Catalytic Carbon Dioxide Sequestration

14.3.1 CO_2 Reduction to Formate

Formate dehydrogenase (FD) is an enzyme capable of catalyzing the oxidation of formate to CO_2. There are various types of FD enzymes. Some of the types contain cofactors such as molybdenum or tungsten, and are capable of reversing the reaction, i.e., reducing carbon dioxide into formate, by the assistance of electrode (Reda et al. 2008). This leads a new way for carbon dioxide sequestration. So far, using FD for carbon dioxide sequestration is still intact. However, Reda et al. (2008) used a tungsten-containing FD enzyme adsorbed onto the electrode that was made of lead, mercury, indium, and thallium to obtain a carbon dioxide reduction rate more than 10 folds higher than other known catalysts used for the same reaction (Jessop et al. 1994; Flores et al. 2003; Tcherkez et al. 2006). The study indicates that FD could be suitable for sequestering carbon dioxide. The mechanism is that CO_2 at the active site of the enzyme is reduced into formate after obtaining electrons from the electrode (Figure 14.2).

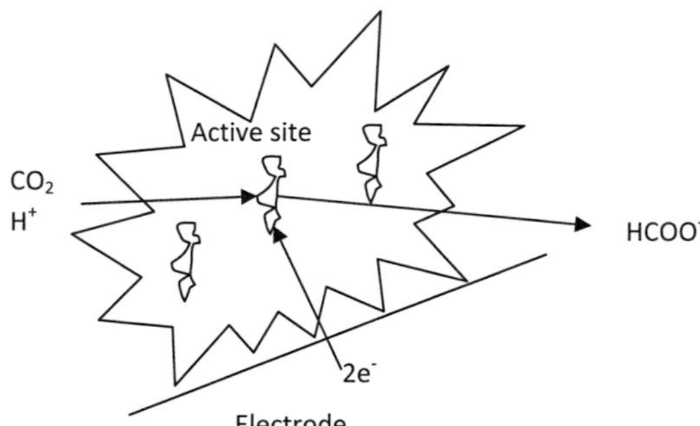

Figure 14.2 Schematic expression of CO2 reduction to fotmate

The factors that may affect carbon dioxide reduction are those that can inhibit the enzyme activity such as pH, temperature, etc. Additionally, the electric potential is also important as the reduction of CO_2 was efficient only at the potential < -1.5 vol (Reda et al. 2008). The FD process would be costly due to the high enzyme cost and a large demand on electricity. Certainly, the cost can be decreased if formate or methanol can be efficiently generated from the process. The cost-effective production of FD would also enhance the application of FD in carbon dioxide sequestration.

In addition, carbon monoxide dehydrogenase (CODH) has also been reported as catalyst for CO_2 reduction to formate. CODH is normally applied in the oxidation of CO to CO_2. While it was found that as CO was oxidized to CO_2, a side reaction of CO_2 to formate was observed (Heo et al. 2002). CO is a toxic gas and required to be converted to a non-harmful gas such as CO_2. When CO is converted to CO_2, the CODH catalytic reaction can be employed to simultaneously reduce CO_2 to formate in order to avoid the emission of CO_2 from the reaction.

14.3.2 CO_2 Reduction to Methane

Nitrogenase is a Mo-dependent enzyme of bacteria. Studies have shown that nitrogenase was capable of reducing small molecular weight and inert compounds with double or triple bond (Rivera-Ortiz and Burris 1975; Burgess and Lowe 1996). Based on these, CO_2 reduction to methane with MoFe nitrogenase was reported (Yang et al. 2012). It was found that methane production was significantly impacted by electron flux where a high electron flux led to a high methanol production. The reduction of CO_2 to methane requires 8 electrons (Eq. 14.6), and the electron was captured to form H_2 by H^+ when the electron flux is low. So far, no other enzymes have been reported to have the capacity of reducing CO_2 to methane.

$$CO_2 + 8H^+ + 8e^- \rightarrow CH_4 + 2H_2O \tag{14.6}$$

14.3.3 CO_2 Reduction to Methanol

CO_2 reduction to methanol is a three step catalytic reaction. The first step is the conversion of CO_2 to formate by formate dehydrogenase ($F_{ate}DH$) catalyzation. In the second step, formate is reduced to formaldehyde by formaldehyde dehydrogenase ($F_{ald}DH$) catalyzation. In the last step, methanol is formed from formaldehyde under alcohol dehydrogenase (ADH) catalyzation. In these reactions, nicotinamide adenine dinucleotide (NADH) is the terminal electron donor.

When $F_{ate}DH$, $F_{ald}DH$ and ADH were encapsulated in the porous silica sol-gel matrix and used for CO_2 reduction in the presence of NADH, methanol production was significantly increased comparing with that from $F_{ate}DH$, $F_{ald}DH$ and ADH solution (Obert and Dave 1999). It was predicated due to the matrix effect. Therefore, to reduce CO_2 to methanol, immobilization of the enzymes should be conducted. As the reaction demands three different enzymes, host materials of immobilization should have reliable stability and reactivity of the enzymes.

14.4 Technical Limitations and Future Perspective

As one of the most promising technologies in carbon dioxide sequestration, CA catalytic carbon dioxide sequestration is a fast, environmentally friendly, and permanent carbon dioxide sequestration method. However, the field-scale application of the technology has not been developed due to the high cost requirement. So far, it still in the research stage due to the difficulties in reducing the CA production cost, producing CA at the industrial scale, maintaining the CA stability after contacting with the environment, and extending the life span of CA after exposing to the sequestration. The high sequestration cost hampers its application. In addition, there is still a great difficulty in keeping the activity and stability of CA with economic methods after CA separated from the CA source. Current immobilization methods have, to some extent, helped to maintain the activity and stability of CA (Bond et al. 2001a; Kanbar 2008). However, very limited materials have been used to immobilize CA. FD could also be a useful enzyme to accomplish carbon dioxide sequestration since it can efficiently convert carbon dioxide into formate. However, to our best knowledge, no detailed studies have been reported so far. Much work is needed before applying the technology in reality.

CA catalytic carbon dioxide sequestration is still considered as a costly alternative due to high enzyme cost. Therefore, if the sequestration can combine with other types of beneficial treatments or production, the cost may be somewhat offset. As previously described, cations which are used to form carbonate for a permanent carbon dioxide sequestration are critical in CA catalytic carbon dioxide sequestration. Desalination is a process which generally produces a large amount of cations (Ca^{2+} and Mg^{2+}) and most of the cations are wasted. The sequestration cooperating with desalination could be an alternative to lower the sequestration cost because of the benefit from freshwater production. Moreover, a large amount of solid carbonate is produced in the sequestration. For instance, around 2.36 g of $CaCO_3$ could be produced when 1 g of carbon dioxide is captured. Therefore, chemical production (e.g., $CaCO_3$ or $MgCO_3$) can be considered in the sequestration in order to reduce the sequestration cost.

In order to develop enzymatic carbon dioxide sequestration, the following aspects should be studied:

(1) Compared to plants and animals, microorganisms have much faster growth rates (around 10 to 50 times faster). Therefore, producing CA from microorganisms would extensively increase CA productivity, and hence, reduces the CA cost.

(2) CA purification takes up a great amount on the cost of CA production. Therefore, the development on the purification would reduce the cost of CA production.

(3) Using raw CA (without purification) for catalyzing carbon dioxide hydration is a cheaper way in compared to using purified CA. Thus, carbon dioxide sequestration using raw CA should be developed.

(4) To seek and use microorganisms for CA enzymatic carbon dioxide sequestration is necessary because it would essentially reduce the sequestration cost. However, the effect of the sequestration condition on the microorganism growth should be considered for building a long term sequestration.

(5) Activity and stability of CA can be inhibited after contacting with reaction solutions. Hence, methods to maintain the activity and stability should be sought.

(6) CA immobilization has given an encouraging result to enhance the stability of CA and to maintain the activity of the CA. However, very few materials have been found suitable for the immobilization, which limits the development. Therefore, more materials suitable for CA immobilization should be studied.

(7) Apart from immobilization, stabilizing CA via modifying the surface of CA is also an alternative (Yadav et al. 2010). However, research on this issue is still on its infancy and more effort is required.

(8) Combining the sequestration with freshwater production from seawater or chemical production ($CaCO_3$ or $MgCO_3$) should be studied to reduce cost.

(9) FD, nitrogenase, $F_{ate}DH$, $F_{ald}DH$ and ADH have provided another choices for sequestering carbon dioxide. Therefore, attention should be paid and related studies should be conducted.

14.5 Summary

CA catalytic carbon dioxide sequestration is taking the major role in enzymatic carbon dioxide sequestration. The carbon dioxide sequestration with CA is achieved by firstly using CA, free CA, immobilized CA, or modified CA, to catalyze CA hydration, which normally occurs 14.1 reactions per second in the absence of CA while around 1,000,000 reactions per second in the presence of CA; and secondly using cations (Ca^{2+} or Mg^{2+}) from the addition of chemicals, brine, produced water, etc. to form solid carbonates ($CaCO_3$ or $MgCO_3$). Studies have revealed that it is a rapid, efficient, and environmentally friendly sequestration method. So far, CA catalytic carbon dioxide sequestration has been tried to capture carbon dioxide from flue gases of power plants or metal-making plants. The fact that the flue gases have complex composition requires studying the factors that may affect the sequestration, such as, pH, temperature, and cations and anions. The application of using CA for catalytic carbon dioxide sequestration from flue gases is still in the study stage due to the cost hamper mainly from the costly CA. Therefore, it is believed that reduction of the cost from CA would enhance the application of the sequestration. Other strategies such as enhancing CA stability and combining the sequestration with other production processes may also be a way to reduce the cost.

FD is another type of enzyme which has capability to reduce carbon dioxide into formate. Even though, no related study has been reported on FD using in carbon dioxide sequestration in our awareness, however, it is a field worth to be studied in order to find a proper method for carbon dioxide sequestration. CO_2 reduction to methane and methanol has also been reported. However, much more efforts are needed to approach the stage of the application of the technologies.

14.6 Acknowledgements

Sincere thanks are due to the Natural Sciences and Engineering Research Council of Canada (Grant A 4984, Canada Research Chair) for their financial support. The views and opinions expressed in this chapter are those of the authors.

14.7 Abbreviations

ADH	Alcohol Dehydrogenase
CA	Carbonic Anhydrase
FD	Formate Dehydrogenase
$F_{ald}DH$	Formaldehyde Dehydrogenase
$F_{ate}DH$	Formate Dehydrogenase
NADH	Nicotinamide Adenine Dinucleotide

14.8 References

Adler, L., Brundell, J., Falkbring, S.O., and Nyman, P.O. (1972). "Carbonic anhydrase from *Neisseria sicca*, strain 6021 I. Bacterial growth and purification of the enzyme." *Biochimica et Biophysica Acta*, 284(1), 298–310.

Bahn, Y.-S., Cox, G. M., Perfect, J.R., and Heitman, J. (2005). "Carbonic anhydrase and CO_2 sensing during cryptococcus neoformans growth, differentiation, and virulence." *Current Biology*, 15(22), 2013–2020.

Bond, G.M., Egeland, G., Brandvold, D.K., Medina, M.G., and Stringer, J. (1999a). "CO_2 sequestration via a biomimetic approach." In Mishra, B. (ed.), *EPD Congress: Proceedings of sessions and symposia*, pp 763–781.

Bond, G.M. Egeland, G., Brandvold, D.K., Medina, M.G., Simsek, F.A., and Stringer, J. (199b). "Enzymatic catalysis and CO_2 sequestration." *World Resource Review*, 11(4), 603–618.

Bond, G.M., Medina, M.-G., and Stringer, J. (2001a). "A biomimetic CO_2 scrubber for safe and permanent CO_2 sequestration." *Eighteenth Annual International Pittsburgh Coal Conference Proceedings (December 3-7, 2001, Newcastle, NSW, Australia).*

Bond, G.M., Stringer, J., Brandvold, D.K., Simsek, F.A., Medina, M.G., and Egeland, G. (2001b). "Development of integrated system for biomimetic CO_2 sequestration using the enzyme carbonic anhydrase." *Energy Fuels*, 15(2), 309–316.

Bond, M.G., Medina, M.-G., Stringer, J., and Simsek-Ege, F.A. (2001c). "CO_2 capture from coal-fired utility generation plant exhausts, and sequestration by a biomimetic route based on enzymatic catalysis-current status." *First National Conference on Carbon Sequestration, 2001.*

Bond, G.M., McPherson, B.J., Abel, A., Lichtner, P., Grigg, R., Liu, N., and Stringer, J. (2003). "Biomimetic and geologic mineralization approaches to carbon sequestration." *Proceedings of the Second National Conference on Carbon Sequestration (May 5–8, 2003, Alexandria, VA), 2003.*

Burgess, B.K., and Lowe, D.J. (1996). "Mechanism of molybdenum nitrogenase." *Chemical Reviews*, 96(7), 2983–3012.

Cheng, L.-H., Zhang, L., Chen, H.-L., and Gao, C.-J. (2008). "Hollow fiber contained hydrogel-CA membrane contactor for carbon dioxide removal from the enclosed spaces." *Journal of Membrane Science*, 324(1–2), 33–43.

Chien, L.-J. Sureshkumar, M., Hsieh, H.-H., and Wang, J.-L. (2013). "Biosequestration of carbon dioxide using a silicified carbonic anhydrase catalyst." *Biotechnology and Bioprocess Engineering*, 18(3), 567–574.

Cioci, F. (1995). "Thermostabilization of erythrocyte carbonic anhydrase by polyhydric additives." *Enzyme and Microbial Technology*, 17(7), 592–600.

Cowan, R.M., Ge, J.J., Qin, Y.J., McGregor, M.L., and Trachtenberg, M.C. (2003). "CO_2 capture by means of an enzyme-based reactor," *Annals of the New York Academy of Sciences*, 984(4), 453–469.

Crumbliss, A.L., McLachlan, K.L., O'Daly, J.P., and Henkens, R.W. (1988). "Preparation and activity of carbonic anhydrase immobilized on porous silica beads and graphite rods." *Biotechnology and Bioengineering*, 31(8), 796–801.

Dilmore, R., Griffith, C., Liu, Z., Soong, Y., Hedges, S.W., Koepsel, R., and Ataai, M. (2009). "Carbonic anhydrase-facilitated CO_2 absorption with polyacrylamide buffering bead capture." *International Journal of Greenhouse Gas Control*, 3(4), 401–410.

Domsic, J.F., and McKenna, R. (2010). "Sequestration of carbon dioxide by the hydrophobic pocket of the carbonic anhydrases." *Biochimica et Biophysica Acta (BBA) - Proteins & Proteomics*, 1804(2), 326–331.

Drevon, G.F., Urbanke, C., and Russell, A.J. (2003). "Enzyme-containing michael-adduct-based coatings." *Biomacromolecules*, 4(3), 675–682.

Favre, N. (2011). *Captage enzymatique de dioxyde de carbone*, Ph.D. dissertation , Report no. 109-2011, Universit´e Claude Bernard-Lyon 1, Lyon, France, 2011.

Figueroa, J.D., Fout, T., Plasynski, S., McIlvried, H., and Srivastava, R.D. (2008). "Advances in CO_2 capture technology–The U.S. Department of Energy's Carbon Sequestration Program." *International Journal of Greenhouse Gas Control*, 2(1), 9–20.

Fisher, S.Z., Maupin, C.M., Budayova-Spano, M., Govindasamy, L., Tu, C., Agbandje-McKenna, M., Silverman, D.N., Voth, G.A., and McKenna, R. (2007). "Atomic crystal and molecular dynamics simulation structures of human carbonic anhydrase II: Insights into the proton transfer mechanism." *Biochemistry*, 46(11), 2930–2937.

Flores, R., Lopez-Castillo, Z.K., Kani, I., Fackler, J.P., and Akgerman, A. (2003). "Kinetics of the homogeneous catalytic hydrogenation of olefins in supercritical carbon dioxide using a fluoroacrylate copolymer grafted rhodium catalyst." *Industrial & Engineering Chemistry Research*, 42(26), 6720–6729.

Franchi, M., Vullo, D., Gallori, E., Antel, J., Wurl, M., Scozzafava, A., and Supuran, C. T. (2003). "Carbonic anhydrase inhibitors: Inhibition of human and murine mitochondrial isozymes V with anions." *Bioorganic & Medicinal Chemistry Letters*, 13(17), 2857–2861.

Ge, J.-J., Cowan, R.M., Tu, C.-K., McGregor, M.L., and Trachtenberg, M.C. (2002). "Enzyme based CO_2 capture for advanced life support." *2002 Life Support & Biosphere Science*, 8, 181–189.

Guilloton, M.B., Korte, J.J., Lamblin, A.F., Fuchs, J.A., and Anderson, P.M. (1992). "Carbonic anhydrase in *Escherichia* coli. A product of the cyn operon." *The Journal of Biological Chemistry*, 267(6), 3731–3734.

Guilloton, M.B., Lamblin, A.F., and Kozliak, E.I. (1993). "A physiological role for cyanate-induced carbonic anhydrase in *Escherichia coli*." *Journal of Bacteriology*, 175(5), 1443–1451.

Hamilton, T. (2007). "Capturing carbon with enzymes." MIT Technology Review, Feb. 22, 2007. Available at <http://www.technologyreview.com/biomedicine/18217/> (access Aug. 2012).

Heck, R.W., Tanhauser, S.M., Manda, R., Tu, C.K., Laipis, P.J., and Silverman, D.N. (1994). "Catalytic properties of mouse carbonic-anhydrase-V." *Journal of Biological Chemistry*, 269(40), 24742–24746.

Heo, J., Skjeldal, L., Staples, C.R., and Ludden, P.W. (2002). "Carbon monoxide dehydrogenase from Rhodospirillum rubrum produces formate." *Journal of Biological Inorganic Chemistry*, 7(7–8), 810–814.

HewettEmmett, D., and Tashian, R.E. (1996). "Functional diversity, conservation, and convergence in the evolution of the alpha-, beta-, and gamma-carbonic anhydrase gene families." *Molecular Phylogenetics and Evolution*, 5(1), 50–77.

Innocenti, A., Pastorekova, S., Pastorek, J., Scozzafava, A., Simone, G.D., and Supuran, C.T. (2009). "The proteoglycan region of the tumor-associated carbonic anhydrase isoform IX acts as anintrinsic buffer optimizing CO_2 hydration at acidic pH values characteristic of solid tumors." *Bioorganic & Medicinal Chemistry Letters*, 19(20), 5825–5828.

Jessop, P.G., Ikariya, T., and Noyori, R. (1994). "Homogeneous catalytic-hydrogenation of supercritical carbon-dioxide." *Nature*, 368(6468), 231–233.

Jovica, D.B., and Kostic, N.M. (1999). "Effect of encapsulation in sol-gel silica glass on esterase activity, conformational stability, and unfolding of bovine carboniic anhydrase II." *Chemical Material*, 11, 3671–3679.

Kanbar, B. (2008). *Enzymatic CO_2 sequestration by carbonic anhydrase.* M.S. Thesis, Izmir Institute of Technology.

Karlsson, J., Clarke, A.K., Chen, Z.Y., Hugghins, S.Y., Park, Y.I., Husic, H.D., Moroney, J.V., and Samuelsson, G. (1998). "A novel alpha-type carbonic anhydrase associated with the thylakoid membrane in Chlamydomonas reinhardtii is required for growth at ambient CO_2." *Embo Journal*, 17(5), 1208–1216.

Keilin, D., and Mann, T. (1939). "Carbonic anhydrase." *Nature*, 144(3644), 442–443.

Keilin, D., and Mann, T. (1940). "Carbonic anhydrase. Purification and nature of the enzyme." *Biochemical Journal*, 34, 1163–1176.

Lavecchia, R., and Zugaro, M. (1991). "Thermal denaturation of erythrocyte carbonic anhydrase." *FEBS Letters*, 292(1–2), 162–164.

Liu, N., Bond, G.M., Abel, A., McPherson, B.J., and Stringer, J. (2005). "Biomimetic sequestration of CO_2 in carbonate form: Role of produced waters and other brines." *Fuel Processing Technology*, 86(14–15), 1615–1625.

Meldrum, N., and Roughton, F.J.W. (1933). "Carbonic anhydrase: its preparation and properties." *The Journal of Physiology*, 80, 113–142.

Mirjafari, P., Asghari K., and Mahinpey, N. (2007). "Investigating the application of enzyme carbonic anhydrase for CO_2 sequestration purposes." *Industrial & Engineering Chemistry Research*, 46(3), 921–926.

Neish, A.C. (1939). "Studies on chloroplasts. Their chemical composition and the distribution of certain metabolites between the chloroplasts and remainder of the leaf." *Biochemical Journal*, 33, 300–3308.

Özensoy, Ö., Nishimori, I., Vullo, D., Puccetti, L., Scozzafava, A., and Supuran, C.T. (2005). "Carbonic anhydrase inhibitors: Inhibition of the human transmembrane isozyme XIV with a library of aromatic/heterocyclic sulfonamides." *Bioorganic & Medicinal Chemistry*, 13(22), 6089–6093.

Premkumar, L., Bageshwar, U.K., Gokhman, I., Zamir, A., Sussmanb, J.L. (2003). "An unusual halotolerant a-type carbonic anhydrase from the alga *Dunaliella salina* functionally expressed in Escherichia coli." *Protein Expression and Purification*, 28(1), 151–157.

Ramanan, R., Kannan, K., Sivanesan, S.D., Mudliar, S., Kaur, S., Tripathi, A.K., and Chakrabarti, T. (2009). "Bio-sequestration of carbon dioxide using carbonic anhydrase enzyme purified from *Citrobacter freundii*." *World Journal of Microbiology & Biotechnology*, 25(6), 981–987.

Reda, T., Plugge, C.M., Abram, N.J., and Hirst, J. (2008). "Reversible interconversion of carbon dioxide and formate by an electroactive enzyme." *Proceedings of National Academy of Sciences of the Unite States of America*, 105(31), 10654–10658.

Rivera-Ortiz, J.M., and Burris, R.H. (1975). "Interactions among substrates and inhibitors of nitrogenase." *Journal of Bacteriology*, 123(2), 537–545.

Rowlett, R.S., and Silverman, D.N. (1982). "Kinetics of the protonation of buffer and hydration of CO_2 catalyzed by human carbonic anhydrase II." *Journal of the American Chemical Society*, 104(24), 6737–6741.

Salmon, S., Saunders, P., and Borchert, M. (2009). "Enzyme technology for carbon dioxide separation from mixed gases." *IOP Conference Series: Earth and Environmental Science*, 6, 172018.

Sharma, A., and Bhattacharya, A. (2010). "Enhanced biomimetic sequestration of CO_2 into $CaCO_3$ using purified carbonic anhydrase from indigenous bacterial strains." *Journal of Molecular Catalysis B: Enzymatic*, 67(1–2), 122–128.

Sharma, A., Bhattacharya, A., Pujari, R., Shrivastava, A. (2008). "Characterization of carbonic anhydrase from diversified genus for biomimetic carbon-dioxide sequestration." *Indian Journal of Microbiology*, 48 (3). 365–371.

Shekh, A.Y., Krishnamurthi, K., and Mudliar, S.N. (2012). "Recent advancements in carbonic anhydrase driven processes for CO_2 sequestration: Minireview." *Critical Reviews in Environmental Science and Technology*, 42(14), 1419–1440.

Silverman, D.N., and Mckenna, R. (2007). "Solvent-mediated proton transfer in catalysis by carbonic anhydrase." *Accounts of Chemical Research*, 40(8), 669–675.

Simsek-Ege, F.A., Bond, G.M., and Stringer, J. (2002). "Polyelectrolyte complex cages for a novel biomimetic CO_2 sequestration system." *Environmental Challenges and Greenhouse Gas Control for Fossil Fuel Utilization in the 21st Centuryedited by Maroto-Valer et al., Kluwer Academic Publishers*, 133–146.

Simsek, F.A., Bond, G.M., and Stringer, J. (2001). "Immobilization of carbonic anhydrase for biomimetic CO_2 sequestration." *World Resource Review* 13, 74–90.

Smith, R.G. (1988). "Inorganic carbon transport in biological systems." *Comparative Biochemistry and Physiology Part B: Comparative Biochemistry*, 90(4), 639-654.

So, A.K.C., Espie, G.S., Williams, E.B., Shively, J.M., Heinhorst, S., and Cannon, G.C. (2004). "A novel evolutionary lineage of carbonic anhydrase (epsilon class) is a component of the carboxysome shell." *Journal of Bacteriology*, 186(3), 623–630.

Supuran, C.T., Scozzafava, A., Mincione, F., Menabuoni, L., Briganti, F., Mincione, G., and Jitianu, M. (1999). "Carbonic anhydrase inhibitors. Part 60. The topical intraocular pressure-lowering properties of metal complexes of a heterocyclic sulfonamide: Influence of the metal ion upon biological activity." *European Journal of Medicinal Chemistry*, 34(7–8), 585–595.

Tcherkez, G.G.B., Farquhar, G.D., and Andrews, T.J. (2006). "Despite slow catalysis and confused substrate specificity, all ribulose bisphosphate carboxylases may be nearly perfectly optimized." *Proceedings of the National Academy of Sciences of the United States of America*, 103(19), 7246–7251.

Trachtenberg, M.C., and Bao, L. (2005). "CO_2 capture: Enzyme vs. amine." *Fourth Annual Conference on Carbon Capture and Sequestration, May 2–5, 2005.*

Trachtenberg, M.C., Cowan, R.M., Smith, D.A., Horazak, A.D., Jensen, D.M., Laumb, D.J., Vucelic, A.P., Chen, H., Wang, L., and Wu, X. (2009). "Membrane-based, enzyme-facilitated, efficient carbon dioxide capture." *Energy Procedia*, 1(6), 353–360.

Veitch, F.P., and Blankenship, L.C. (1963). "Carbonic anhydrase in bacteria." *Nature*, 197(4862), 76–77.

Vinoba, M., Bhagiyalakshmi, M., Jeong, S.K., Nam, S.C., and Yoon, Y. (2012). "Carbonic anhydrase immobilized on encapsulated magnetic nanoparticles for CO_2 sequestration." *Chemistry – A European Journal*, 18(38), 12028–12034.

Vullo, D., Ruusuvuori, E., Kaila, K., Scozzafava, A., and Supuran, C.T. (2006). "Carbonic anhydrase inhibitors: Inhibition of the cytosolic human isozyme VII with anions." *Bioorganic & Medicinal Chemistry Letters*, 16(12), 3139–3143.

Yadav, R., Labhsetwar, N., Kotwal, S., and Rayalu, S. (2010). "Single enzyme nanoparticle for biomimetic CO_2 sequestration." *Journal of Nanoparticle Research*, 13(1), 263–271.

CHAPTER 15

Biochar

Indrani Bhattacharya, J. S. S. Yadav, T. T. More, Song Yan, R. D. Tyagi, R. Y. Surampalli, and Tian C. Zhang

15.1 Introduction

Biochar or black carbon is defined simply as charcoal that has been used for agricultural purposes. Biochar is produced by a process called pyrolysis, which is the direct thermal decomposition of biomass in the absence of oxygen to obtain an array of solid (biochar), liquid (bio-oil), and gas (syngas) products. Biochar usually has high surface area, low density and rich in carbon and has been known to be the key component for a green revolution (Barrow 2012).

Benefits of biochar are multiple and have both direct and indirect effects. Application of biochar in soils improves soil nutrient retention capacity (e.g., increasing NH_4^+ and P concentrations and deceasing NO_3^- in soil, reducing the leaching of nutrients from soil), water holding capacity, soil productivity, soil quality (by removing the contaminants from the soil, making it healthier and cleaner), nutrient cycling, bioavailability for plants and stabilization of solutes, and changes the microbial populations and stimulates the microbial activity of the soil (Mendez et al. 2012; Beesley et al. 2013; Ducey et al. 2013; Masek et al. 2013; Xu et al. 2013).

Biochar is composed of recalcitrant carbon structures, and these properties prevent biochar from decomposition. Therefore, addition of biochar to soil results in long-time carbon storage (e.g., 100-1000 years and more) (Duku et al. 2011; Reid et al. 2013). The burning and natural decomposition of biomass and in particular agricultural waste adds large amounts of CO_2 to the atmosphere. Sustainable use of biochar could reduce the global net emissions of CO_2, methane, and nitrous oxide by up to 1.8 Gt CO_2-C equivalent (CO_2-C_e) per year (= 12% of current anthropogenic CO_2-C_e emissions), and total net emissions over the course of the next century by 130 Gt CO_2-C_e, without endangering food security, habitat, or soil conservation. Moreover, energy is produced as a by-product, which being an alternative energy source, reduces the tendency for further fossil fuel burning (Brown 2009).

Since the 2000s, studies have been accelerated on developing biochar-related technologies for restoring carbon to depleted soils and sequestering significant amounts of CO_2 (Lehmann 2007; Sohi et al. 2010; IBI 2013). This chapter reviews topics related to the biochar for carbon sequestration. Section 15.2 discusses the role of biochar for carbon capture and sequestration (CCS). Section 15.3 describes biochar technology (e.g., biochar feedstocks and yield, transformation principle, production technologies, characteristics of biochar and property optimization). Section 15.4 presents the major benefits of biochar for development of the sustainable society. Section 15.5 describes biochar sustainability (e.g., sustainable biochar industry, social, environmental, economic outcomes). Section 15.6 discusses the concerns and future perspectives, and Section 15.7 summarizes the chapter.

15.2 Role of Biochar for CCS

Carbon has a permanent repository in the plants and soil, and it acts as a natural balance. Over millions of years, coal has obtained the status of pure carbon, which are eventually gathered by plants and further sequestrated through the natural processes. Burning coal means the burning of most of the carbon that nature sequestered. On the other hand, growing plants take CO_2 from the atmosphere and fix it in their cells. However, 99% of the carbon ends up back in the atmosphere as the plant is eventually burned or consumed by living forms, which later on returns the carbon to the atmosphere (Canadian Biochar Initiative 2008; Olah et al. 2009).

Unlike plants, biochar can be used to effectively remove net CO_2 from the atmosphere (Figure 15.1). Biochar is defined as the process that digests the carbon captured by living plants and converts the biomass into the charcoal (Marris 2006). As shown in Fig. 15.1, during the process of pyrolysis, in the charcoal a certain part of the plant carbon (20 -60%) is obtained. When the charcoal rich in carbon is being tilled into soils, it can be sequestered away ($\geq$ 100 -1000 years), thus minimizing carbon dioxide concentrations in the atmosphere. In addition, the co-production of bioenergy and biochar leads to mitigate climate change globally by minimum utilization of fossil fuels. Furthermore, biochar may also reduce emissions of other potential greenhouse gases (GHGs) such as CH_4 and NO_x, which are generated due to speedy decay of biomass (Winsley 2007; Woolf el al. 2010). Finally, the term "Carbon Negative" comes into being, when biochar is buried in the soil and the bioenergy generated during pyrolysis are utilized. Processes are under development to use carbon as a fertilizer that is plowed back into the land, stimulating the growth of crops (Brown 2009). Therefore, biochar is one of the upcoming technologies for CCS and development of the sustainable society.

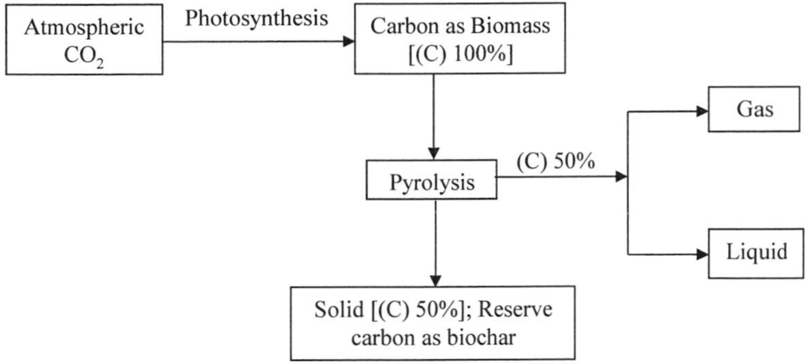

Figure 15.1. Carbon sequestration during transformation of biomass to biochar

15.3 Biochar Technology

In general, biochar technology includes entire integrated biochar systems with the following major elements: i) collection, transport and processing of biomass feedstock; ii) characterization and testing of biochar; iii) biochar production and utilization of energy co-products (e.g., gas, oil, or heat); iv) biochar transport and handling for soil application; v) monitoring of biochar application for carbon accounting, and vi) life cycle assessment and full system monitoring for sustainability assessment (IBI 2010). This section focuses on biochar feedstock and yield, transformation principle, production technologies, characteristics of biochar and property optimization.

15.3.1 Biochar Feedstocks and Yield

Biochar processes take the waste material from food crops, forest debris, and other plant material (Table 15.1), and turn it into a stable form that can be buried away permanently as charcoal. A major source of biomass is provided by agricultural crop residues (Matala et al. 2010). For example, almost every year depending upon the climatic and soil conditions 69.9 million tonnes of crop residues are produced tentatively in India from the six major crops, i.e., rice = 13.1, wheat = 11.4, sugar = 21.6, ground nut = 3.3, mustard = 4.5, and cotton = 11.8 million tonnes (Haefele 2007). As per estimation by Woolf et al. (2010), globally ~2.27 Pg C/year feedstock is available for transformation in to biochar with maximum sustainable technical potential value as presented in Figure 15.2.

Table 15.1. Summary of biochar yield from the different feedstock

Biomass	Process	Biochar Yield (% wt.)	References
Oak wood	Fast pyrolysis at 500 °C	31.2	Novak et al. 2009
Corn husks	Fast pyrolysis at 500 °C	26.0	Purevsuren et al. 2003
Olives stones	Slow pyrolysis at 600 °C	39.7	Mullen et al. 2010
Pine wood	Fast pyrolysis at 800 °C	32.1	Mullen et al. 2010
Olive bagasse	Slow pyrolysis at 500 °C	39.7	Spokas et al. 2010
Palm shell	Slow pyrolysis at 400 °C	24.8	Spokas et al. 2010
Pine saw dust	Slow pyrolysis at 800 °C	24.3	Spokas et al. 2010
Spruce wood	Fast pyrolysis at 600 °C	37.5	Spokas et al. 2010; Sukartono et al. 2011
Eucalyptus wood	Slow pyrolysis at 400 °C	42.2	Spokas et al. 2010
Olive husk	Fast pyrolysis at 800 °C	39.7	Spokas et al. 2010
Beech wood	Slow pyrolysis at 500 °C	26.2	Spokas et al. 2010
Corn cob	Slow pyrolysis at 800 °C	23.2	Tsai et al. 2012; Zhao et al. 2013
Rapeseed stalks	Slow pyrolysis at 400 °C	32.1	Zhao et al. 2013
Pitch pine	Fast pyrolysis at 500 °C	39.7	Zhao et al. 2013
Straw pellets	Slow pyrolysis at 400 °C	24.8	Zhao et al. 2013
Willow pellets	Fast pyrolysis at 700 °C	24.3	Kim et al. 2012
Conocarpus waste	Fast pyrolysis at 500 °C	37.5	Al-Wabel et al. 2013; Masek et al. 2013
Walnut-shell	Slow pyrolysis at 500 °C	21.8	Masek et al. 2013

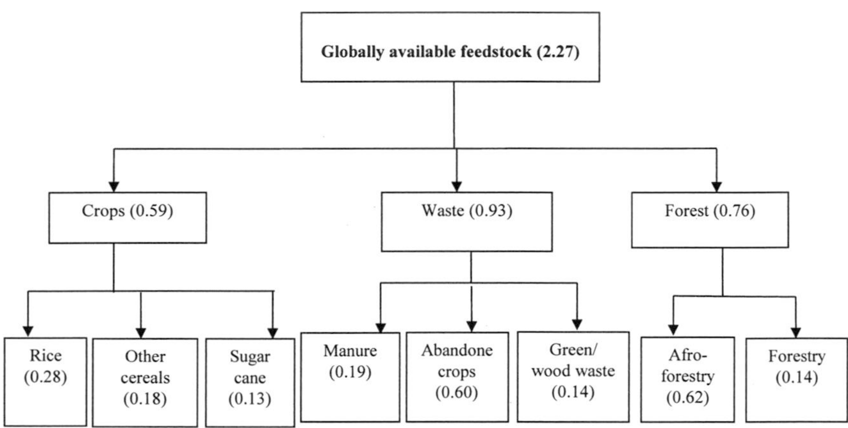

Figure 15.2. Total globally available annual feedstock (value in Pg C/year) and their distribution in different biomass

Several points should be noted concerning feedstocks. First, biochar should be made from biomass waste materials in order to create no competition for land with any other land use option (e.g., food production or leaving the land in its pristine state). Second, using local feedstocks often makes more economic sense as collection; transport and storage usually are costly. Third, feedstocks affect the composition of biochar (the amount of carbon, nitrogen, potassium, calcium, etc.), but pyrolysis conditions also greatly affect nutrient properties contents. Therefore, biochar should be tested on a batch by batch basis to determine specific properties.

The yield of biochar is affected by the process used and applied operational conditions. Moreover, one main factor which affects the yield of biochar is the intrinsic property or composition of feedstock (Lee et al. 2013; Manyà 2012). The yield reported from different types of feedstock is summarized in Table 15.1.

15.3.2 Transformation Principle

The process of biochar initiates with the plant materials (biomass) through the process of pyrolysis. The thermochemical process pyrolysis transforms low density organic materials and biomass into a high density liquid (bio-oil), a high energy density solid (biochar) and comparatively low energy density gas (syngas). Basically during pyrolysis the organic compounds are heated in a closed chamber to temperature more than 400 °C in the absence of oxygen. Due to higher temperature organic compounds decomposes and release a vapour phase and a residual solid phase, i.e. biochar (Laird 2009). Traditionally, pyrolysis technology is employed to produce biochar, bio-oil and syngas. Biochar processing is the similar process as that of preparation of charcoal, which has been followed for years; however, in biochar processes, gases are allowed to escape. Biochar involves no usage of gases. However, a finite volume of the gases are maintained to drive the pyrolysis process (Lehmann et al. 2006). Moreover, the operating conditions used for pyrolysis decide the proportion of the quantity of the final products. The flow diagram (Figure 15.3) shows the pyrolysis process and final product based on the operating condition used (Sohi et al. 2010).

Conventionally, slow pyrolysis is the preferred way to produce biochar while fast pyrolysis is the route for bio-oil (Mohan et al. 2006). In slow pyrolysis biochar is the dominant product; therefore, it is the method of choice for current biochar production techniques over others (Chaiwong et al. 2012). Due to the absence of oxygen during the pyrolysis process, a minimal amount of CO_2 is produced and the biomass melts into carbon with a very fine structure. Apart from that, hydrogen molecules in the plant generate syngas and bio-oil along with heat energy. The bio-oil produced can be used like low-grade diesel fuel for heating and power generation (Lehmann et al. 2006; Blakeslee 2009).

The slow pyrolysis (also called as carbonization pyrolysis) processes apply a long residence time and a low heating rate during the process. Some operational and intrinsic properties of biomass, which play a critical role during the pyrolysis process have been identified. The factors are peak temperature, heating rate, pressure, vapor residence time, particle size and moisture content of the biomass. The highest temperature reached during the process is called as peak temperature. It has been established that peak temperature has the most impact on the yield and characteristics (e.g. surface area and pore size distribution) of the final product (biochar). The yield has been reported to decrease with an increase in temperature. However, as the peak temperature increases; the rise in the amount of fixed carbon of the biochar resulted. The temperature range between 300 to 500 °C has been reported for increment of the fixed carbon (Manyà 2012; Ronsse et al. 2012). The heating rate generally varies from 1 to 30 °C/min. The higher pressure (1.0 to 3.0 MPa) reported to enhance the biochar yield due to the result of the rise in the vapor residence time inside the solid particle (Manyà 2012; Ronsse et al. 2012). However, it is still difficult to predict the yield and properties of the biochar produced. Therefore studies are continued to improve the yield and properties of biochar.

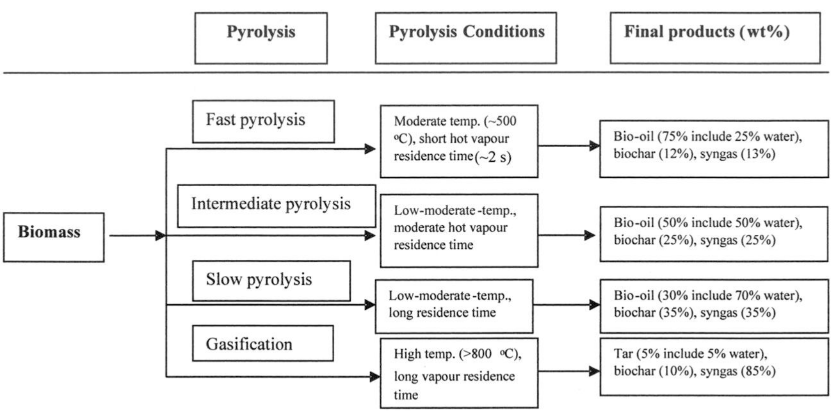

Figure 15.3. Flow diagram of pyrolysis process and final products

The feedstock (biomass) with a moisture content of 42 -62% was reported to enhance the yield of the biochar at higher pressure. Thus, the agricultural residue, with high moisture content could be attractive for biochar purposes. Besides the effect of moisture, the biochar yield from a particular biomass is affected by the compositional characteristics such as hemicellulose, lignin, extractive and other organic matters. The yield value reported for hemicellulose, cellulose and lignin is 23.5%, 19% and 45%, respectively, during slow pyrolysis (Lee et al. 2013; Manyà 2012). Thus, yield is highest from lignin. The inorganic content of the biomass also influence pyrolysis product distribution. The pyrolysis of the biomass with a higher inorganic content, particularly alkali and alkali earth metals, catalyzes biomass disintegration and char making reactions. Lower charcoal yield have been reported by many researchers, with a reduced ash content obtained by pre-treating the biomass with hot water at 80 °C. Others process parameters, which might influence biochar yield are particle size and soak time at peak temperature. For the biomass of a higher particle size, the diffusion rate of the volatiles through the char declines, and thus, the generation of other char due to secondary reaction should be expected (Manyà 2012).

Pyrolysis was used years ago by Brazilian natives to enhance their poor and acidic soil into Terra Preta (Brown 2009), considered as the most productive soils known to man (Blakeslee 2009). Biochar is generated as per the thermal decomposition of the matter, which is organic in nature, and it takes place at lower temperatures (less than 700 °C), at no or limited supply of O_2. What makes biochar different from the charcoal is the production methodology. The overall objective is to improve the soil quality and productivity, carbon storage and filtration of percolating soil water. In relation to the organic carbon (OC) rich biochar, burning of biomass in fire creates ash, which eventually contains minerals such as Ca, Mg and certain inorganic carbonates. A minor portion of the vegetation is quantitatively burned in the areas of the limited supply of O_2. The usage of term char or for that matter charring is used in connection with the making of charcoal or in combination with 'char' deriving from the fire (Brown 2009).

15.3.3 Production Techniques

Methods for producing biochar and advanced biofuels have been reviewed by many researchers (e.g. WSU 2011; IBI 2013). In general, biochar can be produced via either conventional small-scale or current techniques, which are described below.

Conventional Techniques. Conventional or small-scale biochar production techniques are basically founded on the pyrolysis principle. These techniques developed conventionally are not being used on industrial scales. However, these methods are exhibiting good quality biochar production on small-scale basis and still used locally (Brudges 2009). Biochar production with conventional techniques is on a non-domestic scale. Biochar production can be in i) a fixed location; ii) a distributed system, where a lower tech pyrolysis kiln is used by small group of farmers and using the energy generated from such kilns for processing the harvest and producing electricity for local needs; or iii) a mobile pyrolysis plant that could be driven to different sites, stimulating the need to transport the biomass (Whitfield 2008).

Carbon zero experimental biochar kiln. It is a kiln, closed with the help of firebrick for insulation, and the enclosure is at an average of 200 liter barrel of steel. It is stacked with split wood and other feedstock of biomass. These are covered and further heated from beneath. The stove beneath consists of wood fire where it reached maximum pyrolysis temperature of 320 °C and above. The wood for fire is dependent upon the moisture content in the wood; additional heat is needed otherwise, as the exothermic reaction releases heat. Using a pilot model with a retort volume of ~5 m^3, the kiln can process up to 1 tonne of air-dried, chipped woody biomass per hour, with a conversion efficiency of woody biomass to biochar being 30-33% and ~1.4-1.7 MWh of chemical energy in the syngas produced. Straws and grasses have lower lignin content and hence a lower conversion efficiency to biochar (~17%) (Biochar Info 2013).

The advantages to this particular approach are quantitatively numerous. As the enclosure, which keeps on refracting and refraining significant amount of heat, lesser volume of woods are needed for the entire reaction. A support flame, not too big is required to keep the gases ignited. A cylindrical tube-like structure called the blower is mounted at the top of the afterburner, from where injection of the air steam takes place. Apparently, the firebrick plate is placed at a height of 3/4 of the top of the afterburner barrel for the deflection of the gas stream, and it must mix well with the air outside as per the maintenance of the ignition. The probes, which are the thermocouple probes further extend into the retort from the bottom and top for maintaining the temperature (Prins et al. 2011; Biochar Info 2013).

A two barrel charcoal retort. This combination is where the smaller vessels were filled with different biomass. The preferred biomass has to be dried up in order to make the charring process effective. Use of wood chips, dry grass, twigs and sawdust, etc., even some hard bones, are good to put in the retort. The vessel has to be placed upside down. Pair of bricks was laid, upon which filled vessels were placed. Then the vessel is filled to the top. After that, the bigger drum is placed upside down on the smaller vessel containing the biomass. This is to ensure that the vessel is sealed entirely. Holding the inner container tightly to the bottom of the larger container and turning it back making sure that the content doesn't spill out. The entire burning process must for a minimum of 30 -40 min after replenishing the firewood. After that the pyrolysis of the biomass starts. The gases, which escape during the heating process in between the barrel and the vessel emerges along with fire which further heats the vessel (Shackley et al. 2011). The char is ready for use.

A two barrel charcoal retort with afterburner. A container for containing the wood for charring is used in this kind of two barrel charcoal retort with afterburner. The principles are similar throughout; the only difference is the size. Out of the two containers, one is used as the retort, which will be placed at the bottom and the other one, which is placed above will be acting as the afterburner. A bottom and the lid are required for the retort. The 2nd container is used as the afterburner. This barrel has the bottom and the top being removed, cylindrical in shape. The afterburner is placed above the retort (Odesola and Ososeni 2010).

Terra Preta Pot. It is used to make biochar for residues of the crop in spite of the wood. The same pot can be used at home to cook food or warm the home. This is in fact pot inside a pot, which suits the concept of micro-pyrolyzer. It is made from local materials, and can be manufactured by any skilled potter. Thus it is possible to make biochar-generating cooking stoves while maintaining local pottery traditions (Biochar Info 2013).

Current Techniques. In conventional biochar production technologies, all processes (pyrolysis, gasification and combustion) are conducted in a single unit below earthen kiln layers. However, in modern technologies the pyrolysis and combustion process are physically separated by metal obstacle (Meyer 2011). The modern pyrolyzers vary in size, from laboratory scale (can be used to process as minimum as milligrams) to industrial equipment (with a capacity of several tonnes/hour). In North America the largest fast pyrolyzers are efficient to process 250 tonnes and 200 dry tonnes of biomass/day. The design of slow pyrolyzers is less complex than fast pyrolyzers and can be fabricated at smaller scale. Currently, many companies are selling and some are advertising to sell the industrial scale pyrolyzers systems. Scale-up studies from small to large scale pyrolyzer systems are necessary to make a balance between costs of the biomass processing at a large scale central facility over small scale processing facility situated near to biomass source. For a large-scale central processing facility, the additional costs are transportation, storage and handling cost (Laird 2009). The reactors, which are commonly popular, based on the design, are ablative reactor, rotating cone reactors, rotary drum reactor, fixed-bed reactors, fluidised bed system, screw conveyor, microwave, multiple-hearth furnace, belt reactors and compact moving bed. Some companies, which are globally famous to provide pyrolyzer systems, include: Abri-Tech Inc., Quebec, Canada; Adam + Partner, Garmish, Germany; Agri-Tech Producers, Columbia, USA; Agri-Therm Ltd., Canada; Ambient Energy, LLC, Washington, USA; Appropriate Rural Technology Institute, Pune, India; Biocarbo, Brazil; Biochar Industries, Australia; Biochar products, USA; Black Earth Product, Australia; Super Stone Clean International, Japan; waste to Energy Salutations etc. (Knight 2012; IBI 2013). A comprehensive list of biochar companies and organizations are available (Knight 2012).

Research on advanced biochar production focuses on i) continuous feed pyrolyzers to improve energy efficiency and reduce pollution emissions associated with batch kilns; ii) exothermic operation without air infiltration to improve energy efficiency and biochar yields; iii) recovery of co-products to reduce pollution emissions and improve process economics; iv) control of operating conditions to improve biochar properties and allow changes in co-product yields; v) feedstock flexibility allowing both woody and herbaceous biomass (like crop residues or grasses) to be converted to biochar (IBI 2013). Moreover, technology development is continued to expand the utility of biochar. One of such initiative is to synthesize certain metals and their (metal) oxides) (e.g., Fe, Ni, Co) on activated carbon by using electrochemical deposition and traditional impregnation techniques. Although the catalysts are available commercially, yet they can be synthesized on activated carbon used as a support catalyst. Catalyst characteristics are determined based upon the physical properties such as pore column distribution, surface area and porosity (Lehmann et al. 2009). For biochar production, the types of biomass used are dried and pelletized chicken litter and swine manure. These are used as feedstock for the bench-scale carbonization method. Biochar being a type of charcoal, can be burned as a substitute for coal, but using biochar as a soil amendment or directly as a growing medium (e.g., in roof gardens, greenhouses, garden walls) may have many benefits for the environment (Lehmann et al. 2009; IBI 2013).

15.3.4 Characteristics of Biochar and Property Optimization

Physical. The composition of biochar is single and condensed C rings (aromatic), which eventually has a higher surface area per unit mass along with a high density charge (Novak et al. 2009). In micropores surface area of pores present in the biochar ranges from 750-1360 m^2/g and volume ranges from 0.2-0.5 cm^2/g. Similarly the macrospores surface area ranges from 51-138 m^2/g and volume ranges from 0.6-1 cm^2/g (Brown 2009).

The physical properties of biochar are dependent upon the organic biomass (feedstocks) and the carbonization (pyrolysis) methods (Canadian Biochar Initiative 2008; Lehmann et al. 2009). The organic biomass can come from various sources or different feedstocks. The chemical characteristics of different feedstocks have an impact directly on the physical characteristics of the biochar generated. In a comparative study as presented in Figure 15.4, it is observed that coconut shell based biochar and wood-pellet based biochar has the maximum wt% of carbon when compared with others. The biochar containing higher C % along with other aromatic structures would have a more positive land amendment effect to improve the soil fertility (Sukartono et al. 2011). In addition, the portion of inorganic components (ash) also has implications for the physical structure of biochar. Similar processing operations were observed in the technique such as sintering/ash fusion, which may change the physical property of biochar dramatically. Moreover, operating parameters (e.g., heating rate, highest treatment temperature also known as HTT, pressure of reaction, residence time, reaction vessel, pre-treatment and post-treatment) have an influence on the physical properties of the biochar (Brown 2009; Downie et al. 2009; IBI 2010; Yoder et al. 2011). It has been understood that thermal decomposition occurs at temperature above 120 °C of the organic matter. During the pyrolysis process, loosely bound moisture is lost completely. Degradation of hemicelluloses takes places at 200-260 °C, lignin degrades at 280-500 °C and finally cellulose degrades at 240-350 °C.

Chemistry of Biochar. Chemically biochar don't have a defined chemical structure; however, biochar comprises of a range of material which differ as a result of numerous factors such as feedstocks and pyrolysis operating conditions (e.g. highest temperature and pyrolysis time) used. The organic portion of biochar has a very high content of carbon (fixed carbon varies between 50 and 90 wt%), which eventually comprises of aromatic ring structure. Additionally, structures become larger and more condensed with increasing temperature (Preston and Schmidt 2006; Brown 2009).

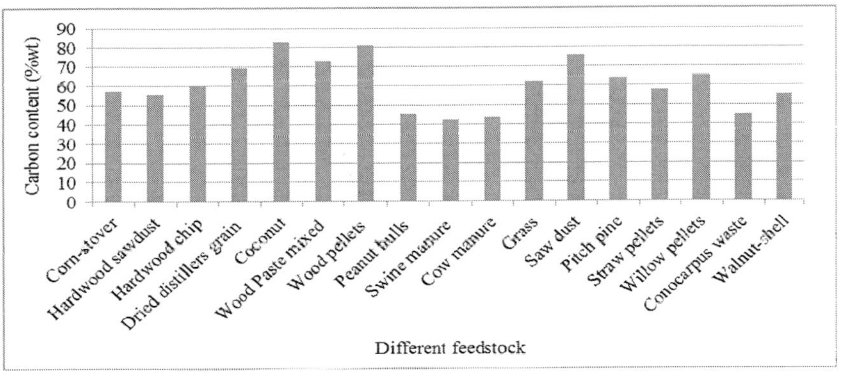

Figure 15.4. The carbon content of the biochar produced from different feedstock

The model structure of biochar (Figure 15.5) was proposed by Bourke (2007) based on the confirmation through different analytical techniques (e.g., SEM, XRD, ESR, ^{13}CNMR and MS). The biochar model structure contains oxygen as heteroatom and free radicals in the conjugated aromatic ring. Moreover, biochar also contain mineral matter (2.4 to 6.1 wt%) with a small fraction of nitrogen and sulphur. The carbonized biochar are microporous carbon having pore half width from 3 to 10 Å and the micropores in the microcrystalline graphite structure are responsible for most of the surface area of the carbon (Bourke et al. 2007).

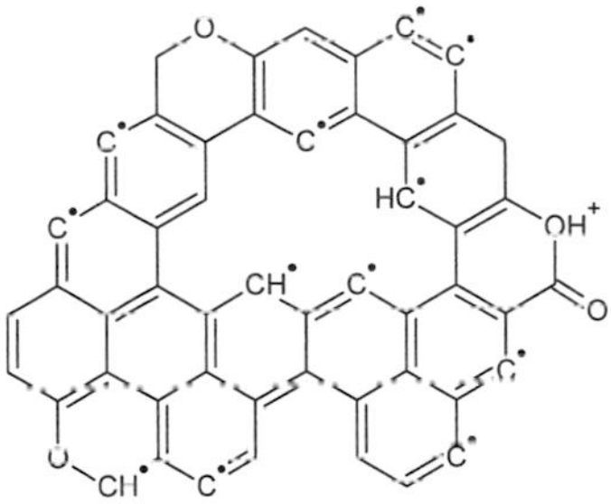

Figure 15.5 Model structure of biochar for a molecular formula $C_{52}H_{13}O_4$

To predict the reactivity and stability of biochar for its potential application (such as soil amendment), it becomes important to understand the biochars organic structural composition. Using the NMR analytical technique for analyzing the terra preta soils and soil structures, it has been derived out that the biochar in these samples is composed of highly heterogeneous mixture of organic structures. The structural form of carbon in biochar certainly depends upon the biogeochemistry of the feedstock of the biomass and the pyrolysis conditions (Sohi et al. 2010). Figure 15.6 represents a diagrammatic arrangement of aromatic rings of biochar at different temperatures during processing.

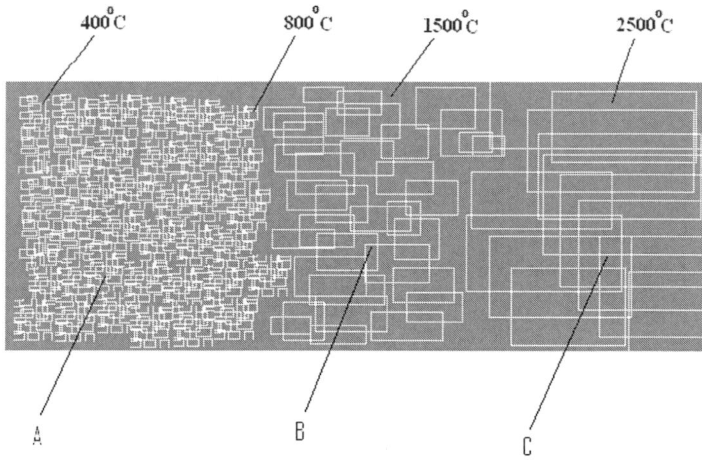

Figure 15.6. Ideal biochar structure development with highest treatment temperature (HTT). A) Increased proportion of aromatic carbon; B) growing sheets of conjugated aromatic carbon; and C) structure becomes graphite three-dimensionally

Internal morphology of pecan shell-based biochar was studied using ^{13}C nuclear magnetic resonance (NMR) correlation spectroscopy. The pecan shell-based biochar has 58% C (in aromatic structures), 29% C (=having single O bond), 13% C (having -COOH). While a few number of C are associated with carbohydrates and acetal C. Pecan shells are primarily composed of lignin and 47% cellulose (Novak et al. 2009). The entire internal morphology can be compared with casein based biomass, which has 52.49% C (Purevsurem 2003) and corn based biomass consisting of corn-cobs (47.35%) and corn-stovers (46.60%) of C, respectively (Mullen et al. 2010). Lignin and cellulose charred at temperatures of 500 °C would result in loss of the aliphatic components by the rapid conversion into ring structures such as aromatic compounds (Rutherford et al. 2004; Novak et al. 2009). In a study by Kim et al. (2012), it is observed that biochars produced at 300 °C to 500 °C are different in wt% of total carbon. Pitch pine biochar at 500 °C has more wt% of carbon than at 300 °C; however the yield is lesser.

The stability of biochar directly related to its structural arrangement. The condensed aromatic carbon of biochar persists in soil environments for millions of years, whereas biochars with higher levels or larger levels of single-ring aromatic and aliphatic carbon will mineralize quickly (Novak et al. 2009). Biochar produced during the pyrolysis of feedstock of different biomass to an infertile form is called Oxisols. Oxisols have shown to have long lasting significance in the fertility of soil due to its stability (Novak et al. 2009; Hale et al. 2013).

In practice, when biochars are being used for the soil fertility amendments, biomass pyrolysis conditions are designed for the carbonization of the material under moist condition and at lower temperatures (Novak et al. 2009). Because of decomposition and oxidation by the microbial communities in the soil, the organic carbon structures of feedstocks are bound to produce by-products during pyrolysis, resulting in the biochar containing higher densities of carboxylate and other oxygen with functional groups like OH, OR, etc., which are capable of serving as the sites for the cation exchange (Ernsting and Smolker 2009). Anhydrocellulose (the dehydrated forms of cellulose) are generally the microbial oxidized compounds and other components such as polysaccharides, alcohols, etc. should ideally exist in biochars, when prepared by pyrolysis from feedstocks at lower temperatures.

A relationship exists between biochar pyrolysis temperature and resistance to soil microbial decomposition. Biochars generated from maize and rye at 350 °C are more susceptible for the soil microbial degradation than the biochar prepared from the oak wood pyrolysis at temperature 800 °C (Novak et al. 2009). Differences in biochar decomposition are based on the C:N ratios. At higher temperatures of pyrolysis, wider C:N ratios are generated (e.g., in oak wood biochar) as the loss of the concentration of N continues with respect to carbon. For example, in pecan biochar, the C:N ratio of the pecan biochar is 244:1. Nitrogen immobilization occurs generally when organic residues possessing a C:N ratio of greater than 32:1 are gradually added to soils. The wide C:N ratio, considering the aromaticity, will enhance slow biochar decomposition. Although biochars (or soil black carbon) will ultimately be degraded via slow chemical and microbial decomposition, the rate of decomposition is very slow that even large additions of biochar to soil will possibly not significantly immobilize N (Lehmann et al. 2006). The high stability of biochar in soil environments is beneficial with regards to carbon sequestration as carbon added to the soil as biochar will be deleted from the atmosphere for ≥ 1000 years. For example, the biochar in Amazonian Black Earth region has a half-life of 6850 year. In comparison to it, the mean residence time of soil organic matter has been speculated at the range of 250-3280 years. Soil organic matter contained sufficient amounts of black carbon, which makes the age of the total SOC pool highly older than the biogenic SOC fraction age (Novak et al. 2009).

Property Optimization. Lehmann (2007) stated that biochar sequestration does not require a fundamental scientific advance, and the underlying production technology is robust and simple. However, it does require studies to optimize biochar properties and to evaluate the economic costs and benefits of large-scale deployment. Currently, one of the major research areas in biochar is about the pyrolysis condition for optimization of biochar yield and property because the final products of pyrolysis are biochar, bio-oil and gases, where fraction of each is determined by the decomposition of key chemical components of the biomass (e.g., cellulose, hemicelluloses and lignin). Another related research area is to understand under what conditions biochar develops its properties that are closely related to its functional groups [e.g., fused-ring aromatic structures and anomeric O-C-O carbons, aliphatic O-alkylated (HCOH) carbons].

Biochar has been generated in various types and qualities (Novak et al. 2009; Steinbeiss et al. 2009). For example, biochar prepared from manure has a higher nutrient value than biochar prepared from wood cuttings. Biochars produced at higher temperatures (e.g., 700 °C) are more porous and more adsorptive than that producd at lower temperature (e.g., 400 °C). For biochar formation, the degree of carbonization during pyrolysis was accelerated from 300-500 °C; whereas formation of aromatic structures of biochar begins at 400 °C (Kim et al. 2012). The presence of oxygen during the carbonization process is detrimental to the production of biochar yield and properties. Zailani et al. (2013) carried out an experiment on several wood species in a fixed-bed pyrolyzer under various fractions of oxygen ranging from 0% to 11% by varying nitrogen and oxygen composition in the pyrolysing gas mixtures at desired compositions. They reported that optimum condition of 15.2% biochar yield of mangrove wood was at pyrolysis temperature of 403 °C, 2.3% oxygen and processing time of two hours. The study shows that oxygen is the unfavourable effect presence in real application, i.e. biochar yield decreases with increasing oxygen composition. Biochar yield was 10.0% and 3.8% at 2.3% and 9.0% oxygen composition, respectively.

Devi and Saroha (2013) used waste sludge generated by the pulp and paper mill as feedstock for pyrolysis, and found biochar produced at higher temperatures (e.g., 600 to 700 °C) are relatively alkaline in nature. Biochar yield decreased with an increase in pyrolysis temperature and maximum yield was obtained at 300 °C. Conversely, surface area of the biochar increased with an increase in temperature and maximum surface area was obtained at 700 °C. Addition of $CaCO_3$ significantly affects the surface area and pore volume of the biochar, although it does not lead to significant change in biochar yield (Devi and Saroha 2014). Usually, biochars with higher specific surface area have immense potential for the adsorption of toxic substances and for rehabilitating contaminated environments (Novak et al. 2009).

Li et al. (2013) reported that the agricultural biomass carbonized to biochars was a dehydroxylation/dehydrogenation and aromatization process, mainly involving the cleavage of O-alkylated carbons and anomeric O-C-O carbons in addition to the production of fused-ring aromatic structures and aromatic C-O groups. The pH and electrical conductivity of rice straw derived biochars were mainly determined by fused-ring aromatic structures and anomeric O-C-O carbons, but the pH of rice bran derived biochars was determined by both fused-ring aromatic structures and aliphatic O-alkylated (HCOH) carbons. In many cases, novel tools (such as 2-D ^{13}C NMR) are needed for characterising the development of functional groups in biochar, and information related to biochar characterization techniques is available in the literature (e.g., Sohi et al. 2010).

15.4 Biochar for Development of Sustainable Society

Utilization of biochar has many benefits for development of the sustainable society, including i) using biochar in soil for carbon sequestration and GHG balance, ii) using biochar to enhance crop productivity and soil performance, and iii) turning bioenergy into a carbon-negative industry by collecting energy generated during pyrolysis processes. This section will describe these benefits and the related issues.

15.4.1 Biochar for Climate Change and Mitigation

The contribution of using biochar to combat climate change is to provide a potential carbon sink of ~1/6 of the net anthropogenic addition of carbon to the atmosphere (Table 15.2). Considering the other possibilities in Table 15.2, this contribution is very significant.

Essentially, the biochar process generates a natural cycle of carbon, which works best on forests and farms. These are the places where waste plant materials are locally gathered. Henceforth, the energy produced is local. The fertilizers which are carbon based are returned for promoting the growth of local crops. These biochar processes or systems are tried and tested; they are small-scale, and cheap to manufacture (Lehmann et al. 2006; Jeffery et al. 2011). More importantly, burying biochar in soil would not only permanently sequestrate atmospheric CO_2 but also contribute to mitigation of climate change by i) simultaneously decreasing associated emission of other GHGs and ii) increasing soil quality and thus, promoting crop production with less consumption of fertilizers, less irrigation costs, and reduced energy requirement in tillage. Brief descriptions of these contributions are as follows.

Carbon in biochar resists degradation, decay and digestion. Biochar remains in soil far longer than other organic matter, such as compost, plant residue or manure that oxidizes quickly. In soil, biochar's carbon-carbon bonds don't break down. So, CO_2 fixed by photosynthesis is now an inert form. Therefore, burying biochar into soils would sequester 20 to 60% of the carbon remains as biochar in soils for hundreds to thousands of years. The literature indicates that corn stalk pyrolysis into biochar on a 250-hectare farm would sequester 1,900 tons of carbon a year; applied worldwide, soil sequestration by biochar can lower CO_2 by ~8 ppm in 50 years (CNN 2013). More importantly, biochar has the potential for increasing the uptake of atmospheric CO_2 by i) stimulating greater growth of crop/plants, and ii) developing critical symbiosis for optimum soil structure and fertility. After applying biochar, the yearly harvest of biomass is larger, allowing more biofuel and biochar which builds a natural positive feedback cycle. In addition, biochar is the preferred habitat of *mycorrhizal* fungi which produces glomalin (Blakeslee 2009). Glomalin accounts for 30-40% of the carbon in soil and forms clumps of soil aggregates. These add structure to soil and keep other stored soil carbon from escaping. Currently, one practice is to use biochar in conjunction with high carbon mulch and with permanent presence of plant roots aiming at providing optimal conditions for *Arbuscular Mycorrhizal* activity for mizimizing glomalin production (Yarrow 2013).

Table 15.2. Carbon balance of different processes and CCS by biochar technology

Processes	Carbon transfer	Reference
Photosynthesis by plants	Draws 120 Gt C/yr from air; ~60 Gt C/yr returns to the atmosphere, and 60 Gt C/yr stored in new plant growth (with 45% of plant biomass = carbon)	DeLucia and Schlesinger 1995
The net primary productivity of managed agricultural and forest ecosystems	12-24 Gt C/yr, of which 20–40% associated with human active management	Vitousek et al. 1997; Sohi et al. 2010
The net anthropogenic addition of carbon to the atmosphere	6.3 Gt C/yr	Houghton, 2003
The potential of interception/stabilization of carbon by biochar technology	1 Gt C/yr	Lehmann 2007
Black carbon produced during wildfire	0.05-0.2 Gt C/yr (as a terrestrial net sink of C)	Kuhlbush 1998
The global soil carbon pool	1500 Gt C	Sohi et al. 2010
The carbon in living plant biomass	560 Gt C	DeLucia and Schlesinger 1995

Biochar contribution to global carbon sequestrations:

1. If 80% of biomass residues are used for making biochar, net biochar production will be 3.2%, and it will retain the carbon concentration alone till the year 2050.
2. Biochar alone and biochar + fossil offsets of the net anthropogenic addition of carbon to the atmosphere (6.3 Gt C/yr above) will be 15% and 31%, respectively, till the year 2050 (IBI 2010).
3. If farmers make biochar from their agricultural wastes using a very low-oxygen chambers and plowing the biochar into soil, they could store 0.5 Gt C/yr (presumably in 2007), rising to 1.75 Gt C/yr by 2060. By 2100, biochar could store 50 Gt C of carbon (Winsley 2007).
4. If biochar additions were applied on 10% of the world's cropland, it could store 8 Gt C.

Biochar can reduce emissions of nitrous oxide by 50-80%. Biochar by modern controlled pyrolysis is an approved Clean Development Mechanism in the UN Framework Convention on Climate Change to avoid methane from biomass decay (Winsley 2007; Sohi et al. 2010; CNN 2013). Biochar can be considered as a support material for microbial inoculants *e.g.* Rhizobium and its gradual application of biochar volumes sufficiently could diminish nitrous oxide emissions and nitrate leaching from soils. Advantages of biochar are that it has a high C:N ratio that would reduce nitrogen-based fertilizers and N_2O emissions (Winsley 2007; Blakeslee 2009; IBI 2010). This is extremely significant because nitrous oxide being a potent and long-lasting GHG and waterways nitrification is another major form of environmental damage from agriculture.

Much synthetic fertilizer is currently produced by using natural gas to synthesize ammonia using nitrogen from the air. However, currently the production and use of 1 ton of fertilizer nitrogen results in a 1.9 ton CO_2-C emission (Mortimer et al. 2003). Using biochar would increase soil quality and thus, promote crop production with less consumption of fertilizers, less irrigation costs, and reduced energy requirement in tillage, which would promote mitigation of climate change.

15.4.2 Improvement in Soil Quality and Crop Productivity

Soil Quality Improvement. Biochar is not another compost type or manure that improves soil properties. Biochar is far efficient in enhancing soil quality other than its organic soil amendments (Canadian Biochar Initiative 2008; Lehmann et al. 2009). The carbon in biochar is not providing nutrients to plants directly. However, it improves soil structure, environment and water retention ability; it also promotes nutrient availability, lowers acidity, and reduces the toxicity of aluminium to plant roots and soil microbiota (Stetiner et al. 2004). Biochar decreases the bioavailability of heavy metals and endocrine disruptors in limited production systems. It has potential in bioremediation. Biochar increases the water holding capacity and reduce soil bulk density. Furthermore, application of biochar in the soil helps to increase soil strength, exchangeable Al, and soluble Fe and increased porosity, organic carbon, soil pH, available P, exchangeable K, and Ca (Masulili et al. 2010). The negative surface charge and the carboxylate groups present on the surface of the biochar are probably the reasons for higher cation exchange capacity (CEC). Biochar additions to the sandy, Coastal Plain soils of the southeastern United States has been observed to eventually increase the SOC content and CEC for improving the fertility status (Brewer et al. 2009; Lehmann et al. 2009; Novak et al. 2009), which is consistent with other studies (Manyà 2012).

Biochar goes through a slow degradation process and the nanoscale molecule looks like an Australian coral reef. It represents an entire ecosystem of soil fungi and bacteria feeding on the roots of plants and also holds soil together (Lehmann et al. 2006). One characteristic of biochar is its macropores, which are very relevant for the vital functions of soil such as aeration and hydrology. These are also relevant for the movement of roots through soil and as habitats, for a vast variety of soil microbes. In biochar, the macropore surface areas are smaller than the micropores surface areas. The volumes of macropores result in bigger and finer functionality in soils and land when compared with the surface area which is narrower (Lehmann et al. 2006; Brown 2009).

Biochar is a source of reduced carbon compounds that may directly or indirectly benefit soil microbial populations (Page-Dumroese et al. 2009). The activated form of charcoal does act as a beneficiary host for the micro-organisms, commonly *mycorrhizal* fungi which further adds to the nutritional value (Brown 2009). Biochar addition to soils returns back major part of the nutrients, which are removed from the soil when biomass was harvested. Biochar enables to maintain in it, bio-available water and is also an ultimate adsorbent of plant nutrients and dissolved organic compounds (Laird 2008a). It is an efficient soil conditioner that reduces soil density and increases the soil capacity to retain, supply nutrients and water for the growing plants (Laird 2008b). Moreover, some biochars have the capacity to hold nutrients either by immobilization, nutrient adsorption or by raising the soil pH at a maximum stage until the nutrients are no longer available.

In the past, carbon sequestration in soil has been practiced by increasing the equilibrium level of active soil organic matter. To do so, carbon (in the form of organic resource or wastes) is forced through the soil system to be permanently increased to a higher rate. However, this practice is slow (often taking decades to be functional) and has its limit as soil has limited capacity to stabilize labile carbon. As the soil is closer to its equilibrium capacity, it becomes more and more difficult to maintain the annual rate of carbon incorporation. After equilibration being reached, extra resources will be needed to keep the organic matter in the soil.

It seems that adding biochar would be a much more efficient strategy for carbon sequestration, especially in fertile soils. According to Yarrow (2013), adding biochar into nutrient-poor, sandy glacial soil would transfer the gritty and granular sand into sponge cake (thick, light, fluffy large chunks soft lump) with the ideal image of perfect "crumb" structure, which is mostly due to the huge biomass of fungal grown in the soil. In addition, *mycorrhizae* and plant roots form close symbiotic associations in which plants secrete sugars to feed the fungi and the fungi share water and nutrients they scavenge from surrounding soil.

After applying biochar, there was increase in soil pH and three major plant nutrients Ca, P and K concentrations (Ernsting and Smolker 2009). The micronutrient concentrations present in the soil were not influenced by the addition of biochar. After water leaching of the biochar, the Norflok soil revealed K enrichment and sorption of most of the multivalent cations, but only net sorption of phosphorous (anions). To improve soil fertility and enhance carbon sequestration, using biochar having more readable oxidized structural groups and a minimal ratio of C:N is more appropriate. After the oxidation of biochar, soil pH will decrease and soil CEC will increase (Mullen et al. 2010; Novak et al. 2009).

Crop Productivity Enhancement. In recent years, there are concerns regarding the potentials of the world's agricultural sector in order to produce sufficient quantities of food for the ever growing world population and diverse effects of diverting a minor portion of the agricultural production for the production of the biomass for bioenergy (e.g., biofuels and biochar). Biochar technology offers a good opportunity for improving the current situation of food shortage.

Biochar is a 4,000 year-old generational practice, which is used for converting agricultural waste into enhancing of soil for the better storage of carbon. For example, in the Amazon basin, anthropogenic dark earth soils recognized as Terra Preta contain large amounts of charred materials, most likely added by pre-Columbian farmers who practiced a form of slash and char agriculture along with disposal of charcoal remains from hearths (Stetiner et al. 2004). Similar soils have been found in Ecuador and Peru, West Africa (Benin, Liberia), and the savanna of South Africa (Sohi et al. 2010). In these soils, the biochar acts as a soil conditioner. These improve soil's physical properties and nutrient use efficiency which on the contrary increases plant growth (Stetiner et al. 2004). After implementation of the practices, which generated certain soils, these Terra Preta soils are of high value for agricultural and horticultural use in the Amazon basin, the initiative would have been taken 500 years ago (Novak et al. 2009).

Studies were conducted in the Amazons, and these lead to the varied application of biochar as a possible soil enhancer (Lehmann et al. 2009). Terra Preta literally means the enriched black soil. Terra Preta is characterized by the presence of low-temperature charcoal in high concentrations, with high quantities of pottery sheds, with organic matter such as plant residues, animal feces, animal bones and other material and also with nutrients such as N, P, Ca, Zn, and Mn. An extreme level of micro level organic activities and certain specific characteristics were carried out within and around the ecosystem. These ecosystems are far lesser adaptable for nutrient leaching, as the latter is susceptible in major rain forests, "Terra Comum" is even called as common soil surrounds Terra Preta zones. Terra Comum is an infertile soil consists of acrisols, farnesols and arenosols (Downie et al. 2011). Terra Preta contains nearly or more than 9% carbon. It is quite productive and being sold as potting soil. Earlier, slash and burn forms of agriculture are the practices previously practiced. This led to soil depletion and spews CO_2, and other pollutants in the air. The terra preta is usually created by slash and char, which induces cutting off oxygen to the burning biomass (Brown 2009; Ernsting and Smolker 2009).

Glaser et al. (2001) reviewed the studies conducted during the 1980s and 1990s, and concluded that low biochar additions (0.5 ton/ha) would enhance productivity of various crops, but inhibition at higher rates were observed. Plants grow well in soil with 9% biochar, at less cost, increased yield, and sustain this greater production longer with less fertilizer. Food from those soils has higher nutritional balance, density and quality (CNN, 2013). Yields 300% greater are common, and some researchers got over 800% more yield from biochar-enriched soils (CNN 2013 and other examples therein). Increase in crop productivity is believed because of increase in soil fertility, water holding capacity, and early warm in sprint due to darker soil, among other reasons.

Masulili et al. (2010) conducted experiments to evaluate the characteristics of biochar prepared from rice husk and its potential as a soil amendment in acidic state of soils. The improvement of rice growth was observed in terms of an increase in plant height, number of tillers, and dry biomass. This enhancement was observed due to significant improvement in some properties of the acid sulfate soil, such as decreasing soil bulk density, soil strength, exchangeable Al, and soluble Fe, and increasing soil pH, soil organic matter, total phosphorous, exchangeable K, and exchangeable Ca. In the U.S. Coastal Plain region, agricultural land has low-value soil fertility characteristics due to i) its acidic pH values, ii) clays which are kaolinitic in nature in sandy soils, iii) low CEC, and iv) lesser soil organic carbon contents (Ernsting and Smolker 2009; Novak et al. 2009). Using column tests, significant fertility improvements were noticed after the addition of biochar into the soil from Coastal Plain region, including the soil pH, soil organic carbon, Ca, K, Mn, and P and decreased exchangeable acidity, S, and Zn content (Novak et al. 2009; Steinbeiss et al. 2009). More information is available in the literature (e.g., Sohi et al. 2010; IBI 2013).

15.4.3 Turning Bioenergy into A Carbon-Negative Industry

As a traditional practice, some farmers are burning crop residue along with weeds and then bury the biochar into the soil. This practice, however, not only wastes the bioenergy stored in the biomass but also contributes to CO_2 emissions. Biomass is the world's third largest fuel, after coal and oil. Current biomass-to-energy technology is at best carbon neutral only if the feedstock is cheap and the biochar (the by-product of pyrolysis) is used to offset fossil-fuel use and reduce costs, which is not sustainable, since harvests deplete nutrients, reducing fertility and productivity.

Pyrolysis making biochar also produces energy. As biomass breaks down into char, hydrogen, methane and other hydrocarbons are released and can be captured to refine into renewable fuels. Energy produced during pyrolysis can be turned into space heat, electricity, reformed into ethanol or ultra-clean diesel. One ton of biomass equals 5.5 barrels of oil. Each gigajoule of hydrogen produced would store 112 kg of CO_2 in soil. Therefore, considering biochar will be put back into soil, the GHG emissions reductions of the pyrolysis process can be 12–84% (Lehmann 2007; Cowie 2010). Moreover, burying biochar in soil would return about half the original carbon and most minerals to the soil and support sustainable, biological fertility. Thus, biochar may be our best chance to turn energy production into a C-negative industry.

Certain companies do claim that biochar production is basically carbon negative, such as Epride, Shell, Brazil's Embrapa, JP Morgan Chase, Biochar Engineering, Dynamotive, Heartland Bioenergy, and Indonesian palm oil association (Lehmann et al. 2006; Blakeslee 2009). Proponents claim that charcoal can not only sequester carbon, on a globally significant scale, but also improve soil fertility, and thereby reduce demand for synthetic fertilizers and emissions of the powerful GHG N_2O, as increased aeration due to biochar application can suppress the conversion and henceforth reduction of N_2O to N_2 (Karhu et al. 2011). Biochar technology also reduces water consumption, dynamically prevents runoff of chemicals from farm lands, reduces emissions of nitrogen oxide (NO_x) and sulfur oxide (SO_x) from coal burning power plants, reduces emissions of black carbon from biomass cooking fires, reduces methane emissions from decomposing organic waste piles and more (Lehmann et al. 2006). Considering Terra Preta as a control, claims are put forward for biochar in a larger extent (Blakeslee 2009).

15.5 Biochar Sustainability

According to the available information, pyrolysis can be cost-effective for a combination of sequestration and energy production when the cost of a CO_2 ton reaches \$37 (Lehmann 2007); it was \$4 in 2007 and \$16.82/ton of CO_2-C in 2010 on the European Climate Exchange (Zhang and Surampalli 2013), and will be rising over the coming years to decades to \$25–85 per tonne (Stern 2007).

Climate change involves the management of carbon, which keeps on flowing in a vicious manner in between the atmosphere, terrestrial and ocean systems. The Kyoto Protocol Article 3.4 has given permission for the recognition of enhanced soil carbon sequestration. According to Kyoto Protocol Article 3.3, a landowner can convert forest plantations to daily pasture to offset carbon losses in the grounds, by sequestrating biochar present in the soil. For every tons of carbon dioxide added as biochar, there will be reducing deforestation liabilities.

The economic impact of developing a significant biochar sector on the wider regional economy and the potential benefits for the farmers and the people involved are: a) increase in yield; b) increased output; c) return back of carbon to the soil; d) low cost of production; e) improved quality of the crops and plants (Collison et al. 2009). Introducing biochar in soil is substantially substituting lime and fertilizers inputs, which are applied when forests are being converted to pastures. Economic benefits are yet to be gathered from the biochar applied soils (Lehmann et al. 2009). Apart from the potential increases in per hectare productivity of the biochar, the biochar market may create a domestic biochar sector, and will affect widely on the regional economy. Potential benefits (e.g. economic growth) consists of creating value from existing wastes in the sectors of agricultural and domestic, municipal and commercial waste streams, etc. The construction and operation of biochar plants depends upon the biochar impact on our society, which is further relied upon agriculture, cost reduction, extensive crop output, raw material processing and a carbon market (Collison et al. 2009).

For an economically viable, socially responsible, and environmentally sound biochar industry, IBI published *the Guiding Principles for a Sustainable Biochar Industry*, with the following principles being endorsed (IBI 2013):

- **Environmental Outcomes**, including: 1) soil health (maintain and enhance soil fertility), 2) climate stability (at least GHG neutral and preferably GHG negative), 3) energy efficiency & conservation (resulting in neutral or preferably net energy export), 4) Feedstocks (prioritizing the use of biomass residuals), 5) biochar production (safe, clean, economical, and efficient, not exceed the environmental standards and regulatory requirements), 6) biochar quality, 7) biological diversity (promoting above- and below-ground biodiversity), 8) water (not polluting nor degrade water resources);
- **Social outcomes**, including: 9) food security (not jeopardizing food security), 10) local communities (involving stakeholders fully and transparently in planning and implementation), 11) biochar knowledge societies (being continuously improved through research, education and the open sharing of scientific and traditional knowledge); and
- **Economic outcomes**, including: 12) labor rights (not violating labor rights), and 13) economic development (contributing to the economic development of local communities).

Sustainable biochar systems are essential to the future of biochar. To establish such a system, the following needs to be done:

- Addressing a wide range of potential environmental/social/economic impacts;
- Sharing experiences and making recommendations on best practices for known sustainability issues;
- Promoting development of new technology will help future projects better encapsulate all aspects of biochar technology including sustainability;
- Developing practical tools and online assessment platforms (rather than depending on absolute standards that require testing and certification) for evaluation of i) biochar technology that is adaptable to different regions, feedstocks, technologies, locations and communities, and ii) sustainability; and
- Incorporating new information and creating new tools for developing bochar industry standards (such as i) biochar characterization standards, ii) biochar sustainability protocols, iii) guidelines for developing and testing a pyrolysis plant, iv) carbon market investment criteria for biochar projects, v) biochar legislation and policy, vi) quantification of the climate change mitigation benefits of biochar, vii) a certification program…).

15.6 Concerns and Future Prospective

15.6.1 Debates and Concerns

Currently, there are some debates about the effects of biochar technology on GHG emissions and improvement of soil quality and crop productivity, which reflects the nature of biochar technology as a relatively new sector. In this section, these debates and concerns are briefly introduced, together with future trends for biochar research.

Debate on Biochar for Carbon Sequestration. There are several scenarios fueling the debate. Scenario I is that reduced plant growth has been observed after the application of biochar. This has been attributed to temporary pH levels, volatile matter and imbalances associated with volatile matter in the freshly-prepared biochar (McClellan et al. 2007). Biochar often can have an initially heavily alkaline, which is perfect when used along with acid-degraded soils. But when soil pH becomes too alkaline, plants may suffer nutrient deficiencies. The mobile matter (i.e., tars, resins, and other short-lived substances) that remain on the biochar surface immediately after production can inhibit plant growth. This is because before microbes decompose and transform these mobile matter into the carbon-rich material for plant nutrients, microbes use them (plus N_2 and other soil elements) for their growth, rendering them temporarily unavailable for uptake by plants and causing transitional imbalances (e.g., decaying, neutralization of pH) (Hunt et al. 2010). A study has shown that substantially growth of *Chlorella vulgaris* (e.g., > 80%) could decrease the negative impact of biochars (Magee et al. 2013); but during this (microbial growth) phase biochar does negatively affect plant growth especially in the growth of certain crops (e.g., soybeans and maize) (Dharmakeerthi et al. 2012). Ernsting and Smolker (2009) reported that biochar has the capacity to suppress plant growth completely after simultaneous harvesting for two times. Graber et al. (2012) has shown that the high surface area of biochar is effective in reducing the potential intake of insecticides and herbicides by plants and thus affecting herbicides and insecticides efficacy. Therefore, such uncertainty needs research to improve the utility of biochar.

In Scenario II, biochar present in the soil stimulates the carbon emission of the soil. Soil microbes can metabolize carbon black, which is the result of carbon emission in the atmosphere. Quantitatively, extensive usage of biochar creates an ecological niche for the microbial community. These microbial community breaks down the carbon black present in the soil which further leads to the release of CO_2. Biochar has the capacity to accelerate the microbial activity in the soil. A study performed at 2008 suggests that if biochar is placed into boreal forest soil, it eventually diminishes substantial amounts of soil organic carbon over a period of ten years period. Primary results of the study showed Colombia having 60% increases in soil carbon losses in the past two years of 'biochar' use as compared to control plots (Ernsting and Smolker 2009). The plant biomass decomposes in a shorter interval of time, whereas biochar is more stable.

In Scenario III, biochar present in the soil becomes a major carbon emission source, rather than the carbon sink. Biochar derived from glucose remained in forest soil for ~12 years longer than that in arable soil. For the biochar derived from yeasts, it takes nearly 6 years to be mineralized in the forest soil (Steinbeiss et al. 2009). However, there is no such straightforward proof to demonstrate that biochar will retain carbon in soils. In certain cases, microorganisms degrade the biochar carbon and soil OC. In the literature, it has been clarified that in certain cases, if not being transported properly, or stored efficiently or being added to the soil carefully, biochar will escape into the air, and thus, further increases the global warming (Ernsting and Smolker 2009). Australian biochar trial suggested that problems were observed by simply placing biochar on the top of soil and vegetation without incorporating it. To avoid the problem of black carbon, biochar need to be tied in the soil, which is a disruptive process resulting in carbon emissions from soil.

Debate on Biochar's Improvement of Soil Quality and Crop Productivity. Certain mechanisms concerning such effects have never been clearly comprehended, and some mechanisms may be related to soil, chemical and microbial properties and processes associated with biochar (Mukherjee and Lal 2013).

Water retention. Terra preta is the form of biochar, which has shown the capacity of water retention in soils, and eventually other forms if biochar leads to same water retention capacity. It has been well accepted that water retention of soils increases due to biochar amendment. This reduces the need for irrigation, henceforth resulting in more of plant growth, decreasing water run-off and thereby reducing soil erosion and leaching of agricultural nutrients (Sopena et al. 2012). It is clearer in the case of sandy soils, and does not appear to hold positive for loamy or clayey soils (Sopena et al. 2012). In loamy soil, biochar does not visibly change water retention, while in the clayey soil, biochar eventually reduces it. Additionally, over the broad period of time and particularly after a fire, biochar could make soils gradually water-repellent (Ernsting and Smolker 2009).

Nutrient loss. Adding Terra Preta to the soil may add an extra building block of biomass, which is humus and charcoal together. However, some researchers argue that biochar has a reverse role in agricultural land forestry residues by efficiently reducing humus form the soil (Blakeslee 2009). On the other hand, practicing swidden agriculture or slash and burn agriculture means farmers clear an area for temporary cultivation by cutting and burning the vegetation. This has been long known as adding some charcoal to the soil to make it temporarily more fertile because 'fresh' charcoal retains nutrients essential for plant growth. However, this should not be considered as fertile Terra Preta. The soil amended with charcoal has had different properties from Terra Preta. Stetiner et al. (2004) reported that it takes 50-100 years of gradual close interactions among soil microbes and charcoal for the creation of soils resembling as Terra Preta.

However, adding Terra Preta with synthetic fertilizers would lead to excessive leaching of nutrients, much more than when these fertilizers are added to the low quality carbon soils. In the similar experiment, biochar was used to replace Terra Preta and used together with synthetic fertilizers. Biochar was found to gradually reduce nutrient leaching caused by synthetic fertilizers, but the results cannot be extrapolated to all different types of soil (Ernsting and Smolker 2009). A field study near Manaus, Brazil results shows that charcoal added with synthetic fertilizers in soil enhances yields more than using the fertilizers alone, but the highest reported yields obtained was using chicken manure only (Blakeslee 2009). Actually, scientific investigations on the soil fertility effects are minimal. Biochar may improve soils of low fertility at the regions of Southeastern U.S. and at the Coastal Plain regions, as indicated in Section 15.4.2 (Novak et al. 2009). However, similar behaviors of the biochar (i.e., increase the SOC content and CEC with improved fertility) were not seen when incubated for longer periods of time; apparently at higher temperatures, biochars are not so effective (Ernsting and Smolker 2009).

Toxic and persistent organic pollutants. The toxic or POPs are due to feedstock characteristics or may be generated during pyrolysis. It has been reported that polycyclic aromatic hydrocarbons (PAHs) are produced during the pyrolysis of biomass (Fabbri et al. 2012). In biochar and biochar-embedded soil, ethylene was observed. Ethylene is produced from the pyrolysis of biomass (Spokas et al. 2010). Ethylene brings a significant plant hormone, and it is also an inhibitor for soil microbes. Ethylene reduces microbial nitrification and has been postulated to affect the spore germination of fungi (Spokas et al. 2010). In a study conducted with twelve different biochars, soils amended with biochar hardly produced any detectable ethylene. On the contrary, five biochars even without any soil or microbial inoculums produced ethylene at the dried state (1.8-18 ng/day-g of char). When the biochar was mixed with soil, six soils out of twelve exhibited an increasing amount of production of ethylene as compared to the soil in which biochar is absent. Therefore, it is important to have the information about characteristics of a biochar for its end use. Research is needed for characterization of biochar to predict it actual property and for development in efficient pyrolyzers systems.

Earthworm reduction. Biochar reduces the population of earthworm. Significant earthworm populations were observed in all types of soil where charcoal generated from the forest fires exists. Topoliantz and Ponge (2003) reported that a difference in community of earthworm has occurred after the addition of biochar. Liesch et al. (2010) observed that the earthworm mortality rate and weight loss was higher with poultry litter biochar than pine chips biochar. Poultry litter biochar was believed to create a more stressful environment, probable due to the existence of ammonia gas in addition to high pH (increased from 7.2 to 8.9 after adding biochar). Additionally, toxic micronutrients (e.g., As, Zn, Cu, Fe and Al) are present in poultry litter biochar. The study was concluded that toxicity and impact of biochar may differ due to difference in feedstock used as well as conditions used for pyrolysis. The researchers recommended that evaluation of toxicity of biochar prior to land application will be helpful to rectify such problems.

15.6.2 Future Research

Intensive biochar research is still needed, particularly in the following areas (Lehmann 2007; Sohi et al. 2010; IBI 2013; Zhang and Surampalli 2013):

- Development of carbon-negative biochar production technology. It is important to i) conduct comparison tests between using biomass for biochar production and that for biofuel; ii) develop large-scale "high tech", and small-scale "low tech" methods for biochar production with all design and operational parameters being understood; iii) produce biochar with different feedstocks or combination of different feedstocks; and iv) develop different biochar-related products (such as slow-released ammonia bound with biochar);
- Characterization of biochar and feedstocks;
- Large-scale field demonstration for carbon sequestration. It is important to i) establish the methods for quantifying the effect of carbon sequestration by biochar in different soils and at different locations; ii) understand the fate and behavior of biochar in the soil and the effects of biochar burying on soil OC and other components; and iii) determine the long-term stability of C-sequestration rate with life cycle analyses;
- Field application for improvement of soil quality and crop production. It is important to understand biochar interaction with i) soil microbes, ii) soil particles, iii) nutrients, iv) plants in soils, biochar impact on v) soil NO_x and CH_4, vi) crop productivity, vii) soil properties, and viii) the evolution of the related mechanisms in time;
- Demonstration of sustainability of biochar technology; and
- Social/economic/environmental impacts of biochar technology. It is important to i) conduct life cycle assessment on environmental sustainability of the technology, and ii) determine the contribution to local communities.

15.7 Summary

As the key element in a new carbon-negative strategy, biochar can mitigate climate change by carbon sequestration and facilitate development of the sustainable society by resolving critical challenges (e.g., food and energy security). In addition, biochar technology can create new local businesses, jobs and financial cycles to raise incomes in rural communities. Development of biochar technology depends upon the biochar impact on our society, the carbon market, and biochar research/practice.

15.8 Acknowledgement

Sincere thanks are due to the Natural Sciences and Engineering Research Council of Canada (Grant A 4984, Canada Research Chair) for their financial support. Views and opinions expressed in this article are those of the authors.

15.9 Abbreviations

Al	Aluminum
Ca	Calcium
CDM	Clean Development Mechanism
CEC	Cation Exchange Capacity
Ce	Carbon Emission
CO_2	Carbon Dioxide
-COOH	Carboxyl group
C: N	Carbon: Nitrogen ratio
CNMR	Carbon-13 Nuclear Magnetic Resonance
Cm^2/g	Square Centimeter per Gram
ESR	Electron Spin Resonance
Fe^{2+}	Ferrous ion
GEO	Geo-ecology Energy Organisation
GHG	Green House Gases
HTT	Highest Treatment Temperature
H_2O_2	Hydrogen Peroxide
IBI	International Biochar Initiative
K	Potassium
M^2/g	Square Meter per Gram
MS	Mass spectrometry
Mn	Manganese
N	Nitrogen
NGO	Non-governmental Organization
NMR	Nuclear Magnetic Resonance
NOx	Nitrogen oxide
N_2O	Nitrous Oxide

NO_3^-	Nitrate
NH_4^+	Ammonium
N_2	Nitrogen gas
N_2O	Nitrous Oxide
Ng	NanoGrams
OC	Organic Carbon
O_3	Ozone
P	Phosphorus
Pg	Petagram (1 Petagram = 10^{15} g C)
PAHs	Polycyclic Aromatic Hydrocarbons
ppm	Parts Per Million
S	Sulphur
SCAD	Social Change And Development
SEM	Scanning Electron Microscopy
SOC	Soil Organic Carbon
SOx	Sulfur Oxide
UNCCD	United Nations Convention to Combat Desertification
UNFCCC	United Nations Framework Convention on Climate Change
Wt/wt	Weight by Weight
Wt%	Weight by percentage
XRD	X-ray Diffraction
Zn	Zinc

15.10 References

Al-Wabel, M.I., Al-Omran, A., El-Naggar, A.H., Nadeem, M., and Usman, A.R.A. (2013). "Pyrolysis temperature induced changes in characteristics and chemical composition of biochar produced from conocarpus wastes." *Bioresource Technology*, 131, 374–379.

Barrow, C.J. (2012). "Biochar: Potential for countering land degradation and for improving agriculture." *Applied Geography*, 34, 21–28.

Beesley, L., Marmiroli, M., Pagano, L., Pigoni, V., Fellet, G., Fresno, T., Vamerali, T., Bandiera, M., and Marmiroli, N. (2013). "Biochar addition to an arsenic contaminated soil increases arsenic concentration in the pore water reduces uptake to tomato plants (*Solanum lycopersicum* L.)." *Science of the Total Environment*, 454, 598–563.

Biochar Info (2013). "Large Scale biochar Production." <http://www.biochar.info/> (accessed Dec. 2013).

Blakeslee, T.R. (2009). "Biochar key to carbon-negative biofuels." Clearlight Foundation, available at *http://www.RenewableEnergyWorld.com*.

Brewer, C.E., Schmidt-Rohr, K., Satrio, J.A., and Brown, R.C. (2009). "Characterization of biochar Report FP6-2004-INCO-DEV-3 from fast pyrolysis and gasification system." *American Institute of Chemical Engineers*, 28(3), 386−396.

Bourke, J., Manley-Harris, M., Fushimi, C., Dowaki, K., Nunoura, T., and Antal, M. J. (2007). "Do all carbonized charcoals have the same chemical structure? 2. A model of the chemical structure of carbonized charcoal." *Industrial & Engineering Chemistry Research*, 46(18), 5954−5967.

Brown, R. (2009). Biochar production technology. Biochar for environmental management: *Science and Technology*, 127−146.

Brudges, J. (2009). "BioChar and Agriculture." The BioChar Debate: Charcoal's potential to reverse climate change and build soil fertility, (4), *The Schumacher Society*, UK, 29–56.

Canadian Biochar Initiative. (2008). Available at *http://www.biochar.ca/*.

Chaiwong, K., Kiatsiriroat, T., Vorayos, N., and Thararax, C. (2013). "Study of bio-oil and bio-char production from algae by slow pyrolysis." *Biomass and Bioenergy*, 56, 600−606.

Carbon-Negative Network (CNN) (2013). "Biochar: Carbon-negative fertility, food & fuel. Frequently asked questions." Available at www.carbon-negative.us/docs/biocharFAQ.pdf (accessed Dec. 2013).

Collison, M., Collison, L., Sakrabani, R., Tofield, B. and Wallage, Z. (2009). *Biochar and Carbon Sequestration: A Regional Perspective*. Report Reference no. 7049, *University of East Anglia, East of England Development Agency (EEDA)*.

Cowie, A. (2010). "Is biochar carbon negative? Quantifying the climate change mitigation benefits of biochar." Presentation given at IBI 2010, Rio de Janeiro, Brazil. Available at <www.biochar-international.org/sustainability> (accessed Dec. 2013).

DeLucia, E.H. and Schlesinger, W.H. (1995). "Photosynthetic rates and nutrient-use efficiency among evergreen and deciduous shrubs in Okefenokee swamp." *Intl. J. Plant Science*, 156, 19−28.

Devi, P., and Saroha, A.K. (2013). "Effect of temperature on biochar properties during paper mill sludge pyrolysis." *International Journal of Chem. Tech. Research*, 5(2), 682−687.

Devi, P., and Saroha, A.K. (2014). "Optimization of pyrolysis conditions to synthesize adsorbent from paper mill sludge." *J. Clean Energy Technol.*, 2(2), 180−182.

Dharmakeerthi, R.S., Chandrasiri, J.A.S., and Edirimanne, V.U. (2012). "Effect of rubber wood biochar on nutrition and growth of nursery plants of *Hevea brasiliensis* established in an Ultisol." *Springer Plus*, 1(1), 1−12.

Downie, A., Crosky, A. and Munroe, P. (2009). "Physical properties of Biochar." *Biochar for Environmental Management: Science and Technology*, Earth Scan, UK and USA, 2, 18−19.

Downie, A.E., Zwieten, L.V., Smernik, R.J., Morris, S., and Munroe, P.R. (2011). "*Terra Preta* Australis: Reassessing the carbon storage capacity of temperate soils." *Agricultural, Ecosystems and Environment*, 140(1–2), 137–147.

Ducey, T.F., Ippolito, J.A., Cantrell, K.B., Novak, J.M., and Lentz, R.D. (2013). "Addition of activated switchgrass biochar to an aridic subsoil increases microbial nitrogen cycling gene abundances." *Applied Soil ecology*, 65, 65–72.

Duku, M.H., Gu, S., and Hagan, E.B. (2011). "Biochar production potential in Ghana–A review." *Renewable and Sustainable Energy Reviews*, 15(8), 3539–3551.

Ernsting, A., and Smolker, R. (2009). "Biochar for Climate Change Mitigation: Fact or fiction." *Agrofuels and the Myth of the Marginal Lands,* Joint Briefing published by *Gaia Foundation,* 1–10.

Fabbri, D., Rombola, A.G., Torri, C. and Spokas, K.A. (2012). "Determination of polycyclic aromatic hydrocarbons in biochar and biochar amended soil." *Journal of Analytical and Applied Physics*, 103, 60–67.

Glaser, B., Haumaier, L., Guggenberger, G., and Zech, W. (2001). "The 'Terra Preta' phenomenon: A model for sustainable agriculture in the humid tropics." *Naturwissenschaften*, 88, 37–41.

Graber, E.R., Tsechansky, L., Gerstl, Z. and Lew, B. (2012). "High surface area biochar negatively impacts herbicide efficacy." *Plant and Soil*, 353(1–2), 95–106.

Hale, S.E., Lehmann, J., Rutherford, D., Zimmerman, A.R., Bachmann, R.T., Shitumbanuma, V., and Cornelissen, G. (2012). "Quantifying the total and bioavailable polycyclic aromatic hydrocarbons and dioxins in biochars." *Environmental Science & Technology*, 46(5), 2830–2838.

Haefele, S.M. (2007). Black Soil Green Rice. *Rice Today*, 6(2), 26–27.

Houghton, R.A. (2003). "Revised estimates of the annual net flux of carbon to the atmosphere from changes in land use and land management 1850–2000." *Tellus B*, 55(2), 378–390.

Hunt, J., DuPonte, M., Sato, D., and Kawabata, A. (2010). *The Basics of Biochar: A Natural Soil Amendment*. SCM-30, Cooperative Extension Service, University of Hawai'i at Manoa.

International BioChar Initiative (IBI) (2010). Available at http://www.biochar-international.org/biochar (accessed Dec. 2013).

IBI (2013). IBI Publications. Available at http://www.biochar-international.org/biochar (accessed Dec. 2013).

Jeffery, S., Verheijen, F.G.A., Velde, M.V., and Bastos, A.C. (2011). "A quantitative review of the effects of biochar application to soils on crop productivity using meta-analysis." *Agriculture, Ecosystems, and Environment*, 144, 175–187.

Karhu, K., Mattila, T., Bergstrom, I., and Regina, K. (2011). "Biochar addition to agricultural soil increased CH_4 uptake and water holding capacity–Results from a short-term pilot field study." *Agriculture, Ecosystems and Environment*, 140(1–2), 309–313.

Kim, K.H., Kim, J.Y., Cho, TS., and Choi J.W. (2012). "Influence of pyrolysis temperature on physiochemical properties of biochar obtained from the fast pyrolysis of pitch pine (*Pinus rigida*)." *Bioresources Technology*, 118(158–162), 158–162.

Knight, E. (2012). *Biochar Companies & Organizations*, complied September 2012 by Knight, E., available at http://www.biochar-international.org/company_list (accessed Dec. 2013).

Kuhlbush, T.A.J. (1998). "Black carbon and the carbon cycle." *Science*, 280, 1903–1904.

Laird, D.A. (2008a). "The charcoal vision: A win–win–win scenario for simultaneously producing bioenergy, permanently sequestering carbon, while improving soil and water quality." *Agronomy Journal*, 100(1), 178–181.

Laird, D.A. (2008b). United States Department of Agriculture, available at http://www.ars.usda.gov/research/publications.

Laird, D.A. (2009). United States Department of Agriculture, available at http://www.ars.usda.gov/research/publications.

Lee, Y., Park, J., Ryu, C., Gang, K.S., Yang, W., Park, Y.K., Jung, J. and Hyun, S. (2013). "Comparison of biochar properties from biomass residues produced by slow pyrolysis at 500 °C." *Bioresource Technology*, 148, 196–201.

Lehmann, J., Gaunt, J., and Marco, R. (2006). "Bio-char sequestration in terrestrial ecosystems – A Review." *Mitigation and Adaptation Strategies for Global Change, Springer*, 11, 403–427.

Lehmann, J. (2007). A handful of carbon. *Nature*, 447, 143–144.

Lehmann, J., Czimczik, C., Laird, D., and Sohi, S. (2009). United States Department of Agriculture (USAD), available at <http://www.ars.usda.gov/research/publications/publications.htm?seq_no_115=230704> (accessed Dec. 2013).

Li, X., Shen, Q., Zhang, D., Mei, X., Ren, Z., Xu, Y., and Yu, G. (2013). "Functional groups determine biochar properties (pH and EC) as studied by two-dimensional ^{13}C NMR correlation spectroscopy." *PLOS ONE*, 8(6), e65949. DOI: 10.1371/journal.pone.0065949.

Liesch, A.M., Weyers, S.L., Gaskin, J.W. and Das, K.C. (2010). "Impact of two different biochars on earthworm growth and survival." *Annals of Environmental Science*, 4(1), 1.

Magee, E., Zhou, W., Yang, H., and Zhang, D. (2013). "The Effect of Biochar Application in Microalgal Culture on the Biomass Yield and Cellular Lipids of *Chlorella vulgaris*." Available at <http://www.biochar-international.org/node/4445> (accessed Dec. 2013).

Manyà, J.J. (2012). "Pyrolysis for biochar purposes: A review to establish current knowledge gaps and research needs." *Environmental Science & Technology*, 46(15), 7939–7954.

Marris, E. (2006). "Putting the carbon back: Black is the new green." *Nature*, 442(10), 624–626.

Masek, O., Budarin, V., Gronnow, M., Crombie, K., Brownsort, P., Fitzpatrick, E., and Hurst, P. (2013). "Microwave and slow pyrolysis biochar-Comparison of physical and functional properties." *Journal of Analytical and Applied Pyrolysis*, 100, 41–48.

Masulili, A., Utomo, W., Hadi, S., and Kom Y. (2010). "Characteristics of rice husk biochar and its influence on the properties of acid sulfate soils and rice growth in West Kalimantan." *Agricultural Science*, 2(1), 39–47.

Matala, P.Z., Ajay, O.C., Oduol, P.A., and Agumya, A. (2010). "Socio-economic factors influencing adoption of improved fallow practices among small holder farmers in western Tanzania." *African Journal of Agricultural Research*, 5(8), 818–823.

McClellan, A.T., Deenik, J., Uehara, G., and Antal, M. (2007). "Effects of flashed Carbonized© Macadamia nutshell charcoal on plant growth and soil chemical properties." *AA*, 80(100), 120.

Mendez, A., Gomez, A., Paz-Ferreiro, J., and Gasco, G. (2012). "Effects of sewage sludge biochar on plant metal availability after application to a Mediterranean soil." *Chemosphere*, 89, 1354–1359.

Meyer, S., Glaser, B., and Quicker, P. (2011). "Technical, economical, and climate-related aspects of biochar production technologies: a literature review." *Environmental Science & Technology*, 45(22), 9473–9483.

Mohan, D., Pittman, C.U., and Steele, P.H. (2006). "Pyrolysis of wood/biomass for bio-oil: a critical review." *Energy & Fuels*, 20(3), 848–889.

Mortimer, N. D., Cormack, P., Elsayed, M. A., and Horne, R. E. (2003). *Evaluation of the Comparative Energy, Global Warming and Socio-Economic Costs and Benefits of Biodiesel.* Report to the Department for Environment, Food and Rural Affairs Contract Reference No. CSA, 5982.

Mukherjee, A., and Lal, R. (2013). "Biochar impacts on soil physical properties and greenhouse gas emissions." *Agronomy*, 3(2), 313–339.

Mullen, C.A., Botaeng, A.A., Goldberg, N.M., Lima, I.M., Laird, D.A., and Hicks, K.B. (2010), "Bio-oil and biochar-production from corn-cobs and stover by fast pyrolysis." *Biomass and Bioenergy, Science,* 34, 67–74.

Novak, J.M., Busscher, W.J., Laird, D.L., Ahmedna, M., Watts, D.W., and Niandou, M.A.S. (2009). "Impact of biochar amendment on fertility of a southeastern coastal plain soil." *Soil Science*, 174(2), 105–112.

Odesola, I.F., and Ososeni, T.A. (2010). "Small scale biochar production technologies: A review." *Journal of Emerging Trends in Engineering and Applied Sciences*, 1(2), 151–156.

Olah, G.A., Goeppert, A., and Urya Prakash, G.K. (2009). "Chemical recyclying of carbon diaoxide to methanol and dimethyl ether: from greenhouse gas to renewable, environmentally carbon neutral fuels and synthetic hydrocarbons." *The Journal of organic chemistry*, 74(2), 487–498.

Page-Dumroese, D., Coleman, M., Jones, G., Venn, T., Dumroese, R.K., Anderson, N.l, Chung, W., Loeffler, D., Archuleta, J., Kimsey, M., Badger, P., Shaw, T., and McElligott, K. (2009). "Portable in-woods pyrolysis: Using forest biomass to reduce forest fuels, increase soil productivity and sequester carbon." *Rocky Mountain Research Station*, Moscow, 1–13.

Preston, C.M., and Schmidt, M.W.I. (2006). "Black (pyrogenic) carbon: A synthesis of current knowledge and uncertainties with special consideration of boreal regions." *Biogeosciences*, 3(4), 397–420.

Prins, R., Teel, W., Marier, J., Austin, G., Clark, T., and Dick, B. (2011). "Design, construction, and analysis of a farm-scale biochar production system." *Catalyzing Innovation*, Washington DC, March 24–26, 2011.

Purevsuren, B., Avid, B., Tesche, B., and Davaajav, Y.A. (2003). "A biochar from casein and its properties." *Journal of materials, Science*, 38, 2347–2351.

Reid, B.J., Pickering, F.L., Freddo, A., Whelan, M.J., and Coulon, F. (2013). "Influence of biochar on isoproturon portioning and bioaccessibility in soil." *Environmental Pollution*, 181, 44–50.

Ronsse, F., Van Hecke, S., Dickinson, D., and Prins, W. (2012). "Production and characterization of slow pyrolysis biochar: Influence of feedstock type and pyrolysis conditions." *GCB Bioenergy*, 5, 104–115

Rutherford, D.W., Wershaw, R.L., and Cox. L.G. (2004). *Changes in Composition and Porosity during the Thermal Degradation of Wood and Wood Components*. USGS Sci. Invest. Rep. 2004–5292, available at website http://pubs.usgs.gov/sir/2004/5292/. (Accessed Dec. 2013).

Shackley, S., Hammond, J., Gaunt, J., and Ibarrola, R. (2011). "The feasibility and costs of biochar deployment in the UK." *Carbon Management*, 2(3), 335–356.

Sohi, S.P., Krull, E., Lopez-Capel, E., and Bol, R. (2010). "A review of Biochar and its use and function in soil." *Advances in Agronomy*, 105, 47–82.

Sopena, F., Semple, K., Sohi, S., and Bending, G. (2012). "Assessing the chemical and biological accessibility of the herbicide isoproturon in soil amended with biochar." *Chemosphere*, 88(1), 77–83.

Spokas, K.A., Baker, J.M. and Reicosky, D.C. (2010). "Ethylene: Potential key for biochar amendment impacts." *Plant Soil*, 333, 443–452.

Steinbeiss, S., Gleixner, G., and Antonietti, M. (2009). "Effect of biochar amendment on soil carbon balance and soil microbial activity." *Soil Biology and Biochemistry*, 41(6), 1301–1310.

Stern, N. (2007). *The Economics of Climate Change: The Stern Review*. Cambridge Univ. Press, Cambridge, 2007.

Stetiner, C., Teixeira, W.G., and Zech, W. (2004). "Slash and char: An alternative to slash and burn practiced in the Amazon Basin." *Amazonian Dark Earths: Explorations in Space and Time. Springer*, 631, 183–193.

Sukartono, Utomo, W.H., Nugroho, W.H., and Kusuma, Z. (2011). "Simple biochar production generated from cattle dung and coconut shell." *Journal of Basic and Applied Scientific Research*, 1(10), 1680–1685.

Topoliantz, S., and Ponge, J.F. (2003). "Burrowing activity of the geophagous earthworm *Pontoscolex corethrurus* (Oligochaeta: Glossoscolecidae) in the presence of charcoal." *Applied Soil Ecology*, 23(3), 267–271.

Tsai, W.T., Liu, S.C., Chen, H.R., Chang, Y.M., and Tsai, Y.L. (2012). "Textural and chemical properties of swine-manure-derived biochar pertinent to its potential use as a soil amendment." *Chemosphere*, 89(2), 198–203.

Vitousek, P.M., Aber, J.D., Howarth, R.W., Likens, G.E., Matson, P.A., Schindler, D.W., and Tilman, D.G. (1997). "Human alteration of the global nitrogen cycle: sources and consequences." *Ecological applications*, *7*(3), 737–750.

Washington State University (WSU) (2011). *Methods for Producing Biochar and Advanced Biofuels in Washington State Part 1: Literature Review of pyrolysis Reactors*. WSU and Dept. of Ecology, State of Washington, Ecology Publication # 11-07-017. April 2011.

Whitfield, R. (2008) "Biochar: A way forward for India and the world." *Action for a Global Climate Community*, Paper II.3.

Winsley, P. (2007). "Biochar and bioenergy production for climate change mitigation." *New Zealand Science Review*, 64(1), 5–10.

Woolf, D., Amonette, J.E., Street-Perrott, F.A., Lehmann, J. and Joseph, S. (2010). "Sustainable biochar to mitigate global climate change." *Nature communications*, 1, 56.

Xu, G., Wei, L.L., Sun, J.N., Shao, H.B., and Chang, S.X. (2013). "What is more important for enhancing nutrient bioavailability with biochar application into a sandy soil: Direct or indirect mechanism?" *Ecological Engineering*, 52, 119–124.

Yarrow, D. (2013). "Biochar and mycorrhizae critical symbiosis for optimum soil structure & fertility." Available at http://dyarrow.blogspot.com/ (accessed Dec. 2013).

Yoder, J., Galinato, S., Granatstein, D., and Garcia-Perez, M. (2011). "Economics tradeoff between biochar and bio-oil production via pyrolysis." *Biomass and Bioenergy*, 35(5), 1851–1862.

Zailani, R., Ghafar, H., and So'aib, M.S. (2013). "Effect of oxygen on biochar yield and properties." *World Academy of Science, Engineering and Technology*, 73.

Zhang, T.C., and Surampalli, R.Y. (2013). "Carbon capture and storage for mitigating climate changes." Chapter 20 in *Climate Change Modeling, Mitigation and Adaptation,* Surampalli, R., Zhang, T.C., Ojha, C.S.P., Gurjar, B.R., Tyagi, R.D., and Kao, C.M. (eds.), ASCE, Reston, Virginia, February 2013.

Zhao, L., Cao, X., Masek, O., and Zimmerman, A. (2013). "Heterogeneity of biochar properties as a function of feedstock sources and production temperatures." *Journal of Hazardous Materials*, 256–257, 1–9.

CHAPTER 16

Enhanced Carbon Sequestration in Ocean: Principles, Strategies, Impacts and Future Perspectives

P. N. Mariyamma, S. Yan, R. D. Tyagi, R. Y. Surampalli, and Tian C. Zhang

The primary components of global carbon cycle are ocean, atmosphere, plants and soil, and they actively exchange carbon (Prentice et al. 2001). One of the possible ways to mitigate the greenhouse gas (GHG) effect is to make changes to the global carbon cycle. Among the various sequestration methods available, oceans have been found to be the most technologically feasible, immediately available and low cost technique for carbon sequestration. The vastness of the ocean and the property of CO_2 to dissolve in sea water forming various ionic species make it capable of storing a large quantity of CO_2 (Table 16.1). On average, the ocean absorbs 2% more carbon than they emit each year; the net ocean uptake is 2 GtC per year which accounts to 30 percent of total anthropogenic emissions (Herzog et al. 2001).

This chapter describes enhanced ocean sequestration of CO_2 and the related impact on the ecosystems and marine food supply. The issue of providing food security is as important as that of managing climate change as 800 million people are undernourished at present and the population is rising rapidly (Jones and Young 2009). The chapter starts with an introduction to the general concepts of CO_2sequestration in ocean, followed by major methods for ocean sequestration of CO_2. The chapter then moves into its focused topic, ocean fertilization for sequestering carbon, followed by evaluation on potential impacts of ocean carbon sequestration, future trend/perspective and summary.

16.1 Background of CO_2 Sequestration in Ocean

Ocean Sequestration Capacity and Strategies. Oceans, which occupy 70 percent of earth's surface with an average depth of 3,800 m, offer the most powerful long-term buffer against the increase in temperature and atmospheric CO_2. The vastness of the ocean provides no practical physical limit to the amount of anthropogenic CO_2 that can be stored in the ocean (see Table 16.1). It is estimated that oceans contain 40,000 GtC compared to a storage capacity of 750 GtC in atmosphere and 2200 GtC in terrestrial biosphere (Sabine et al. 2004). About 80 percent of CO_2 released into the atmosphere will be sequestered in ocean though it may take years to equilibrate with carbonate sediments.

As shown in Table 16.1, the most dominant strategies of CO_2sequestration in ocean include: i) direct release of CO_2 in different forms into the ocean; ii) carbonate mineral dissolution; and iii) ocean nourishment (IPCC 2005; Zhang and Surampalli 2013).

Table 16.1. Ocean sequestration capacity, strategies and general concerns[a]

Capacity (Gt):

- Ocean: 38,000–40000 and Marine sediments & sedimentary rocks: 66–100) x10^6
- Atmosphere: 578 (as of year 1700) and 766 (as of year 1999); Terrestrial plants: 540–610; soil organic matter 1,500–1,600; fossil fuel deposits: 4,000

Strategies and Capacity:

- Direct release of CO_2 in different forms into the ocean (Cost (for the year 2000) = \$65 /tonne CO_2) (Jones and Young 2009)
  - CO_2 dispersal in a very dilute form at depths of 1000–2000 m, a most promising option in the short-term. The cost of capturing the CO_2 + transporting it to a distance of 500 km + storing it = ~ \$70/tonne CO_2
  - Injecting CO_2 directly into the sea at > 3000 m to for a lake of liquid CO_2 on the seabed
  - Formation of a sinking plume (e.g., bicarbonate) to carry most of the CO_2 into deeper water
  - Release of solid CO_2 at depth
- Release of carbonate minerals to accelerate carbonate neutralization. Cost = (for the year 2000) \$18 /tonne CO_2 sequestered under the most favorable conditions (Jones and Young 2009)
- Ocean nourishment, including iron fertilization (US\$5/tonne CO_2), ocean nourishment (US\$15/tonne CO_2) (Jones and Young 2009)

Concerns and needs:

- Concerns: a) unknown impact on ecosystems (e.g., ocean acidification, wildlife, oxygen supply); b) difficult to certify the dissolution, leakage and location of CO_2; c) unknown impact on microbial carbon pump and biological carbon pump.
- Needs: a) making reliable predications of the technical feasibility and storage times; b) understand how to predict and minimize any environmental impact; and c) making reliable cost estimates and assess the net benefit.

[a]Zhang and Surampalli (2013).

It had been observed that 80% of CO_2could be sequestered permanently within a residence time of 1000 years and up to 150-300 GtC can be absorbed with a pH change of 0.2–0.4. The outgoing of remaining 20% of injected carbon would occur within a time period of 300–1000 years (Cole et al. 1993). The lateral transport of re-mineralized CO_2originating from shallow layers of ocean represent a powerful route for carbon sequestration in deep sea which needs to be investigated (Hoppemma 2004).

Major Mechanisms. CO_2 is absorbed by the ocean as per the reactions below (IPCC 2005):

$$CO_2\,(g) + H_2O \leftrightarrow H_2CO_3\,(aq) \leftrightarrow HCO_3^- + H^+ \leftrightarrow CO_3^{2-} + 2H^+ \quad (16.1)$$

In ocean dissolved inorganic carbon can be present in any of four forms: dissolved carbon dioxide (CO_2), carbonic acid (H_2CO_3), bicarbonate ions (HCO_3^-) and carbonate ions (CO_3^{2-}). Addition of CO_2 to seawater, lead to an increase in dissolved CO_2 (Eq. 16.1), which reacts with seawater to form carbonic acid, which rapidly dissociates to form bicarbonate ions and further to form carbonate ions (Eq. 16.1). At a typical seawater pH of 8.1 and salinity of 35 the dominant dissolved inorganic carbon species is HCO_3^- with only 1% in the form of dissolved CO_2. It is the relative proportions of the dissolved inorganic carbon species that control the pH of seawater on short-to-medium timescales.

Once in the ocean, CO_2 is transported and/or transformed in two major mechanisms (Zhang and Surampalli 2013):

a) Physical pump. Cold water holds more CO_2 than warm water. Because cold water is denser than warm water, this cold, CO_2-rich water is pumped down by vertical mixing to lower depths. Depending on the density of the CO_2 in relation to the surrounding water, injected CO_2 can either move upward or downward. Drag forces aid in transferring the momentum from CO_2 droplets to surrounding water creating motion in the direction of droplet motion. Eventually CO_2 dissolves making surrounding water denser and then sinks. As the CO_2-enriched water moves, it gets mixed with less CO_2 enriched surrounding water creating additional dilution and diminishing the density contrast between the CO_2-enriched water and the surrounding water. CO_2 transported by ocean currents undergo mixing and dilution with other water masses along surfaces of constant density, whereas in a stratified fluid, buoyancy forces inhibit vertical mixing (Alendal and Drange 2001).

b) Biological carbon pump (BCP) forcing CO_2 going through the food chain. This is a process whereby CO_2 in the upper ocean is fixed by primary producers and transported to the deep ocean as sinking biogenic particles or as dissolved organic matter. The fate of most of this exported material is remineralization to CO_2, which accumulates in deep waters until it is eventually ventilated again at the sea surface. However, a proportion of the fixed carbon is not mineralized but is instead stored for millennia as recalcitrant dissolved organic matter.

The consequence of pathways a) and b) are that ocean surface waters are supersaturated with respect to $CaCO_3$, allowing the growth of corals and other organisms that produce shells or skeletons of carbonate minerals. In contrast, the deepest ocean waters have lower pH and lower CO_3^{2-} concentrations, and are thus undersaturated with respect to $CaCO_3$. The net effect of pathway b) is that a large amount of carbon is suspended in the water column as dissolved organic carbon (DOC). For example, green, photosynthesizing plankton converts as much as 60 GtC/yr into organic carbon–roughly the same amount fixed by land plants and almost 10 times the amount emitted by human activity. Even though most of DOC is only stored for a short period of time, marine organisms are capable to convert immense amounts of bioavailable organic carbon into difficult-to-digest forms known as *refractory* DOC; this organisms driven conversion has been named the "jelly pump" (Hoffman 2009) and the microbial carbon pump (MCP) (Jiao et al. 2010). Once transformed into "inedible" forms, these DOCs may settle in undersaturated regions of the deep oceans and remain out of circulation for thousands of years, effectively sequestering the carbon by removing it from the ocean food chain (Hoffman 2010). As shown in Table 16.1, there is a tremendous amount of CO_2 storage capacity in marine sediments and sedimentary rocks. However, what is the contribution due to the inedible forms of DOCs or carbonate compounds that are formed by biological pumps and the related mechanisms are not fully understood yet, rending more studies about the real contribution of these mechanisms to CO_2 storage.

16.2 Major Strategies for Ocean Sequestration of CO_2

Direct Release of CO_2. Direct ocean disposal of CO_2 refers to the injection of solid, liquid or gaseous CO_2 into the mid and deep ocean waters (Table 16.1). It was proposed for the first time by Marchetti (1977), aiming at the artificial acceleration of the natural process of CO_2 absorption. Here, we will introduce the methods and conditions required for direct release of CO_2.

Methods. There are several techniques available for the implementation of direct injection like medium-depth sequestration which takes place at depths between 1000–2000 m, high-depth sequestration zone at depths greater than over 3000 m, sequestration on the bottom of the ocean, or sequestration at the undersea earth's layer. The process of switching industrial CO_2 emissions directly to the oceanic column below 800 m was studied by Ametistova et al. (2002). The most preferred injection capture would be the dissolution of liquid CO_2 using fixed pipeline at depths between 1000–1500 m. Liquid CO_2 will be diffused as droplets at this depth. The advantages of this method are that CO_2 will be transferred close to the carbonate dissolution boundary at very slow release rates and with reduced environmental impacts. A pure stream of CO_2 can be either directly injected to the ocean or deposited on the sea floor. It can also be loaded on ships or transported to fixed platforms and dispersed from a towed pipe to ocean forming a CO_2 lake on sea floor (Nakashiki 1997).

CO_2 can be sequestered more effectively and for a longer period of time if CO_2 is stored in liquid form on the sea floor or in hydrate form below 3000 m depths (Shindo et al. 1995). CO_2 hydrate could be designed to produce a hydrate pile or pool on the sea floor. CO_2 released onto the sea floor deeper than 3 km is denser than surrounding sea water and is expected to fill topographic depressions, accumulating as a lake of CO_2 over which a thin hydrate layer would form. This hydrate layer would retard dissolution, but it would not insulate the lake from the overlying water. Although hydrate layer will isolate CO_2 for a longer period from the contact of atmosphere, absence of physical barrier will permit CO_2 to dissolve in the overlying water (Haugan and Alendal 2005). The hydrate would dissolve into the overlying water (or sink to the bottom of the CO_2 lake), but the hydrate layer would be continuously renewed through the formation of new crystals (Mori 1998). Laboratory experiments (Aya et al. 1995) and small deep ocean experiments (Brewer et al. 1999) show that deep-sea storage of CO_2 would lead to CO_2 hydrate formation and subsequent dissolution. The time taken for complete dissolution of CO_2 in a CO_2 lake with an initial depth of 50 m, varies from 30 to 400 years, depending on the local sea and sea floor environment. The dissolution time also depends on mechanism of CO_2 dissolution, properties of CO_2 in solution, turbulence characteristics and dynamics of the ocean bottom layers and the depth and complexity of the ocean lake.

Properties of CO_2 and conditions required for CO_2 injection. The most important factor for carbon sequestration is the depth (as described above) that is sufficient to keep carbon from surface ocean, and it depends on various factors like ocean current, temperature, weather, patch dissolution and grazing activity (De Baar et al. 2005). The behavior and the forms of injected CO_2 will depend on the physical properties of CO_2, location and method of release (Song et al. 2005). CO_2 can be injected either as gas, liquid, solid or solid hydrate. Irrespective of the form in which CO_2 is injected, it gets dissolved in the sea water with time. The dissolution rate of CO_2 is highly variable and depends on certain factors such as form of CO_2, depth and temperature of disposal and local water velocities.

CO_2 can be potentially released as a gas at depths shallower than ~500 m, and gas bubbles being less dense than the surrounding area will rise to the surface at a radial speed of 0.1 cm/hr (Teng et al. 1996). CO_2 can exist as a liquid in ocean at depths roughly below 500 m. At depths < ~2500 m, the density of CO_2 (usually < 1.038 g/cm^3) is less dense than that of sea water (~1.038-1.039 g/cm^3), and hence liquid CO_2 released shallower than 2500 m would tend to rise towards the surface. It was observed that a 0.9-cm diameter droplet would rise at a rate of ~400 m/hr before dissolving completely, and 90 percent of its mass would be lost in the first 200 m (Brewer 2004).

Solid CO_2 (dry ice) has a density between 1.4 and 1.6 g/cm^3. Being denser than sea water, solid CO_2 will dissolve in sea water at a speed of about 0.2 cm/hr and sink (Aya et al. 1999). Proportionately small quantities of solid CO_2 would dissolve completely before reaching the sea floor whereas large masses could potentially reach the sea floor before complete dissolution. CO_2 hydrate refers to a form of CO_2 in which water molecules surrounds each molecule of CO_2. It is normally formed in ocean waters below about 400-m depth. A fully formed crystalline CO_2 hydrate is denser than sea water and dissolves at a speed about 0.2 cm/hr, similar to that of solid CO_2 (Teng et al. 1999; Rehder et al. 2004). In water colder than 9 °C and at greater depths, a CO_2 hydrate film will be formed on the droplet wall, which induces the droplet to diminish to reduce at a speed of 0.5 cm/hr. Fully formed crystalline CO_2 hydrate being denser than sea water will sink. Liquid CO_2 being negatively buoyant at greater depths than 2600 m forms a hydrate skin on water droplet due to ambient temperature and pressure, possess the potential to remove CO_2 from the atmospheric reservoir (Haugane and Drange 1992). Pure CO_2 hydrate is a hard crystalline solid and will not flow through a pipe; however a paste-like composite of hydrate and sea water may be extruded, and this will have a dissolution rate intermediate between those of CO_2 droplets and a pure CO_2 (Tsouris et al. 2004). Although formation of solid CO_2 hydrate is a dynamic process, the nature of hydrate nucleation in these systems are not fully understood (Sloan 1998).

CO_2 diffusers can produce droplets that will dissolve within 100 m of the depth of release. Alternatively CO_2 diffusers can be engineered with nozzles that can produce mm scale droplets that would produce CO_2 plumes that would rise less than 100 m. Hence droplets could be produced that either dissolve completely in the sea water or sink to the sea floor.

Carbonate Mineral Dissolution. This strategy is to change the pH and alkalinity of the surface water such that more carbon can be dissolved in the ocean (for the same partial pressure of atmospheric CO_2). This can be done by the natural dissolution of carbonate mineral in sea floor sediments and lands. Broecker and Takahashi (1977) pointed out that the calcium carbonate stored in the marine sediments will, on a very long time scale, play a role in adjusting the oceanic partial pressure of CO_2. Dissolution of marine $CaCO_3$ sediments will decrease the partial pressure of CO_2 in the ocean. At the pH and temperature of the ocean, the rate of dissolution of calcium carbonate to form bicarbonate is low. If CO_2 is dissolved in sea water to decrease the pH, the reaction rate can be increased. Larger surface area to volume also accelerates the process (Jones and Young 2009). Carbonate neutralization approaches involves the reactions in which limestone reacts with CO_2 and water forming calcium and bicarbonate ions in the solution (Eqs. 16.2–16.5):

$$CO_{2\,(gas)} \rightarrow CO_{2\,(aq)} \tag{16.2}$$

$$CO_{2\,(aq)} + H_2O \rightarrow H_2CO_{3\,(aq)} \tag{16.3}$$

$$H_2CO_{3\,(aq)} + CaCO_{3\,(solid)} \rightarrow Ca^{2+}_{(aq)} + 2HCO_3^-{}_{(aq)} \quad (16.4)$$

$$\text{Net reaction: } CO_{2\,(gas)} + CaCO_{3\,(solid)} + H_2O \rightarrow Ca^{2+}_{(aq)} + 2HCO_3^-{}_{(aq)} \quad (16.5)$$

According to the speciation of dissolved inorganic carbon in sea water, for each mole of calcium carbonate dissolved, there would be 0.8 moles of additional CO_2 sequestered in equilibrium with fixed CO_2 partial pressure. In other words, for a reactor able to neutralize 1 tonne of CO_2 per day, 2.3 tonnes of calcium carbonate would need to be supplied. Adding alkalinity would increase ocean carbon storage in short term and long term time periods. Sea water acidity caused by the CO_2 addition for over thousands of years can be neutralized by this way, which in turn allows oceans to sequester more CO_2 from the atmosphere without significant change in the ocean pH and carbonate ion concentration (Archer et al. 1998). Carbonate minerals have been considered as the primary source of alkalinity for the neutralization of CO_2 acidity. Kheshgi (1995) suggested promoting the reaction of calcining limestone to form readily soluble CaO since ocean surface waters being over saturated with carbonate minerals. It had been observed that enhanced mineral weathering reactions occur in environments with elevated CO_2 like decomposing organic rich soils and in deep ocean.

Ocean Nourishment. The third strategy is ocean nourishment by enhancing BCP. Ocean nourishment refers to the introduction of nutrients to the upper ocean so as to stimulate the marine food chain and to sequester CO_2 from the atmosphere. It belongs to geoengineering techniques, which intentionally alters the environment on a planetary scale to mitigate the global warming. Ocean nourishment offers the prospect of reducing the concentration of atmospheric GHGs and also increasing the primary production in ocean. Primary production refers to the process of producing organic compounds from atmospheric or aquatic CO_2 through the process of photosynthesis. Majority of the primary production is carried out by microscopic organisms called phytoplankton and algae. Horiuchi et al. (1995) studied that inorganic carbon concentration in deep ocean is not in equilibrium with the atmospheric CO_2 partial pressure. The surface ocean is considered to be deficient in nutrients since they are consumed rapidly by phytoplankton. Fertilizing with nutrients would promote propagation of phytoplankton and assimilate organic carbon. Consequently this leads to a decrease in partial pressure of CO_2 on the ocean surface, and resulting in drawing more CO_2 from the atmosphere. Once essential nutrients such as N, P and Fe are used up, algal bloom die, and they sink and thereby sequester carbon. Dead phytoplanktons and marine organisms act as CO_2 vessels, during the natural BCP as they sink towards the bottom of the ocean (Fertiq 2004). Microorganisms that feed on this particulate organic matter produces CO_2, of which some portion dissolves in the ocean and the rest ends up as detritus.

In general, nutrients such as N, P, Fe can be added for ocean fertilization. The use of macronutrients created by the Haber-Bosch process has been termed Ocean Nourishment. The addition of iron to macronutrient rich zones has been termed iron fertilization, while the biological fixing of nitrogen by cyanobacteria using iron, phosphate and other nutrients has yet to be seriously considered (Jones and Young 2009). Currently, it is not clear what the variety of nutrients needed for ocean fertilization is, and whether trace nutrients are enough to allow increased primary production or must they also be added.

Judd et al. (2008) and Jones (2001) concluded that the enhanced BCP can result in ~30% more net GHG emission reduction. A reasonable cost estimate for iron fertilization is US$5 per tonne of organic carbon exported from the ocean surface. Shoji and Jones (2001) suggested that, for energy costs in the year 2000, Ocean Nourishment would require US$15 per tonne of CO_2 avoided (reduced); the results are supported by Matear and Elliott (2001).

16.3 Ocean Nourishment

Ocean fertilization may be a way to create low cost protein in sufficient quantity to supply the needs of the additional two billion people expected to populate the earth before the population stabilizes at values near eight billion. Global wild fish catch has leveled off at 90 Mt/yr. One ocean nourishment plant can provide a minimum protein needed by an extra 38 million people. Many Ocean Nourishment plant could reduce starvation and control climate change. While manipulation of the land ecosystem in support of agriculture for the benefit of humans has long been accepted, it is a new concept to enhance the large scale ocean productivity and so creates some apprehension. In this section, we introduce 3 different methods for ocean nourishment for carbon sequestration and the related approaches.

16.3.1 Major Methods

Ocean Iron Fertilization. This refers to the addition of iron artificially to the water to promote the phytoplankton growth in ocean, which in turn will help in enhancing oceanic CO_2 uptake and reducing CO_2 in the atmosphere (Buesseler et al. 2004; Denman 2008). Iron is one of the major limiting factors for primary production in oceans. Martin (1990) published the *Iron Hypothesis*, suggesting that Fe could be the limiting factor for photosynthesis in the ocean, where the concentration of macronutrients is high, but chlorophyll is low. The availability of light, nutrients and trace elements are the factors that influence carbon cycling and growth of phytoplankton in the ocean. In the oceanic region where there is deficiency of Fe, all the macronutrients cannot be used for photosynthesis, and hence, fertilizing the surface ocean in these regions increases the amount of CO_2 used by phytoplanktons for photosynthesis and will increase the primary production and carbon sequestration in deep ocean. All the carbon taken up by the phytoplankton is not sequestered in the ocean, some portion returns to the atmosphere within short time scales. Being the lowest in hierarchy of the food chain, phytoplanktons are grazed upon by zooplanktons, which in turn are taken by fish and other higher animals. A fraction of carbon goes back to ocean as dissolved inorganic or organic carbon due to the physiological activities of higher animals. Moreover, bacteria also remineralize much of the organic carbon into inorganic carbonate and bicarbonate ions (Denman 2008). It was observed that 50% of exported organic carbon get remineralized during 100 m of sinking, another 2–25% reaches to depths of 1000–1500 m and 1–15% of carbon sinks below 500 m (Powell 2008).

Ocean Urea Fertilization. Ocean urea fertilization refers to the process of fertilizing the ocean with urea, the nitrogen rich substance, so as to boost the growth of CO_2 absorbing phytoplankton, as a means to combat climate change. It had been proved that the efficiency of urea fertilization is dependent on the efficiency of carbon burial and species composition of stimulated bloom (Glibert et al. 2008). The production of higher phytoplankton biomass can be stimulated by nitrogen fertilization. The desired amount of nutrients required to offset the rising concentration of CO_2 in the atmosphere is based on the redfield ratio (PNC ratio: Phosphorous: Nitrogen: Carbon ratio) of the phytoplankton in the ocean. Typical chemical composition of an algal cell is 106 C: 16 N: 1P: 0.0001 Fe (ECOR 2011):

$$106CO_2 + 16NO_3^- + H_2PO_4^- + 17H^+ + 122H_2O \Leftrightarrow (CH_2O)_{106}(NH_3)_{16}(H_3PO_4) + 138O_2 \qquad (16.6)$$

Hence for each unit of Fe added, 1,000,000 units of carbon biomass can be produced. For each unit of nitrogen that is added to a nitrogen limited region, seven units of carbon biomass can be produced. It was observed that urea uptake was positively correlated with the proportion of phytoplankton composed of cyanobacteria in water. Lucas et al (2007) studied rates of phytoplankton production in an iron fertilized region in oceans and also the sensitivity of cell size to iron availability through the size fractionated measurements of nitrate, ammonium and urea uptake. In contrast, nitrate uptake is positively correlated with diatom biomass and negatively correlated with urea uptake (Glibert et al. 2004; Heil et al. 2007). The cellular enzyme urease hydrolyzes urea to ammonium, and the enzyme activity is positively correlated with temperature. Diatoms do not excrete inorganic nitrogen, and hence, the catabolic end products of urea cycle are returned to the anabolic pathways that produce glutamine and glutamate. Urea enrichment leads to enhanced production of cyanobacteria, and picoeukaryotes rather than diatoms (Berg et al. 2001).

Cyanobacteria Nourishment. Cyanobacteria, also known as Cyanophyta, is a phylum of bacteria that obtain their energy through photosynthesis. Many cyanobacteria are able to reduce nitrogen and CO_2 under aerobic conditions (http://en.wikipedia.org/wiki/Ocean_fertilization#cite_note-4). The information on cyanobacteria nourishment has been recently reviewed by ECOR (2011) and Jones (2011).

Macronutrient Fertilization. There is another method to add macronutrients such as nitrogen and phosphorus to the surface ocean to stimulate phytoplankton production. By this way, it is expected to increase photosynthesis and remove CO_2 from the atmosphere. Phytoplankton consumes these growth-limiting nutrients until one or more are exhausted (Matear and Elliott 2004). By increasing the available macronutrients in the low and mid-latitudes, where there is ample sunlight, the ocean primary production can be enhanced (ECOR 2011).

16.3.2 Approaches

The nutrients (nitrogen and/or phosphorus) are distributed into to the surface water of the ocean by pipeline or by ship. The point of release is at the edge of the continental shelf. Carbon dioxide is taken up over the deep ocean. Monitoring of the process is carried out by ocean color satellites. Satellite monitoring of the increase in phytoplankton from the ocean nourishment site ensures that the change in the productivity of the ocean downstream of the nutrient release point is always small and within safe limits. The nutrient concentration and initial uptake efficiency of nutrient are the engineering challenge (Judd et al. 2008; ECOR 2011).

16.3.3 Case Studies–Patch Fertilization of Iron in Ocean

Patch fertilization refers to fertilizing an ocean area with iron and then measuring a few hundred kilometers for a period of one month to several years. Bakker et al. (2005) explored the changes in the biological carbon uptake and surface water fugacity of CO_2 in the iron fertility experiment in Southern Ocean. They studied the effect of CO_2 air-sea transfer on dissolved inorganic carbon for patch iron fertilization. It has been observed that algal carbon uptake reduced surface CO_2 from the 4th day onwards. Surface water CO_2 decreased at the rate of 3–8 micro atoms per day, thus making iron enriched water a potential sink for atmospheric CO_2. The surface water CO_2 and dissolved inorganic carbon decreased at the rate of 32–38 micro atoms after thirteen days. The studies revealed that iron addition made ocean water sinks for atmospheric CO_2, and replenishment of CO_2 by air-sea exchange was less in comparison to algal carbon uptake. IPCC has predicted that accumulated CO_2 emission until the year 2100 would be in the range of 770–2540 Gt. The potential of large scale iron fertilization could be 26–70 GtC for a period of one month. Large scale iron fertilization could share to mitigate CO_2 concentration whereas potential for patch fertilization would have only a relatively small impact. In-situ iron fertility experiments have demonstrated that it will promote the development of algal bloom, build-up of biomass and uptake of inorganic carbon (Boyd et al. 2000; Watson et al. 2000). Four Lagrangian in-situ iron fertilization experiments conducted in Southern Ocean proved that iron addition promoted uptake of inorganic carbon, algal bloom and build-up of biomass (Coale et al. 2004).

16.4 Impact of Ocean Sequestration of Carbon Dioxide

It had been predicted that, with the consumption of fossil fuels at the present rate, by the year 2030, pH of ocean surface water will decrease to 7.8 as atmospheric CO_2 doubles to 700 μatm. However, injection of CO_2 accounting to 1,300 GtC would decrease the pH by 0.3 units, with a corresponding decrease of pH by 0.5 units in the deep ocean. Ocean general circulation models have been used to predict the changes in ocean chemistry as a result of the dispersion of injected CO_2. It had been predicted that injection of 0.37 Gt CO_2/yr for 100 years would produce a pH change of 0.3 units over a volume of sea water equivalent to 0.01 percent (Wickett et al. 2003).

The most obvious consequence of injection of CO_2 is that CO_2 alters the food web by changing the partitioning of energy between metabolic processes (Angel 1992). The magnitude of ocean sequestration and its impact on the environment depends on the duration of exposure, the organism's compensatory mechanism, energy requirement and mode of life (Adams et al. 1997). The low pH is harmful to zooplankton, bacteria, bottom dwelling plants and animals due to limited mobility. The potential effects of liquid CO_2 injection on deep sea foraminiferal assemblage on California margin was studied by Ricketts et al. (2009). Results suggested that foraminiferal diversity decreased due to CO_2 emplacement. Release of liquid CO_2 caused an increase in the dissolution of calcareous taxa in sediments directly below the CO_2 pool. Liquid CO_2 injection also caused significant mortality in benthic foraminifera, since increased CO_2 concentrations caused metabolic changes such as intracellular acidosis and respiration stress.

The effect of CO_2 injection to the physiological changes in the marine organisms and to the ecosystems should be taken into consideration while studying the impacts of ocean carbon sequestration. The adverse effect on the diverse fauna that resides in the deep ocean and in sediments is also one of the most alarming consequences of ocean CO_2 sequestration, which can lead to changes in ecosystem composition and functioning. The dissolution of CO_2 can lead to dissolution of calcium carbonate present in the sediments or in the shells of the microorganisms. Changes in the productivity pattern of algal/heterotrophic bacterial species, biological calcification or decalcification and metabolic impacts on zooplankton species are the observed consequences of the lowered sea water pH. Changes in the pH of the marine environment will affect the carbonate system, nitrification, speciation and uptake of nutrients (Huesemann et al. 2002). Although it had been stated that CO_2 injection has ecosystem consequences, no controlled ecosystem experiments have been performed in deep ocean, and hence, no environmental thresholds have been identified.

Ocean fertilization has drawbacks in the sense that it can affect the ocean ecosystem in the long run and change the plankton structure. The potential for ocean nourishment experiments for mitigation of GHGs is controversial since the magnitude and direction of carbon stored remains uncertain, and the verification of carbon stored is impossible (Gnanadesikan et al. 2003). The most crucial limitation is the production of methane gas triggered by the sinking of organic matter as a result of large scale iron fertilization. Moreover, it can induce the production of GHGs such as nitrous oxide, methane, dimethylsulphide, alkyl nitrates and halocarbons (Turner et al. 2004; Jin and Gruber 2003). Production of other GHGs and their outgassing could partially offset the carbon drawdown from the atmosphere into the ocean. Changes in the marine ecology and biogeochemical changes are induced by large-scale iron fertilization experiments (Chisholun et al. 2001). Experimental studies have revealed that significant microbial community structural modifications occur in response to 7% increase in CO_2 concentration (Sugimori et al. 2001). A firm response of heterotrophic bacteria to phytoplankton bloom, in terms of biomass production and respiration can be induced by natural iron fertilization (Obernosterer et al. 2008). Ocean fertilization also alters the partitioning of energy between metabolic processes, induces ocean acidification and alters the physical properties of the ocean.

Limitations of ocean sequestration include distraction of energy usage in an efficient way, alternate energy generation from renewable sources and tampering the ecological processes. Urea fertilization is likely to cause eutrophication impacts, which include development of hypoxic or anoxic zones and alteration of species leading to harmful algal blooms (Anderson 2004). Sinking of algae to deep water ocean cause hypoxia upon their decomposition and hence is responsible for fish kills. Nitrogen loading can bring forth a shift in the marine community of coral reef directed towards algal overgrowth of corals and ecosystem disruption (McCook et al. 2001). Ocean iron fertilization has some negative effects such as the development of toxic algal blooms, unforeseeable changes in food web and ecosystems, anoxia due to remineralization sinking organic matter, increased production of nitrous oxide which can lead to the death of marine life (Denman 2008). More research has to be undertaken before using iron fertilization on a large scale due to substantial uncertainties and effectiveness associated to it (Buesseler et al. 2004).

16.5 Future Perspectives

The opportunity to produce food (protein) while managing climate change justifies future research on ocean fertilization for carbon sequestration. Determining the feasibility, efficiency, and environmental consequences of this process involves significant scientific, technological, economic, and legal investigation. The oceans have a far greater capacity to absorb carbon than all other sinks combined. However, the effects of ocean storage are even more uncertain, raising additional environmental concerns. Although scientists have explored possible physical, chemical, and biological methods of carbon storage in oceans for a few decades, there is an urgent need on research with large-scale field experiments, and the computational models with embedded biogeochemistry into the effects of ocean fertilization. In addition, it is imperative to develop platforms and instruments to gather sufficient data on fertilization, direct CO_2 injection, and other methods for ocean carbon sequestration. Furthermore, the legal framework of intentional carbon storage in the ocean needs to be established, and public acceptance needs to be promoted of the deliberate storage of CO_2 in the ocean as part of a climate change mitigation strategy.

16.6 Summary

In order to stabilize the increasing GHG emissions, it is always advisable to adopt a combination of mitigation strategies. Ocean carbon sequestration has been suggested as a scientifically and ecologically sound method for reducing atmospheric CO_2. On a global scale, ocean sequestration of CO_2 will help to lower the atmospheric CO_2 content, their rate of increase and in turn will reduce the detrimental effects of climate change and chance of catastrophic events. The physical capacity for ocean storage of CO_2 is large compared to the fossil fuel resource, and the utilization of this capacity to its full range depends upon cost, equilibrium pCO_2 and environmental consequences. Ocean nourishment can address both the increasing demand for food (protein) and the reduction of atmospheric CO_2 levels. One of the main knowledge gaps for ocean sequestration is the environmental impact that might pose to the marine biota due to the injection of CO_2. Almost all the data available and predictions made are based on the model. Alterations in the biogeochemical cycles will have large sequences, which may be secondary, yet difficult to predict. Since oceans play a pivotal role in maintaining the ecosystem balance, any change in the oceanic environment should be dealt with seriously. Hence, detailed research is needed to develop techniques to monitor the CO_2 plumes, their biological and geochemical behavior in terms of long duration and on a large scale.

16. 7 Acknowledgements

Sincere thanks are due to the Natural Sciences and Engineering Research Council of Canada (Grant A 4984, and Canada Research Chair) for their financial support. The views and opinions expressed in this chapter are those of the authors.

16.8 Abbreviations

atm atmosphere

BCP	biological carbon pump
C	carbon
cm	centimeter
GHGs	Greenhouse Gases
GT	Gigatonne (10^9 tonnes)
IEA	International Energy Agency
IPCC	Intergovernmental Planet on Climate Change
km	kilometer
md	millidarcies
Mha	million hectare
mm	millimeter
MMT	million metric tonnes (10^9 Kilograms)
p CO_2	partial pressure of CO_2
Pg	Picogram (10^{-12} grams)
Ppmv	Parts per million by volume
Tg	Teragram (10^{12} grams)
yr	year

16.9 References

Alendal, G., and Drange, H. (2001). "Two-phase near-field modeling of purposefully released CO_2 in the ocean." *J. Geophysical Research*, 106(C1), 1085–1096.

Anderson, D.M. (2004). "The growing problem of harmful algae." *Oceanus,* 43(1), 1–5.

Angel, M.V. (1992). "Managing biodiversity in the oceans." p. 23–62. In *Diversity of Oceanic Life: An Evaluative Review* (ed. Peterson, M.N.A.). The Center for Strategic and International Studies, Washington D.C.

Ametistova, J., Twidell, J., and Briden, J. (2002). "The sequestration switch: removing industrial CO_2 by direct ocean absorption." *Science of the Total Environment*, 289, 213–223.

Archer, D.E., Kheshgi, H., and Maier-Reimer, E. (1998). "Dynamics of fossil fuel neutralization by Marine $CaCO_3$." *Global Biogeochemical Cycles*, 12(2), 259–276.

Archer, D., Kheshgi, H., and Maier-Reimer, E. (1997). "Multiple timescales for neutralization of fossil fuel CO_2." *Geophysical Research Letters*, 24, 405–408.

Aya, I., Yamane, K., and Shiozaki, K. (1999). "Proposal of self sinking CO_2 sending system." *COSMOS*. Elsevier Science Ltd, Pergamon, 269–274.

Aya, I., Yamane, K., and Yamada, N. (1995). "Simulation experiment of CO_2 storage in the basin of deep-ocean." *Energy Conversion and Management,* 36(6–9), 485–488.

Bakker, D.C.E., Bozec, Y., Nightingale, P.D., Goldson, L.E., Messias, M.J., de Baar, H.J.W., Liddicoat, M.I., Skjelvan, I., Strass, V., and Watson, A.J. (2005). "Iron and mixing affect biological carbon uptake in SOIREE and EisenEx, two Southern Ocean iron fertilisation experiments." *Deep-Sea Research*, Part I, 52, 1001–1019.

Bariteau, L. (2008). "Shake and Bake!" https://sogasex.wordpress.com/page/2/ (accessed May 2013).

Berg, G.M., Glibert, P.M., Jorgensen, N.O.G., Balode, M., and Purina, I. (2001). "Variability in inorganic and organic nitrogen uptake associated with riverine nutrient input in the Gulf of Riga, Baltic Sea." *Estuaries,* 24, 176–186.

Boyd, P.W., Watson , J.A., Law, C.A., Abraham, E.R., Trull, T., Murdoch, R., Bakker, D.C.E., Bowie, A.R., Buesseler, K.O., Chang, H, O., Charette, M., Croot, P., Downing, K., Frew, R., Gall, M., Hadfield, M., Hall, J., Harvey, M., Jameson, G., LaRoche, J., Liddicoat, M., Ling, R., Maldonado, M.T., McKay, R.M., Nodder, S., Pickmere, S., Pridmore, R., Rintoul, S., Safi, K., Sutton, P., Strzepek, R., Tanneberger, K., Turner, S., Waite, A., and Zeldis, J. (2000). "A mesoscale phytoplankton bloom in the polar Southern Ocean stimulated by iron fertilization." *Nature*, 407, 695–702.

Brewer, P.G. (2004). "Dissolution rates of pure methane hydrate and carbon dioxide hydrate in under-saturated sea water at 1000 m depth." *Geochimica et Cosmochimica Acta,* 68(2), 285–292.

Brewer, P.G., Friederich, G., Peltzer, E.T., and Orr, F.M. (1999). "Direct experiments on the ocean disposal of fossil fuel CO_2." *Science*, 284(5416), 943–945.

Broecker, W.S., and Takahashi, T. (1977). "Neutralization of fossil fuel CO_2 by marine calcium carbonate." In Anderson, R., and Malahoff, A. (eds.) *The Fate of Fossil Fuel CO_2 in the Ocean*, New York: Plenum Press, pp.213–241.

Buesseler, K.O., Andrews, J.E., Pike, S.M., and Charette, M.A. (2004). "The effects of iron fertilization on carbon sequestration in the southern ocean." *Science*, 304 (5669), 414–417.

Chisholm, S.W. (2000). "Stirring times in the Southern Ocean." *Nature*, 407, 685–687

Chisholm, S.W., Falkowski, P.G., and Cullen, J.J. (2001). "Discrediting ocean fertilization." *Science*, 294, 309–310.

Coale, K.H., Johnson, K.S., Chavez, F.P., Buesseler, K.O., Barber, R.T., Brzezinski, M.A., Cochlan, W.P., Millero, F.J., Falkowski, P.G., Bauer, J.E., Wanninkhof, R.H., Kudela, R.M., Altabet, M.A., Hales, B.E., Takahashi, T., Landry, M.R., Bidigare, R.R., Wang, X., Chase, Z., Strutton, P.G., Friederich, G.E., Gorbunov, M.Y., Lance, V.P., Hilting, A.K., Hiscock, M.R., Demarest, M., Hiscock, W.T., Sullivan, K.F., Tanner, S.J., Gordon, R.M., Hunter, C.N., Elrod, V.A., Fitzwater, S.E., Jones, J.L., Tozzi, S., Koblizek, M., Roberts, A.E., Herndon, J., Brewster, J., Ladizinsky, N., Smith, G., Cooper, D., Timothy, D., Brown, S.L., Selph, K.E., Sheridan, C.C., Twining, B.S., and Johnson, Z.I. (2004). "Southern Ocean iron enrichment experiment: carbon cycling in high- and low-Si waters." *Science,* 304, 408–414.

Cole, K.H., Stegen, G.R., and Spencer, D. (1993). "The capacity of the deep oceans to absorb carbon-dioxide." *Proceedings of the International Energy Agency Carbon*

Dioxide Disposal Removal Symposium, Energy Conversion and Management, special issue, 34(9–11), 991–998.

de Baar, H.J.W., Boyd, P.W, Coale, K.H., Landry, M.R., Tsuda, A., Assmy, P., Bakker, D.C.E., Bozec, Y., Barber, R.T., Brzezinski, M.A., Buesseler, K.O., Boyé, M., Croot, P.L., Gervais, F., Gorbunov, M.Y., Harrison, P.J., Hiscock, W.T., Laan, P., Lancelot, C., Law, C.S., M.Levasseur, Marchetti, A., Millero, F.J., Nishioka, J., Nojiri Y., Van Oijen, T., Riebesell, U., Rijkenberg, M.J.A., Saito, H., Takeda, S., Timmermans, K.R., Veldhuis, M.J.W., Waite, A.M., and Wong, C.-S. (2005). "Synthesis of eight in-situ iron fertilization in high nutrient low chlorophyll waters confirms the control by wind mixed layer depth of phytoplankton blooms." *Journal of Geophysical Research,* 110, doi: 10.1029/2004JC002601.

Denman, K.L. (2008). "Climate change, ocean processes and ocean iron fertilization." *Marine Ecology Progress Series,* 364, 219–225.

ECOR (Engineering Committee for Oceanic Resources) (2011). *Enhanced Carbon Storage in the Ocean.* ECOR Report, Version 4.6.

Fertiq, B. (2004). "*Ocean gardening using iron fertilization.*" Available http://www.csa.com/discoveryguides/oceangard/overview.php (accessed Dec. 2013).

Finney, B., and Jacobs, M. (2011). "Phase diagram of CO_2." Available at https://enwikipedia.org/wiki/File:Carbon-Dixide... (accessed May 2013).

Glibert ,P.M., Heil, C.A., Hollander, D., Revilla, M., Hoare, A., Alexander, J., and Murasko, S. (2004). "Evidence for dissolved organic nitrogen and phosphorus uptake during a cyanobacterial bloom in Florida Bay." *Marine Ecology Progress Series,* 280, 73–83.

Glibert P.M., Azanza, R., Burford, M., Furuya, K., Abal, E., and Al-Azri, A., (2008). "Ocean urea fertilization for carbon credits poses high ecological risks." *Marine Pollution Bulletin*, doi·:10.1016/j.marpolbul.2008.03.010).

Gnanadesikan, A., Sarmiento, J.L., and Slater, R.D. (2003). "Effects of patchy ocean fertilization on atmospheric carbon dioxide and biological production." *Global Biogeochemical Cycles,* 17, 1050, doi: 10.1029/2002GB001940.

Haugan, P.M., and Drange, H. (1992). "Sequestration of CO_2 in the deep ocean by shallow injection." *Nature,* 357, 318–320.

Haugan, P.M., and Alendal, G. (2005). "Turbulent diffusion and transport from a CO_2 lake in the deep ocean." *Journal of Geophysical Research-Oceans*, 110, C09S14, doi:10.1029/2004JC002583.

Heil, C.A., Revilla, M., Glibert, P.M., and Murasko, S. (2007). "Nutrient quality drives phytoplankton community composition on the West Florida Shelf." *Limnology and Oceanography*, 52, 1067–1078.

Herzog, H., Caldeira, K., and Adams, E. (2001). "Carbon sequestration via direct injection." Steele, J.H., Thorpe, S.A., and Turekian, K.K. (ed.), *In: Encyclopedia of Ocean Sciences,* Vol. 1, London, UK: Academic Press, 408–414.

Herzog, H., Drake, E., and Adams, E. (1997). *CO_2 capture, reuse, and storage technologies for mitigating global climate change. A White Paper Final Report*, Order No. DE-AF22- 96PC01257, U.S. Department of Energy, Washington, D.C.

Hoffman, D.L. (2009). "New Jelly pump rewrites carbon cycle." Available at <http://theresilientearth.com/?q=content/> (accessed Feb. 2012).

Hoffman, D.L. (2010). "Ocean CO_2 storage revised." Available at <http://www.theresilientearth.com/?q=content/> (accessed Feb. 2012).

Hoppemma, M. (2004). "Weddell Sea is a globally significant contributor to deep-sea sequestration of natural carbon dioxide." *Deep Sea Research Part I: Oceanographic Research Papers,* 51(9), 1169–1177.

Horiuchi, K., Kojima, T., and Inaba, A. (1995). "Evaluation of fertilization of nutrients to the ocean as a measure for CO_2 problem." *Energy Conversion and Management*, 36, 915–918.

Huesemann, M.H., Skillman, A.D., and Crecelius, E.A. (2002). "The inhibition of marine nitrification by ocean disposal of carbon dioxide." *Marine Pollution Bulletin*, 44(2), 142–148.

IPCC (Intergovernmental Panel on Climate Change) (2005). *IPCC Special Report on Carbon Dioxide Capture and Storage.* IPCC Working Group III. Cambridge Univ. Press, Dec., 19, 2005, 431 pp.

Jiao, N., Herndl, G.J., Hansell, D.A., Benner, R., Kattner, G., Wilhelm, S.W., Kirchman, D.L., Weinbauer, M.G., Luo, T., Chen, F., and Azam, F. (2010). "Microbial production of recalcitrant dissolved organic matter: long-term carbon storage in the global ocean." *Nature Reviews Microbiology*, 8, 593–599.

Jin, X., and Gruber, N. (2003). "Offsetting the radiative benefit of ocean iron fertilization by enhancing ocean N_2O emissions." *Geophysical Research Letters*, 30 (24), 2249, doi:10.1029/2003GL018458.

Jones, I.S.F. (2001). "The global impact of ocean nourishment." *Proc. of the 1st Nat. Con. Carbon Sequestration*, Washington. <www.netl.doe.gov/publications/proceedings/01/carbon_seq/6b2.pdf> (accessed 2009).

Jones, I.S.F. (2011). *Engineering Strategies for Greenhouse Gas Mitigation*, Cambridge University Press, May 12, 184 pages. ISBN: 9780521731591. Cambridge, UK.

Jones, I.S.F., and Caldeira, K. (2003) "Long-term ocean carbon sequestration with macronutrient addition." *Jc03y.doc, Second Annual Conference on Carbon Sequestration*, Washington, May.

Jones, I.S.F., and Young, H.E. (2009). "The potential of the ocean for the management of global warming." *Int. J. Global Warming,* 1(1/2/3), 43–56.

Judd, B.J., Harrison, D.P., and Jones, I.S.F. (2008). "Engineering ocean nourishment." *Proceedings of the World Congress on Engineering,* July 2–4, London, U.K. http://hdl.handle.net/2123/2664.

Kheshgi, H.S. (1995)."Sequestering atmospheric carbon dioxide by increasing ocean alkalinity." *Energy*, 20(9), 915–922.

Lucas, M.I., Seeyave, S., Sanders, S., Moore, M.C., Williamson, R., and Stinchcombe, M. (2007). "Nitrogen uptake response to a naturally Fe fertilised phytoplankton bloom during the 2004/2005 CROZEX study." *Deep-Sea Res. Pt. II*, 54, 2138–2173.

McCook, L.J., Jompa, J., and Diaz-Pulido, G. (2001). "Competition between corals and algae on coral reefs: a review of evidence and mechanisms." *Coral Reefs*, 19, 400–417.

Marchetti, C. (1977). "On geoengineering and the CO_2 problem." *Climate Change*, 1977, 1, 59–68.

Martin, J.H. (1990). "Glacial-interglacial CO_2 change: The iron hypothesis." *Paleoceanography*, 5, 1–13.

Matear, R.J., and Elliott, B. (2001). "Enhancement of oceanic uptake of anthropogenic CO_2 by macro nutrient fertilization." In Durie, R.A., McMullan, P, Paulson, C.A.J., Smith, A.Y., and Williams, D.J. (Eds.) *Greenhouse Gas Control Technologies*, CSIRO, Syd., 451–456, ISBN: 0643066721.

Matear, R.J., and Elliott, B. (2004). "Enhancement of oceanic uptake of anthropogenic CO_2 by macronutrient fertilization." *Journal of Geophysical Research*, Oceans (1978–2012).109(C4). DOI: 10.1029/2000JC000321.

Mori, Y.H. (1998). "Clathrate hydrate formation at the interface between liquid CO_2 and water phases-a review of rival models characterizing "hydrate films"." *Energy Conversion Management,* 39, 1537–1557.

Nakashiki, N. (1997). "Lake-type storage concepts for CO_2 disposal option." *Waste Management*, 17 (5–6), 361–367.

Obernosterer, I., Christaki, U., Lefevre, D., Catala, P., Van Wambeke, F., Lebaron, P., (2008). "Rapid bacterial remineralization of organic carbon produced during a phytoplankton bloom induced by natural iron fertilization in the Southern Ocean." *Deep-Sea Research II*, 55, 777–789. [doi:10.1016/j.dsr2.2007.12.034].

Powell, H. (2008). "Will ocean iron fertilization work?" *Oceanus Magazine,* 46 (1), 10–13.

Prentice, I.C., Farquhar, G.D., Fasham, M.J.R., Goulden, M.L., Heimann, M., Jaramillo, V.J., Kheshgi, H., S., Le Quéré, C., Scholes, R.J., and Wallace, D.W.R. (2001). "The carbon cycle and atmospheric CO_2." J.T. Houghton (ed.), *Climate Change 2001: The Scientific Basis: Contribution of WGI to the Third Assessment Report of the IPCC.* Cambridge University Press, New York, 183–237.

Rau, G.H., and Caldeira, K. (1999). "Enhanced carbonate dissolution: A means of sequestering waste CO_2 as ocean bicarbonate." *Energy Conversion & Management*, 40, 1803–1813.

Rehder, G., Kirby, S.H., Durham, W.B., Stern, L.A., Peltzer, E.T., Pinkston, J., and Brewer, P.G. (2004). "Dissolution rates of pure methane hydrate and carbon dioxide hydrate in under-saturated sea water at 1000 m depth." *Geochimica et Cosmochimica Acta*, 68(2), 285–292.

Ricketts, E.R., Kennett, J.P., Hill, T.M., and Barry, J.P. (2009). "Effects of carbon dioxide sequestration on California margin deep-sea foraminiferal assemblages." *Marine Micropaleontology*, 72(3–4), 165–175.

Sabine, C.L., Feely, R.A., Gruber, N., Key, R.M., Lee, K., Bullister, J.L., Wanninkhof, R., Wong, C.S., Wallace, D.W.R., Tilbrook, B., Millero, F.J., Peng, T.H., Kozyr, A., Ono, T., and Rios, A.F. (2004). "The oceanic sink for anthropogenic CO_2." *Science*, 305, 367–371.

Shoji, K., and Jones, I.S.F. (2001). "The costing of carbon credits from ocean nourishment plants." *The Science of the Total Environment*, 277, 27–31.

Shindo, Y., Fujioka, Y., and Komiyama, H. (1995). "Dissolution and dispersion of CO_2 from a liquid CO_2 pool in the deep ocean." *International Journal of Chemical Kinetics*, 27(11), 1089–1095.

Sloan, E.D. (1998). *Clathrate Hydrates of Natural Gases.* Marcel Dekker Inc., New York, 705.

Song, Y., Chen, B., Nishio, M., and Akai, M. (2005). "The study on density change of carbon dioxide seawater solution at high pressure and low temperature." *Energy,* 30(11–12), 2298–2307.

Sugimori, M., Takeuchi, K., Ozaki, M., Fujioka, Y., and Ishizaka, J. (2001) "Responses of marine biological communities to different concentrations of CO_2 in a mesocosm experiment." In: Williams, D.J., Duire, R.A., McMullan, P., Paulson, C.A.J., and Smith, A.Y. [eds.] *Proceedings of the Fifth International Conference on Greenhouse Gas Control Technologies.* Publ. CSIRO Publishing, Collingwood, Australia, pp 511–522.

Teng, F., and Tondeur, D. (2005)."Efficiency of carbon storage with leakage: Physical and economical approaches." *Energy,* 32, 540–548.

Teng, H., Yamasaki, A., and Shindo, Y. (1996). "The fate of liquid CO_2 disposed in the ocean." *International Energy,* 21(9), 765–774.

Teng, H., Yamasaki, A., and Shindo, Y. (1999). "The fate of CO_2 hydrate released in the ocean." *International Journal of Energy Research,* 23(4), 295–302.

Tsouris, C., Brewer, P.G., Peltzer, E., Walz, P., Riestenberg, D., Liang, L., West, O.R. (2004). "Hydrate composite particles for ocean carbon sequestration: Field verification." *Environment Science and Technology*, 38, 2470–2475.

Turner, S.M., Harvey, M.J., Law, C.S., Nightingale, P.D., and Liss, P.S. (2004). "Iron induced changes in oceanic sulfur biogeochemistry." *Geophysical Research Letters*, 31, L14307, doi: 10.1029/2004GL020296.

Watson, A.J., Bakker, D.C.E., Ridgwell, A.J., Boyd, P.W., and Law, C.S. (2000). "Effect of iron supply on Southern Ocean CO_2 uptake and implications for glacial atmospheric CO_2." *Nature*, 407, 730–733.

Wickett, M.E., Caldeira, K., and Duffy, P.B. (2003). "Effect of horizontal grid resolution on simulations of oceanic CFC-11 uptake and direct injection of anthropogenic CO_2." *Journal of Geophysical Research,* 108. Doi: 10.1029/ 2001JC00130.

Zhang, T.C., and Surampalli, R.Y. (2013). "Carbon capture and storage for mitigating climate changes." Chapter 20 in *Climate Change Modeling, Mitigation and Adaptation,* Surampalli, R., Zhang, T.C., Ojha, C.S.P., Gurjar, B.R., Tyagi, R.D., and Kao, C.M. (eds.), ASCE, Reston, Virginia, February 2013.

CHAPTER 17

Modeling and Uncertainty Analysis of Transport and Geological Sequestration of CO_2

Munish K. Chandel, B. R. Gurjar, C. S. P. Ojha, and Rao Y. Surampalli

17.1 Introduction

Carbon dioxide emitted from large point sources (e.g., large fossil fuel or biomass energy facilities, major CO_2-emitting industries, natural gas production, synthetic fuel plants and fossil fuel-based hydrogen production plants) could be captured, compressed and transported for storage in i) geological formations (such as oil and gas fields, unmineable coal beds and deep saline formations), ii) the ocean (e.g., direct release into the ocean water column or onto the deep seafloor), and iii) mineral carbonates, or for use in industrial processes (e.g., industrial fixation of CO_2 into inorganic carbonates) (IPCC 2005). Among different techniques of CO_2 storage, geologic CO_2 sequestration (GCS), including the stable carbonate mineral formation, is one of the sequestration mechanisms that ensure long-term storage of CO_2 (Lackner 2003; Matter and Kelemen 2009).

Underground GCS is the most widely acknowledged mode for large CO_2 sequestration. As CO_2 is naturally found worldwide trapped in natural geologic formation, on the same lines engineered setups for injecting CO_2 into deep geological formations with some modification and supervision can be accomplished. It is estimated that 99% or more of the injected CO_2 will be retained for 1000 years. Due to the widespread formation of carbonate minerals in marine and sedimentary environments, the chemical process of precipitation of these minerals warrants intensive research (Riding 2000). The extreme physico-chemical conditions present in the subsurface reservoirs make prediction of the mineralization processes difficult. A combination of factors related thermodynamic and kinetic properties contributes to this uncertainty (Fernandez et al. 2013).

Among the different geologic CO_2 sequestration systems, saline formations have much larger potential. However, the upper limit estimates are uncertain due to lack of information and an agreed methodology. On the other hand, the capacity of oil and gas reservoirs is better known, but technical storage capacity in coal beds is much smaller and less well known. Model computations for the capacity to store CO_2 in the oceans indicate that the estimated capacity could be of the order of thousands of $GtCO_2$, depending on the assumed stabilization level in the atmosphere and on certain environmental constraints (e.g., ocean pH change). The extent to which mineral carbonation may be used is hard to determine at present. This is because it depends on the unknown amount of silicate reserves that can be technically exploited and on some important environmental issues (such as the volume of product disposal). The CO_2 emissions and capture ranges not only reflect the inherent uncertainties of scenario and modelling analyses but also the technical limitations of applying carbon capture and storage (CCS) (IPCC 2005). Thus, it is very much relevant to address issues related to modelling and uncertainty analysis of CCS technologies and their performance.

The objective of this chapter is to give an overview of the models used for the analysis of CO_2 transport and geological sequestration. CO_2 transport models are described, and the level of understanding of CO_2 transport through onshore and offshore pipes, and ships is explained. The formulation of different models used to estimate CO_2 storage capacity for different reservoirs and the assumptions used to formulate these models is described. The uncertainty involved in the estimates of CO_2 storage and risks associated with the leakage of geologically sequestrated CO2 is elucidated. This chapter also explains briefly how sequestrated CO_2 might leak into the freshwater aquifers and how the water chemistry might be affected.

17.2 Modeling CO_2 Transport to Sequestration Site

Carbon dioxide transport is an essential component of CCS. As suitable sequestration sites may not be at the same place where CO_2 is captured, hence CO_2 needs to be transported in between capture and sequestration sites. CO_2can be transported through pipes and tanks onshore and through ships and pipes offshore. Onshore CO_2 transport is well understood as the 1st large onshore CO_2 transport pipe in the U.S. was the Canyon Reef Carriers, which was built in 1970 to transport 4.4 million metric tonne of CO_2 per year (IPCC 2005). As in 2012, 2500 km of CO_2 pipeline, the largest in world, is in the U.S. alone. Shute Creek is the biggest CO_2 pipe with a diameter of 0.76 m and a maximum flow capacity of 19.3 million metric tonnes/year (Gale and Davidson 2004).

The experience of the offshore CO_2 transport is limited. The only existing offshore pipe is the subsea Snohvit pipeline (8", 153 km) with a capacity of 0.7 million tonne/yr (Serpa et al. 2011). However, several offshore pipes are used to transport oil and natural gas. The deepest pipe is of ~650m (26 inches) and laid at the seawater depth of 2200 m (IPCC 2005).

Pipe Transport Models. Although, theoretically CO_2 can be transported in any of its state (solid, liquid or gas), practically CO_2 transport through long distance pipes is carried out in dense phase (CO_2 at pressure higher than critical pressure of CO_2). The dense phase transport is economical as high density at the dense phase requires smaller diameter pipes as compared to the gaseous phase.

CO_2 can be transported at high-pressure supercritical phase which has low viscosity and high density. For the flow of CO_2 in the supercritical phase, the critical pressure (7.36 MPa) and critical temperature (31°C) should be maintained. It is reasonable to maintain the high pressure in a pipe flow. The CO_2 temperature, however, would depend to a certain extent upon the initial temperature of the flue gases and the ambient temperature. It may not be economic in all cases to maintain the temperature above 31°C. It may be more practical and economical to allow the CO_2 temperature to be equal to the ambient temperature.

As per the Department of Transportation regulation for "Transportation of hazardous liquids by pipeline (part 195.248)" (CFR 2008), the pipes must be buried under the soil, except for some exempted subparts. Since soil temperatures remain more uniform than surface temperatures, buried pipelines would remain at a relatively stable temperature. If the pressure above the supercritical pressure is maintained for all possible soil temperatures, CO_2 will remain in the dense phase–either as liquid or supercritical fluid. The most important concern is to avoid two-phase flow (gas and liquid) in the pipe, which could create problems for compressors and other equipment and possibly lead to pipe failure (IPCC 2005; WRI 2008). Because even a small change in pressure near the critical pressure could lead to high variations in density, the pressure should be maintained well above the critical pressure.

Design of the Pipe System. As CO_2 pipes are being existed for more than three decades now, the design methodology is well understood. The necessary conditions for the design of a pipe line system is to make sure that the dense phase of CO_2 is maintained throughout the pipe flow. The phase change and two-phase systems make it difficult to operate compressors and other transport equipment and reduce efficiency of the system (IPCC 2005; Serpa et al. 2011). The design of the pipeline system consists of the following components:

- Design of pipe[s] size based on the mass of CO_2 to be transported;
- Design of the booster pumps/compressors required to maintain CO_2 in the dense phase; and
- Other mechanical components such as operating and control valves and overall monitoring system

Pipes are designed based upon the hydraulic equation of flow in which pipe diameter is estimated using head loss equation due to frictional resistance (IEA GHG

2002; Heddle et al. 2003). Some models design the diameter of the pipe on the basis of a velocity flow equation which assumes uniform gas velocity in the pipe in the range of 1-3 m/s (IEA GHG 2005; Chandel et al. 2010).The internal diameter of the pipe is calculated from the standard velocity-flow equation of the fluid as:

$$d = \sqrt{\frac{4Q}{\pi U}} \quad (17.1)$$

where d = the piper internal diameter; U = the velocity of the fluid, a variable depending upon flow characteristics and fluid density; and Q = the flow in the pipe. Q is calculated from the mass flow-density relation. Required pipe thickness depends upon the maximum pipe pressure and pipe material characteristics. Pipe thickness can be determined by re-arranging Barlow's formula:

$$t = \frac{Pd}{2(S-P)} \quad (17.2)$$

where t = the pipe thickness; S = the specified minimum yield strength of the pipe, which is a function of the pipe material grade; and P = the designed maximum pressure. The pipe thickness would depend upon the pipe material grade and the design pressure.

The head loss occurs during the CO_2 transport mainly due to frictional losses. To maintain the pressure above critical pressure, booster pumps are required to be designed. The spacing of the pumps along the pipes would depend upon the head loss, which can be calculated by using Darcy-Weisbach equation:

$$h_f = f.\frac{L}{d}\frac{U^2}{2g} \quad (17.3)$$

where h_f = head loss due to friction; f = friction factor; L = length of the pipe; d = inner diameter of pipe; and U = velocity in a pipe. Based upon head loss due to friction and neglecting other minor losses, the pressure drop across the length of the pipe is:

$$\Delta P = \rho g f.\frac{L}{d}\frac{U^2}{2g} \quad (17.4)$$

where ΔP = pressure drop across the length of the pipe. Once we decide what the allowable pressure drop in the pipe is, the spacing of the booster pumps could be calculated from the above equation. The friction factor (f) depends upon the type of flow, pipe diameter, velocity of flow, fluid characteristics (i.e., density and viscosity), and pipe material. The flow in the pipe is assumed to be a turbulent flow.

Offshore Pipes. Offshore pipes can also be designed on the same line as the onshore pipe systems. Additional factor required to be handled is the water pressure on the pipes. Depending upon the wellhead pressure requirement and the distance of the sequestration site from the onshore, booster pumps to enhance/maintain pressure may or may not be required. If the booster pumps are not required, CO_2 can be distributed by a subsea facility, and the platforms on the sequestration site may be small and simple-only required to operate and maintain the injection and distribution facility (Pershad and Slater 2007). However, if booster pumps are required at the sequestration site, then additional platform may be required which should be able to sustain the booster station. Subsea booster stations may be required if the water depth is more than 3000m.

Ships. As of now due to limited commercial demand the use of ships for transporting CO_2 is in a rudimentary stage with potential for further development in the future when CCS technology might get widely implemented and captured carbon gets transported from far off places across the globe. So far, four ships are used to transport CO_2 worldwide (IPCC 2005). The amount of CO_2 transported by these ships is relatively small as compared to liquefied petroleum gas (LPG) and liquefied natural gas (LNG) transported by ships. Three ships operated by the Yara in the North Sea basin for CO_2 transport are in the size range from 1000–1500 m^3 and are pressurized in the range of 14–20 bar (Barrio et al. 2004). However, large size tankers which are used for LPG transport can be used for CO_2 transport. These large size tankers are designed for a pressure range of 5–7 bar and temperature range of -50 to -48°C with a capacity of 20000–22000 m^3 (Golomb 1997; Barrio et al. 2004; IPCC 2005). However, CO_2 is in gaseous phase under these conditions. To utilize the design criteria for LPG ships and use the existing ships, the temperature and pressure identified for economic transport are 6.5 bar and -52°C (Bario et al. 2004). However, new ships of higher capacity dedicated to CO_2 transport could be built in the future (Golomb 1997). Decarre et al. (2010) proposed a ship of size 30,000 m^3 and compared two temperature and pressure conditions: -30°C, 15 bar and -50°C, 7 bar. The study concluded that the ship operated at -30°C, 15 bar are economic.

17.3. CO_2 Storage Capacity and Injectivity

17.3.1 Storage Capacity of Different Sites

CO_2 can be stored in oceans, geological media and mineral carbonation. Ocean storage has different concerns including imbalancing of ocean ecosystems, acidification of oceans and leakage of CO_2 back to the atmosphere. Mineral carbonation is expensive as compared to the other two and also has significant environmental footprint (IPCC 2005; Bachu et al. 2007). Hence, of all the options, geological media storage is most significant. The following sites are identified as potential candidates for sequestration activity (IPCC 2005):

i) Depleted oil and gas reservoirs;
ii) Coal formations; and
iii) Saline formations, i.e. deep underground porous reservoir rocks saturated with brackish water or brine.

Also, CO_2 can be stored in underground cavities such as salt caverns in similar fashion as natural gas is stored in some places. Although the storage capacity of such cavities is small, nevertheless they can act as temporary storage. Basalt has been explored as possible storage sites for the places where sedimentary basins do not exist. Of all the geological storage sites, deep saline aquifers are most significant because of the storage potential of these aquifers is thought to be the largest among all the classes of CO_2 storage (IPCC 2005; Bachu et al. 2007).

In this section, CO_2 storage estimates are provided for oil and gas reservoirs, deep saline aquifers, and unmineable coal beds, together with uncertainty in storage. The estimates are adopted from the U.S. DOE methodology that was used to develop Carbon Sequestration Atlas of the United States and Canada (Goodman et al. 2011). These estimates are based on the volumetric approach. In addition, several uncertainties are associated in the estimation of the storage capacity of an aquifer. The uncertainties factors include nature of the reservoir, sweep efficiency and injection process (Shafeen et al. 2004).

Oil and Gas Reservoirs. Oil and Gas reservoirs characteristics are relatively better known as compared to coal beds and saline aquifers (Bachu et al. 2007). Storage estimates of oil and gas fields are based on the assumption that the volume of oil and gas produced is replaced by the equivalent volume of CO_2. This basically implies that the reservoir pressure after CO_2 storage could be equal or less than the original pressure of the reservoir (Goodman et al. 2011). The volumetric based CO_2 storage estimate is given by the following equation (Goodman et al. 2011):

$$G_{CO_2} = A h_n \varphi_e (1 - S_{wi}) B \rho_{CO_2 std} E_{oil/gas} \tag{17.5}$$

where A= area of the reservoir; h_n = net thickness of the reservoir; φ_e = average effective porosity; S_{wi} = initial water saturation; B= initial oil/gas formation volume factor that converts standard oil/gas volume to subsurface volume (at reservoir pressure and temperature); $\rho_{CO_2 std}$= standard CO_2 density; and $E_{oil/gas}$ = storage efficiency factor. Storage capacity estimates for oil and gas reservoir are based on the standard methods used by oil and gas industry to estimate original oil or gas in the fields (Goodman et al. 2011). Storage efficiency factor could be estimated from the local oil/gas reservoir experience or simulation.

Deep Saline Aquifer. Saline aquifers are those aquifers that are saturated with saline water. They are particularly significant because the storage capacity of these aquifers is thought to be the largest among all the classes of CO_2 storage (Bachu et al. 2007). Saline formations with following conditions could be suitable for CO_2 storage (Goodman et al. 2011):

- Temperature and pressure of the saline formation should be such that CO_2 is either in supercritical or liquid form in the formation. Generally a saline formation at $\geq$ 800 m depth keeps CO_2 supercritical or liquid;
- The formation should be with suitable caprock consisting of low permeability sealing rocks, such as shales, anhydrites, and other evaporates, for the structural trapping; and
- The formation is with hydrogeological conditions to keep CO_2 in the formation.

The equation to estimate the CO_2 storage in saline aquifer is (Goodman et al. 2011):

$$G_{CO_2} = A_t h_g \varphi_{tot} \rho E_{saline} \tag{17.6}$$

where A_t= the total area; h_g = formation thickness; φ_{tot} = total porosity; ρ= density of CO_2 at reservoir temperature and pressure; and E_{saline} = CO_2 storage efficiency factor. The storage efficiency factor would be a function of several formation characteristics such as gross thickness, porosity and different displacement efficiency components. Storage efficiency factors for saline formations, estimated using statistical analysis, vary from 0.4–5.5% (Goodman et al. 2011).

Unmineable Coal Beds. Unmineable coal beds can be used for CO_2 sequestration. Coalbed permeability should be > 1 mD for the possible storage of CO_2 in the coal beds. The coal bed permeability generally decreases with depth and 1 mD is generally at the depths of 1300–1500m. These depth ranges hence sets a limit for the coal bed CO_2 storage (IPCC 2005; Bachu et al. 2007). The equation to estimate the CO_2 storage in unmineable coal beds is given as (Goodman et al. 2011):

$$G_{CO_2} = A h_g \, C_{s,max} \varphi_{tot} \rho \rho_{CO_2\,std} E_{coal} \tag{17.7}$$

where A_t= the total area; h_g = formation thickness; $C_{s,max}$ = maximum volume of CO_2 at standard conditions that can be sorbed per volume of coal. $C_{s,max}$ depends upon the coal characteristics and to a degree on the temperature; and E_{coal}= storage efficiency factor which tells the extent of the CO_2 stored with respect to the bulk volume of the coal. The storage efficiency factor of coal seam would depend upon the basin's volume in which coal is actually present, portion of the in situ volume for which CO_2 is accessible, degree of CO_2 saturation within the CO_2-accesble deposit. Storage efficiency factors for coal seams are estimated to vary from 21–48% (Goodman et al. 2011)

17.3.2 CO_2 Injectivity

Storage capacity is the volumetric storage potential of a storage site whereas injectivity signifies the ease with which CO_2 can be injected into it. For CO_2 storage, both are important as some sites may have good storage capacity but low injectivity, hence cannot be considered as a good storage site. Operators of a storage site would like to inject CO_2 at the rate maximum which do not fracture the reservoir caprock and with a sustainable rate for a considerably longer time. As an example, a 500 MW coal power plant would generate 4–5 million tonnes of CO_2 in a year. Given such high levels of CO_2 emission, good injectivity is required so that CO_2 can be stored from an industrial scale CO_2 source. Injectivity would also decide the number of wells required to inject CO_2 in a storage site within a reasonable time. Low injectivity means more wells and dense spacing of the wells (Kobos et al. 2011).

Injectivity would primarily depend upon the permeability and thickness of the formation but also upon a number of other parameters including porosity, fluid viscosity, pressure, temperature, brine salinity and relative permeability to brine and CO_2 and amount of CO_2 injected (Bachu et al. 2007; Burton et al. 2009; Ghaderiet al. 2009; Mathias et al. 2011; Mathias et al. 2013). Lack of information of these different parameters could lead to the uncertainty in the estimation of injectivity. Information about many of these parameters could be available especially for those formations which are in the regions of oil and gas fields. However, information about some of the parameters such as CO_2-brine relative permeability could be limited because of no historical interest in the data. As an example, injectivity uncertainty associated with relative permeability could be as high as 57% (Mathias et al., 2013).

Semi-analytical models are developed to study injectivity with different level of complexities and assumptions. Burton et al. (2009) developed a one-dimensional model of injection at constant pressure using Darcy's Law and a modified form of Buckley-Leverett fractional flow theory. The model accounts for partial solubility of CO_2 and H_2O in each phase which could be an important factor in CO_2 injectivity. Mathias et al. (2011) developed a semi-analytical solution for pressure build-up due to a constant rate of injection. Different simulators used to study the performance of CO_2 (described in next section) have capabilities to analyse the CO_2 injectivity. For example simulators, GEM , TOUGH2 , etc. could be used to analyse the CO_2 injectivity.

17.4 Modeling of Sink Performance

The objective of the sink modeling is to understand the fate of CO_2 in the sequestrated formations. Since data available of subsurface geology, where CO_2 is to be sequestrated, is limited so the numerical simulation of suitable formations is of paramount significance. Fate of CO_2 will vary with time and spatial scale. Different multi-dimensional models with different degree of complexities and accuracy are used to study the CO_2 sequestration in the reservoirs. The modelling philosophy could be either to include as many processes as possible with very fine scales of time and space or to use model that is as simple as possible. The first approach will give accurate results but are limited by the computation capabilities and the data availability. Second approach would be highly restrictive in assumptions, but would provide insight of the overall system. Fully coupled 3-dimensional models are based upon the first approach. These models include maximum possible processes and parameters such as governing equations for mass balance of each component with suitable expressions for permeability, capillary pressure, diffusion, dispersion, equations of state, energy transport and geochemical reaction equations with associated parameters. Although these models are fairly accurate but are complicated and sometimes defining parameters over spatial domains is a tedious task (Court et al. 2012).

Suitable assumptions can be applied to simplify these fully coupled 3-dimensional models. Several of the equations can be simplified by taking suitable assumptions and the results would be of reasonable accuracy. The equations with simplifying assumptions are better manageable than that of fully coupled models. These 3-dimensional models with simplified assumptions are in wide use. TOUGH2, ECLIPSE, STOMP, NUFT, LLNL are a few examples of simplified 3-dimensional models (Court et al. 2012).

In a CO_2 storage reservoir, the thickness of the reservoir is much smaller as compared to the lateral dimensions. The 3-dimensional analysis can be converted into 2-dimensional modelling approach by integrating full 3-dimensional equations over the direction perpendicular to the top and bottom boundaries of the formation. These vertically integrated models, usually macro-scale models, are useful in answering the practical questions as the scale we are mostly interested in is the macroscale. These vertically integrated models can be further simplified by making additional assumptions (e.g., the entire formation is homogeneous and the top and bottom boundaries are horizontal) (Court et al. 2012).

A hybrid modelling approach can also be applied where detailed numerical models are applied where needed and simpler models are applied in other regions. In the grids where injection wells, leaky wells and faults are present, a grid block analytical solutions are applied locally to analyse the local behaviour of these features whereas coarser grid blocks are used otherwise. The grid blocks are discretized on the macro scale whereas the local well behaviours are analysed by local analytical solutions. These models have been constructed successfully to model multiple formations with multiple wells. In this way, very challenging computational problems can be solved quantitatively by a simplified approach of hybrid models (Court et al. 2012).

Different multi-dimensional models used to study the CO_2 sequestration in the reservoirs are based upon several assumptions and have different computational capabilities. A brief summary of different models is presented in Table 17.1. Most of the models solve conservation equations for mass and energy using equations of state for multiphase flow. Several simplifying assumptions are used to solve the equations. Different phase parameters are incorporated using simplified correlations and standard tables. Different models are used to analyse the CO_2 fate with time and spatially, leakage potentials and pathway, storage capacity, trapping mechanisms, etc.

As an example, TOUGH (Transport of Unsaturated Groundwater and Heat) and the associated codes developed by Earth Sciences Division of Lawrence Berkeley National Laboratory of the U.S. DOE are used by the parent laboratory and other organisations to study the reservoir dynamics, storage capacity, CO_2 leakage through caprock and faults, mineral trapping and caprock integrity (LBNL 2012). TOUGHREACT/ECO2 simulates reactions between gas-aqueous-solid phases and is used to study mineral trapping, caprock integrity and CO_2 reservoirs. TOUGH2/ECO2 is a multiphase flows simulator for non-isothermal flows for mixture of water/CO_2/NaCl in porous and fractured media and is used to study of reservoir dynamics, storage capacity and CO_2 leakage. The modular architecture of TOUGH2/ECO2 is shown in Fig. 17.1.

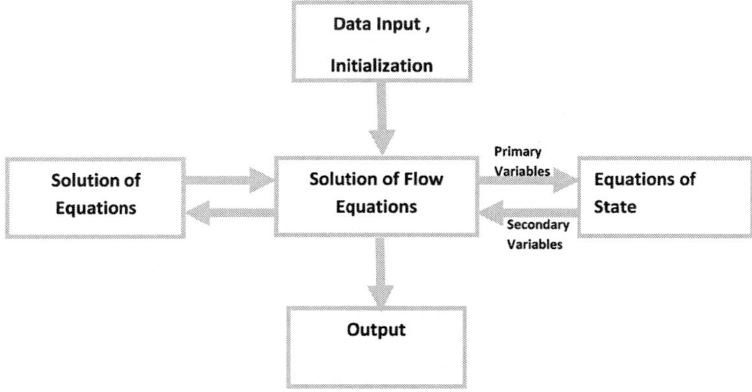

Figure 17.1. Modular architecture of TOUGH2/ECO2 (adopted from García 2003)

The Following section describes the model architecture, model equations, and different assumptions used to develop the TOUGH2/ECO2 simulator and is adopted from the model descriptions of the simulator elsewhere (Pruess and Garcia 2002; García 2003; LBNL 2012). The objective is to give the readers an overview of the model methodology.

Table 17.1. Models used to describe CO_2 sequestration in geological formations

Model & software	Model origin	Model application[s]	Description	Reference
CO_2-PENS	Earth and Environmental Sciences, Los Alamos National Laboratory, Los Almos, NM, USA	Simulates following CO_2 capture from a power plant, transport to the injection site, injection into geologic reservoirs, potential leakage from the reservoir, and migration of escaped CO_2.	• A hybrid system and process level models, integrates different modules described by analytical/semi analytical/detailed numerical models. • Supports a science based quantitative risk assessment.	Viswanathan et al. (2008)
$DuMu^X$	Univ. of Stuttgart, Institute for Water & Environ. Systems, Stuttgart, Germany	Simulates CO_2 transport and flow in reservoirs.	• Multi-Phase, multi-Component, multi-Scale, non-isothermal model for flow and transport in porous media. • Open-source simulator based on the Distributed and Unified Numerics Environment (DUNE). • Use Finite differences approach for the time discretization and BOX method for space discretization	Flemisch et al. (2010)
FEHM (Finite Element Heat and Mass)	Los Alamos National Laboratory, Los Alamos, NM, USA	• Simulates complex coupled subsurface processes as well flow in large and geologically complex basins • Non-isothermal, multi-phase flow of CO_2 and water	• Non-isothermal, multiphase flow and transport code that simulates the transport of heat and contaminants in both saturated and partially saturated heterogeneous porous media. • The code includes comprehensive reactive geochemistry and transport modules and a particle tracking capability. • Uses a control volume finite element method to form the discrete equations that represent the conservation of mass and energy. • These discrete nonlinear equations are solved with a Newton-Raphson method for the outer nonlinear iterations, and a preconditioned Krylov method for the inner linear iterations.	FEHM (2012)
GEM	Commercial Computer Modeling Group Ltd.	To determine CO_2 injectivity in aquifers, CO_2 sequestered in solution, residual gas saturation, and mineral precipitates.	• Flow of three-phase, multi-component fluids. • Modeling of gas solubility in the aqueous phase, intra aqueous reactions, mineral dissolution and precipitation. • Use equation of states to calculate phase equilibrium and the mass transfer of components between phases.	GEM (2012)
General Purpose Research Simulator (GPRS)	School of Earth Sciences, Stanford University, USA	CO_2 sequestration simulation for complex reservoirs	• Modeling of flow in porous media for mixtures with arbitrary number of phases. • Unifies different type of phases typical for flow of multi-component at equilibrium conditions.	GPRS (2012)
PNLCARB	Pacific Northwest National Laboratory, Hydrology Group Richland, WA, USA	Can simulate multiphase, radial injection of CO_2 and the growth of its area around the injector, buoyancy- driven migration of CO_2 toward the top confining layer, and dissolution of CO_2 during injection and vertical migration.	Semi-analytical model based on equations governing the radial injection of an immiscible CO_2 phase into saturated confined formations representing deep saline aquifers and reservoirs.	Saripalli et al. (2012)
Subsurface Transport over Multiple Phases (STOMP-CO_2 and STOMP-CO_2e)	Pacific Northwest National Laboratory, Hydrology Group Richland, WA, USA	Solves flow and transport problems for deep saline formations	• Solves the partial-differential equations that describe the conservation of mass or energy quantities by employing integrated-volume finite-difference discretization to the physical domain and backward Euler discretization to the time domain. • The resulting nonlinear coupled algebraic equations are solved using Newton-Raphson iteration. • STOMP-CO_2 is the isothermal version and STOMP-CO_2e is the nonisothermal version.	STOMP (2012)
1) TOUGH2 2) TOUGH2/ECO2 3) TOUGHREACT/ECO2 4) TOUGH2 coupled to commercial FLAC3D	Earth Sciences Division, Lawrence Berkeley National Laboratory Berkeley, CA, USA	• Studies reservoir dynamics, storage capacity, CO_2 leakage • Study mineral trapping, caprock integrity, natural CO_2Reservoirs • Reactions between gas - aqueous - solid phases. • Analyze leakage through caprock and Faults; stress-strain analysis	• Multi-dimensional numerical models for simulating multiphase flow in fractured porous media. • Equations of states property module for mixtures of water, NaCl, and CO_2 used for the analysis of geologic carbon sequestration processes. • Nonisothermal, Multi-phase approach to fluid and heat flow is used, which fully accounts for the movement of gaseous and liquid phases, their transport of latent and sensible heat, and phase transitions between liquid and vapor	TOUGH (2012)

The major assumptions of TOUGH2/ECO2 simulator include (García, 2003):

- Darcy's law is applicable for each phase of the multi-phase system;
- No chemical reaction is assumed to take place;
- Different phases are in local thermal and chemical equilibrium;
- Effects of mechanical stress are assumed to be negligible; and
- The dependable variables are the volumetric average of the representative volume.

A general conservation equation to represent multiphase, multicomponent, non-isothermal system is given as follows:

$$\frac{d}{dt}\int_{V_n} M^k \, dV_n = \int_{\Gamma_n} F^k . n \, d\Gamma_n + \int_{V_n} q^k dV_n \quad (17.8)$$

where Vn = arbitrary volume of the flow system over a closed surface, Tn; M^k = the mass per unit reservoir volume of k component; k corresponds to water, NaCl and CO_2 and heat; F^k = mass flux for water, NaCl and CO_2, and heat flux for the heat term; q^k= sinks and sources; and n= an unit normal vector on surface element pointing inwards into V_n. Sum of mass terms over all fluid phases is given as:

$$M^k = \phi \sum_\beta S_\beta \, \rho_\beta X_\beta^k \quad (17.9)$$

where ϕ = porosity; S_β = the saturation of phase β; ρ_β = density of phase β; and X_β^k= the mass fraction of component k phase β. Similarly, M^k for heat term is given as:

$$M^k = \phi \sum_\beta \rho_\beta S_\beta \, u_\beta + (1 - \phi)\rho_R C_R T \quad (17.10)$$

where ρ_R and C_R= density and specific heat of the rock grains respectively; T = temperature; and u_β= the specific internal energy of phase β. The mass flux includes advective and diffusive-dispersive components of all mobile phases. The advective mass flux is the sum of the individual mass fluxes, F_β and is given as follows:

$$F^k|_{adv} = \sum_\beta X_\beta^k \, F_\beta \quad (17.11)$$

The individual mass fluxes are given by the Darcy's law:

$$F_\beta = \rho_\beta u_\beta = -k \frac{k_{r\beta}\rho_\beta}{\mu_\beta} (\nabla P_\beta - \rho_\beta g \;) \quad (17.12)$$

where u_β=the Darcy velocity and k=absolute permeability; $k_{r\beta}$= relative permeability to phase β; μ_β = viscosity; and P_β = the fluid pressure in phase β. The diffusive-dispersive flux is given by the following equation:

$$F^k|_{dis} = -\sum_\beta \rho_\beta \, \overline{D}_\beta^k \nabla K_\beta^k \quad (17.13)$$

where $\overline{D}_\beta^k$= the hydrodynamic dispersion tensor and depends upon longitudinal and transverse dispersion coefficients. Similarly, the heat flux term is given as:

$$F^k = -K\nabla T + \sum_\beta h_\beta F_\beta \tag{17.14}$$

where K = the overall porous media thermal conductivity; and h_β is specific enthalpy.

The conservation equation is spatially discretized by using integral finite difference method which provides flexibility in representing irregular geological surfaces. The mass term is presented as follows:

$$\int_{V_n} M dV = V_n M_n \tag{17.15}$$

Similarly, surface integrals are discretized as the sum of averages over surface segments A_{nm}:

$$\int_{F_n} F.n d\Gamma = \sum_m A_{nm} F_{nm} \tag{17.16}$$

Time is discretized as a first order backward finite-difference. The space and time discretized form of Eq.17.16 can be written as the following non-linear equation:

$$R_n^{k,k+1} = M_n^{k,k+1} - M_n^{k,k} - \frac{\Delta t}{V_n}\left\{\left\{\sum_m A_{nm} F_{nm}^{k,k+1} + V_n q_n^{k,k+1}\right\} = 0 \tag{17.17}$$

The above equation is solved for each NaCl aqueous and CO_2, and for energy by using Newton-Raphson iteration such that the residual value is reduced to a preset convergence tolerance. To solve the above equation, values of several unknowns are required. This is achieved by using equations of state for multiphase flow and using several simplifying assumptions (García 2003). All thermo-physical properties are calculated as functions of basic primary thermodynamic variables. Several fluid properties are needed in the mass balance equations, e.g., density and viscosity of different phases; vapour pressure; salinity of aqueous phase; solubility of CO_2 as a function of temperature, salinity and partial pressure, etc. (Pruess and Garcia 2002). Various tabular equations and correlations are used in the software to present the thermo-physical properties numerically.

17.5 Leakage Potential and its Mitigation for Geological Storage of Carbon Dioxide in Saline Aquifer

Once CO_2 is injected in the formation with an impermeable caprocks, CO_2 would be sequestrated by dissolution in formation fluids, mineralization in the surrounding rock or by trapping in small pores. All these mechanisms are slow processes, and the time scale could be of the range of thousand years (LeNeveu 2008). During the CO_2

storage, it can leak through different pathways, including concentrated leakage through natural faults and fractures, diffuse leakage through caprocks and through manmade wellbores. In the area where extensive oil and gas exploration has been occurred in the past, CO_2 plume stored in the formation could leak through these abandoned wells, which could be a major leakage pathway. The leaked CO_2 may or may not reach the biosphere. In some formations, downward fluid pressure could prevent upward leakage, and it is also possible that CO_2 could be dissolved and would never reach the surface (LeNeveu 2008).

Analysis of leakage has a high level of uncertainty because of different possible leakage pathways. Analysis of leakage through abandoned wells would require specific data of the spatial and time-scale position of the wells, material used for the closures of these wells, and the state of the material and wells. This means that the risk analysis would require many simulations with different parameters associated with individual wells. In addition, the unknown factors of the material properties of the geological formations add further to the uncertainty in the risk analysis (Celia et al. 2004). The leakage estimation would need site specific analysis and would be a costly affair.

The different models used to model the CO_2 sequestrations sites can be used to model the leakage of CO_2. As an example, CQUESTRA, a semi-analytical model, can be used for the assessment of leakage through abandoned well bores and fractures. The model considers the trapping mechanisms of dissolution, mineralisation and capillary and assumes that the CO_2 pool under the caprock is uniform and immobile. The geological sequestration system is modelled as a series of layers beginning below the caprock and extending upto the surface. The properties of each layer such as pressure, salinity and Darcy's velocity of the fluid are considered in the model. The different processes which are included in the model are (LeNeveu 2008):

- Potential failure of conventional wellbore seals;
- Corrosion of the wellbore casing;
- Dissolution and mineralization of CO_2;
- Movement of the buoyant phase into the annular space around the wellbore; and
- Potential degradation of the central wellbore plug seal and annular cement seal at specific times of leakage.

The results of CQUESTRA model, based upon simulation for the Williston basin near Weyburn, Saskatchewan, indicate that in the event of CO_2 leak from a wellbore, a significant amount of the CO_2 could dissolve into the formations above the caprock. Also, once the transient pumping pressure is dissipated, the reservoir can be under-pressurized and a downward hydraulic gradient around a leaking wellbore could develop, counteracting buoyancy such that the CO_2 mixture would not rise into the wellbore or annulus.

17.5.1 CO_2 Leakage to Fresh Water Aquifers

One important risk associated with the CO_2 sequestration is the possibility of its leakage from the saline aquifer into the groundwater and also to the atmosphere. It is very unlikely that CO_2 would leak if geological storage sites are chosen carefully. However, some leaks may be possible due to heterogeneity of the subsurface and due to some unknown geological factors which are difficult to know due to the complex geology of the subsurface. Possible leaks may be via abandoned wellbores, faults and fractures and due to diffused leaks in permeable caprocks and micro-fractures in the caprock (Siirilaet al. 2012). If CO_2 escapes to freshwater aquifers, it may form carbonic acids, which may be buffered only to an extent by carbonate dissolutions. Lowered pH may dissolve heavy metals into the groundwater and increase their concentrations above the acceptable levels. The extent of groundwater contamination due to the dissolution of contaminants would depend upon the degree of leakage and on the geology of the groundwater aquifers as well.

Little and Jackson (2010) performed laboratory incubations of CO_2 infiltrations on 17 sediment samples collected from four freshwater aquifers for >300 days to understand the impact of CO_2 leakage on the water quality. The samples were collected from the sites which overlie the potential CO_2 sequestration sites in the U.S. The results revealed that water pH declines by 1–2 units for all aquifers, and the concentrations of the alkali and alkaline earths and manganese, cobalt, nickel, and iron increased by more than two orders of magnitude. In some samples the results showed the increase of uranium and barium as well.

A batch experiment for 2-weeks on samples of representative aquifers of the Gulf Coast region of the U.S.A. shows that the elevated CO_2 concentration in the groundwater aquifers increased the concentration of Ca, Mg, Si, K, Sr, Mn, Ba, Co, B and Zn initially and then became stable. The concentration of Fe, Al, Mo, U, V, As, Cr, Cs, Rb, Ni and Cu increased initially but declined before the end of the experiment, in most cases, even lower than the pre-CO_2 concentration (Lu et al. 2010).

A numerical modeling study in France was carried out to see the geochemical impact of CO_2 leak into a groundwater aquifer (Albian sandstone, a potable water source), though a leaky well from a potential CO_2 sequestration site, the Jurassic Dogger formation in the Paris Basin. A multiphase reactive transport model used for the study shows that CO_2 dissolution leads to a decrease in pH from 7.3 to 4.9. Glauconite present in the Albian aquifer dissolved and increased the concentration of silicon and aluminium in the solution. The calcium concentration increased from 1.3 to 2500 mmol/kg of water; the iron concentration increased from 1.3 to 2500 μmol/ kg of water at the CO_2 intrusion point (Humezet al. 2011).

17.5.2 Mitigation Strategies for Sealing Geologically Stored CO_2

CO_2 is kept in storage formation predominantly by stratigraphic or structural trapping. However, it has a potential to leak or move out of its place as a result of typical situations. For example, fractures may occur in the cap rocks if the CO_2 is stored at potentially high pressure. Also, earlier unidentified pathways such as fractures or faults in the cap rock, or badly designed wells or poorly plugged abandoned wells may cause leakages. The comprehensive mitigation and remedial strategies are required and available in case of such unlikely situations of leakage.

Strategies. There are various leakage mitigation and remediation strategies such as: i) pressure reduction in storage reservoir; ii) extraction of the gas plume from the reservoir before leaking and reaching out of the storage structure; iii) pressure increase in the formation from where the leakage outbreaks; and iv) separation of identified and accessible locations where leaks occur. It is of utmost importance to design storage projects because mitigation and remediation strategies depend on site specific conditions. Also, it is very difficult to restore the affected sites because of hydro-geological limitations. Generally, the combined use or application of groundwater pressure increase and saline aquifer pressure reduction is considered to be the best mitigation option for CO_2 leakage. In some cases, the number of wells may be increased, but this can be considered with due consideration of cost aspects (IEAGHG 2011). The comprehensive strategy for CO_2 leak prevention and remediation is comprised of the following five elements:

a) The selection of the site should be accomplished in such a way that it has very low risk of CO_2 leakage;
b) More emphasis should be given to well integrity;
c) A complete and well maintained monitoring system should be installed for the concerned CO_2 storage project;
d) A simulation based modeling should be conducted for reservoir in phased manner to track and cite the location of CO_2 plume; and
e) A "ready to use" contingency plan should be established for remediation purpose.

Leakage mitigation strategies are mainly associated with operational activities. Accordingly, some strategies can be adopted immediately, while others require more time. If the migration occurs through any of the unrestraint or active well, then above listed remedial options may be assisted with other strategies like repairing of wellbore leakages with the help of cement plugs or appropriate chemical sealants in casing; using well recompletion techniques; capping external leaks with the help of cement or chemical sealant; replacement of corroded pipe; repairing liner pipe patches; and plugging or rejection of not repairable wells.

Materials and Methods. As stated by Philiips et al. (2013), ideally speaking, potential CO_2 storage sites are not connected to freshwater aquifers and low permeability cap rocks; these sites are generally used to separate functional aquifers from target storage reservoirs. Nevertheless, there are concerns related to the possible impacts of potential CO_2 leakage into overlying aquifers and the atmosphere, resulting into risk to public health and carbon credit forfeiture under a carbon trading system and wastage of resources/energy associated with the CO_2 injection. Thus, prevention or mitigation of leakage is essential to managing risks associated with GCS.

The legal framework may be well advanced in countries where extensive oil and gas production activities take place, because of the similarity of CCS to those activities (IEAGHG 2007). Relatively less of the legal framework may be in place in other countries. The project operator is responsible for safely operating the injection facilities and closing it properly after completing the injection period. Especially post-injection responsibilities may vary related to monitoring and remediation. Parties with post-injection responsibility may include the operator, governments, a third party brought in under contract, or some combinations, all subject to the prevailing legal framework subjected to change over time.

In general, use of CO_2 resistant cements and ultrafine cements are considered as developed technologies. Whereas, use of low viscosity fluids have substantial merit of having the ability to stop small aperture leaks like fractures or delamination interfaces, to reduce the necessary injection pressures, and to increase the radius of influence around injection wells. Also, rock formation pore space can be properly pugged around the wellbore in particularly challenging situations. Nowadays, there is potential in research in progress related to use of microbial biofilms capable of inducing the precipitation of crystalline calcium carbonate using the process of ureolysis. The reduction in permeability of well bore, CO_2–related corrosion, and lowering the risk of unwanted upward CO_2 movement are some potential advantages of this method. Based on this aspect, currently the research is underway at the Center for Biofilm Engineering (CBE) at Montana State University (MSU) in the area of ureolytic biomineralization sealing for reducing CO_2 leakage risk. This research program aims at developing and verifying biomineralization sealing technologies and strategies which can successfully be applied at the field scale for carbon capture and geological storage (CCGS) projects (Cunningham et al. 2013).

In the same context, Phillips et al. (2013) described aqueous solutions and suspensions can effectively promote microbially induced mineral precipitation and can also prove to be an effective technology that can decrease the permeability between 2 and 4 orders of magnitude. They experimentally found that after fracture sealing, the sandstone core withstood three times higher well bore pressure than during the initial fracturing event that occurred prior to biofilm-induced mineralization of $CaCO_3$.

Summary. One can summarize that depleted oil and gas reservoirs are one of the best storage alternatives for GCS wherein it is quite often to have large numbers of abandoned wells. Nevertheless, the leakage of CO_2 through these abandoned wells poses a high potential risk and hence calls for methods of leakage mitigation such as re-plugging abandoned wells before exposure to CO_2. The procedures for abandonment of oil, natural gas and other mineral extraction wells are established by many countries that can be applied to CO2 injection.

Wherever possible, plugging the source of leak (e.g., wellbore/fracture) would be the best leakage mitigation strategy. In case if it is not possible, the following three basic steps can be taken for mitigating CO_2 leakage from the reservoir:

a) The pressure should be reduced from the reservoir in which leak occurs;
b) The pressure should be increased in the geological interval into which leak occurs; and
c) The CO_2 plume should be captured and CO_2 should be extracted before leak outbreaks and injected back into another formation, wherever possible.

Finally, a barrier should be created by injecting water upstream or otherwise, such as a barrier can be created by using a chemical sealant. Generally, these steps are considered as initial steps to be taken for mitigation of leakage of CO_2 from storage reservoir. But there are following nine essential steps that are considered as detailed response procedures, technologies and actions so as to give response to any CO_2 leakage (Kuuskraa, 2007):

1) Stop CO_2 Injection;
2) Notification;
3) Identify Source of Leak;
4) Remediate Wellbore Leak;
5) Remediate Caprock or Spill-Point Leak;
6) Conduct Integrated Leakage and CO_2 Accumulation Study;
7) Create Pressure Boundaries;
8) Drill Shallow CO_2 Recovery Wells; and
9) Remediate or Reconfigure Storage Site.

17.6 Conclusion

CO_2 transport through pipes has existed for more than three decades now. Thus, the modeling of CO_2 transport systems is reasonably understood. CO_2 pipe transport could be modeled by using standard hydraulic equation of flow in which CO_2 is mostly assumed to be transported in dense phase. To maintain the dense phase, pressure and temperature of $CO_2$is required to be maintained. Hence booster pumps may be required through the transport line. Offshore pipes could also be designed similar to the onshore lines but additionally, water pressure on the pipes should be considered.

Experience of CO_2 transport through ships is limited so far but a good understanding of the pressure vessels used to transport LPG could be applied to model the ships required for CO_2 transport.

Among different potential methods of CO_2 sequestration, sequestration in geological formations such as oil and gas fields, unminable coal beds and deep saline formation could be significant. Of all the geological formations, deep saline aquifers are understood to have the maximum storage potential. To understand CO_2 sequestration in geological formations, storage capacity, injectivity, fate of CO_2 in the formations, potential leakage through caprock and faults, mineral trapping and caprock integrity are required to be analysed. Due to the large spatial and time scale of CO_2 sequestration and variation in characteristics of the formations, modeling of these systems is a huge challenge.

Generalised storage estimates are made by several researchers using the volumetric approach. Storage capacity estimates for oil and gas reservoir are based on the standard methods used by oil and gas industry to estimate original oil or gas in the fields. Since a better understanding is developed for oil and gas reservoirs, better estimates are possible for oil and gas fields than for the saline formations and coal beds. Different multi-dimensional models (e.g. TOUGH2, ECLIPSE, STOMP, NUFT, LLNL) have been used to study the CO_2 sequestration in the reservoirs. These models are based upon several assumptions and have different computational capabilities. The modeling philosophy could be either to include as many processes as possible with very fine scales of time and space or to use the model that is as simple as possible. The first approach will give accurate results but are limited by the computation capabilities and the data availability. The second approach would be highly restrictive in assumptions; however, it would provide insight of the overall system. A hybrid modeling approach can also be applied where detailed numerical models are applied where needed and simpler models are applied in other regions.

One important risk associated with the CO_2 sequestration is the possibility of its leakage from the saline aquifers into the groundwater and also to the atmosphere. CO_2 leakage to freshwater aquifers may form carbonic acids, which may be buffered only to an extent by carbonate dissolutions. Lowered pH may dissolve heavy metals into the groundwater and increase their concentrations above the acceptable levels. The extent of groundwater contamination due to the dissolution of contaminants would depend upon the degree of leakage and on the geology of the groundwater aquifers as well.

17.7 References

Aspelund, A., Molvnik, M.J., and Koeijer, G. (2006). "Ship transport of CO_2 technical solutions and analysis of costs, energy utilization, energy efficiency and CO_2 emissions." *Chemical Engineering Research and Design,* 84(A9) Special Issue Carbon Capture and Storage. 847–856.

Bachu, S., Bonijoly, D., Bradshaw, J., Burruss, R., Holloway, S., Christensen, N.P., and Mathiassen, O.M. (2007). "CO_2 storage capacity estimation: methodology and gaps." *International Journal of Greenhouse Gas Control,* 1,430–443.

Barrio, M., Aspelund, A., Weydahl, T., Sandwik, T.E., Wongraven, L.R., Krogstad, H., Henningsen, R., Mølnvik, M., and Eide, S.I. (2004). "Ship-based transport of CO_2." *Proceeding of GHGT7,* Vancouver, Canada.

Burton, M., Kumar, N., and Bryant, S.L. (2009). "CO_2 injectivity into brine aquifers: why relative permeability matters as much as absolute permeability." *Energy Procedia* , 1, 3091–3098.

Celia, M., Bachu, S., Nordbotten, J.M., Gasda, S.E., and Dahle, H.K. (2004). "Quantitative estimation of CO_2 leakage from geological storage: analytical models, numerical models, and data needs." *Proceeding of GHGT7,* Vancouver, Canada, 663– 671.

CFR (2008). "Electronic code of federal regulations (e-CFR)." *Transportation, title 49, part 195. CFR.*http://ecfr.gpoaccess.gov/cgi/t/text/text-idx?c=ecfr&tpl=%2 Findex.tpl (retrieved Oct. 11, 2012).

Chandel, M.K., Pratson, L.F., and Williams, E. (2010). "Potential economies of scale in CO_2 transport through use of a trunk pipeline." *Energy Conversion and Management,* 51(12), 2825–2834.

Court, B., Celia, M.A, Nordbotten, J.M., Dobossy, M., Elliot, T.R, and Bandilla, K. (2011). "Modeling options to answer practical questions for CO_2 sequestration operations." <http://arks.princeton.edu/ark:/88435/dsp01rf55z769f.> (retrieved Oct., 11, 2012).

Cquestra (2008). "A risk and performance assessment code for geological sequestration of carbon dioxide." *Energy Conversion and Management,* 49(1), 32–46.

Cunningham, A.B., Lauchnor, E., Eldring, J., Esposito, R., Mitchell, A.C., Gerlach, R., Phillips, A.J., Ebigbo, A., and Spangle, L.H. (2013). "Abandoned well CO_2 leakage mitigation using biologically induced mineralization: current progress and future directions." In: Greenhouse Gases: Science and Technology, special issue, 11th US annual conference on Carbon Capture, Utilization, and Sequestration, 3(1), 40–49.

Decarre, S., Berthiaud, J., Butin, N., and Guillaume-Combecave J.L. (2010). "CO_2 maritime transportation." *International Journal of Greenhouse Gas Control,* 4(5), 857–864.

FEHM (2012). *Finite Element Heat and Mass Code (FEHM).*http://ner.com/analysis-software-2/FEHM.html.Retrieved Dec. 22, 2012.

Fernandez-Martinez, A., Yandi, H., Lee, B., Jun, Y.S., and Waychunas, G.A. (2013). "In situ determination of interfacial energies between heterogeneously nucleated $CaCO_3$ and quartz substrates thermodynamics of CO_2 mineral trapping." *Environ. Sci. Technol.*, 47, 102–109.

Flemisch, B., Darcis, M., Erbertseder, K., Faigle, B., Lauser, A., Mosthaf, K., Muthing, S., Nuske, P., Tatomir, A., Wolff, M., and Helmig, R. (2010). *DuMux: DUNE for Multi- {Phase, Component, Scale, Physics} Flow and Transport in Porous.* http://www.simtech.uni-stuttgart.de/publikationen/prints.php?ID=206. Retrieved Dec. 22, 2012.

Gale, J., and Davison, J. (2004). "Transmission of CO_2–safety and economic considerations." *Energy*, 29, 1319–1328.

García, J.E. (2003). *Fluid Dynamics of Carbon Dioxide Disposal into Saline Aquifers.* Ph.D dissertation, Civil and Environmental Engineering, University of California-Berkeley. Fall 2003. Lawrence Berkeley National Laboratory. <http://www.escholarship.org/uc/item/2g11g3t9.> Retrieved Dec. 28, 2012.

GEM (2012). *Generalized Equation of State Model Reservoir Simulator.* http://science.uwaterloo.ca/~mauriced/earth691duss/CO2_General%20CO2%20Sequestration%20materilas/CMG%20GEM.pdf. Retrieved Dec. 27, 2012.

Ghaderi, S. M., Keith, D. W., and Leonenko, Y. (2009). "Feasibility of injecting large volumes of CO_2 into aquifers." *GHGT-9, Energy Procedia*, 1, 3113–20.

Golomb, D. (1997). "Transport systems for ocean disposal of CO_2 and their environmental effects." *Energy conversion and management,* 38, Suppl., 279–286.

Goodman, A., Hakala, A., Bromhal, G., Deel, D., Rodosta, T., Frailey, S., Small, M., Allen, D., Romanov, V., Fazio, J., Huerta, N., McIntyre, D., Kutchko, B., and Guthrie, G. (2011). "DOE methodology for the development of geologic storage potential for carbon dioxide at the national and regional scale." *International Journal of Greenhouse Gas Control,* 5.952–965.

GPRS (2012). *General Purpose Research Simulator.* https://pangea.stanford.edu/researchgroups/supri-b/research/research-areas/gprs, Retrieved Dec. 11, 2012.

Heddle, G., Herzog, H., and Klett, M. (2003). "The economics of CO_2 storage." Massachusetts Institute of Technology, Laboratory for Energy and the Environment, publication no. LFEE 2003-003RP.

Humez, P., Audigane, P., Lions, J., Chiaberge, C., and Bellenfant, G. (2011). "Modeling of CO_2 leakage up through an abandoned well from deep saline aquifer to shallow fresh groundwater." *Transport in Porous Media,* 90, 153–181.

IEA GHG (2002). *Transmission of CO_2 and Energy.* IEA greenhouse gas R&D programme, *Report no.*4/6.

IEA GHG (2005). *Building the Cost Curves for CO_2 Storage: European sector.*IEA greenhouse gas R&D programme, Report no. 2005/2.

IEA GHG (2007). *Geological Storage of Carbon Dioxide: Staying Safely Underground.* IEA Greenhouse Gas R&D Programme, Cheltenham, GL52 7RZ, UK.

IEA GHG (2011). *Potential Impacts on Groundwater Resources of CO_2 Storage. 2011/11*, Cheltenham, GL52 7RZ, UK.

IPCC (2005). *IPCC Special Report on Carbon Dioxide Capture and Storage.* Cambridge University Press, Cambridge.ISBN-13 978-0-521-86643-9.

Kobos, P.H., Roach, J.D., Heath, J.E., Dewers, T.A., McKenna, S.A., Klise, G.T., Krumhansl, J.L., Borns, D.J., Gutierrez, K.A., and McNemar, A. (2011) "Economic uncertainty in subsurface CO_2 storage: Geological injection limits and consequences for carbon management costs." *30th USAEE/IAEE North American Conference, SAND2011-5975C*, Washington, D.C., October 9–12, 2011.

Kuuskraa, V.A. (2007). "Overview of mitigation and remediation options for geological storage of CO_2." In: AB1925 Staff Workshop California Institute for Energy and Environment, University of California, Oakland, CA.

Lackner, K.S. (2003). "A guide to CO_2 sequestration." *Science,* 300, 1677−1678.

LBNL (2012). *TOUGH: Suite of Simulators for Nonisothermal Multiphase Flow and Transport in Fractured Porous Media.* Lawrence Berkeley National Laboratory, Earth Sciences Division. http://esd.lbl.gov/research/projects/tough/. Retrieved Dec. 11, 2012.

LeNeveu, D.M. (2008). "CQUESTRA, a risk and performance assessment code for geological sequestration of carbon dioxide." *Energy Convers. Manage.*, 49, 32–46.

Lu, J., Partin, J.W., Hovorka, S.D., and Wong, C. (2010). "Potential risks to freshwater resources as a result of leakage from CO_2 geological storage: a batch-reaction experiment." *Environ. Earth Sci.,* 60, 335–348.

Little, M.G., and Jackson, R.B. (2010). "Potential impacts of leakage from deep CO_2 geosequestration on overlying freshwater aquifers." *Environ. Sci. Technol.*, 44, 9225–9232.

Matter, J.M., and Kelemen, P.B. (2009). "Permanent storage of carbon dioxide in geological reservoirs by mineral carbonation." *Nature Geoscience*, 2, 837−841.

Mathias, S.A., Gluyas, J.G., Gonzalez Martinez de Miguel, G.J., and Hosseini, S.A., (2011). "Role of partial miscibility on pressure buildup due to injection of CO_2 into closed and open brine aquifers." *Water Resources Research,* 47, W12525, http://dx.doi.org/10.1029/2011WR011051.

Mathias, S.A., Gluyas, J.G., Gonzalez Martinez de Miguel, G.J., Bryant, S.L., Wilson, D. (2013). "On relative permeability data uncertainty and CO_2 injectivity estimation for brine aquifers." *International Journal of Greenhouse Gas Control,* 12, 200–212.

Pershad, H., and Slater. S. (2007). *Development of a CO_2 Transport and Storage Network in the North Sea. Report to the North Sea Basin Task Force.* Department of Business Enterprise and Regulatory Reform U.K. <http://webarchive.nationalarchives.gov.uk/+/http://www.berr.gov.uk/files/file42476.pdf> (Retrieved Dec. 22, 2012).

Phillips, A.J., Lauchnor, E., Eldring, J.J., Esposito, R., Mitchell, A.C., Gerlach, R., Cunningham, A.B., and Spangler, L.H. (2013). "Potential CO_2 leakage reduction through biofilm-induced calcium carbonate precipitation." *Environ. Sci. Technol.,* 47, 142–149.

Pruess, K., and García, J. (2002). "Multiphase flow dynamics during CO_2 disposal into saline aquifers." *Environmental Geology,* 42, 282–295.

Riding, R. (2000). "Microbial carbonates: the geological record of calcified bacterial-algal mats and biofilms." *Sedimentology*, 47, 179–214.

Saripalli, K.P., McGrail, B.P., and White M.D. (2012). "Modeling the sequestration of CO_2 in deep geological formations." Pacific Northwest National Laboratory, Richland, Washington 9352.http://www.netl.doe.gov/publications/ proceedings/ 01/carbon_seq/p36.pdf. (Retrieved Dec. 22, 2012).

Serpa, J., Morbee, J., and Tzimas, E. (2011). *Technical and Economic Characteristics of A CO_2 Transmission Pipeline Infrastructure.* JRC Scientific and Technical reports. EUR 24731EN. <http://publications.jrc.ec.europa.eu/repository/ bitstream/111111111/16038/1/reqno_jrc62502_aspublished.pdf> (retrieved Dec. 2012).

Shafeen, A., Croiset, E., Douglas, P.L., and Chatzis, I. (2004). "CO_2 sequestration in Ontario, Canada. Part I: storage evaluation of potential reservoirs." *Energy Conversion and Management,* 45. 2645–2659.

Siirila, E.R., Navarre-Sitchler, A.K., Maxwell, R.M., and McCray, J.E. (2012). "A quantitative methodology to assess the risks to human health from CO_2 leakage into groundwater." *Advances in Water Resources,* 36,146–164.

STOMP (2012). *Subsurface transport over multiple phases.* Pacific Northwest National Laboratory, Hydrology Group. http://stomp.pnnl.gov/documentation/guides/ theory.pdf. (Retrieved Dec. 19, 2012).

TOUGH (2012). *TOUGH: Suite of Simulators for Nonisothermal Multiphase Flow and Transport in Fractured Porous Media.* Earth Sciences Division, Lawrence Berkeley National Laboratory Berkeley, CA, USA http://esd.lbl.gov/research/ projects/tough/. (Retrieved Dec. 22, 2012).

Viswanathan, H.S., Pawar, R.J., Stauffer, P.H., Kaszuba,P.J., Carey, J.W., Olsen, S.C., Keating, G.N., Kavetski, D., and Guthrie, D.J. (2008). "Development of a hybrid process and system model for the assessment of wellbore leakage at a geologic CO_2 sequestration site." *Environ. Sci. Technol.*, 42 (19), 7280–7286.

World Resources Institute (WRI) (2008). *CCS Guidelines: Guidelines for Carbon Dioxide Capture, Transport and Storage.* Washington, DC: WRI. http:// pdf.wri.org/ ccs_guidelines.pdf. (Retrieved Nov. 10, 2012).

CHAPTER 18

Carbon Capture and Storage: Major Issues, Challenges and the Path Forward

Tian C. Zhang, Rao Y. Surampalli, C. M. Kao, and W. S. Huang

18.1 Introduction

In the portfolio of climate change mitigation strategies, three options are being explored: a) increasing energy efficiency, b) switching to less carbon-intensive sources of energy, and c) carbon capture, storage and sequestration (CCS) (White et al. 2003). Here, we define the term CCS (= Carbon Capture, Storage, and Sequestration) as any technologies/methods that are to a) capture, transport, and store carbon (CO_2), b) monitor, verify, and account the status/progress of the CCS technologies employed, and c) advance development/uptake of low-carbon technologies and/or promote beneficial reuse of CO_2. While our society is embracing options a) and b), deployment of CCS seems to be inhibited due to several challenges.

Major public concerns about CCS include: a) limitations of CCS for power plants, b) cost of CCS and limitation of CCS because of its energy penalty, c) mandating CO_2 emission reductions at power plants, d) regulating the long-term storage of CO_2, and e) concerns related to health, safety, and environmental impacts (Zhang and Surampalli 2013). The pros and cons of CCS have been discussed recently. The opponents of CCS believe that CCS has several constrains. The *first* constrain is about CCS efficacy. CCS delays inevitable transition to clean energy; CCS distracts attention and resources form clean energy; CCS is not feasible; CCS will take far too long to implement for climate change. The *second* is about risks involved. The potential problems associated with CCS are not fully understood. Leakage of CO_2 from CCS facilities is a risk and a burden of taxpayers and our children. The *third* is about economics. The estimated costs for CO_2 transportation (\$1–3/t-100 km) and sequestration (\$4–8/t-CO_2) are small compared to that for CO_2 capture (\$35–55/t CO_2 capture) (Li et al. 2009). In general, CCS is less cost-effective than renewable energy; CCS raises costs and energy prices, and requires significant water (e.g., power plants with CCS technology needs 90% more freshwater than those without CCS); without a price on carbon, CCS will not fly. These concerns and constrains can be classified into the following four issues: a) costs and economics, b) legal and regulatory frameworks, c) social and acceptability, and d) uncertainty and scalability. Each of them is associated with different challenges.

In the past, these issues/challenges have been evaluated and reviewed (e.g., Herzog 2001; IPCC 2005; de Coninck et al. 2006; IEA 2009, 2010; CCCSRP 2010; Surampalli et al. 2013). Debate over the regional climate change policies and the timing of CCS deployment has been continuing. This chapter provides a non-exhaustive analysis of these reviews and debate, with the intention of introducing the basic principles and major frameworks related to climate change policies and CCS. The chapter also introduces intensive discussions on the challenges associated with each issue and future perspectives. Understanding these issues would better prepare us to overcome the associated challenges and to address economic, technical, regulatory and social implications of implementing CCS technologies in the world.

18.2 Cost and Economics Issues

It is very difficult to estimate exactly what the cost of CCS is. One can understand this difficulty by visiting the document published by CMU et al. (2011). Many factors affect CCS costs, such as a) choice of power plant and CCS technology, b) process design and operating variables, c) economic and financial parameters, d) choice of system boundaries (e.g., one facility vs. multi-plant system; GHG gases considered (CO_2 only vs. all GHGs); power plant only vs. partial or complete fuel cycle), and e) time frame of interest (e.g., first-of-a-kind plant vs. n^{th} plant; current technology vs. future systems; consideration of technological "learning") (Rubin 2011). In general, the economics of CCS are often discussed in terms of a) mitigation costs (i.e., how much it costs to avoid/capture a tonne of CO_2), b) increase in the electricity cost, c) capital and operation/maintenance (O&M) costs, and d) comparison to the costs of other mitigation options. Since it is almost impossible to summarize the different methods for cost estimation of CCS in a reasonable page limit, we only provide some general ranges of and discussions on CCS costs here.

Table 18.1 shows a range of CCS component costs. Estimation of the costs for capturing CO_2 from mobile/distributed point or non-point sources (e.g., Table 18.2) has more uncertainty because we don't have enough data for meaningful evaluation. When estimating CO_2 avoidance costs for a complete CCS system for an industrial sector (e.g., electricity generation), one needs to add the cost of CCS component together. In this case, the cost of CCS can be significantly higher than the cost of individual components (Table 18.3). Table 18.4 shows that the estimates of investment costs (in 2006) vary significantly over the models and different research groups. The increase in investment costs as a consequence of CO_2 capture is on the order of 30% for coal-based integrated gasification combined cycle (IGCC) power plants, but up to 100% for gas-fired plants (de Coninck et al. 2006). Table 18.5 shows that electricity prices will be increased due to the deployment of CCS, but the estimation can be very different, depending strongly on the base price of electricity. The relative price changes shown in Table 18.5 may affect the acceptability of CCS to both the general public and the private sector. The public in regions with low electricity prices might be more reluctant to accept CCS (or other mitigation options) than elsewhere (de Coninck et al. 2006). However, the public acceptability of CCS may be affected by many other factors (see Section 18.4). One example is based on the competition between CCS deployment and development of other low-carbon technologies. Table 18.6 compares the cost of low-carbon technologies with that of conventional power generation. CCS may cost a lot, but then, so do all the other near zero carbon options. Once the relatively low-cost technology options (e.g., hydropower and onshore wind technologies) are fully exploited, CCS becomes very competitive. As shown in Fig. 18.1, continuous increase in the CO_2 abatement by CCS starts in 2020 as compared with renewables and nuclear technologies. It should be noted that, without CCS, abatement costs in the electricity sector could be higher by more than 70% if energy-related emissions are to be halved by 2050 (IEA 2010). In addition, the costs of new technologies that have not yet reached full maturity (e.g., CCS) will become lower in the future. Furthermore, the future cost of CCS (and other low-carbon technologies) will depend on the carbon price and related regulations (see Section 18.3).

Table 18.1. CCS component costs

CCS component	Cost range
Capture from power plants	Prices vary: \$4.660/kW real vs. 2.600/kW estimated[4]
Pre-combustion from IGCC	43–55[3] US\$/tCO_2 avoided
Oxy-combustion capture	52[3] US\$/tCO_2 avoided
Capture from a coal-fired power plant	15–75[1]; 60–95[2]; 43–58[3] US\$/tCO_2 avoided
Capture from gas processing or NH_3 production	5–55 US\$/tCO_2 avoided[1]
Post-combustion capture using amines	58[3] US\$/tCO_2 avoided
Capture from other industrial sources	25–115 US\$/tCO_2 avoided[1]
Transport in general	1–8 US\$/tCO_2 transported per 250 km[1]
Transport cost for the complete CCS chain	7–12% of capture costs[4]
Geological storage	0.5–8 US\$/tCO_2 injected[1]
Storage in general	1 to 20 € (2011)/tCO_2[4]
Enhanced Oil Recovery	- (20–30)[b] US\$/tCO_2 injected[5]

[a]IGCC = integrated gasification combined cycle. [b] - (mines) = subtracting 20–30 US\$/tCO_2 injected from the total cost. [1] = de Coninck (2006); [2] = USDOE (2010); [3] = IEA (2011); [4] = CMU et al. (2011); [5] = de Coninck et al. (2006)

One limitation of CCS is its energy and cost penalty. Wide-scale application of CCS would reduce CO_2 emissions from flue stacks of coal power plants by 85–90% with an increase in resource consumption by one third. As shown in Fig. 18.2, there is a need to invest over ~US\$5 trillion for CCS deployment from 2010 to 2050 to achieve a 50% reduction in GHG emissions by 2050 (IEA 2009; Lippone 2012). However, the true costs of different CCS technologies for different applications are still unknown because our experience is still limited (we don't have enough full-scale applications of CCS). This demonstrates the challenge for commonality in generation of CCS cost estimates so they may be used in a consistent manner (see Table 18.7). This will ultimately lead to a reduction in the uncertainty, variability, and bias of CCS costs estimates (CMU et al. 2011).

Table 18.2. Costs of capturing CO_2 from mobile/distributed point or non-point source[a]

CCS method	Cost range
Trees/organisms	0.03–8 US$/t$CO_2$ avoided
Biomass-fueled power plant, bio-oil and biochar	41 US$/t$CO_2$ avoided
Steel slag & waste concrete used as CO_2 sorbents	2–8 US$/t$CO_2$ avoided
Liquid sorbents (synthetic trees): NaOH scheme/Ca(OH)$_2$	7–20/20 US$/t$CO_2$ avoided

[a]Adapted from Zhang and Surampalli 2012.

Table 18.3. CCS costs for the complete CCS system for electricity generation[a]

Power plant with CO_2 capture and geological storage[b]	Type of power plant with CCS	
	IGCC[c] ref. plat (US$/t$CO_2$ avoided)	Pulverized coal ref. plat (US$/t$CO_2$ avoided)
Natural gas combined cycle	40–90	20–60
Pulverized coal	70–270	30–70
Integrated gasification combined cycle	40–220	20–70

[a]de Coninck et al. (2006). [b]Cost of enhanced oil recovery can be obtained by substracting 20 to 30 US$/t$CO_2$. [c]IGCC = integrated gasification combined cycle.

Table 18.4. Initial investment costs of CCS

Plant type	Cost range (€/kW)
Natural gas plants w/capture	515–724 €/kW[1]; 600–1150 €/kW[2]; 1700 US$/kW[3]
IGCC plant w/ capture	1169–1565€/kW[1]; 800–2100 €/kW ± 21% deviation[2]
Pulverized coal baseline w/ capture	3800US$/kW[3]

[1] = IPCC (2005); [2] = de Coninck et al. (2006); [3] = Lippone 2012.

Table 18.5. Incremental electricity costs of CCS

Estimation basis	Cost range
An engineering costs analysis for large-scale deployment	0.02–0.03 US$/kWh (IPCC 2005)
Completing the cycle of CCS in the US	~0.06 US$/kWh (IPCC 2005)
Country/electricity price (US$/kWh)	**Increase in electricity price (%)**
Germany/0.20 for household and 0.08 for industry	10–15 for household and 25–60 for industry (IPCC 2005)
USA/0.09 for household and 0.05 for industry	20–35% for household and 25–60 for industry (IPCC 2005); 50% for household and ~100% for industry (Charles 2009)

Table 7.6. Costs of low-carbon technologies vs. conventional power generation[a]

Technologies	Cost range		
	Capital investment US\$/kW[b]	**Levelized electricity US\$/MWh[b]**	**CO_2 avoided US\$/kW[b]**
Natural gas-fired plant (as baseline 1)	~900	~88	-
Coal-fired plant (as baseline 2)	~2600	~75	-
CSC: natural gas/coal	1700–2200/4100–4800	107–119/89–139	67–106/23–92
Wind onshore	1700–2700	67–86	-8–16
Hydropower	2000–2600	52–60	-27–0
Geothermal	2000–3500	43–61	-38–0
Biomass	3400–4400	81–113	9–49
Solar: PV/thermal	4000–5100/5100–6800	220–265/185–265	182–239/139–203
Wind offshore	4200–6800	146–215	90–176
Nuclear	4300–6700	68–94	-7–25

[a]Abellera and Short (2011). [b] US\$ in 2011.

Table 18.7. Major variables and uncertainties in cost estimation of CCS[a]

Variables/uncertainty	Concerns and Challenges
• Reference plants (RPs)	• Results are highly sensitive to the RPs used; some RPs do not exist.
• Different ways to report a singular measure	• Different parameters and the measures used generate different results.
• Cost elements at different levels	• A consistent and complete set of cost elements are not identified yet. System-wide costs, singular plant costs, costs of different technical options are often mixed used.
• Terms related to costs	• Many terms related to costs (e.g., owner's costs) are not defined in a consistent set of categories, but reported as the same in different studies.
• Interest and year-of-currency used	• Audience often confused without such info.
• Technology development	• A moving target as elements of the technology are in development
• Cost estimation methods	• Improvement is needed for the reporting and transparency of these methods (e.g., assumptions)

[a]CMU et al. (2011).

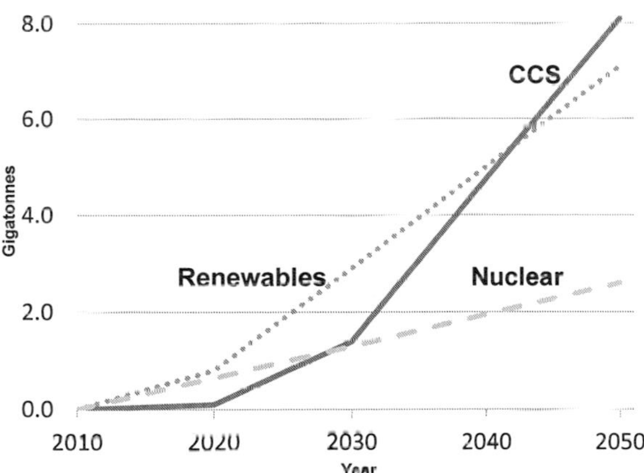

Figure 18.1. Evolution of CO_2 abatement by different low-carbon technologies grouping in the IEA Blue Map Scenario (adapted from Abellera and Short 2011)

18.3 Legal and Regulatory Issues

Legal and regulatory issues surrounding CCS are very complicated, and have been studies by many researchers and organizations for a long time. Currently, the legal and policy framework for CCS is under the umbrella of the international law that is related to the framework of climate changes policies and incentivizing carbon management (ICM). These frameworks are not established yet, and they are tangling with each other. In this section, we will review and discuss the following: a) international legal policies related to climate changes policies, ICM and CCS; b) domestic legal and policy framework for CCS; and c) key legal issues and uncertainties related to CCS implementation.

International Legal Policies Related to ICM and CCS. The international law consists of a diffuse patchwork of agreements, regulations and customs. As shown in Table 18.8, there is currently no comprehensive regulatory framework in the world to deal specifically with CCS or even ICM. The existing public international law closely related to ICM and CCS is in the marine protection treaties. However, clarification and amendment of several provisions in these treaties are needed (de Coninck et al., 2006).

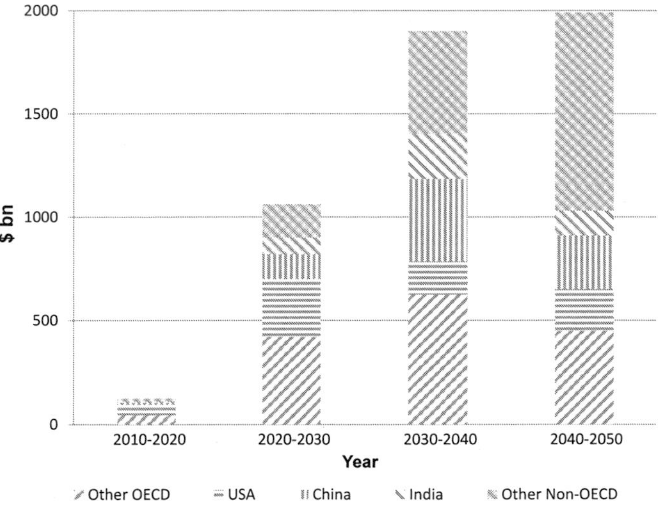

Figure 18.2. Predicted CCS investment between 2010 and 2050 to meet the IEA CCS Roadmap ambitions (Lippone 2012). OECD = the Organization for Economic Co-operation and Development

Table 18.8. International legal and policy framework related to ICM and CCS

Law or Convention/ Year established	Implications to ICM and CCS
The General Agreement on Tariffs and Trade (GATT)/after WW II	To avoid protectionism so domestic trade policy and unilateral measures for ICM and CCS are vulnerable to challenge if they discriminate between domestic and foreign products or between imports from two different countries.
The World Trade Organization (WTO)/1995	Under the WTO, the GATT brings uniformity and certainty to international trade law by providing a conflict resolution mechanism for international trade disputes.
• The WTO's Agreement on Subsidies and Countervailing Measures (ASCM)	• Targets trade-distorting subsidies.
• The WTO's Agreement on Technical Barriers to Trade (TBT)	• Targets trade-distorting technical regulations and standards.
• Doha Development Round/ Commenced in 2001	• This round put trade law reforms, including a uniform approach to the growing number of multilateral environmental agreements (including climate change policy) among WTO countries, on the agenda. Future unclear due to the disparate interests held by participating countries.
The London Convention (LC)/1972 and The London Protocol (LP)/1996 (came into force 2006)	87 States are parties to LC. To control waste dumping into the sea. CO_2 is not in the blacklist and reverse list, but its disposal into the sea would violate the LC. It is not clear if storage belongs to dumping (deliberate disposal) at sea, which is prohibited. In 1996, with 42 parties, the LP was agreed to further modernize and eventually replace the LC. It prohibits the storage of CO_2 in the water column and sub-seabed repositories.
The UN Convention on the Law of the Sea (UNCLOS)/1982 (came into force in 1994)	Provides a framework for all areas, including marine protection (e.g., prevent, reduce and control pollution), applying to the seabed and its subsoil (and thus, CCS beneath it). Uncertainty exists about disposal of CO_2 via pipeline into the exclusive economic zone or the continental shelf and via marine storage. CO_2 disposal is acceptable to the high seas.
The OSPAR Convention/1992	Includes 15 Northern European Member States and the European Community. It allows offshore-derived CO_2 disposal/placement. There is a basic lack of certainty as to the applicability of OSPAR to CCS.
The UN Framework Convention on Climate Change (UNFCCC)/1992 (came into force 1994)	Mentioned the sustainable management, conservation and enhancement of sinks and reservoirs of all GHGs. Annex I Parties are obliged to enhance GHG sinks/reservoirs. The Conference of the Parties (COP) is the "supreme body" of the Convention. The Convention established two permanent subsidiary bodies: the Subsidiary Body for Scientific and Technological Advice (SBSTA) and that for Implementation (SBI).
• The Kyoto Protocol	• Agreement to reduce GHG emissions by average 5% below 1990 levels by 2012; all parties take action on mitigation and adaptation. Requiring research/promotion/ development and increase use of CCS; current commitments expire in 2012.
• Bali Action Plan (BAP)/2007	• This is COP13. It designed a two-year process to finalize a binding agreement at the COP15 in Copenhagen in 2009.The BAP identified five key building blocks required (shared vision, mitigation, adaptation, technology and financial resources) for a strengthened future response to climate change and to enable the full, effective and sustained implementation of the Convention, now, up to and beyond 2012
• Copenhagen Climate Conference (COP15)/2009	• The Copenhagen Accord recognizes the scientific case for keeping temperature rises below 2 °C, but does not contain a baseline for this target, nor commitments for reduced emissions to achieve the target. One part of the agreement pledges US$ 30 billion to the developing world over the next three years, rising to US$100 billion per year by 2020, to help poor countries adapt to climate change.
• COP18/CMP8/2012	• COP18 and CMP8 took place on 11/26–12/07/2012 in Doha, Qatar, with negotiations being focused on ensuring the implementation of agreements reached at previous conferences and on amendments to the Kyoto Protocol to establish its second commitment period (ENB 2012).

Domestic Legal and Policy Framework for CCS. On the domestic front, domestic legal and policy framework for CCS remains ambiguous. Currently no comprehensive regulatory framework for CCS exists in the U.S., European Union (EU) or any other counties although in some countries and regions, there is a developing understanding of how to apply (or extend) current regulatory regimes to CCS. For example, in 2010 the US Environment Protection Agency (USEPA) finalized the regulations of using Class V wells for CO_2 geological sequestration (USEPA, 2012a). Table 18.9 shows the existing regulations or regulatory analogs in the US that may be closely related to CCS components. Table 18.10 shows the similar information in EU. Many countries, such as the United Kingdom (UK), have existing regulations or regulatory analogs related to CCS and ICM, but they are not included here for the simplicity purpose. These regulatory analogs may be modified toward (or at least provide insight into) a future CCS regulatory framework.

Table 18.9. Existing regulations or analogs related to CCS and ICM in the US

Regulation or analogs	Description and relationship to GHG capture and CCS
USEPA's cap & trade programs[1]	Focused on the pollutants (SO_2, NO_x, and mercury) from the power sector, including: a) the clean air interstate rule; b) clean air visibility rule; c) the acid rain program; d) the NO_x budget trading program; and e) other programs. Emissions trading became law as part of the Clean Air Act of 1990; cap and trade takes effect in 1995. The program can be extended to other sectors for GHG capture.
• Regional Greenhouse Gas Initiative (RGGI)[1] • Western Climate Initiative[1]	• RGGI: 10 northeastern and Mid-Atlantic States will cap and then reduce CO_2 emissions from the power sector by 10% by 2018. • Launched in February 2007, it includes 7 western states and four Canadian provinces, has established a regional target for reducing heat-trapping emissions of 15% below 2005 levels by 2020. It requires participants to implement California's Clean Car Standard, and recommends other policies and best practices for states to adopt and to achieve regional goals for cutting emissions.
• Assembly Bill 32 (AB32)[1]	• AB32: In 2006, California passed AB 32, the Global Warming Solutions Act of 2006, which set the 2020 GHG emissions reduction goal into law. Later, California's Climate Change Scoping Plan was created to mitigate and reduce greenhouse gas emissions in California to 1990 levels by 2020.
Enhanced oil recovery (EOR) Enhanced coal-bed methane (ECBM)[2]	The most commonly used analog for geological carbon sequestration; both EOR and ECBM have tax incentive and will grow tremendously under a carbon-constrained situation; important difference between EOR or ECBM and CO_2 storage exists; using wells as storage sites bring new challenges and calls for changes in regulatory framework (e.g., USEPA UIC program) and industry practices.
Energy storage: storage of natural gas, liquefied natural gas, and petroleum reserves[2]	These are regulated with a) monitoring protocols to avoid leaks and potential human health or ecosystem impacts and b) siting and operations guidelines; similar regulations and guidelines are needed for carbon storage; however, energy storage is temporary while carbon storage is permanent.
Waste disposal: • Ocean dumping • incineration	Historically went through a range of regulatory challenges. • Ocean dumping is controlled by OSPAC and London Conventions • Incineration is drastically restricted. Waste prevention/minimization/land disposal/ underground storage is encouraged.
• USEPA's UIC program[3]	• Five classes of wells for waste injection; Class V is for CO_2 geological sequestration

[1] = USEPA (2012b); [2] = Forbes (2002); [3] = USEPA (2012a). UIC = Underground injection and control.

Tables 18.9 and 18.10 indicate that the climate change policies and regulations related to CCS are different between the US and EU. This is mainly because that the EU and the US disagree with the certainty of global climate change and, thus, have different tone of the strategies (Carlarne 2006). For example, the EU have approved and ratified the Kyoto Protocol, but the US has publicly repudiated the Protocol. Thus, the US is not obligated to comply with internationally agreed baselines or to meet internationally negotiated commitments. Under the Kyoto Protocol, the EU and the UK are obligated to monitor GHG emissions and to develop increasingly ambitious climate change policies (see Table 18.10). On contrary, the US lacks a comprehensive regulatory regime and has a limited scope to develop complimentary enforcement mechanisms. The US strategy relies majorly on market-base programs and voluntary collaboration with the privet sector, leaving much of the onus on individual states to regulate and monitor GHG emissions and CCS (Carlarne 2006). Table 18.11 shows the comparison of the political and legal responses to climate change between the US and EU and factors shaping these differences. It is important for policymakers to understanding them in order to formulate effective climate change and CCS policies.

Table 18.10. EU's existing regulations and analogs related to CCS and ICM

Regulation or analogs	Description and relationship to GHG capture and CCS
The European Union's Emission Trading Scheme (EU-ETS) established by Directive 2003/87/EC[1]	The first cap-and-trade program for GHG emissions and cap to meet commitments to Kyoto Protocol. It includes 27 countries and all large industrial facilities (power sector, refine petroleum, and produce iron, steel, cement, glass, and paper). Phase 1: 2005–2007; Phase 2: 2008–2012, and Phase 3: 2013–2020. Key info includes: • Starting 01/2013, new sectors (petrochemicals, ammonia and aluminum, nitrous oxide and perfluorocarbons and aviation) will be included. • A single EU-wide cap on total allowances will replace nationally-determined caps. • A principle of full auctioning for the allocation of allowances will begin in 2013 with power stations. Transitional allocations will see auctioning phased in gradually for other sectors; at least 50% of auctioning proceeds should be used for climate-related adaptation and mitigation. • CO_2 captured and safely stored will be considered as 'not emitted'; and smaller emitters (<25,000 tCO_2/year) may opt out of the EU ETS. • It is targeting 10–12 full-scale demonstrations in 2015.
Geological storage (GS) of CO_2 (Directive 2009/31/EC)[1]	• Is the legal framework for safe GS of CO_2 by permanent containment in the ground. • Has permit regime for exploration and storage, and selection criteria for storage sites. • Covers operation, closure and post-closure obligations, CO_2 acceptance criteria, monitoring and reporting obligations, inspections, measures in case of irregularities and/or leakage and provision of financial security.
Regulation EC/443/2009: Emission performance standards for reducing CO_2 from new passenger cars[1]	• Contains CO_2 emissions performance requirements for new passenger cars. • Car manufacturers must ensure their annual CO_2 average emissions (AEs) < 130g CO_2/km. From 2020 onwards AEs for the new car fleet must be 95g CO_2/km. • Manufacturers' AEs are determined based on a proportion of their new passenger cars registered that year: 65% in 2012 rising to 100% by 2015. • If targets are exceeded manufacturers must pay an excess emissions premium.
Reducing GHG emissions from transport fuels (Directive 2009/30/EC)[1]	• Binding target for the reduction of life cycle GHG emissions. • Requires suppliers to reduce life cycle GHG emissions per unit of energy from fuel and energy supplied by up to 10% by Dec. 31, 2020. • 'Life cycle GHG emissions' means all net emissions of CO_2, CH_4 and N_2O that can be assigned to the fuel or energy supplied. This includes emissions from extraction or cultivation (taking account of land use changes), transport and distribution, processing and combustion.
Other relevant EU directives[2]	The following directives were created without consideration of CCS, but they may be extended to CCS: a) water (2000/60/EC), b) waste (75/442/EEC), c) landfill (1999/31/EC), d) pollution (1996/61/EC), e) environmental impact assessment (85/337/EEC), and f) strategic environmental assessment (2001/42/EC). CO2 will be classified as either 'waste' or 'special category' under these Directives[3].

[1] = IEEP (2012); [2] = de Coninck et al. (2006); [3] = de Figueiredo et al. (2007).

Key Legal Issues and Uncertainties raised by CCS. CCS and any related carbon policies would raise many issues and uncertainties at both the international and domestic legal level, and these issues and uncertainties are delineated below.

Table 18.11. Comparison of US and EU's regulation framework and shaping factors[a]

US	EU
Differences in regulation framework	
• Utilizes a GHG intensity standard to relate GHG with economic prosperity.	• Calculate GHG emissions based on absolute discharges regardless of economic activity.
• Not a signatory to the Kyoto Protocol. No specific GHG emission reduction objective. Its GHG intensity is projected to decrease by 18% while projecting an overall GHG increase of 14%.	• a signatory to the Kyoto Protocol. Have firm GHG emission reduction goals (e.g., 8% below 1990 levels by 2012) and established trading scheme. Must establish how CCS fits into CO2 trading and accounting system.
• Its strategy relies on market-based programs and voluntary collaboration with the private sector without the supporting backbone of a strong regulatory program.	• Its strategies embody mandatory programs and obligations that are complimented by market-based programs and voluntary agreements.
• State and local policy-makers are adopting more stringent legislation, pushing the federal government to change.	• Actively support regional/local programs that compliment and extend the centralized climate change strategies.
• Not bounded by Kyoto Protocol. Not obligated the monitoring, enforcement and progressive development of related polices.	• Bounded by Kyoto Protocol. Obligated to the monitoring, enforcement and progressive development of related polices.
Key factors shaping the differences	
• Social factors:	• Social factors:
o Low levels of public awareness and less concerned about pressuring the government to respond to international environmental issues.	o High levels of public awareness and pressure for action on domestic and international environmental issues.
o In direct contradiction to the precautionary principle[b], the US has relied on a perceived lack of scientific certainty as a reason for postponing Kyoto-style measures to prevent climate change.	o The EU has promoted the precautionary principle as a key underlying factor for enacting the UNFCCC and the Kyoto Protocol and for aggressively addressing climate change.
o A less risk adverse society	o A more risk adverse society
• Political factors:	• Political factors:
o Big influence of different interests groups. Not easy to placate large energy and oil companies	o Politicians are more united concerning the certainty of climate change.
o Everything involves in party politics	o Climate change is beyond party politics
• Legal, economic and technological factors:	• Legal, economic and technological factors:
o On private lands, mineral rights and surface/pore space ownership are held by different parties, making the legal framework for pore space acquisition difficult.	o Mineral rights and surface/pore space owned by central government, making the legal framework for pore space acquisition more straightforward.
o Significant on-shore geological storage capacity.	o Off-shore locations are important.
o CCS costs a lot, but the US is not obligated to the Kyoto Protocol, so it does not need to comply with it.	o CCS costs a lot, and the EU is obligated to the Hyoto Protocol, so it must comply with it

[a] Mainly based on Carlarne (2006) and de Figueiredo et al. (2007). [b]The precautionary principle is a commonly applied principle of international environmental law. It states that "where there are threats of serious or irreversible damage, lack of full scientific certainty shall not be used as a reason for postponing cost-effective measures to prevent environmental degradation." (UN 1992).

International legal level. From a legal point of view, international trade law (e.g., WTO's agreements) can act to constrain domestic policy; this has not come up before the WTO, but serious issues could be raised with respect to a) subsidization of CCS, b) performance standard, and c) carbon pricing due to the different domestic legal policies (see Table 18.12) (WP 2012). For example, Canadian steel producers were granted a subsidy for using electricity produced with CCS technology, which has displaced steel imports from the US, and thus, is a WTO challenge. Currently, it is very uncertain about where to set the baseline and the extent that a CCS technology and its associated subsidy can be justified under GATT.

Applying carbon standard on the domestic production of a certain good (e.g., steel) or a certain industry (e.g., cement) is often coupled with a requirement that imports of that good meet the same standard, which may have several effects. For example, it often provokes the trading partners at the WTO because the trading partners may have different climate initiatives and level of development. Also, it stimulates changes in production methods in either domestic or foreign trading partners, resulting in in-effect discrimination, which then can be used by domestic or trading partners to mount a challenge at the WTO. More importantly, it is often very difficult to set emissions standards based on some baseline scenario of emissions for a certain industry. Take cement-making industry as an example, cement manufacturing is a heterogeneous one regarding GHG emissions. To set emissions standards based on some baseline scenario of emissions for cement-making is to invite challenge at the WTO.

As shown in Table 18.12, polices related to carbon tax and cap and trade also are associated with many potential issues. Carbon tax means that a government determines what emitters of GHG would pay for their emissions. Cap and trade means that the government sets a limit on the total amount of carbon that may be emitted in the country (the cap) and requires companies to bid on emission rights; unused emission rights can then be resold (trade). While both methods have pros and cons (see Table 18.13), cap and trade may provide more flexibility than carbon tax. A carbon tax requires a firm to decide, each year, how much to reduce its emissions and how much tax to pay. Under a cap-and-trade system, borrowing, banking and extended compliance periods allow firms the flexibility to make compliance planning decisions on a multi-year basis. In addition, historically, it proved cap and trade works. For example, implementation of a cap-and-trade system for sulfur dioxide reduction achieved the long-term reduction targets three years ahead of schedule and at a cost ($1.1 and $1.8 billion annually once fully implemented) significantly lower than expected ($6 billion). Anyway both methods may create real problem under the current WTO trade frame. One example is that in September 2012 EU recommended the suspension of the continent's carbon emission fees for airlines to avert a trade war with other countries, allowing time to forge a global agreement on climate charges for the aviation industry, which will delay the implementation of the EU-ETS Phase 3 plan concerning airline emission change (Table 18.10).

Therefore, there is a need for a uniform and effective approach to international trade law. This uniform approach has implications for policy intended to promote ICM and CCS. As an international environmental treaty, UNFCCC carries the big hope to push different countries or regions to build the regulatory framework concerning ICM and CCS. For example, The US develops the national GHG inventory each year to track the US trend in emissions to comply with its obligations under the UNFCCC, even though the US has not accepted the Kyoto Protocol (USEPA 2012c).

Domestic level. To make CCS economically viable requires considerable governmental intervention, resources and coordination powers. At the domestic level, CCS raised issues and uncertainties include: a) making the high costs of on-site CCS units affordable; b) investing infrastructure (e.g., a vast pipe network) to transport captured carbon to storage sites; and c) developing a transparent and effective legal and regulatory regime to promote CCS. For example, laws regarding mining, oil and gas operations, pollution control, waste disposal, nuclear waste storage, pipelines, property and liability may be relevant and extendable to CCS (de Coninck et al. 2006). The existing standards were not designed with long-term carbon sequestration in mind; thus carbon sequestration may require new standards and increased cooperation between federal and state agencies.

Table 18.12. Key international legal issues and uncertainties raised by CCS and ICM[b]

Policy	Key issues and considerations
Subsidies (from government/ private backers). Examples: • Grants on specific infrastructure • Tax exemptions • Emission permits • Transportation/storage	• Any subsidies may be considered as potential forms of "aids granted by states" that violate WTO's fair-trading standards under ASCM o May cause serious prejudice (e.g., significant price depression) o May displace imports or increase in the market share of the subsidizing member • Failing to equalize marginal costs across different emission source • Failing to provide appropriate price incentives to reduce consumption • Reduce emissions at a cost > a broadly-based cap-and-trade scheme or carbon tax • Requirement of funds, which requires increasing other taxes
Performance Standards: • Standards require use of CCS tech. in production • Standards require level of carbon emission only achievable w/ CCS tech. • Standards promotes low carbon emissions	• The imposition of an emissions performance standard needs to conform to or be justified under GATT (e.g., prohibits import bans or quotas, avoids protectionism, and prohibits discrimination between importing countries) o Must not discriminate, e.g., a) a standard applied to some countries and not others or b) banning the import of non-CCS products or high emission products, without a parallel domestic regulation, would violate GATT o May affect the conditions of competition • It is difficult to determine an acceptable performance standard that accounts for a country's climate initiatives and level of development • It is not easy to implement regulations without having an in-law or in-effect discriminatory effect on the conditions of competition to "no less favorable states" • Imposing these standards may lose comparative advantage in emission-intensive industries, which may move to less stringently-regulated places (states) • Unilateral imposition of performance standards may cause 5–20% emission leakage[b]
Carbon Pricing: • Domestic carbon tax • Cap-and-trade scheme	• A carbon price may make foreign companies more competitive while eroding market share for the businesses of the country (or region) with such a policy • Any restriction a country imposes on imports can also just as easily be turned around and imposed on the country's exports • Carbon pricing coupled with a border adjustment is a solution, but calculation of carbon footprints on an industry level is a big administrative burden • One byproduct of cap and trade is "leakage," by which investment and jobs are driven to nations that have looser or nonexistent climate regimes and therefore lower costs • Method of distributing allowances could have mixed resuls

[a] GATT administered by WTO, WTO's ASCM, and WTO's TBT are the relevant provisions. GATT = The General Agreement on Tariffs and Trade; WTO = The World Trade Organization; ASCE = Agreement on Subsidies and Countervailing Measures; TBT = Agreement on Technical Barriers to Trade. [b] 5-20% leakage = if EU emissions are reduced by 100 units, rest-of-world emission increases by 5–20 units.

Currently, it is commonly believed that establishment of comprehensive climate change legislation is most important, including GHG emission lists, carbon credit trading/carbon price). The next urgent issues are related to legal and regulatory clarity, authority and support for safe and effective CCS deployment. Nowadays, it is not clear (at least in the US) about what agency is for issuing CCS permits; who owns the right to use geological formation for CO_2 storage; how to prevent significant environmental impacts of CCS project; how to regulate the safety and operation of CCS projects; how to develop and improve long-term liability and stewardship framework; what are the procedures for aggregating and adjudicating the use of and compensation for pore space for CCS projects; and what are the standard protocols of monitoring, verification and accounting (MVA) for CO_2 storage. Governments (state and federal) must work together to address these issues and uncertainties.

Table 18.13. Comparison between carbon tax and cap-and-trade[1–4]

Criterion	Carbon tax	Cap-and-trade
• Mechanisms/Principle	• Government sets the carbon price; the market sets the quantity emitted[1]	• Government sets the quantity of carbon emitted, and the market sets the price[1]
• Potential for developing carbon market network	• Low potential to link all related components together for network development[1]	• Allows us to link state, regional, and national carbon permit markets with each other and with international ones, which may contain the costs of climate solutions[1]
• Providing the certainty for reducing GHG emissions	• Emissions levels automatically trigger tax-rate adjustments; thus, the certainty for reducing GHG emissions exists[1]	• Overall emissions of GHG would be capped; reducing the allowed emissions over time as technology allows for greater reductions[2]
• Responsive to economic conditions	• Very limited. it only be changed by an act of Congress	• It is a market-based system, allowing the price to emit GHGs change with the economy[2]
• Historical/current practices and cases	• Some cases, e.g. Finland, Norway, the Netherlands, Sweden, Denmark, the UK, Quebec, British Columbia, Boulder Colorado and the Bay Area of California have the programs[4]	• SO_2 cap and trade program for acid rain control works[2]. All member States in EU + Iceland, Liechtenstein and Norway, Australia, New Zealand and ten Northeastern US States have cap-and-trade programs[4]
• Complexity for administration	• Simpler to initiate and administer quickly[1]	• Complex and takes time for the system to evolve into a mature one
• Resources/Revenues distribution	• Controlled by the government[2], could generate revenues to be used for the public benefit and may have built-in protections for low-income families[1]	• Creates its own durable political constituency. Businesses will protect actions to keep the value of carbon permits and uniform distribution of the resources and revenues[1]
• Potential for gaming (manipulation through collusion or fraud	• Low. However, as with other taxes, a carbon tax can be rendered ineffective through loopholes and exemptions[1]	• High. Gaming could lower confidence in the carbon market, decrease its liquidity, and reduce the economic efficiency of the market[3]

[1] = Durning et al. (2009); [2] = PNM (2012); [3] = Taylor (2012); and [4] = WP (2012).

18.4 Social Acceptability Issues

Working toward public acceptance of CCS is urgent and necessary in order to a) obtain public subsidies for early demonstration projects, b) negotiate property rights issues to create large and legal storage units, c) secure siting approvals, and d) resolve issues related to long-term post-closure liability. The social acceptability of CCS includes the responses of the lay public and of stakeholders (de Coninck et al. 2006), and they are described separately here.

Lay Public Perceptions. Studies have been conducted to study public perceptions of CCS (de Coninck et al. 2006), and the main findings are as follows:

- Most of the lay public (70–96%) are not familiar with the terms CCS, and other power generation technologies (e.g., nuclear power, renewables).
- Most of the lay public support the concept of CCS (limited to offshore geological storage) or nuclear power and are away from renewables. They are slightly positive about CCS in general terms, but neutral to negative about storage in the immediate vicinity of the neighborhood.
- The negative effects that are thought of by most of the lay public are leakage, potential impacts upon ecosystems and health risk. Some people believe that the risk and drawbacks of CCS are larger than the benefits to the environmental and society.
- The lay public has not regarded anthropogenic global climate change as a relatively serious problem and has not accepted the need for large scale reduction of CO_2 to reduce the threat of global warming.
- Levels of trust in key institutions and the role of the media have a major influence on how CCS is received by the lay public.

Stakeholder Perceptions. Stakeholders are agents that have professional interest in CCS via employment or personal engagement, such as industry, industry associations, private or governmental organizations, etc. Usually, stakeholders have a defined agenda or *a priori* viewpoint when evaluating CCS, whereas the lay public has no well-formed opinions on most issues associated with CCS. Some stakeholders are directly involved in the formulation and design of CCS policies, programs and projects (PPPs). Many stakeholders of coal and lignite industries often support CCS as they are trying to stay in business, whereas it is much more difficult to characterize the position of nongovernment organizations (NGOs). Perceptions of CCS by stakeholders are often tangled with that of other energy technologies and carbon policies, such as nuclear power, carbon pricing, etc. Usually, if the CCS is developed at the cost of renewable energy, the NGOs would be against it. The CCS technologies are more acceptable by the NGOs if they are bridging options to help strong renewables development. The comparison with lack of acceptance in other energy technologies could lead to interesting insights and lessons. Further research is needed.

Studies indicate that there is a long way to go before CCS can be widely accepted by the lay public. Major public concerns about CCS originate from constrains of CCS, such as its efficacy, associated risks, the costs and its competition with other low-carbon options. To this end, expanding government education and engagement efforts is imperative, such as development of well-thought-out and well-funded public outreach programs to educate the public about climate change, low-carbon technologies and the risks and benefits of CCS. In addition, a reliable government with a trustworthy regulatory framework for CCS and carbon pricing is likely to increase acceptance. Moreover, conducting demonstration CCS projects is also vital to positively influence public perception. These projects must have broad science and technology components to answer key regulatory questions and their results must be publicly available (de Figueiredo et al. 2007).

18.5 Technical Issues: Uncertainty and Scalability

In Sections 18.2–18.4, issues are described and discussed related to cost and economics, legal and regulatory, and social acceptability of CCS. In this section, we mainly focus on technical issues related to CCS development. To provide a comprehensive picture about this topic, we divide the CCS processes and/or technologies into several different areas as shown in Table 18.14.

In general, source identification is easy, but quantification for carbon footprint is very challenge, particularly on the basis of factory-by-factory. However, this calculation is critical as carbon pricing will be based on such information. One big concern under the category of source identification is that several different sources are not targeted by CCS even though they make major contribution to global GHG emissions (Herzog et al. 2009; Zhang and Surampulli 2012). This could be because technologies (e.g., air capture systems, synthetic trees) for capturing CO_2 from small mobile/distributed point source are not mature (Lackner 2009; Lackner and Brennan 2009; Plasynski et al 2009; Zhang and Surampulli 2012).

CCS involves 4 major systems: a) capture and compression; b) transportation; c) injection; and d) storage reservoir. Each system can leak CO_2 and thus, should be treated as a source for emissions. As shown in Table 18.14, many issues and unknowns are linked with these systems. Currently, most of public concerns originate from MVA of CO_2 storage in different settings due to complexity of the system and the related environmental issues and health risks. However, beneficial uses of captured CO_2 and the relationships among different low-carbon options are also critical for CCS implementation.

It is imperative to improve CCS technologies to lower the cost, demonstrate the feasibility of CCS, solve fundamental problems, and reuse CO_2 in a beneficial way. From technical point of view, the future path of CCS depends on how much efforts will be made to: a) promote government support to establish framework for CCS deployment; b) foster the success of CCS projects, particularly commercial-scale demonstrations; c) conduct cutting-edge research to establish CCS, such as (i) technologies and the related fundamentals (e.g., long-term strategies for CO_2 source clusters and CO_2 pipeline networks, mapping CO_2 storage potential of deep saline formations, value-added CO_2 reuse pathways and (ii) standards and consistent requirements to ensure the safe and effective operation of CCS, MVA and reporting; and d) support international collaboration to facilitate the global deployment of CCS (Zhang and Surampalli 2013).

Table 18.14. Key technical issues for CCS development (Zhang and Surampalli 2013)

System/technology	Issue/unknown and their implications
Source identification	**Calculation of carbon footprint is a big burden to regulatory officers**
• Concentrated point sources	• Calculate carbon footprint on a factory-by-factory base is formidable
• Mobile/distributed point sources	• Contributing 22% of global GHG emissions, but not targeted for CCS
• Non-point sources	• Contributing 35% of global GHG emissions, but not targeted for CCS
Carbon capture technologies	**Capture reliability, cost, energy penalty are unknown**
• Conc./mobile/diffused point sources	• Effects of varying purity of CO_2 streams are unknown
• Non-point sources	• Poor capture of CO_2 from small mobile/distributed point source
Transport of CO_2	• Issues: a) regulatory classification of CO_2 itself (commodity or pollutant), b) economic regulation, c) utility cost recovery, d) pipeline right-of-ways, e) pipeline safety, f) environmental impact • Optimization of pipeline network in concert with sophisticated CCS (e.g., zero-emission power generation plants), CO_2 inventory (what, where, when), renewable energy technologies, and CO_2 reuse technologies is very challenging
Long-term storage of CO_2	**Many unknowns and challenges exist**
• Geological storage	• Issues: a) little is known about geological performance in a variety of geological settings and reservoir types (e.g., saline aquifers, consequence of overpressuring and acidification of the reservoir); b) new technologies to ensure CO_2 stays in place forever; and c) site appraisal studies to reduce harmful effects
• Mineral storage	• Issues: a) the kinetics of natural mineral carbonation is slow; b) the resulting carbonated solids must be stored at environmental suitable locations; c) how to: i) reduce cost and energy requirement and ii) integrate/optimize power generation, mining, carbonation reaction, carbonates' disposal, material transport, and energy in a site-specific manner
• Ocean storage	• Issues: a) unknown impact on ecosystems (e.g., ocean acidification, wildlife, oxygen supply) and on microbial carbon pump and biological carbon pump; b) difficult to certify the dissolution, leakage and location of CO_2; c) how to: i) make reliable predications of the technical feasibility and storage times, ii) predict and minimize any environmental impact, and iii) make reliable cost estimates and assess the net benefit.
MVA and LCRM of CCS[a]	**Leakage of CO_2 is possible. The CO_2 with concentrations > 5–10% of the air volume is lethal**
• MVA	• Unknowns: standard procedures for a) site performance assessment, b) regulatory compliance; and c) health, safety, and environmental impact assessment are not available
• LCRM	• Issues/risks: a) for pre-operation, i) problems with licensing/permitting, ii) poor conditions of the existing well bores, and iii) lower-than-expected injection rates; b) for operation, i) vertical CO_2 migration with significant rates, ii) activation of the pre-existing faults/fractures, iii) substantial damage to the formation/caprock, iv) failure of the well bores, v) lower-than-expected injection rates, vi) damage to adjacent fields/producing horizons; and c) for post-injection, leakage i) through pre-existing faults or fractures, ii) through the wellbores, iii) due to inadequate caprock characterization, iv) due to inconsistent or inadequate monitoring
Beneficial uses of CO_2	• Information is lack on large-scale CO_2 beneficial uses for value-added products or beneficial activities • Pathways and novel approaches for beneficial uses of captured CO_2 are not fully understood
Relationships among low-C options	• Advances of CCS depends on future carbon restrictions and price and their interactions with other low-C options • There exists an implicit competition between the development of CCS and that of other low-C technologies; CCS may be less cost-effective than renewable energy; CCS raises costs and energy prices, and requires significant water • It is unknown what an optimal portfolio of a range of energy sources contains and how to achieve it

[a] MVA = monitoring, verification and accounting; LCRM = life cycle risk management of CCS, including development and quality CCS technology → Propose site → Prepare site → operate site → close site → post closure liability.

18.6 Conclusions

CCS can play a central role in the mitigation of GHG emissions. Currently, there's a huge gap between what can technically do and what we are doing. High costs, inadequate economic drivers, remaining uncertainties in the regulatory and legal frameworks for CCS deployment, and uncertainties regarding public acceptance are barriers to large-scale applications of CCS technologies in the world.

Wide-scale application of CCS would reduce CO_2 emissions from flue stacks of coal power plants by 85–90% with an increase in resource consumption by one third. Completing the cycle of carbon capture and storage may double the US industrial electricity price (i.e., from 6 to 12 ¢/kWh) or increase the typical retail residential electricity price by ~50%. There is a need to invest over ~US$5 trillion for CCS deployment from 2010 to 2050 to achieve a 50% reduction in GHG emissions by 2050. However, the true costs of different CCS technologies for different applications are still unknown because of our limited experience. Overcoming the challenge for commonality in generation of CCS cost estimates will ultimately lead to a reduction in the uncertainty, variability, and bias of CCS costs estimates.

Currently, the legal and policy framework for CCS is under the umbrella of the international law that is related to the framework of climate changes policies and incentivizing carbon management. These frameworks are not established, with international legal policies tangling with domestic ones. Adopting any type of carbon policy or promoting low-carbon technologies (e.g., CCS) may trigger the design and implementation of some measures to offset the perceived competitive disadvantage that might be imposed to domestic firms. These measures raise international trade issues as they may increase tariffs and other barriers to trade. Therefore, domestic trade policy and unilateral measures for CCS are vulnerable to challenge if they discriminate between domestic and foreign products or between imports from two different countries.

The social acceptability of CCS includes the responses of the lay public and of stakeholders. Currently, government education and engagement efforts (e.g., support field-scale demonstration projects) are far behind. Major public concerns about CCS originate from constrains of CCS, such as its efficacy, associated risks, the costs and its competition with other low-carbon options. There is an urgent need to improve CCS technologies to lower the cost, demonstrate the feasibility of CCS, solve fundamental problems, and reuse CO_2 in a beneficial way.

It is imperative to overcome the technical, regulatory, financial and social barriers. Deployment of large-scale demonstration CCS projects within a few years will be critical to gain the experience necessary to reduce cost, improve efficiency, remove uncertainties, and win public acceptances of CCS. Because CCS is expensive, regulatory frameworks should help in establishing and promulgating best practices and allowing regulated utilities to make investments in capture technologies (Landcar and Brennan 2009). In addition, government should be more involved in clarifying legislation barriers, the management of safe and permanent carbon storage, and supporting international collaboration to facilitate the global deployment of CCS. Furthermore, considerable research is needed in the future for CCS development.

18.7 References

Abellera, C., and Short, C. (2011). "The costs of CCS and other low-carbon technologies." Global CCS Institute, *Issues Brief* 2011, No 2, 1–12.

Carlarne, C.P. (2006). "Climate change policies an ocean apart: EU & US climate change policies compared." *Pennsylvania State Environmental Law Review,* 14(3), 435–482.

CCCSRP (California Carbon Capture and Storage Review Panel) (2010). *Findings and Recommendations by the California Carbon Capture and Storage Review Panel.* CCCSRP, Dec. 2010. Available at <www.climatechange.ca.gov/carbon_capture_review_panel/.../2011-01-14...> (accessed Feb. 2012).

CMU (Carnegie Mellon University), Electric Power Research Institute (EPRI), Global CCS Institute, IEA Greenhouse Gas R&D Programme (IEAGHG), International Energy Agency (IEA), Massachusetts Institute of Technology (MIT), Vattenfall (2011). *Proceedings from a CCS Cost Workshop*, CCS Cost Workshop, Paris, France. March 22, 2011. Available at <http://www.global ccsinstitute.com/publications/proceedings-ccs-cost-workshop> (accessed Oct. 2012).

Rubin, E.S. (2011). "Methods and measures for CCS Costs." Presentation given at CCS Cost Workshop, Paris, France. March 22, 2011. *Proceedings from a CCS Cost Workshop* by Carnegie Mellon University CMU), Electric Power Research Institute (EPRI), Global CCS Institute, IEA Greenhouse Gas R&D Programme (IEAGHG), International Energy Agency (IEA), Massachusetts Institute of Technology (MIT), Vattenfall. Available at <http://www.global ccsinstitute.com/publications/proceedings-ccs-cost-workshop> (accessed Oct. 2012).

de Coninck, H. (2006). "Costs and economics of CO_2 capture and storage." Presentation given at Near Zero Emissions Coal Workshop, July 5, 2006, Beijing, China.

de Figueiredo, M.A., Herzog, H.J., Joskow, P.L., Oye, K.A., and Reiner, D.M. (2007). *Regulation of Carbon Capture and Storage*. Policy Brief written for International Risk Governance Council (IRGC) workshop on regulation of geological storage of CO_2, Washington, D.C., March, 2007. Available at <http://www.irgc.org/Expert-contributions-and-workshop.html> (accessed Oct. 2012).

de Coninck, H., Anderson, J., Curnow, P., Flach, T., Flagstad, O.-A. Groenenberg, H., Norton, C., Reiner, D., and Shackley, S. (2006). *Acceptability of CO_2 Capture* *and Storage.* Energy Research Centre of the Netherlands (ECN), ECN-C-06-026, 42 p.

de Richter, R. (2012). "Optimizing geoengineering schemes for CO_2 capture from Air." PPT slides Available at <http://data.tour-solaire.fr/optimized-Cabon-Capture% 20RKR% 20final.pps> (accessed Feb. 2012).

Durning, A., Fahey, A., de Place, E., Stiffler, L., and Williams-Derry, C. (2009). *Cap and trade 101: A federal climate policy primer.* Sightline Report, Sightline Institute, Seattle, Washington 98101, 2009.

ENB (Earth Negotiations Bulletin). (2012). "Summary of the Doha climate change conference." *ENB*, 12(567), Dec., 11, 2012.

Forbes, S.M. (2002). "Regulatory barriers for carbon capture, storage and sequestration." National Energy Technology Laboratory. Available at <http:// www.netl.doe.gov/technologies/carbon_seq/refshelf/reg-issues/capture.pdf> (accessed Oct. 2012).

Herzog, H.J. (2001). "What future for carbon capture and sequestration?" *Environ. Sci. Technol.*, 35(7), 148A–153A.

Herzog, H.J., Meldon, J., and Hatton, A. (2009). *Advance Post-Combustion CO_2 Capture.* Technical Report prepared for the Clean Air Task Force. <http://web. mit.edu/mitei/docs/reports/herzog-meldon-hatton.pdf> (accessed Feb. 2012).

IEA (International Energy Agency) (2009). *Technology Roadmap: Carbon Capture and Storage.* An IEA report, Available at <http://www.iea.org/papers/ 2009/CCS_ Roadmap.pdf> (accessed Feb. 2012).

IEA (International Energy Agency) (2010). *Energy Technology Perspectives 2010: Scenarios and Strategies to 2050.* OECD/IEA, Paris, France.

IEA (International Energy Agency) (2011). *Cost and Performance of Carbon Dioxide Capture from Power Generation.* OECD Publishing at 22–24, 27, 31.

IEEP (Institute for European Environmental Policy) (2012). "Key EU climate change legislation." Available at <http://www.ieep.eu/minisite_assets/ briefings-on-climate/pdfs/key_legislation .pdf> (accessed Oct. 2012).

IPCC (Intergovernmental Panel on Climate Change) (2005). *IPCC Special Report on Carbon Dioxide Capture and Storage.* IPCC Working Group III. Cambridge Univ. Press, Dec., 19, 2005, 431 pp.

Lackner, K.S. (2009). "Carbon dioxide capture from ambient air." In: Blum, K., Keilhacker, M., Platt, U., and Roether, W. (eds.) *The physics perspective on energy supply and climate change–a critical assessment.* Eur. Phys. J. Special Topics, Springer Verlag, Bad Honnef, 2009.

Lackner, K.S., and Brennan, S. (2009). "Envisioning carbon capture and storage: expanded possibilities due to air capture, leakage insurance, and C-14 monitoring." *Climatic Change*, 96, 357–378.

Li, Z.-S., Fang, F., and Cai, N.-S. (2009). "CO_2 capture from flue gases using three ca-based sorbents in a fluidized bed reactor." *J. Environ. Eng.*, 135(6), 418–425.

Lippone, J. (2012). "Incentive policy strategy for CCS." Power-point presentation given to Carbon Capture and Storage: Regaining Momentum (an international conference), Sydney, Feb. 21, 2012.

Plasynski, S.I., Litynski, J.T., McIlvried, H.G. and Srivastava, R.D. (2009). "Progress and New Developments in Carbon Capture and Storage." *Critical Reviews in Plant Sciences*, 28(3), 123–138.

PNM (2012). "Cap and trade vs. carbon tax." Available at <http://F:\CO2\References\CCS-Cost-Policy-Trading\C trading\Cap and Trade vs_Carbon Tax PNM.htm> (accessed Nov. 2012).

Surampalli, R., Zhang, T.C., Ojha, C.S.P., Tyagi, R.D., and Kao, C.M. (2013). *Climate Change Modeling, Mitigation and Adaptation,* ASCE, Reston, Virginia, 2013.

Taylor, M. (2012). *Evaluating the policy trade-offs in ARB's cap-and-trade program.* A Legislative Analyst's Office Report. Feb. 2012. Available at <http://www.lao.ca.gov> (accessed Nob. 2012).

UN (United Nations) (1992). *Rio Declaration on Environment and Development*, U.N. Conference on Env. & Dev., U.N. Doc. A/CONF.151/5/Rev.1, 31 I.L.M. 874 (1992).

USDOE (US Department of Energy) (2010). *Report of the Interagency Task Force on Carbon Capture and Storage.* USDOE Interagency Report, Aug., 2010. <http://www.fe.doe.gov/programs/sequestration/ccstf/CCSTaskForceReport2010.pdf> (accessed July 2012).

USEPA (U.S. Environmental Protection Agency) (2012a). Class V wells. Available at <http://water.epa.gov/type/groundwater/uic/class5/index.cfm> (assessed Oct. 2012).

USEPA (U.S. Environmental Protection Agency) (2012b). Cap and Trade Programs Available at <http://www.epa.gov/captrade/> (assessed Oct. 2012).

USEPA (U.S. Environmental Protection Agency) (2012c). U.S. Greenhouse Gas Inventory Report. Available at <httphttp://www.epa.gov/climatechange/ghgemissions/usinventoryreport.html> (assessed Oct. 2012).

White, C.M., Strazisar, B.R., Granite, E.J., Hoffman, J.S., and Pennline, H.W. (2003). "Separation and capture of CO_2 from large stationary sources and sequestration in geological formatons: Coalbeds and deep saline aquifers." *J. Air & Waste Manag. Asso.*, 53(6), 645–715.

WP (White Paper) (2012). "An international trade and investment framework for carbon management technologies." A white paper for a workshop on July 12–13, 2012 in Vancouver, Canada. Available at <http://www.cmc-nce.ca/wp-content/uploads/ 2012/08/2012_07_18_white_paper.pdf> (accessed Oct. 2012).

Zhang, T.C., and Surampalli, R.Y. (2013). "Carbon capture and storage for mitigating climate changes." Chapter 20 in *Climate Change Modeling, Mitigation and Adaptation,* Surampalli, R., Zhang, T.C., Ojha, C.S.P., Gurjar, B.R., Tyagi, R.D., and Kao, C.M. (eds.), ASCE, Reston, Virginia, February 2013.

Index

Page numbers followed by e, f, and t indicate equations, figures, and tables, respectively.

Editor Biographies

Dr. Rao Y. Surampalli, P.E., Dist.M.ASCE, received his M.S. and Ph.D. degrees in Environmental Engineering from Oklahoma State University and Iowa State University, respectively. He is a Registered Professional Engineer in the branches of Civil and Environmental Engineering, and also a Board Certified Environmental Engineer (BCEE) of the American Academy of Environmental Engineers (AAEE) and a Diplomate of the American Academy of Water Resources Engineers (DWRE). He is an Adjunct Professor in seven universities and distinguished/honorary professor in four universities. Currently, he serves, or has served on 66 national and international committees, review panels, or advisory boards including the ASCE National Committee on Energy, Environment and Water Policy. He is a Distinguished Engineering Alumnus of both the Oklahoma State and Iowa State Universities, and is an elected Fellow of the American Association for the Advancement of Science, an elected Member of the European Academy of Sciences and Arts, an elected Member of the Russian Academy of Engineering, an elected Fellow of the Water Environment Federation and International Water Association, and a Distinguished Member of the American Society of Civil Engineers. He also is Editor-in-Chief of the ASCE Journal of Hazardous, Toxic and Radioactive Waste, and past Vice-Chair of Editorial Board of Water Environment Research Journal. He has authored over 600 technical publications in journals and conference proceedings, including 235 refereed journal articles, 15 patents, 16 books and 109 book chapters.

Dr. Tian C. Zhang, P.E., is a Professor in the department of Civil Engineering at the University of Nebraska-Lincoln (UNL), USA. He received his Ph.D. in environmental engineering from the University of Cincinnati in 1994. He joined the UNL faculty in August 1994. Professor Zhang teaches courses related to water/wastewater treatment, remediation of hazardous wastes, and non-point pollution control. Professor Zhang's research involves fundamentals and applications of nanotechnology and conventional technology for water, wastewater, and stormwater treatment and management, remediation of contaminated environments, and detection/control of emerging contaminants in the environment. Professor Zhang has published more than 84 peer-reviewed journal papers, 48 book chapters and 8 books since 1994. Professor Zhang is a member of American Association for the Advancement of Science (AAAS), the Water Environmental Federation (WEF), and Association of Environmental Engineering and Science Professors (AEESP). Professor Zhang is a Diplomate of Water Resources Engineer (D.WRE) of the American Academy of Water Resources Engineers, and Board Certified Environmental Engineers (BCEE) of the American Academy of Environmental Engineers, Fellow of American Society of Civil Engineers (F. ASCE), and Academician of European Academy of Sciences and Arts. Professor Zhang is the Associate Editor of *Journal of Environmental Engineering* (since 2007), *Journal of*

Hazardous, Toxic, and Radioactive Waste (since 2006), and *Water Environment Research* (since 2008). He has been a registered professional engineer in Nebraska, USA since 2000.

Dr. R. D. Tyagi is an internationally recognized Professor with 'Institut national de la recherche scientifique – Eau, terre, et environment', (INRS-ETE), University of Québec, Canada. He holds Canada Research Chair on, 'Bioconversion of wastewater and wastewater sludge to value added products.' He conducts research on hazardous/solids waste management, water/wastewater treatment, sludge treatment/disposal, and bioconversion of wastewater and wastewater sludge into value added products. He has developed the novel technologies of simultaneous sewage sludge digestion and metal leaching, bioconversion of wastewater sludge (biosolids) into *Bacillus thuringiensis* based biopesticides, bioplastics, biofertilisers and biocontrol agents. He is a recipient of the ASCE State-of-the Art of Civil Engineering Award, ASCE Rudolph Hering Medal, ASCE Wesley Horner Medal and ASCE Best Practice Oriented Paper Award. He also received 2010 Global Honor Award (Applied Research) of International Water Association, 2010 Grand Prize (University research) of American Academy of Environmental Engineers, and Excellence Award of Natural Sciences and Engineering Research Council of Canada for Industry-University collaborative research. Dr. Tyagi has published/presented over 600 papers in refereed journals, conferences proceedings and is the author of eleven books, seventy book chapters, ten research reports and nine patents.

Professor Ravi Naidu received his PhD in soil science from Massey University, New Zealand and his MS was jointly awarded by the University of Aberdeen, UK, and the University of the South Pacific, Fiji. He is the Chief Executive Officer and Managing Director of CRC CARE and the Founding Director of the University of South Australia's Centre for Environmental Risk Assessment and Remediation. His work focuses on the remediation of contaminated soil, water and air, and the potential impacts of contaminants upon environmental and human health at local, national and global levels. He has helped revolutionise contamination science by leading the move to a risk-based approach to managing contaminated sites. He has also been a leader in the shift to *in situ* remediation – cleaning up contamination where it lies, rather than the traditional 'dig and dump' approach. He is an elected Fellow of the Soil Science Societies of America (2000), New Zealand (2004) and Agronomy Society of America (2006). In 2012 he was chosen as a winner of the Soil Science Society of America's International Soil Science Award, and in 2013 was elected a Fellow of the American Association for the Advancement of Science. He is Chair of the International Committee on Bioavailability and Risk Assessment and was Chair of the Standards Australia Technical Committee on Sampling and Analyses of Contaminated Soils (1999-2000), Chair of the International Union of Soil Sciences Commission for Soil Degradation Control, Remediation and Reclamation (2002-10), and President of the International Society on Trace Element Biogeochemistry (2005-07). He has authored or co-authored over 600 research articles and technical publications as well as six patents, and co-edited 10 books and over 55 book chapters in the field of soil and environmental sciences. He has also supervised over 30 PhD completions.

Dr. B. R. Gurjar is on the faculty of Environmental Engineering in the Civil Engineering department of Indian Institute of Technology – IIT Roorkee. He is also Core Faculty in IIT Roorkee's Centre for Transportation Systems, and Centre of Excellence in Disaster Mitigation & Management, and heading Max Planck Partner Group for Megacities and Global Change. He holds a PhD in Environmental Risk Analysis from IIT Delhi. He is fellow and life member of several professional bodies and organizations and Honorary Secretary of Roorkee Local Centre of The Institution of Engineers (India). He is co-author/co-editor of six books and a number of research papers. He has received several awards and fellowships including the prestigious Advanced Postdoctoral Research Fellowship of the Max Planck Society (Germany) to support his research tenure (2002-2005) at the Max Planck Institute for Chemistry (MPIC) in Mainz, Germany. The Scientific Steering Committee of the global change System for Analysis, Research and Training (START), Washington, D.C., U.S.A., honored him with the 2004 START Young Scientist Award. He is also co-recipient of The Nawab Zain Yar Jung Bahadur Memorial Medal (best research paper award) from Environmental Engineering Division of The Institution of Engineers (India), Kolkata, for the year 1995-96. The Institution of Engineers (India)'s National Design and Research Forum (NDRF) has recently awarded him the National Design Award for Environmental Engineering 2011.

Dr. C.S.P. Ojha is a Professor in Civil Engineering Department of Indian Institute of Technology, Roorkee. He holds a Ph.D. in Civil Engineering from Imperial College of Science, Technology and Medicine, London, U.K. He has published more than 150 research papers in peer-reviewed journals and supervised/co-supervised 34 PhD theses and 85 M.Tech. dissertations. His research interests include modeling of water resources and environmental engineering systems with a focus on Food and water security related problems. He has worked as Editor for Journal of Indian Water Resources Society and also as guest editor for couple of special issues. Dr. Ojha has been a Commonwealth Research Scholar at Imperial College of Science, Technology and Medicine, London, U.K. (Oct. 1990-Sept. 1993); Visiting Scholar at Louisiana State University, USA. (April-July 2000); Alexander Von Humboldt Fellow at Water Technology Center, Karlsruhe, Germany (Dec. 2001-July, 2002); Guest AvH Fellow, Institute for Hydromechanics, University of Karlsruhe, Germany (Jan. 2002-July, 2002); and Visiting Professor, Civil Engineering, AIT Bangkok (Aug. 2004-Nov. 2004). He has coordinated/co-coordinated several international research projects involving EU, Germany, France, Canada, U.K. and USA. He was conferred BCEEM by American Academy of Environmental Engineers in 2011 and E.M. Curtis Visiting Professorship by Purdue University in 2012.

Dr. Song Yan is currently working as a research associate at Institut National de la Recherche Scientifique (INRS), Eau, Terre et Environnement, Université du Québec, Canada. Dr. Yan's research focuses on: the production of bioplastics, biodiesel, value added products using wastewater and sludge; greenhouse gas emissions from wastewater and sludge (biosolids) treatment processes; and wastewater and sludge treatment process control. Dr. Yan has published 38 papers in refereed journals and14 papers in conference proceedings. She is author of 40 book chapters and 2 scientific reports. She has attended 33 national and international conferences, and also has submitted 99 gene sequences to GenBank, USA. She was selected as an Outstanding Reviewer of the ASCE's Journal of Environmental Engineering in 2012, and is the member of Canadian Association for Water Quality (CAWQ) and International Water Association (IWA).

Dr. S. K. Brar is an Associate Professor at Institut National de la Recherche Scientifique (Eau, Terre et Environnement, INRS-ETE). She graduated in Master's in Organic Chemistry from National Chemical Laboratory, Pune, India with Master's in Technology in Environmental Sciences and Engineering from Indian Institute of Technology, Bombay, Mumbai, India and a Ph.D. in Environmental Biotechnology from INRS, Quebec, Canada. Dr. Brar is a recipient of the ASCE State-of-the-Art of Civil Engineering award (2007) for her article titled, "Bioremediation of Hazardous Wastes - A Review," which was published in the Practice Periodical of Hazardous, Toxic & Radioactive Waste Management – Special issue on Bioremediation. She has also received the Rudolf gold medal (2008) for her originality of the article published in Practice Periodical of Hazardous, Toxic & Radioactive Waste Management. Her research interests lie in the development of finished products (formulations) of wastewater and wastewater sludge based value-added bioproducts, such as enzymes, organic acids, platform chemicals, biocontrol agents, biopesticides, butanol and biohydrogen. She is also interested in the fate of endocrine disrupter compounds, pharmaceuticals, nanoparticles and other toxic organic compounds during value-addition of wastewater and wastewater sludge in turn finding suitable biological detoxification technologies. She is on the editorial board of Brazilian Archives of Biology and Technology Journal and associate editor of two international repute journals. She has won several accolades through her professional career through awards, such as outstanding young scientist in India and several others. She has more than 162 research publications which include three books, 30 book chapters, 80 original research papers, 50 research communications in international and national conferences and has registered 2 patents to her credit.

Dr. Anushuya Ramakrishnan is a Research Fellow at the "South West Center for Occupational and Environmental Health of UT-Health Science Center, Houston, Texas. She conducts research on hazardous and criteria air pollutants, human health effects of air pollutants, water/wastewater treatment, hazardous/solid waste management and anaerobic biodegradation of hazardous waste to generate bio-energy. She has developed novel microorganisms for the generation of methane from coal waste. She has also arrived at the energy economics of coal wastewater treatment using hybrid bioreactor comparing it to UASB. She is a recipient of the University Grants Commission-Junior Research Fellowship from Government of India at Indian Institute of Technology, Bombay. She has published/presented 36 papers in refereed journals, conference proceedings and 11 book chapters. She serves as an Associate Editor for ASCE-Journal of Hazardous, Toxic and Radioactive Waste and also in the editorial board of Macrothink- Journal of applied Biotechnology.

Dr. C.M. Kao, P.E., is a distinguished professor with the Institute of Environmental Engineering at National Sun Yat-Sen University, Taiwan. He received his MS and Ph.D. degrees in Environmental Engineering from North Carolina State University. Prof. Kao is a Fellow of the American Society of Civil Engineers and American Association for the Advancement of Science, and a Diplomate of the American Academy of Environmental Engineers and American Academy of Water Resources Engineers. He is also the Associate Editor of the ASCE journal Hazardous, Toxic, and Radioactive Waste and WEF journal, Water Environment Research. He serves as the executive committee and program chair of the Taiwan Association of Soil and Groundwater Environmental Protection and Taiwan Environmental Engineering Association. Dr. Kao has published more than 180 refereed journal articles including 15 patents, 6 books and numerous book chapters.